Universe
Based on
Aether

Universe
Based on
Aether

Bahram Esmailzadeh, M. Sc.

To order additional copies of this book, contact:
Xlibris
1-888-795-4274
www.Xlibris.com
Orders@Xlibris.com
710838

This book is dedicated,

to Ms. Scharback who patiently and open-mindedly reviewed most of the material presented here and also acted as a responsive soundboard, all along, and also

to those individuals who have an open mind about new ideas and theories.

Bahram Esmailzadeh

Acknowledgement

The author wishes to acknowledge that he is totally indebted to the Creator, who provided him with not only his existence, but also his sense of curiosity, his steadfast enthusiasm and interest, the required time as well as guidance to understand the lessons which were set before him.

The author also wishes to thank the Creator and all of the others, visible or not, for their direct guidance all through this particular research.

Table of Contents

Foreword

The contents of this book and its companion books "'The Evolution of Spirits', What is the Reason for Being Here, in This Creation?" (2012) and "The Formation and Evolution of Living Beings" (2015) are direct results of the author's personal thoughts and contemplations.

These three books introduce new ideas / theories which provide reasonable explanations for fundamental concepts in their respective fields. Yet, they are complimentary to each other and together they intend to open up new paths towards a better understanding of the physical world, the spiritual world and their interplay, respectively. In short, they provide an overall picture of what simply exists in this universe.

In this book, the author is concentrating on the physical universe and the development of its contents.

This book and its companion books are written for the general public. However, possessing some knowledge of the topics covered will be quite helpful in appreciating the simplicity of the theories presented, and the straightforwardness of the explanations they provide for various physical and spiritual phenomena.

Preface

In order to truly appreciate the great expanse of this universe, one needs to get out of the populated areas and on a clear and moonless night, look at the sky. He/she will see a carpet of lights that are in fact stars, stars that make up a very small portion of only one galaxy, namely the Milky Way Galaxy. Using a variety of telescopes, hundreds of thousands of such galaxies have been observed. It is estimated that there are over 10 billion galaxies in this universe.

Man has only begun to understand and appreciate a few of the laws, rules and regulations that dictate how this universe functions, on different scales. So far, he has become aware of some of the physical contents of this universe, and is only guessing about the rest.

Even though man is learning quickly, his knowledge regarding the formation of this universe and the development of its contents are still in their infancies. Therefore, he needs to have an open mind, as he is introduced to reasonable extensions (or even detours) to his current understanding of different aspects of this physical world.

Here, the author is presenting his own personal views and theories. Through their applications, he provides consistent explanations for a variety of phenomena such as 'time', 'light', 'space', 'gravity', 'electricity', 'electric field', 'magnetic field', 'black holes', 'energy', 'dark energy', 'vacuum energy' and 'matter' and 'dark matter', which are also shown to be directly interdependent.

All of the theories presented in this book are based on the existence of aether, the very same aether that physicists in the nineteenth century had believed to exist, everywhere. However, contrary to what was believed back then, aether medium is demonstrated to be quite an active, fluid-like and dynamic medium.

The first chapter provides an overview of how this universe has come into existence and how its contents have evolved so far. It also forecasts how the contents of this universe will evolve from now on. The information presented is solely based on the theories presented in the following chapters.

The second chapter provides detailed information about aether, its properties as well as how it directly relates to various physical phenomena.

Chapters three through seven concentrate on specific phenomena, namely 'space', 'time', 'light', 'gravity' and 'black holes', respectively.

Chapters eight through eleven, cover 'magnetic field', 'electric field', 'electricity' and "natural disasters", respectively.

Five independent methods of locating the birthplace of this universe are introduced and described in detail in chapter twelve. The information necessary to apply three of these methods is readily available to the general public through the internet. The final results of these three methods are presented in chapter thirteen. They are shown to agree with each other on the physical location of the birthplace of this universe. In other words,

<u>The birthplace of this universe has been discovered.</u>

The appendices provide,

- List of nominations,
- List of currently unresolved issues which are consistently explained in this book,
- List of predictions which are made possible by the theories presented in this book,
- List of experiments which are proposed and described in detail in this book, experiments that will confirm the validity of the newly proposed theories, and

- List of 25 fundamental breakthroughs, each one potentially deserving a Nobel Prize

As it is demonstrated throughout this book, once the existence of a dynamic aether is accepted and its effects are properly taken into account, many unresolved issues regarding various basic physical phenomena can be explained, understood and readily expected.

The contents of each chapter are presented in such a way that they are as independent, complete and self-contained as possible. Therefore, certain parts are necessarily repeated due to the interconnectedness of the topics presented. Also, some of the chapters are written in a point by point style, as needed, to focus on one specific issue at a time. This method will also encourage the reader to stay focused, as well.

This book is written for the general public. However, possessing some knowledge of the topics covered will be quite helpful in appreciating the simplicity of the theories presented, and the straightforwardness of the explanations they provide for various physical phenomena.

Introduction

In 1865, Mr. Maxwell proposed his theory on electromagnetism, the very same theory that is still valid to this day. Mr. Maxwell had based his theory on the existence of some kind of stationary medium through which light and other electromagnetic waves were assumed to propagate. At that time, this medium was referred to as 'Aether'. What aether was made of or what kinds of properties it had were not known. The medium of aether to electromagnetic waves was thought to be as a medium such as air is to sound waves.

Over time, scientists became curious about aether and based on the assumption that it is stationary in space and all waves and even planets are moving through it, they tried to detect its drift effect on the surface of earth. The assumed motion of earth in the stationary medium of aether is shown in the following figure.

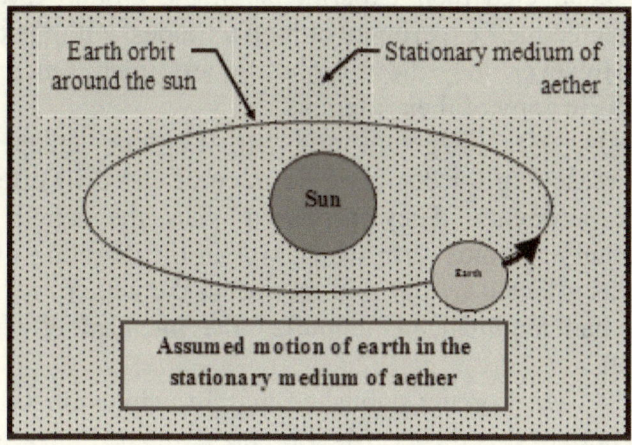

Amongst the multitude of experiments that were proposed, designed and conducted was the one designed and performed by Mr. Michelson and Mr. Morley in 1887. As it is shown below, their experiment was based on reuniting the wave patterns of two light beams from the same monochromatic light source, after they were made to follow two paths that were orthogonal to each other, one in the direction of earth's motion around the sun and the other at 90 degrees to the said direction.

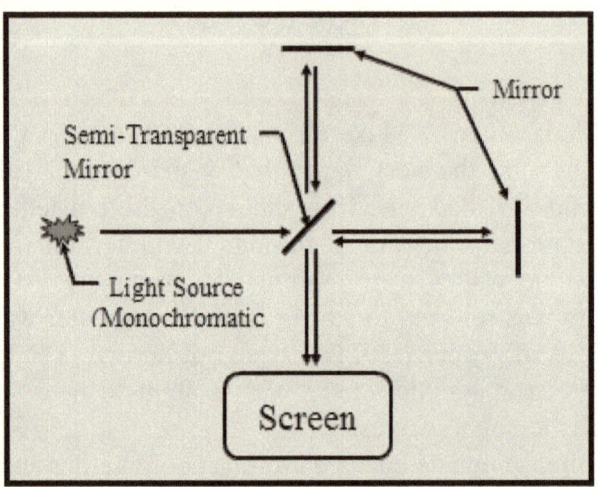

By reflecting these two light branches back and superimposing them on each other, Mr. Michelson and Mr. Morley were expecting to see the formation of some sort of interference pattern.

They conducted their experiment on the ground level. They also conducted their experiment at high altitudes, with the help of a balloon. However, they did not observe any interference patterns forming during any of their attempts.

In 1904, Mr. Fitzgerald and Mr. Lorentz, independently, provided reasoning for detecting no trace of any drift effect in all of the experiments. They proposed that, the length of the experimental apparatus' arm which was in the direction of earth's motion was affected by the said motion in the aether medium and therefore had automatically annulled the intended effect of the aether drift.

In 1904, Mr. Lorentz published his "Principles of Relativity". With the help of this theory, Mr. Lorentz explained the effects due to moving at relatively high velocities. To this day, effects such as the shortening of the length of an object in the direction of its motion, as well as the slowing down of the rate at which time is experienced and also the increase in the mass of an object, as its speed approaches that of light in a vacuum, are calculated using Mr. Lorentz's transformation equations.

It should be noted that, the existence of aether was commonly accepted by the then-current scientific community.

In 1905, Mr. Einstein used Mr. Lorentz's "Principles of Relativity" and added two postulates to it, namely:

- Light is made of particles (photons) which do not need a medium to travel, and

- The speed of light in a vacuum is the maximum possible speed and it is independent of the observer's state of motion relative to the source of light.

Note that, the experiments performed by Mr. Michelson and Mr. Morley, by no means had indicated that there was no aether in space, or that the speed of light was the maximum speed possible.

A revised version of the experiment performed by Mr. Michelson and Mr. Morley is proposed and presented in detail in the chapter titled "What is Light?".

It will be demonstrated throughout this book that, once the existence of a dynamic aether medium is accepted and its effects are properly taken into account, a variety of unresolved issues regarding different basic physical phenomena become explicable, understood and readily expected.

Therefore, **it is very crucial that, the present-day scientific community accepts the existence of aether and takes its various effects into account**.

Formation of the Universe
and
Development of its Contents

Introduction

This chapter provides an overview of how this universe has come into existence and how its contents have formed and developed so far. It also provides reasonable extrapolations on the future development of the contents of this universe, as well as the future of the universe, as a whole. The overview presented here is made possible as different theories presented in the following chapters are allowed to play their respective roles.

All of the theories introduced in this book are based on the existence of aether which is literally occupying the whole universe.

The history of this universe and the development of its contents can be divided into several (in some cases overlapping) periods:

- The beginning (The initial rapid expansion of space)

- Formation of matter and anti-matter particles (The slowing down of the rapid expansion of space)

- Formation and development of galaxies

- Formation and development of physically living beings

- Current conditions in this universe (Acceleration of the expansion of space)
- Current conditions in the accompanying universe

- The future / destiny of this universe and its contents

But, first and foremost, one needs to define what **'Space'** is.

Space

Space owes its existence to aether. Aether, as it is described in greater detail in the chapter titled "What is Aether?", is an elastic, compressible, non-viscous fluid that occupies the whole universe and more. The relation between space and aether is just like that of an ocean and water. The very existence of water gives meaning to an ocean. If all of the water that is inside an ocean is somehow taken out, what remains can no longer be called an ocean. The very same is also true about the direct relation between space and aether, since for space to exist it must be occupied by aether.

The internal structure of space can be analyzed on the following two extreme scales:

- **On the macroscopic scale**

  Space is not limited or restricted to the three spatial dimensions that jointly define what is commonly referred to as the physical universe. Space also includes three hidden dimensions, among others, which are accessible under certain conditions. These three hidden dimensions form a universe which is accompanying this universe.

  Even though the accompanying universe is a parallel universe to this one, and it is apparently hosting a duplicate set of the matter and anti-matter particles that exist in this universe, it is quite a different universe. For example, currently, the density and the pressure of aether that is in the accompanying universe are much, much lower than the density and the pressure of aether that is in this universe.

- **On the microscopic scale**

The internal structure of space in this universe is quite like a sponge which has countless number of tiny holes in its volume. These holes, as explained in greater details in the chapter titled "What is Gravity?" are the matter (and anti-matter) particles which act as drain holes (not as black holes, but as drain holes) allowing aether to escape from this universe into the accompanying universe. The two universes are also connected through black holes which are in fact large aggregates of matter particles that by merging (unifying) together have literally formed much bigger openings (gateways) for aether to flow through.

Therefore, as viewed from this universe, matter (as well as anti-matter) particles are like drain holes at the bottom of a swimming pool through which aether is leaking out. While, as viewed from the accompanying universe, matter (as well as anti-matter) particles are like water spring jets in a swimming pool through which aether is flowing into and joining the aether that is already there. An example of such a connection point, namely a matter particle, between this universe and the accompanying universe is shown below.

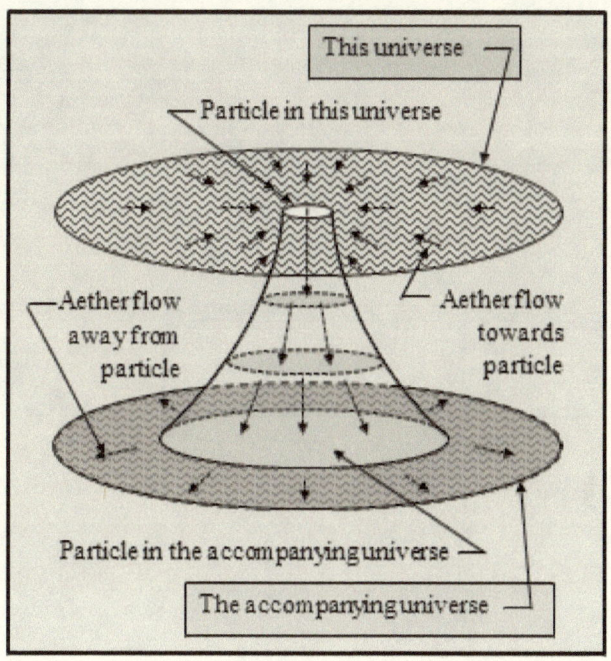

Note that, the difference in the physical size of the particle in the two universes is due to the difference that exists between the pressure of aether that is in this universe and the pressure of aether that is in the accompanying universe.

For more details, please refer to the chapters titled "What is Space?" and "What is Gravity".

The Beginning
(The initial rapid expansion of space)

In the beginning, the fluid aether was in a compressed state having a very small volume. In other words, space was quite limited in its overall size. Aether's density and pressure were enormous. One could say, they were at their critical points.

Somehow a ripple effect was introduced into the aether medium, just like the waves that are formed on the surface of a pond as a pebble is dropped in the middle of it. The ripples caused the over excitation of the aether medium, encouraging it to expand (not to explode, but to expand) in all directions.

Due to not facing any kind of resistive force such as gravity, the fluid aether medium that defines 'space' in this universe went through an unrestricted expansion process. In other words,

"The very lack of the force of gravity, allowed the aether medium, hence space in this universe, to experience an enormous rapid growth burst."

This growth burst, in a relatively short period of time (**based on then-current time scale**) spread the fluid aether and hence expanded the overall size of space from a very small region to a volume that was equivalent to millions of **present-time light years** across. In other words,

"The very introduction of the ripples in the fluid aether medium, <u>marked the beginning of the</u>

**rapid expansion era of that medium and hence
the rapid inflation of space in this universe.”**

The internal pressure of the aether medium was the driving force for its expansion. Due to aether's density and pressure being so high, when its expansion period started, the propagation speed of all types of vibrations, such as the ripples that were introduced into that medium was at an all-time low. This simply meant that,

**“In the beginning, the ripples spread more
rapidly due to being carried by the expanding
aether medium, rather than due to propagating
in that medium.”**

In other words,

**“As the aether medium and hence space
expanded, the ripples simultaneously expanded
with and spread in the aether medium.”**

The induced ripples were in fact some type of phase vibrations in the medium of fluid aether. The expansion of the aether medium as a whole, and the simultaneous propagation of phase vibrations in that medium are shown in the following figure.

Note that, both frontiers (the outer limits), namely that of the aether medium and that of the initial phase vibrations in that medium, were spherical in their overall geometries.

Even though the outer perimeter of the aether medium (which defines 'space' in this universe) was expanding at super-phase vibration (super-light) speeds, phase vibrations were propagating in that medium at their relatively slower speed which was solely depended on the density and the pressure of the local aether medium.

As the aether medium was expanding and its pressure and density were decreasing, the expansion rate of the outer perimeter of the aether medium, as a whole, was decreasing while the speed at which phase vibrations were propagating in that medium was increasing.

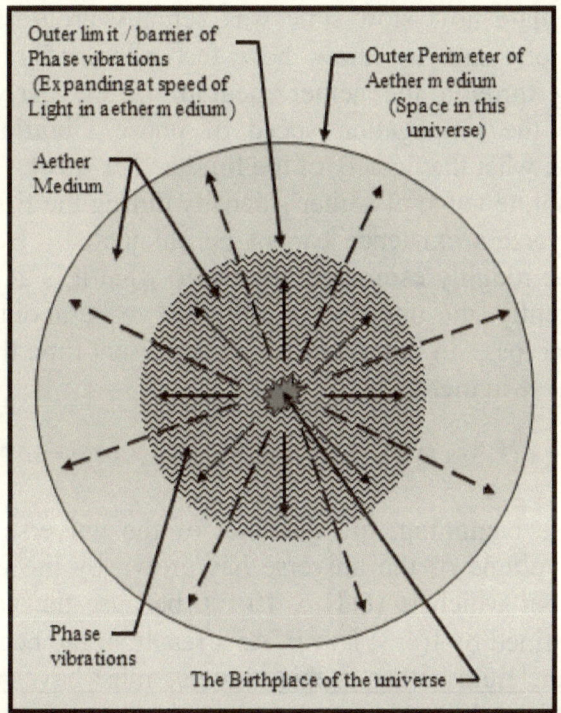

The following labels appear in the figure:

Outer limit / barrier of Phase vibrations (Expanding at speed of Light in aether medium)

Outer Perimeter of Aether medium (Space in this universe)

Aether Medium

Phase vibrations

The Birthplace of the universe

As it is explained in great detail in the chapter titled "What is Time?", the rate at which 'time' is experienced by any entity (inanimate objects as well as living beings) directly depends on the speed of that entity relative to its local aether medium, as compared to the speed at which phase vibrations propagate in that medium. As the speed of the entity relative to its local aether approaches that of phase vibrations in that medium the entity experiences 'time' at an ever slower pace.

The slow speed of ripples (phase vibrations) spreading in the aether medium of the early universe automatically meant that, back then, **'time' was progressing at a much, much slower pace, as compared to the rate it is progressing and it is experienced at the present time.** Since, the slower the propagation speed of phase vibrations in the aether medium simply meant that the closer were the speed of objects (even though there were none existing at that time) to the speed of phase vibrations in that medium. Hence, the effect on the pace at which 'time' was experienced was more pronounced.

To calculate how slow time was actually progressing in the beginning, one has to know how fast phase vibrations were propagating through the aether medium of the early universe. Back then, the propagation speed of phase vibrations in turn depended on what the density of the fluid aether was.

The absolute value of aether's density during the first moments of this universe's existence cannot be calculated. However, its value can be roughly estimated relative to what it is at the present time. Currently, the universe is estimated to be about 93 billion light years across. In other words, at the present time the diameter of the universe in meters is,

$$(9.3 \times 10^{10}) \times (365) \times (24) \times (3600) \times (300,000) \times (1,000) = 8.80 \times 10^{26} \, m$$

If, in the beginning, the diameter of the universe was ONE meter, the volume of the universe has grown by the cube of the above number which is (6.81×10^{80}), because the volume of a sphere is defined by $[(4/3) \pi r^3]$. As a result, at the beginning, <u>the density of the fluid aether in this universe must have been higher than what it is at the present time</u> **at least** <u>by the same ratio.</u> Since, over billions of years, quite a bit of the fluid aether has also leaked out of this universe through the matter and anti-matter particles, let alone through the vast number of black holes.

Based on the aether density ratio alone, one can readily estimate that the speed of phase vibrations (such as light) in the medium of aether back then, must have been billions of billions of billions of times slower, as compared to what it is at the present time. Consequently, one can confidently make the following statement regarding the rate at which the passage of 'time' was experienced, back then.

"The duration of the very first 'SECOND', when this universe had just started its existence, must have been equivalent to millions if not billions of years, based on today's time scale."

Formation of
Matter and Anti-Matter Particles
(The slowing down of the rapid expansion of space)

As aether expanded, the ripples spread and stabilized in their patterns. These ripples were in fact a type of phase vibrations in the medium of fluid aether. Such phase vibrations were just like regular sine waves. They had a positive half and a negative half, during every complete cycle.

Under certain conditions, phase vibrations in the medium of fluid aether can form full-wave or even half-wave spikes (resonances). Simple presentations of such spikes are shown below.

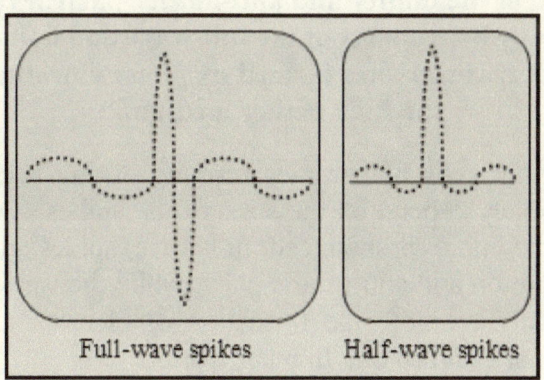

Full-wave spikes   Half-wave spikes

When spiked, the positive halves of the waves lead to the formation of particles that are of positive (+) charge and the negative halves lead to the formation of particles that are of negative (-) charge.

According to the currently accepted definitions in the field of particle physics, the only difference between matter particles and their anti-matter counterparts is that their electric charges are opposite of each other. For example, a proton is a positively (+) charged particle, while an anti-proton which has the same amount of mass is a negatively (-) charged particle.

However, as far as the relations between aether and matter and anti-matter particles are concerned, all of the positively (+) charged particles belong to one group and all of the negatively (-) charged particles belong to another group, regardless of normally being defined to be matter or anti-matter particles. In other words,

Based on this newly proposed matter particle categorization strategy, protons (+) and anti-electrons (positrons) (+) are grouped together, just as anti-protons (-) and electrons (-) are grouped together.

Therefore, full-wave spikes result in the formation of particle pairs (such as proton and anti-proton, or electron and positron) that are of opposite electric charges. Correspondingly, half-wave spikes result in the formation of single particles that are of either positive or negative charge, depending on their respective half-wave spikes being positive or negative. In short,

"All of the matter and anti-matter particles are basically byproducts of the full-wave and half-wave spikes that were/are formed by phase vibrations in the fluid aether medium."

Note that, the types of particles forming during such processes solely depend on the sizes of the spikes. For example, the more pronounced full-wave spikes give birth to proton and anti-proton pairs, while the weaker ones can only promote the formation of electron and positron pairs, and so on. In other words,

"Phase vibration spikes in the fluid aether medium manifest as matter and anti-matter particles."

To be more precise,

"Matter and anti-matter particles are like bubbles that are formed as a result of induced 'cavitation' in the very fabric of the fluid aether medium, as the local phase vibrations resonate into sufficiently high amplitudes"

The formation of matter and anti-matter particles (bubbles) in the aether medium by resonating phase vibrations is analogous to the formation of bubbles in a fluid medium such as water, due to cavitation, when that medium is exposed to ultrasound waves which are phase vibrations in the medium of water.

Bubbles (particles) formed by cavitation in the medium of fluid aether are basically openings into the accompanying universe. They automatically act as drain holes (openings / tubes) through which fluid aether flows from this universe into the accompanying universe.

The following picture shows the entrance of a water-shoot inside the reservoir of a dam. Such water-shoots prevent the water that is behind the dam to rise beyond certain safe level. They perform their important task by simply allowing the excess water to fall in them and get transferred downstream, bypassing the dam structure altogether. The flow of aether towards and through matter and anti-matter particles is quite the same as the flow of water towards and through such openings.

However, the flow of aether towards a given particle is from the whole volume of space surrounding that particle, a volume of space which is three dimensional, while the flow of water towards the inlet of such a water-shoot is from a planar surface which is two dimensional. Yet, such water intakes clearly demonstrate how the flow of the fluid water changes direction, and it literally disappears from the two dimensional plane in which it existed a moment earlier. The very same type of a process takes place with the fluid aether. Since, as aether reaches a given matter or anti-matter particle it literally disappears from the familiar three dimensional volume of space that forms this universe and enters the dimensions corresponding to the accompanying universe.

Note that, in the two-dimensional surface of water, the middle part of such a water-shoot is observed as a void.

The very same is also the case for the matter and anti-matter particles in the aether medium, since they too are bubbles, voids, in the three-dimensional space which is this universe.

In the beginning, phase vibrations were distributed uniformly in the aether medium, except for its outer regions which were (and still are) void of any kind of phase vibrations. Therefore, as the pressure and the density of the fluid aether medium were decreasing, due to its expansion, the same circumstance was simultaneously provided almost throughout its volume. Consequently, the phase vibrations present in the fluid aether resulted in the formation of matter and anti-matter particles everywhere, at about the same time. In other words, one can say,

"The whole universe literally bloomed with matter and anti-matter particles across its volume (the portion that was hosting phase vibrations), at about the same time."

Neutral particles, such as neutrons, where formed from the joining of electrons and protons (also the joining of anti-protons and positrons), as they attracted each other due to having unlike charges.

Note that, full-wave and half-wave spikes can also form randomly, as resonances or harmonics are continuously forming, due to the ongoing interactions between the varieties of phase vibrations that readily exist anywhere and everywhere in the medium of fluid aether.

Such events lead to the **random production and hence random appearance of particles,** either as in single particles or as in particle pairs.

Such events can also cause **the random disappearance of existing particles,** without any direct connection to other particles in this universe. Because, by forming resonances at the right place and at the right time, the existing phase vibrations can nullify the wave functions of locally existing particles.

Such interactions between phase vibrations in the fluid aether medium and matter (as well as anti-matter) particles are analogous to the formation and destruction of bubbles in a medium such as water by ultrasound waves.

Ultrasound waves (phase vibrations) of certain frequency and amplitude introduced in the medium of water can cause cavitation, hence the formation of bubbles, in that medium. The very same type of phase vibrations can also cause the existing bubbles to collapse and literally disappear in that medium.

It must be mentioned and emphasized at this point that,

"Individual particles such as electrons, protons and neutrons, as well as their corresponding anti-particles, are simply bubbles in the medium of aether, bubbles which are not composed of any smaller constituents."

In other words,

"There are no such things as quarks in this universe."

Therefore,

"There is no such a force as the strong nuclear force in this universe."

Also, as it is explained in the chapter titled "What is Gravity", according to the aether based theory of gravity,

"The gravitational attraction forces that protons and neutrons exert on each other override the repelling forces generated between protons by their positive electrical charges."

In other words,

"The gravitational attraction forces that protons and neutrons exert on each other ARE the forces responsible for holding the nucleons together in the atomic nuclei.

Hence, according to the aether-based theory of gravity,

"There is no such a force as the weak nuclear force in this universe."

As it is shown in the following figure, matter and anti-matter particles that were simultaneously formed across the central region of the aether medium fell behind relative to the phase vibrations that had generated them.

The sudden abundance of matter and anti-matters particles in the central region of the aether medium (which was occupied by phase vibrations) allowed nearly unrestricted flow of aether out of this universe. Basically, a huge aether leak was initiated by the formation of matter and anti-matter particles in the central region of the aether medium, in this universe.

As a consequence, the aether medium, as a whole, experienced a relief in its pressure due to leakage of aether in its central region, as well as its expansion towards its then-current outer perimeters. Hence, up to certain midway point, the outward (spreading) flow of aether was partially diverted towards the central region to compensate for such a huge rate of aether loss in that region.

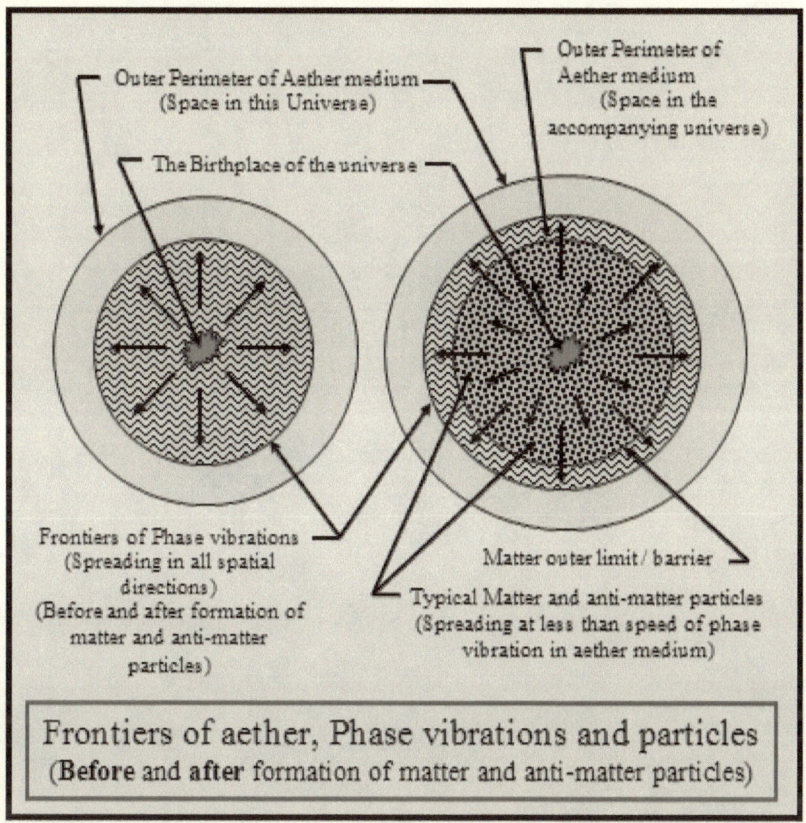

Frontiers of aether, Phase vibrations and particles
(**Before** and **after** formation of matter and anti-matter particles)

Therefore, at some distance, between the outer limits of the aether medium, as a whole, and the outer limits of where matter (and anti-matter) particles were formed, a narrow region of aether automatically experienced a stall situation, as it had no tendency to flow towards either directions. Yet, its pressure and hence its density kept on decreasing with the rest of the aether medium. Basically a neutral, stationary, spherical zone/halo was formed surrounding the central region where matter (and anti-matter) particles existed in the fluid aether medium. This neutral aether flow zone is shown in the figure, below.

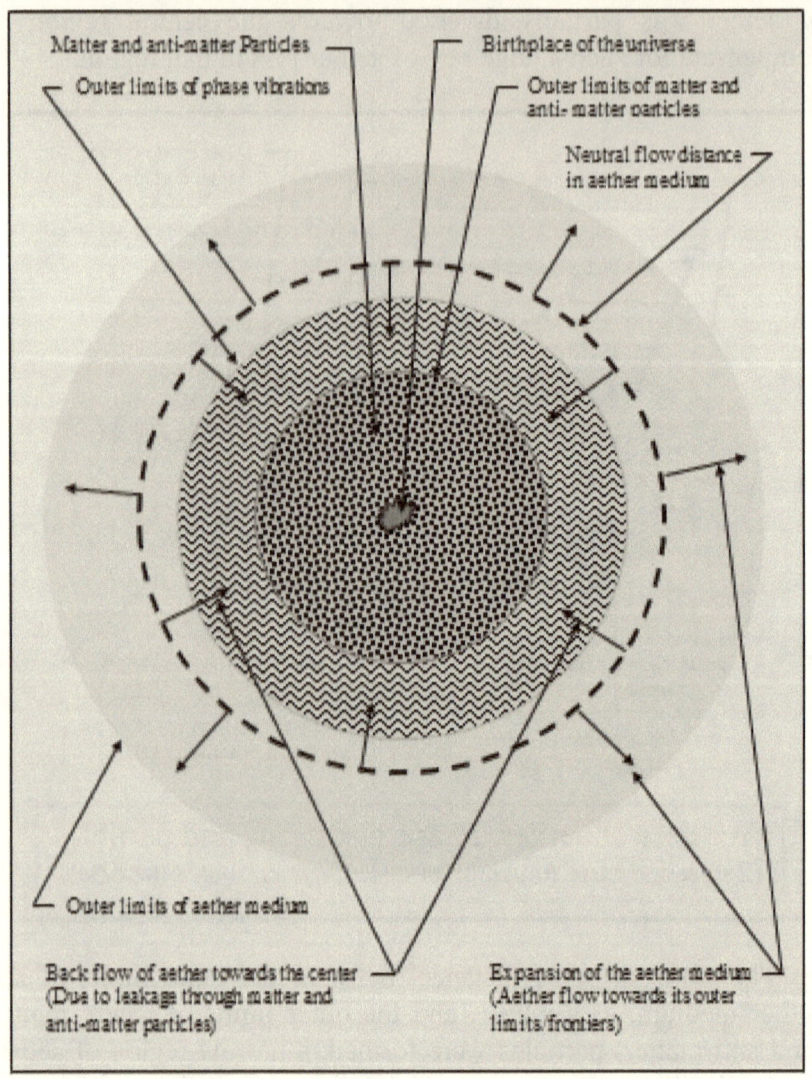

Matter and anti-matter particles basically acted as pressure relief valves. The rapid leakage of aether through the matter and anti-matter particles which were formed in the central region of the aether medium caused six very important effects:

1- By their very formation, matter and anti-matter particles allowed aether to flow from this universe into an adjacent environment and hence gave meaning to 'Space' in the accompanying universe. In other words,

"The very formation of matter and anti-matter particles (bubbles) in the fluid aether medium in this universe <u>initiated the existence of the accompanying universe</u>."

2- The rapid leakage of aether from this universe encouraged a very quick drop in both pressure and density of aether in this universe.

3- The accelerated flow of aether towards each of the matter and anti-matter particles, as shown below, exerted a drag force on the neighboring particles and eventually on all existing particles in this universe.

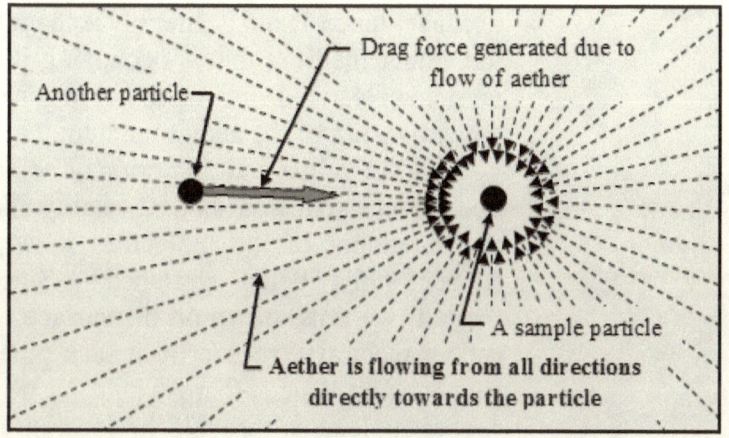

This very action of aether, namely its accelerated flow towards particles and inducing a drag force on other particles that happened to be in its path, literally gave meaning to what is known as the "Force of Gravity". In other words,

"The very formation of matter and anti-matter particles marked <u>the birth of the Force of Gravity</u> in this universe."

Note that, One could state that, the very expansion process of the aether medium must also give rise to a force of gravity of some sort which is actually acting opposite of the normal force of gravity. Since, instead of encouraging particles (as well as their aggregates) to come closer together it is pushing them apart from each other, as it is expanding the domain of this space.

However, the term gravity as it is defined in this book refers to **the drag force induced by the flow of aether which is accelerating through any given location in this universe**, an accelerated motion that is towards particles of matter or anti-matter. The motion of aether that is due to its expansion process, at any given location in this universe, does not contribute to any flow of aether at that location (in space), because space itself is literally expanding with the aether medium that is occupying it. Hence, no net accelerated flow of aether is generated by the expansion of the aether medium.

The spreading action/motion of aether that is due to its expansion can be readily likened to the spreading of the molecules in a piece of rubber while being stretched. Since, even though two dots drawn on the surface of such a rubber piece will recede from each other, as the rubber is stretched, there would be no flow of rubber molecules through the location of either one of those two dots. Please, refer to the chapter titled "What is Gravity?" for detailed information on gravity.

4- The slower motions of the matter and anti-matter particles relative to the local aether, as compared to the speed of phase vibrations in that medium, gave meaning to "TIME". As it is explained in greater details in the chapter titled "What is Time?", 'time' is experienced only by any and all entities that move (relative to their local aether) at a speed that is slower than the speed of phase vibrations in their respective local aether mediums. In other words,

"'Time' had commenced as the first ripples were formed in the fluid aether medium,

However,

<u>"The meaningful startup point for actually experiencing 'Time' was when matter and anti-matter particles were formed.</u>

Since, by falling behind relative to the ripples that had formed them, they were the very first entities that actually experienced the passage of 'time'."

5- Also, the very formation and subsequent presence of the charged particles in this universe gave rise to the electric field.

<u>"Charged particles,</u> with their very presence in this universe, <u>gave birth to Electric Field</u>."

For detailed information on electric field, please refer to the chapter titled "What is Electric Field?".

6- The very motion of charged particles in the aether medium gave rise to the magnetic field. In other words,

<u>"Charged particles,</u> with their very motion in this universe, <u>gave birth to Magnetic Field</u>."

For detailed information on magnetic field, please refer to the chapter titled "What is Magnetic Field?".

By their very formation in the central region of the aether medium, matter and anti-matter particles allowed the unrestricted flow of aether from this universe into the accompanying universe. In doing so, they:

- Drastically slowed down the spreading of the particles in this universe, due to their mutual force of gravity.

- Drastically slowed down the expansion rate of the aether medium, as a whole, by expediting the rate at which its internal pressure was reduced.

Therefore, one can say,

"The very formation of matter and anti-matter particles was like the activation of an effective <u>braking system</u> on a universal scale, which almost brought the rapid initial expansion of the matter contents of this universe to a complete halt."

Shortly after matter and anti-matter particles had slowed down, due to their mutual force of gravity, most of the matter and anti-matter particles joined together and annihilated each other. Their annihilation processes which were in fact the cancelation of their wave functions generated phase vibrations which then spread in the surrounding aether medium as shock waves. The following figure shows the end result of such a process in a very simple fashion.

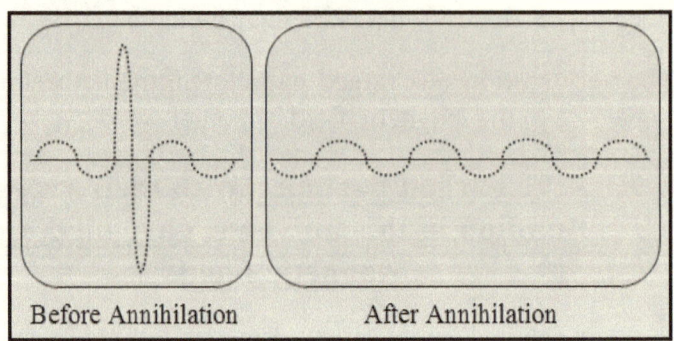

| Before Annihilation | After Annihilation |

Note that, only charged matter particles and their charged anti-matter particles, such as protons and antiprotons (or electrons and positrons), could (can) annihilate each other's wave forms by merging, since they were (are) a match (180 degrees out of phase).

However, two oppositely charged particles that were different in their wave forms could not (cannot) annihilate each other. A good example of such a mismatch in wave forms was the case of electrons and protons. As such mismatched particles merged together, they only neutralized each other's charge but did not result in any overall mass reduction. They actually complemented each other and formed a more massive particle. That was how neutral particles such as neutrons were formed.

The phase vibrations generated due to the annihilation of the matter and anti-matter particles were further spread and flattened by the ongoing expansion of the aether medium, as a whole.

Over time, these phase vibrations not only have blended together by crossing each other's paths but also have blended with the existing phase vibrations in that medium, phase vibrations (the original ripples) that were introduced into the medium of aether and had prompted the birth of this universe. Eventually, they have formed a uniform background noise in the medium of the fluid aether.

The process of phase vibrations mixing and blending with each other in the fluid aether medium can be likened to how waves formed on the surface of a pond mix and blend together, as multiple pebbles are thrown in the middle of the pond at about the same time. Because, as the newly formed waves spread outwards, they not only blend together but also blend with waves that already exist on the surface of the water. Eventually, they become a uniform disturbance throughout the whole surface of the pond.

This uniform background noise (phase vibrations in the medium of aether) has been detected, already. It is known as the **Cosmic Microwave Background Radiation, CMBR**.

However, the detected cosmic microwave background radiation consists of two general types of phase vibrations. One type/group of phase vibrations are the remnants of the initial phase vibrations which were introduced into the aether medium, while the other type/group of phase vibrations were due to the annihilation of matter and anti-matter particles. Since both groups of vibrations were phase vibrations in the medium of aether, they propagated in the aether medium at the very same speed.

Therefore, the outer limit or frontiers of phase vibrations due to matter and anti-matter annihilations were and in fact still are basically chasing the frontiers of the phase vibrations that prompted this universe's birth.

It should be emphasized that,

"The outer regions of the phase vibrations' domain in this universe which do not host phase vibrations that are due to matter and anti-matter annihilations and the inner regions that do host them, must have two totally different textures."

The variations in their textures can be likened to the differences that can exist between the textures of waves on the surface of the outer regions of a vast lake as compared to the textures of waves formed in its inner regions in which many small pebbles are dropped, at about the same time. As the newly generated waves propagate and spread towards the boundaries of the lake, they mix/blend with the existing waves which are already covering the surface of the lake.

In this case, the boundaries of the lake are fixed in their positions. Therefore, after a certain amount of time (which depends on the overall size of the lake surface area) nearly the entire surface of the lake will exhibit the very same wavy texture. Only the very outer regions will be slightly different, due to waves being reflected off of the outer edges.

However, in the case of the phase vibrations in the aether medium and their intermixing in this universe, **the outer regions will always be void of any kind of phase vibrations that are due to matter and anti-matter annihilations.** This is due to the fact that, both frontiers are expanding / spreading at the very same rate of speed which is equivalent to the speed of phase vibrations in the medium of aether.

This is analogous to the propagation of the complex sound waves generated by various instruments in a symphony orchestra through the medium of air. These complex sound waves propagate at the speed of sound (speed of phase vibrations) in the medium of air. Therefore, they cannot catch up with other sound waves that were generated before the conductor had prompted the musicians to play their respective pieces.

In the case of the phase vibrations in the aether medium, in this universe, the overall modifications imposed on the texture of the existing / initial phase vibrations in the inner regions of the aether medium depended on the frequencies of those two groups of phase vibrations. Since, their frequencies dictated how they could affect each other, as they strived for reaching a compromise in their energies.

The following figure shows the frontiers of the cosmic microwave background radiation in relation to the frontier associated with the aether medium (in this universe).

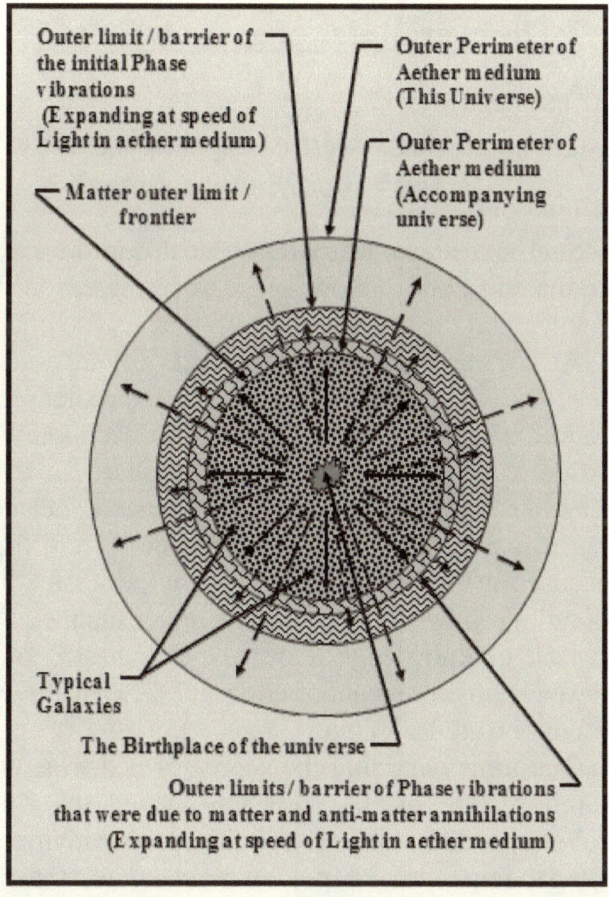

The frontier of the original phase vibrations and the frontier of the region of space hosting matter particles in this universe, which is slightly smaller than the frontier of the region of space hosting the accompanying universe, are also shown in the above figure.

Note that, the outer limit designating the matter frontier is expanding at a much slower speed as compared to the other outer limits / frontiers.

It should be emphasized that, all of the outer limits / frontiers shown in the above figure are spherical in their overall geometries, since aether's expansion has been and still is symmetrical in all spatial directions. Aether's outer limit or its overall frontline is a sphere which is expanding just like a ball. This automatically implies that,

"The overall size of space is finite."

- One may ask:

What is aether expanding in?
Isn't that the true space?

Such questions are philosophical questions. They are like the questions that one may ask in regards to the source of aether and the source of that source, and so on.

At the present time, any and all answers to such questions can only be based on pure speculation. This is due to the simple fact that, in order to even know what the outside of this universe (or even a house, for that matter) looks like, let alone where this universe is located, or into what it is expanding, one needs to see it from the outside. Because, even if the whole of the interior of a house (for example) can be surveyed and precisely mapped out, due to the lack of knowledge about the thicknesses of different exterior walls, one cannot even guess what the outside of the house truly looks like.

Questions regarding the source of aether as well as the overall picture of the **void** into which the fluid aether (along with this universe and the accompanying universe) is expanding must be put on hold, at least for the time being. Since, **the human race has only begun to have a glimpse at the structure of this universe and has become aware of only a few of the laws governing its contents' interactions with each other. Hence, its members have a long journey ahead of them before they can expect to have reasonable answers for such questions.**

After most of the matter and anti-matter particles had annihilated each other, the aether medium became a relatively calm and hospitable environment for the surviving matter particles. The remaining matter particles continued with their expansion process, but at a much, much slower pace. In other words,

This period marked the end of the rapid expansion era.

Note that, the very process of temporary rapid expansion that the infant universe experienced can be likened to what is experienced by the **critical mass inside a nuclear bomb**, as it explodes in the open atmosphere. In the case of a nuclear bomb, as an external neutron source is introduced into its center, the critical mass suddenly gets the urge to start a progressively more violent nuclear chain reaction. The material forming the critical mass experiences a sudden expansion rate that is nearly equivalent to the speed of light. In the meantime, a shock wave forms at the very center of the critical mass which starts to propagate through the expanding material.

As it is viewed from a distance, a nuclear explosion exhibits a very sudden but temporary growth burst, at nearly the speed of light. Then, as the initial phase vibrations keep on expanding at the speed of light, the byproducts of the explosive material continue to expand but at an ever slower pace. Because, they face a tremendous drag force, which is due to collisions with other matter particles forming the surrounding atmosphere, a force which is analogous to the mutual force of gravity that matter and anti-matter particles exerted on each other after their formation, in the early universe.

The initial, temporary rapid expansion period of this universe can also be likened to the very operation of **airbags**, installed in a variety of passenger vehicles. Because, as an airbag is deployed, the internal gas pressure causes the volume of the airbag to increase quite rapidly and hence cause its cover fabric to stretch. The expansion process of the airbag continues only to a certain point when holes appear in its cover fabric. The pressurized gas automatically leaks through these holes, and consequently its internal pressure drops drastically, in a fairly short period of time. The following figure shows the basic operation of an airbag.

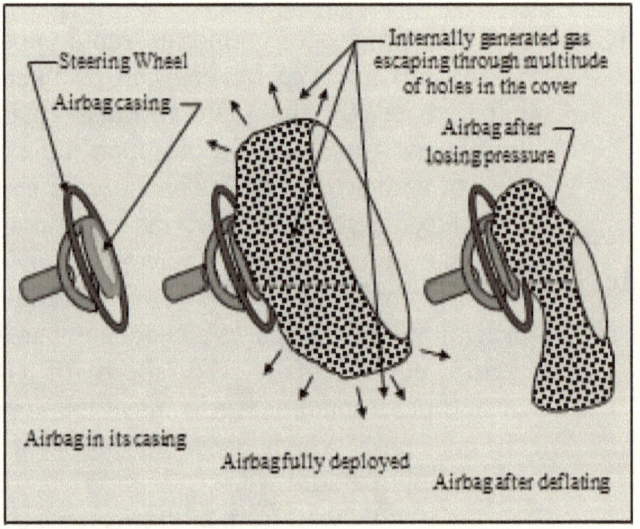

One can say that, airbags experience a miniaturized inflationary process of their own, as compared to what the whole universe experienced on a grand scale, during its infancy.

However, it must be particularly noted that, according to the information presented in the chapter titled "What is Time?", when universe was just born, "time" was experienced at a much, much slower pace, as compared to the rate it is experienced at the present time. In fact, **the first moments of the expansion of the universe had taken equivalent of possibly millions of years, based on today's time scales.**

Matter particles that had survived the annihilation period had lost most of their initial outward momentums, due to the tremendous force of gravity that existed while matter and anti-matter particles were still in abundance. Since their speeds were lower than the speed of phase vibrations in that medium, matter particles fell behind from the phase vibrations (ripples) that had formed them. Therefore, as they were still being dragged by the expansion of the aether medium (space), matter and anti-matter particles also had their own movements / momentums within the aether medium. In other words,

"As matter and anti-matter particles came into existence they had semi-independent motions of their own <u>IN</u> the aether medium, due to the outward momentum (energy) of the phase vibrations that had led to their formations."

It is stated as 'semi-independent motions', because the direction of expansion of the phase vibrations was away from the birthplace of this universe. Therefore, due to conservation of momentum, the <u>general direction</u> of the motion of matter and anti-matter particles that were formed was also away from the center, from where the expansion of aether medium had started.

What is referred to as 'Mass' of a particle, is actually the resistance to any change in the motion of that particle by the fluid aether medium in its immediate vicinity. Mass phenomenon manifests only as a particle is encouraged to move or stop, or it is forced to change the magnitude of its speed in any particular direction, or any combination of different directions. Once a given particle is in motion, it continues to do so at a constant rate of speed forever, unless it is acted upon by another external force of some kind. In other words,

"As the variety of forces act upon matter and anti-matter particles, the local fluid aether generates a resistance

to any changes in their motions, a resistance which is due to the formation of a wake-like wave in that medium. <u>This aether resistance manifests itself as what is commonly referred to as the</u> 'MASS' of particles."

Since the density of aether in this universe is gradually decreasing due to its expansion and leakage, the resistance it generates in response to changes in the motion of particles (bubbles in that medium) is gradually weakening. In other words,

"Mass of any and all particles, in this universe, is gradually decreasing."

Therefore, **in the earlier times, any and all existing particles were much more massive**, since the aether medium was much denser back then.

It should be particularly emphasized that,

"The fluid aether is not made of any kind of particles. Hence, it does not have any mass associated with it."

Note that, <u>the artificial generation of massive particles in various particle accelerators</u> is only indicative of the formation of ever more pronounced (higher amplitude) full-wave as well as half-wave spikes in the phase vibrations that either already exist or are artificially generated in the local aether medium.

However, by no means, such particles are representatives of the previous generations of particles which by splitting have supposedly led to the formation of the present-day particles.

The formation of heavier particles in particle accelerators can be simply looked at as if someone is turning up the 'volume' on a stereo system or as if more speakers are used in sync to generate specific vibrations that are of higher amplitudes.

The newly formed heavier particles (artificially amplified spikes in phase vibrations in the aether medium) have nothing to do with what have naturally existed in the past or will naturally exist in the future, in this universe. In short,

"The massive particles that are artificially generated in various particle accelerators do not represent what naturally have existed in the past or will exist in this universe."

Formation and Development of Galaxies

Each and every one of the matter particles that survived the annihilation period was moving at a different speed relative to its local aether medium, as well as relative to the other particles. However, their general direction of motion was away from the center, from where the aether expansion had started.

Over millions of years following the end of the rapid expansion era, due to the gravitational forces that the remaining matter particles exerted on each other, they formed countless number of huge cloud-like gatherings. The following figure shows several such cloud-like matter particle gatherings.

Since individual particles that were present in each of these cloud-like gatherings were moving at different speeds, by tugging each other, due to their mutual force of gravity, gradually they caused their cloud-like collectives to exhibit an overall rotational motion.

Note that, the orientation of the spin axis and the direction of rotation of each of the cloud-like gatherings were dictated by the overall momentum of their respective constituents.

The only major force governing the motion of matter particles in such gatherings was their mutual force of gravity. Their electric and magnetic fields were coming into play only if they were both charged particles, or charged particles were joining an existing group of particles that already had a net charge, as a whole.

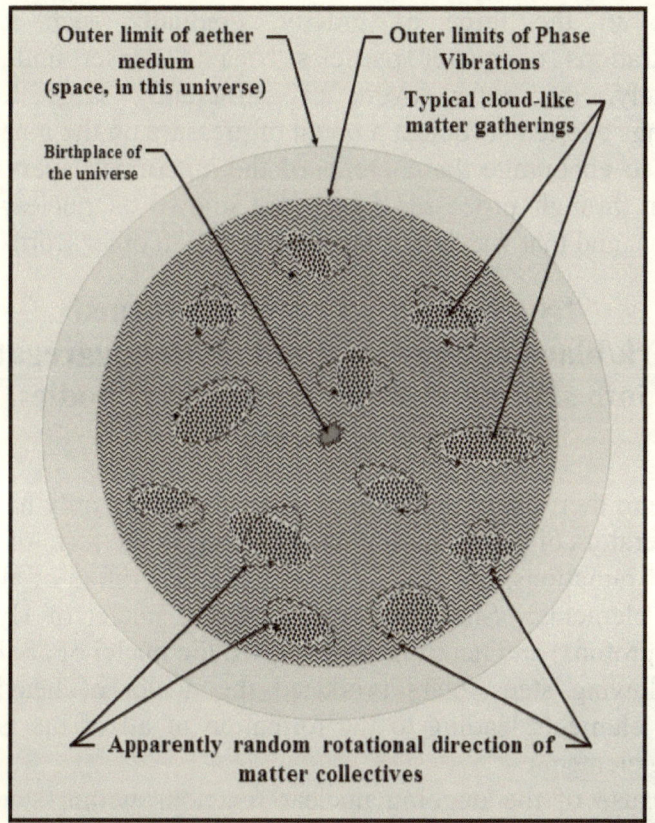

Over time, particles' force of gravity also encouraged the formation of numerous even more localized concentrations of matter particles within each of these giant cloud-like gatherings. Again, the differences between the speeds of the individual particles that step by step joined together and formed such gatherings caused their collectives to possess overall rotational motions. The joining process of particles with different speeds relative to each other is shown below for the simplest case that involves a proton and a neutron.

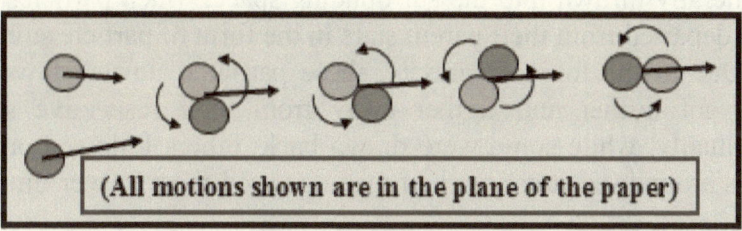

Due to the force of gravity, gradually such localized concentrations of matter particles became denser and denser. Ultimately, the outer layers of sufficiently large localized gatherings exerted sufficient amount of pressure on the inner layers needed to encourage the contents of the innermost layers to fuse together through processes which are known as nuclear fusion reactions, and that was how stars were born. In other words,

"Nuclear fusion reactions turned dark/black/opaque localized matter aggregates into shining, light giving heavenly bodies, namely stars."

Due to their ongoing fusion reactions, stars not only have been the generators of enormous amounts of heat, in the form of various types of radiations, but also have been the generators of all of the natural elements. Since, by combining the nuclei of Hydrogen (single protons) and neutrons they formed the nuclei of Helium. In the following steps, stars produced the nuclei of heavier and heavier elements leading to the formation of all of the naturally existing elements.

Because of the ongoing nuclear reactions within stars, their surface temperatures were/are in the range of thousands of degrees Kelvin and their inner core temperatures were/are in the range of millions of degrees Kelvin.

In the meantime, due to the enormous size of the nuclear reactions which were/are continuously in progress inside stars, strong magnetic and electric fields were/are generated inside and around them.

The violent interactions between their electric and magnetic fields that occur relatively close to the surface of stars have resulted in all kinds of particles and ions (elemental byproducts) to be literally thrown into the surrounding space. Such particles and ions departed from their parent stars in the form of particle storms.

Due to the force of gravity, these particles slowed down, as they got farther and farther away from their respective stars. Eventually, while some were drawn back, others followed orbital paths around the stars that had manufactured them. Over time, as these orbiting particles exerted their own mutual force of gravity on each other, they formed localized groups of their own.

The formation of the planets has been due to such gatherings of more and more particles that have attracted each other and over billions of years have managed to form localized collectives of their own, while orbiting their respective stars.

Stars are continuously throwing particles and elemental byproducts into their surrounding spaces, in all spatial directions. Most of such particles and elemental byproducts possess a certain amount of momentum in the same direction as the star is rotating. Having such common momentums affects their final orbital paths around the star, as they form aggregates. Because, as aggregates of particles are formed, their overall direction of motion around the star will automatically be in the same direction as the portion of the momentum that most, if not all, of them share, a momentum that is in the same direction as the rotation of their source star.

Of course, the final orbital path of any newly formed matter aggregate such as a planet around a star is dictated by where from on the surface of the star the majority of its matter particles have originated. Because, the farther the originating location on the surface of the star is from that star's equator, the more inclined will be the final orbital path of the aggregate that is formed. This is due to the fact that, most of the particles thrown out by stars start their journeys into the surrounding space by following a nearly vertical climb away from the surface of the star. However, <u>only if they also possess sufficient horizontal (diagonal) momentum they can remain in some kind of orbit and not crash back down onto the star.</u> Their required inclined momentum needs to be in the direction that will allow them to go around the center of mass of the star. In other words, <u>their orbital paths have to cross the equatorial plane of the star.</u>

If a matter aggregate (a planet) is formed of roughly equal amounts of matter particles originating from the Northern and the Southern hemispheres of a given star in such a way that they counter each other's inclination momentums, as a whole, that matter aggregate (planet) will follow an orbital path that is superimposed on the equatorial plane of the source star. In such a case, the spin axis of the matter aggregate will also be nearly, if not exactly, perpendicular to its orbital plane. In other words,

"In the absence of any imbalance in the overall inclination momentum of the incoming particles that form a planet, that planet's equatorial plane, if extended, will pass through its source star."

However, if a matter aggregate is formed of unequal amounts of matter particles that have originated from the Northern and the Southern hemispheres of the star in such a way that they do not cancel each other's inclination momentums, as a whole, that matter aggregate will follow an orbital path that is inclined with respect to the equatorial plane of the source star. In such a case, the spin axis of the matter aggregate may or may not be perpendicular to its orbital plane.

Of course, over time, planets that are close to each other, or pass by each other, gravitationally affect one another's orbital paths. Orbital modifications due to such close encounters provide the planets with the needed opportunities to share their momentums, particularly the portions that are in an inclined direction with respect to their common star's equatorial plane.

Note that, over an infinite length of time, the gravitational exchanges between planets orbiting a given star will cause the final orbital paths of all of the planets to become superimposed on a single plane. Because, by exerting their mutual force of gravity on each other they eventually normalize the differences that exist between their orbital inclinations with respect to their common star's equatorial plane.

If all of the matter particles forming all of the planets put together cancel each other's inclined momentums (with respect to their common star's equatorial plane) the final orbital paths of all of the planets will be superimposed on their star's equatorial plane. The following figure shows a simple presentation of such a long term effect.

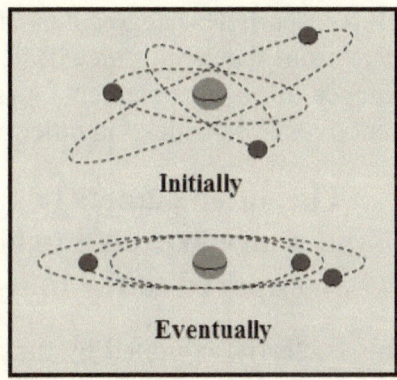

Correspondingly, if all of the matter particles forming all of the planets put together do not cancel each other's inclined momentums (with respect to their common star's equatorial plane) their final orbital paths will have an inclination with respect to their star's equatorial plane.

Since planets are formed from particles and elemental byproducts that are thrown out by stars over billions of years, the overall inclination momentums associated with all of the planets in any given solar system will cancel each other, in most cases.

Notes,

1- As time goes on, due to absorbing all of the particles that are thrown into their paths, every planet is in fact continuously gaining mass.

2- Due to the differences that exist between the speeds of the freely floating particles which join together and form a union such as a planet, their collectives exhibit rotational motions.

3- The orientation of the rotational plane of individual planets is also affected by the interactions between their magnetic fields and the magnetic field of the star around which they are orbiting.

4- The lighter elements and particles, in general, are thrown farther away by the stars, as compared to the heavier ones.

Therefore, particle aggregates forming at different distances from the same star will be composed of different percentages of various elements and correspondingly have different overall densities. In other words,

"The outer planets in any given star system are expected to be less dense as compared to the inner planets."

This is clearly exhibited by the overall density profiles of the planets that exist in our solar system. The overall densities of the three inner planets are around 5.5 g/cm^3, while the overall densities of the outer planets are about 1.5 g/cm^3 and that of Mars, which is located in between the two groups of planets, is about 4.15 g/cm^3.

Note that, any and all of the planets discovered in other star systems, **regardless of their overall physical sizes**, are expected to follow the very same rules as the ones obeyed by the planets in our solar system, particularly the rule that dictates their overall densities. Therefore,

"The closer any newly discovered planet is found to orbit its respective star, the denser that planet is expected / predicted to be."

Correspondingly,

"The farther any newly discovered planet is found to orbit its respective star, the less dense that planet is expected / predicted to be."

Over time, billions and billions of stars and star systems have formed through such processes. And, as many, many nearby stars and star systems have become organized in their relative motions, they have formed galaxies.

As shown in the following figure, just as the mutual force of gravity between planets affects their final orbital planes in a solar system, the mutual force of gravity between stars, in any given galaxy affects the final orbital paths of stars around the center of mass of that particular galaxy. The following figure shows the gradual change in the overall shape / geometry of a typical galaxy, as it starts from a cloud-like gathering of matter particles, then forms a spherical / elliptical galaxy and over time evolves into a galaxy with a more or less well-defined flat geometry.

Therefore, as galaxies take shape, they mostly exhibit a more or less spherical shape. But, as time goes on, they become more and more flat in their geometries. Because, as their stellar constituents affect each other gravitationally, they share part of their momentums as they compromise the orientations of their orbital paths around the center of mass of their mutual collectives.

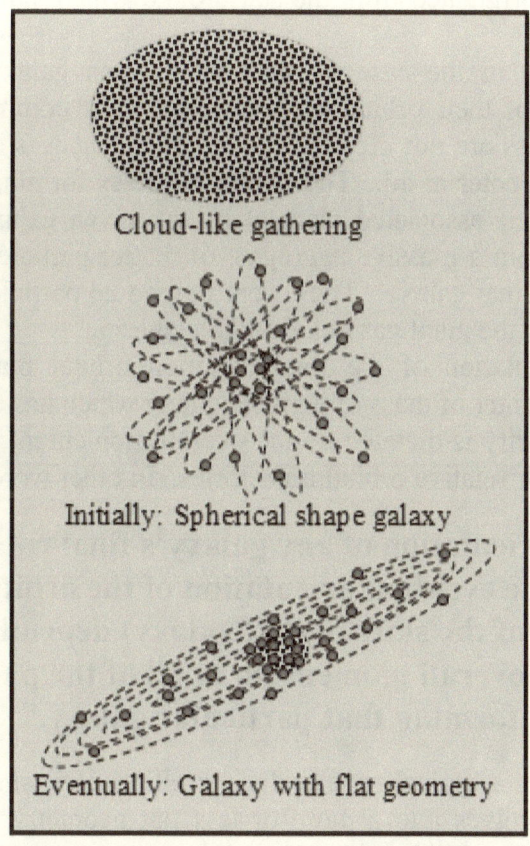

Cloud-like gathering

Initially: Spherical shape galaxy

Eventually: Galaxy with flat geometry

Eventually, most if not all of the stars in any given galaxy will follow orbital paths that are in a relatively common plane around the galactic center of mass.

In short, galaxies are the final results of the evolutionary processes that the giant cloud-like gatherings of matter particles have gone through, over billions of years. They all started from a totally chaotic state. However, due to the gravitational interactions between their constituents, they have managed to develop into well behaved systems of celestial bodies which continue to hold on to each other with the help of the force of gravity that they continue to exert on one another.

In the case of planets, in any given star system, the final orientations of their orbital paths around their common star are dictated by that star's spin direction and the orientation of its equatorial plane, since nearly all of the matter contents of any planet have originated from the star around which that planet is in orbit.

However, in the case of stars in a given galaxy, the final orientations of their orbital planes around their common galactic center of mass are not dictated by anything that was/is located at the galactic center at all. The matter particles forming individual stars (and their associated planets) in any given galaxy have not originated from a massive aggregate of matter particles located at the center of that galaxy. They were individual particles that were distributed in the giant gas / cloud-like gathering.

The orientation of the final common orbital plane of stars around the center of mass of a given galaxy which has evolved into its flat geometry is dictated by the overall momentum of all of the stars and their relative orbital trajectories. In other words,

"The orientation of any galaxy's final rotational plane (the eventual orientation of the orbital paths of most of the stars in that galaxy) depends solely on the overall momentum of all of the particles forming that particular galaxy."

Note that, as a typical spherically (or elliptically) shaped galaxy evolves into a more or less flat geometry many of its stars collide with each other, as they cross each other's orbital paths, as they go around their common galactic center of mass. During such collisions, some of the matter particles that used to form these stars lose some of their orbital momentums and are drawn towards the galactic center, while some other particles gain some momentum and are thrown outwards due to their excess velocities.

The very same applies to stars that experience a Nova moment of their own, since parts of their outer layers get thrown towards the galactic center where they get gravitationally trapped, while other parts get thrown away from the galactic center where they become widely spread.

That is why, as shown in the previous figure, the overall diameter of a galaxy that has evolved into its final flat geometry is quite a bit larger than its diameter when it was spherical (or elliptical).

Also, matter particles that are thrown towards the galactic center contribute towards the formation of more new stars there. Hence, as shown in the figure, the galactic centers continuously become more populated with newly formed stars, while the outer regions become thinner and more spread out.

Due to the direction of the momentums of the particles thrown towards the galactic centers being random, newly formed stars in the galactic centers follow randomly oriented orbital path around the galactic center of mass. Hence, they form a spherical zone (a bulge) near the galactic centers which are quite densely populated with stars.

Even though, at a first glance, it seems like as if the rotational planes of galaxies are oriented in random directions. Yet, when they are studied in some detail, it becomes clear that they do abide by certain rules, the very same type of rules that are followed by individual planets in any star system.

In the case of a given solar system, the source of matter particles forming planets is a star which is located at a specific location relative to the planets. But, in the case of this universe, the matter particles forming individual galaxies were not formed at one specific location in this universe. Matter particles were simultaneously formed in a huge volume of space, billions of light years across, which was occupied by the phase vibrations.

However, the general direction of propagation of the phase vibrations and hence **the initial general direction of the momentum that was transferred over to the matter and anti-matter particles (as they were formed) was pointing directly away from the birthplace of the universe.** In other words,

"As matter and anti-matter particles were formed, they already possessed an initial momentum that was (in general) directed away from a common region of space, as if they had literally started their individual journeys from a common center."

Matter particles were also moving at different rates of speed with respect to their local aether, as well as relative to each other. Therefore, the same methodology that was used to explain the orientation of the underlined rotational plane of individual planets in star systems can also be applied to determine the preferred orientation of the final, flattened geometry of individual galaxies in this universe. Because, as constituents of planets had originated from a direction that was pointing towards their stars, so were the constituents of galaxies since they were also moving away from a common region of space, namely where this universe was born. Hence,

"The preferred orientations that the final, flattened geometries of galaxies can possess would be those orientations that, if their flattened planes were spread out (extended), they would point towards (pass through) their common ultimate source, namely the birthplace of this universe."

However,

The direction of rotation of individual galaxies
is not related to the source of their constituents.
Because, the phase vibrations that led to the
formation of matter and anti-matter particles
were only SPREADING away from the birth
center. They were not ROTATING around it.
Therefore, the direction of rotation of any given
galaxy is simply dictated solely by the overall
momentum its own constituents.

In other words, when examined closely, **the extensions of the rotational planes of most galaxies** (particularly those of the spiral type galaxies, which are well developed and are readily verifiable), regardless of their distances to the observer, **must pass through a common center, or the birthplace of this universe.**

As it is shown in the following figure, most of the spiral galaxies that happen to be along any given straight line that is drawn between the birthplace of the universe and the outer perimeters of the universe, their planes of rotation must be viewed as being edge-on or nearly edge-on to each other's line of sight. However, in all other directions, one can expect to see the galactic planes to be relatively randomly oriented.

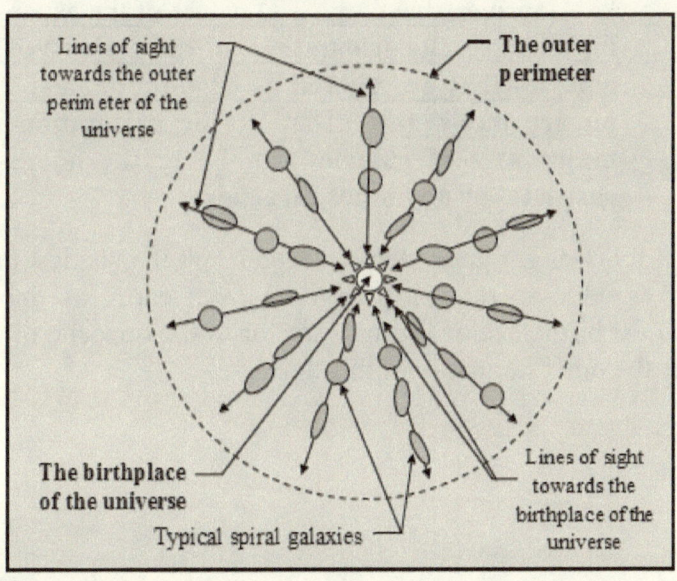

Therefore, in order to narrow down the direction (the galactic coordinates) pointing towards the birthplace of this universe, one must only pay close attention to the apparent angles of the spiral galaxies' rotational planes with respect to his/her line of sight from Earth. The easiest and simplest way to achieve this task is by comparing the length (major axis) and width (minor axis) of galaxies, as they appear to one's line of sight.

Samples of edge-on, full-faced and inclined (near edge-on) views of a typical spiral galaxy's plane of rotation, with respect to the line of sight of an arbitrary observer, are shown below.

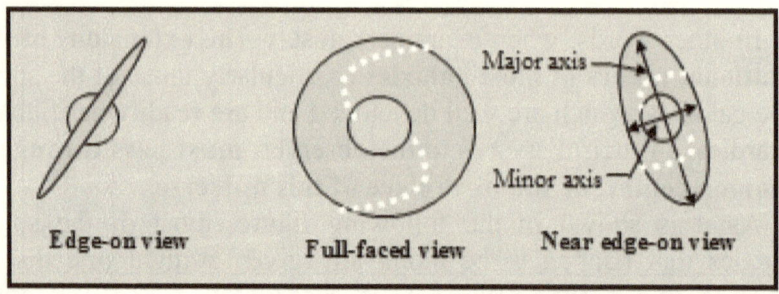

Note that, if the universe was expanding in a flat plane (two-dimension), the orientation of the rotational planes of all of the galaxies would have been expected to be superimposed on that plane.

However, <u>due to the 3-D nature of the expansion of the universe,</u> **the orientations of the major axis of the galaxies that are viewed as edge-on or nearly edge-on are irrelevant.** **That is, the orientation of the major axis of galaxies can be in the north-south, east-west or any other direction.**

The following figure shows several spiral galaxies that are viewed as edge-on or nearly edge-on from earth, as one looks towards the birthplace of this universe or in the opposite direction, towards the outer perimeter of this universe.

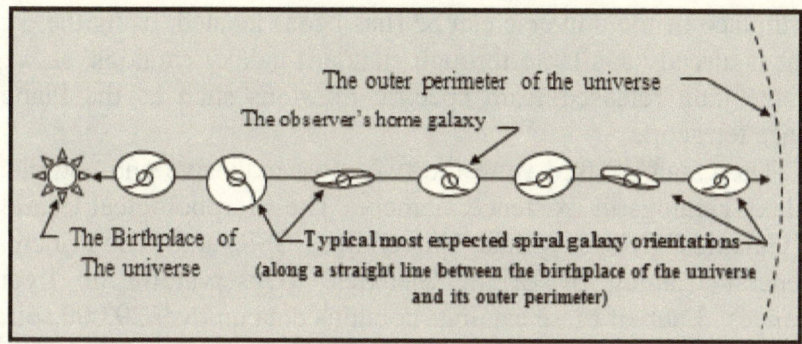

The outer perimeter of the universe

The observer's home galaxy

The Birthplace of
The universe

Typical most expected spiral galaxy orientations
(along a straight line between the birthplace of the universe
and its outer perimeter)

It must be noted that, the mutual gravitational forces that particularly nearby galaxies exert on each other affect the orientations of their rotational planes. Therefore, over billions of years, the orientations of most of the galaxies have changed, and will continue to change in the future. Eventually, the orientations of their rotational planes will become truly random, as viewed from the position of any given galaxy in this universe.

Therefore, **the orientation of the rotational planes of the farther away flattened galaxies, which are actually viewed as they were farther in the past, must be more-or-less their own original directions, before they were seriously affected by the gravitational forces of their neighboring galaxies.**

If sufficient amount of time has not passed already, for the orientations of the rotational planes of the galaxies to become totally random, **one should still be able to use orientations of spiral galaxies' rotational planes to determine the coordinates of a very unique direction in the sky, a direction that points towards the birthplace of this universe.** Correspondingly, the coordinates pointing in the opposite direction will be directly towards the outer perimeter of this universe.

The details needed to perform such a task (for **five different methods**) are described in the chapter titled "Locating the Birthplace of the Universe". Using the data already available to the general public through the internet, the author has already performed this very task, by applying three of these methods. They were found to agree in their final results and therefore confirmed each other's findings. The results of this undertaking are presented in the chapter titled "The Birthplace of the Universe (Discovered)". In that chapter, it is clearly shown that, the birthplace of the universe can be (has been) located, using the data that is already available through standard galaxy catalogs, as well as the data released from specific missions such as the Planck space telescope.

The catalogs used were the two most extensive and complete galaxy catalogs in existence, namely "The Morphological Catalog of Galaxies" (Moscow State University, USSR) and "The Principal General Catalog of Bright Galaxies" (Observatoire de Lyon, France). Each of these catalogs contains data on over 29,000 spiral galaxies. According to the results obtained,

"The direction pointing towards the birthplace of the universe has been found to be at <u>about 120 degrees Longitude East and about 30 degrees Latitude North,</u> according to the Galactic Coordinate System. These coordinates also correspond to the Northern Equatorial Pole."

The following figure shows the direction pointing towards the birthplace of this universe, in relation to earth and sun.

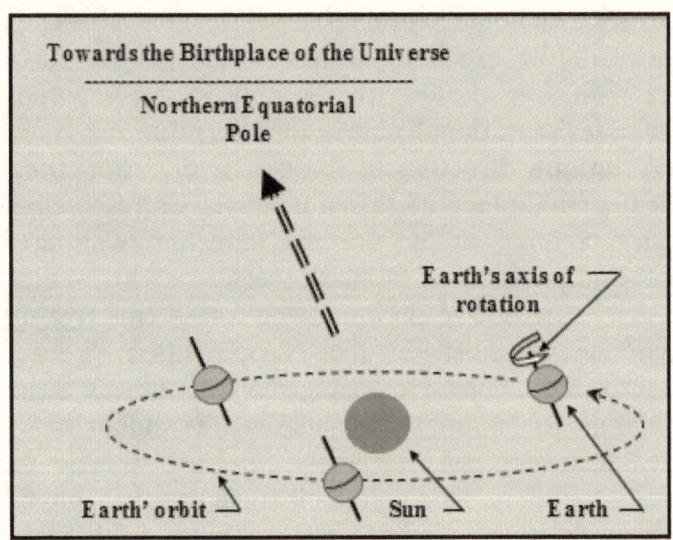

Note that, since there is a noticeable ratio of galaxies with rotational planes that are still oriented towards the birthplace of the universe, the following statements can be made with definite certainty,

"This universe has begun with a Big Bang."

Also,

"This universe is finite."

Since, if it were infinite, the orientations of the rotational planes of all galaxies would have been totally in random, but as it is shown, they are not.

Formation and Evolution
of
Physically Living Beings

Physically living beings owe their existences to the cooperation of the contents of two worlds, namely those of the physical world and those of the spiritual world. Since the beginning of time, both worlds as well as their respective constituents have gone through many changes.

The physical world performed its task by allowing the formation of stars and planets. Planets, due to their surface temperatures being much lower than those of the stars, allowed the formation of atoms and molecules which in turn led to the formation of gases, liquids, and in some cases even solids, each with a variety of compositions. Also, as various types of gasses gradually accumulated around most of the planets, they formed their respective atmospheres.

Planets that were at a certain range of distances from their respective stars had surface temperatures as well as overall surface and subterranean activities which were moderate. Some of the planets also had the proper gaseous composition forming their atmospheres and/or had the proper liquid or even solid compositions forming their surface materials that were necessary environments in which the physical bodies of living beings could form and survive / thrive.

Over time, surface and atmospheric conditions on some of the planets led to the formation of water and other chemical compounds, such as the amino acids. Such chemical compounds were formed due to **being directly controlled / guided by the**

contents of the spiritual world, namely spirits. Gradually, while being manipulated by spirits, such compounds led to the formation of various simple types of physical life-forms which manifested as the varieties of primitive types of plants as well as animals.

The spiritual world has performed its function by allowing the individual tiny incapable spirits to join together and step by step form more complex and more capable spirits that could take over evermore complex physical bodies and perform evermore complex tasks.

In order to properly understand the joining process of spirits together and their residing in individual living beings (both plant and animal types of life-forms), one needs to be familiar with the spiritual world and particularly the internal structure of complex spirits.

The author has presented such information in great details, in a separate book titled "The Evolution of Spirits" (2012). A summary of which is also presented in a different book titled "Purpose of Life in This Universe" (2012). Also, a detailed description of how physically living beings have formed and have gradually evolved in this universe is presented in a separate book titled "The Formation and Evolution of Physically Living Beings" (2015). A brief and yet combined description is presented below.

All of the spirits started their existences at the very same instant, as an original unified spirit split into infinite number of tiny pieces. The tiny spirits have been and still are evolving through various stages, as they reside in progressively more complex physical bodies and simulate more challenging life-forms. They start with plant type of life-form simulations. Later on, they progress into simulating animal life-forms.

In between physical life-simulations, they unite with one or more other tiny spirits that used to be their immediate neighbors in the original unified spirit, just like the pieces of a jigsaw puzzle fit together. In so doing, they form evermore complex spirits.

As a given complex spirit starts a new physical life simulation, different parts of it take over the operations of different body parts. In the case of plant type of life-form simulations, each complex spirit stays intact and occupies a single leaf or a single petal in a plant's flower (not the whole plant or the whole tree), and takes over its internal operations. In other words, each plant or tree can be considered to be like a village and the individual leaves are the residents of that village.

However, in the case of animal life-forms, each complex spirit divides itself into two major parts, upon taking over a physical body which is forming either inside a womb or inside an egg. Each major part is composed of several simpler spirits.

- The constituents of one part (which is referred to as the "Soul" of that animal) take over the internal functions of its physical body, as they try to control the development and operation of individual body parts in a coordinated fashion. The proper coordination of the tasks performed by different body parts (just like different departments in a complex organization / factory) ensures the proper health of that physical body.

- The constituents of the second part (which is referred to as the "Spirit" of that animal) take over the external functions of the living being, including communication with the outside world and analyzing the information gathered through various senses.

The overall abilities, as well as limitations of any given complex spirit are dictated by the previous life experiences of its constituents / spiritual members. However, depending on which spiritual constituent is selected as the leader/organizer in a complex spirit, the complex spirit (the living being) will demonstrate different sets of abilities and limitations, as it starts its new life experience. It will even have a specific personality, since the personality of the leading spiritual constituent will be the dominant one.

Even though some of such abilities and even limitations will be quite obvious from the early stages in the life of a given living being, some others can stay hidden or dormant, due to various reasons. However, they are there and can manifest at any time later, as the proper circumstances arise.

The spiritual components of any complex spirit always try to preserve their configurations. However, as they enter a physical body, due to the conditions and situations they encounter in the body and also how the body can be guided to develop into its final form, they have to improvise in their configuration so that they can adapt to the potential abilities as well as limitations of the physical body in which they are residing. These changes in the configuration of the complex spirits can even include choosing a different constituent as the leader, which automatically implies that a different personality and a different set of characteristics will be exhibited by the individual, later on in life.

Therefore, the very same complex spirit can and will demonstrate possessing different abilities and limitations, as well as personalities, as it resides in different bodies or even the same body but under different circumstances.

This is exactly like when a chef is asked to bake a certain kind of pie under different conditions which may include using different types of cookware, different sources of heat and even within different amounts of time. Even though, the raw materials can be nearly exactly the same, but due to different combinations of other variables involved in the process of baking, the finished pie can have a surprisingly different flavor, every time.

Such potential variations in the outcome of the union between a given complex spirit and a physical body can also be likened to different scenarios that can arise as an individual is allowed to drive a bicycle, a motorcycle, a car, a pickup truck, a semi-truck or a dump truck, on roads that might be in quite different conditions.

Spirits that simulate animal life-forms are very complex spirits that are formed from the unification of many simpler spirits. During their previous lives, these simpler spirits had identified each other as adjacent neighbors in the original unified spirit and had managed to unite, step by step (a few at a time), after they had finished their concurrent physical lives. By joining together, step by step they have formed more complex and more complete spirits that are more capable than each and every one of the spiritual constituents in the group. Any newly formed more complex spirit has to adapt itself to the next animal body in which it is going to reside, so that it may benefit from all of its potential abilities while coping with its potential limitations.

Even though, it is going to be limited to the physical body, the complex spirit must try to benefit from all of the knowledge that it has access to, due to each and every one of its spiritual components. This is just like an individual who has lots of knowledge and experience in regards to the construction of various types of vehicles, yet he/she is placed in a situation where he/she has access to limited materials / components / parts. He/she will still try to use the available information and expertise to better his/her own life, as well as the lives of the others who are around.

In short, the internal structure of any complex spirit that resides in a physical body depends on:

- The previous experiences and abilities of individual spiritual components involved in its collective, and

- The internal structure as well as the abilities of the physical body in which the complex spirit resides.

As the spiritual components join together and form a complex spirit, they choose one spiritual component to act as the leader. The leader has to follow certain spiritual rules of conduct, as it manages the group.

The newly formed complex spirit will wander in the spiritual world for a while, as the spiritual components coordinate their positions in the complex spirit. This is exactly like several individuals who get together to plan for a joint trip to some faraway destination. As the abilities of each and every spiritual component are identified and their respective duties are specified, with their overall agreement a specific trip is picked for the group, a trip which is in fact experiencing another physical life in this physical world.

From the time it enters a new physical body, the complex spirit has to try to benefit from any and all of that body's abilities. Again, this is exactly like the scenario in which a group of friends decide to go for a distant trip together, and depending on the type of vehicle they are provided with, which may be a car, a truck, a plane, a train, a ship or even a bicycle, they have to make the best of it (including choosing the right member in their collective to act as the leader or the driver/operator of the vehicle), so that they can get themselves to their desired destination.

After accepting the vehicle, which is the new physical body, all of the spiritual components must cooperate with each other and try to perform their assigned tasks in a coordinated fashion, so that they may succeed in achieving their common goals.

Note that, the involvement of spirits in the formation of the amino acids, as well as the following steps leading to the development of living organisms, can be readily likened to the involvement of humans (intelligence) in the proper mixing of the raw materials such as soil, water and so on, as they are made to take the form of bricks, etc. Then, step by step, such bricks, along with other similarly manufactured components, are put together and are encouraged to take the shape of the needed walls and so on, and finally form buildings, in which the individuals (humans) can reside, as shown below.

If not assisted by spirits (humans or any other creative living being), raw materials can and will continue on sitting where they are, as they may have already been for millennia, and nothing will ever become of them, as shown below.

Yet, once intelligent beings such as humans get involved and organize them in certain proper orders, the very same raw materials can and have taken the shapes of various small and large scale structures such as skyscrapers, bridges and so on, samples of which are shown below.

The very same can also be said about the involvement of many other types of animals in such processes. This is particularly obvious in the case of birds, since by manipulating the locally available raw materials in different ways, they construct their desired/particular styles of nests. In other words,

"For raw materials to take the shapes, or form compositions, which are necessary to perform higher functions (than just being raw materials), they must be manipulated / controlled by some form of intelligence which, in one way or another, is provided by spirits."

For instance, in the case of the amino acids, it was the direct involvement of spirits that led to the proper bindings of various elemental constituents which were necessary to form them.

Over time, as more capable spirits start their respective physical lives, by becoming superimposed on physical bodies, they make certain changes to their host bodies. Such changes are made during different periods in the lives of the physical bodies. Some are implemented as the eggs are forming / fertilized, while others are implemented as the embryos are developing into their final forms before being born. Yet, some changes are gradually implemented as the living beings are faced with different living conditions in their immediate environments and/or due to the availability of food. These changes are necessary so that those bodies can either perform certain tasks or be able to cope with certain physical limitations and/or environmental conditions.

The following are a few of the types of changes in the overall construct of various physical bodies of plant type and/or animal type of life-forms (as well as actions taken by them) that are made possible only with the direct involvement of spirits.

- Plants and animals knowingly have developed a variety of defense mechanisms to protect themselves against local entities which may not be so pleasant or friendly in their interactions. Development of claws, thorns, poisons and camouflage are good examples of such defense mechanisms.

- Physically living beings have developed various external and internal capabilities that enable them to survive. The development of the immune systems has been to ensure the ongoing health of the physical body by fighting any harmful substance that may attempt to interfere in the proper operation of any of its compartments.

- Various types of digestive systems were developed according to the desired dietary needs of different animal bodies. Some animals are vegetarian, while some others are carnivores. Yet, some others can munch on both kinds of food.

- Instinctive information is also indicative of the presence of spirits within physically living beings.

- The development of various body parts such as muscles, arms, legs, neck, head, skeleton, wings, tail, various internal organs, and so on are emphasized as the particular physical body is further developed, generation after generation.

This aspect of the development of physical bodies can be likened to the manufacturing of various types of cars, trucks and so on. Since, every year their designs are fine-tuned and improved upon, so that they can perform certain desired tasks better, over longer operational life, while preferably requiring less maintenance / repair.

- The sensory system is also developed according to necessity. In some animals the hearing is more important, in others the sense of taste, smell or touch is of prime importance, yet in some others the acuteness of the visual sense is taken as priority. In each case, the proper organs were developed (invented) and perfected so that the physical body could acquire its needs and also become fully aware of what is going on in its immediate vicinity.

- The mental capabilities manifested by different types of animal life-forms are of a variety. Since, mental capabilities were developed according to the specific needs of each particular life-form, as they needed to perform their everyday chores.

- Various animal types have developed different ways of communicating either with their own kinds or with other types of animals. In all cases, the body language has been the most needed/used, yet the verbal languages are also well developed in the case of many of the animal types. In some cases, the verbal languages are developed so well that they are well structured.

- The various techniques used to gather or obtain food are developed according to the type of physical body that a particular animal has. Their skills in performing such tasks are quite obvious in many animals. Even plants readily demonstrate their abilities in performing such tasks.

- Different types of feelings are also well developed in most animals. They can readily express their reactions, as they face different scenarios or situations. Plants also demonstrate their reactions to music and other phenomena, as well.

• Many of the animal types also clearly demonstrate possessing some form of extra-sensory perception or premonition in regards to events that are going to take place in the near future, let alone what is taking place at the present time and is not quite obvious to the other types of animals.

Some animals have developed mentally to the point that their spirits can wander around in the spiritual world and basically experience dreaming, while their souls stay in their physical bodies and keep them functioning.

Note that, spirits residing in leaves of plants cannot experience dreaming, since if they leave their respective host leaves, those leaves will die.

• The development of flaps, tail, wings, legs and other body parts have enabled animals become mobile in different ways. The development of such body parts in various animals can be likened to the development of various modes of transportation by humans, modes of transportations that include walking, carts pulled by animals, bicycles, motorcycles, cars, trucks, ships, airplanes and even rockets.

• The self-healing capability of various body parts is well demonstrated by just about all animals. In most cases, the healing is complete, depending on the external effects such as germs and so on not interfering. While, in some cases scars are left behind indicating where the old injuries had occurred. In some cases, the cut-off body part is replaced by a newly re-grown one.

• The overall operation of the various body parts in any of the animal (as well as plant) types of life-forms are clear indications that they are coordinated / managed by some sort of intelligence, since otherwise their operations can easily jeopardize the very health of the physical body they are part of.

The very formation of more complex life-forms from simpler ones with the direct help and guidance of spirits can be likened to the manufacturing of more complex and more capable vehicles by humans. Step by step, as the need arises, due to a variety of requirements and availability of resources and different technologies, many parts are added, taken away or modified. The result is a new vehicle that is capable of performing the desired tasks while exposed to the expected environmental conditions.

Particularly, as more stringent requirements / conditions are imposed, various special vehicles are designed and built using available resources and newly proposed scientific knowledge and technology to perform specific tasks that are literally of extreme in certain ways. Such scenarios include (but not limited to) mining trucks, ships, airplanes, rockets and weapons, let alone rovers crawling on the surface of the other planets, and so on. In other words,

"Vehicles have not evolved and do not evolve on their own, but rather it is their individual users, in fact spirits residing in their users' bodies, that are the originators of variations in their designs and are responsible for their manufacturing, as well as implementing new modifications in their overall constructions, year after year."

The very same applies to the case of physically living beings in this universe. Since, new species start as the residing spirits induce the needed changes in the genetics of already existing species. The following generations simply copy what has been done and so a new species of animals forms. This is already shown in practice by humans, through genetic engineering of both plants and animals. Yet,

"Even though genetic engineering is done by external spirits (from outside, for instance by spirits residing in humans) and evolutionary genetics are done by the internal spirits (from

**inside, by resident spirits), both are done by
certain controlling entities called spirits."**

Note that, this method of introducing new animal species can be readily likened to the invention of new types of vehicles such as varieties of bicycles, motorcycles, cars, trucks, trains as well as ships and airplanes. Since, once one company designs and manufactures one demo, which demonstrates to have the desired features, such as a sedan, a station wagon, a hatchback, a van, a pickup truck, an SUV (Sub Urban Vehicle) and so on, other manufacturers jump on the band wagon and flood the market by copying the general idea (the overall design) and mass producing the same.

The very same can be stated in regards to the way varieties of bicycles, motorcycles, sports cars, trucks, boats, ships, airplanes, trains and even rockets have come into existence.

Even the very peculiar activities which are demonstrated by various types of animals during their mating seasons, or the way they take care of their young ones, as well as how they communicate either through their body language or by generating different types of sounds, all are due to the existence of spirits in them, individually. In other words,

**"The spiritual components residing in a
given living being not only develop the
physical body and decorate it as their home,
but also based on the local circumstances
and the goals in mind, conduct its actions,
such as its mating ritual, communication,
acquiring nutritional supplies, protection
and so on, accordingly."**

The efforts contributed by the contents of the physical world and the contents of the spiritual world, leading to the formation of living beings in this universe, can also be likened to the combination of the physical efforts put into the manufacturing of various types of vehicles and the teaching efforts put into the training of qualified drivers for them. Since, as properly trained drivers operate such vehicles, they enable those vehicles to manifest as if they are alive, in a sense.

Anyone who disputes the validity of such a comparison is encouraged to visually compare the motion of vehicles on city traffic routes or even on busy highways, as viewed from high above the ground, to either the movements of humans walking on sidewalks or the movements of ants along their busy food trails. Even though at first glance, vehicles demonstrate randomness in their movements, yet individual vehicles, even human bodies as well as those of ants are controlled by spirits who by acting as their operators are knowingly trying to get to certain destinations. In other words, none of their movements are performed based on chance or at random. Samples of such cases which clearly demonstrate their similarities are shown in the following pictures.

It is due to being controlled and guided by spirits that physical bodies express themselves in a lively fashion. That is why, as a complex spirit due to any reason leaves its host body, that body loses its living status and becomes just like the many driverless vehicles that are parked on the sides of the streets or are parked in parking lots, let alone the vehicles that are not properly maintained or operated by their now-missing operators and have ended up in various auto wreckers / junkyards.

In short,

"Just as various evermore capable means of transportation have been designed, manufactured and operated by humans, the evermore complex physical life-forms have been and still are being formed / developed, step by step, by their correspondingly evermore capable guest spirits directly controlling their developments."

Current Conditions in This Universe
(Acceleration of the expansion of space)

From the time when the initial rapid expansion of the universe slowed down, the overall distances between the remaining matter particles have been (and still are) increasing both due to their individual movements IN the aether medium and also due to being carried by the expansion of the aether medium (space), as a whole. These two types of motions and the gradual decreases in the density and the pressure of aether are causing the gradual weakening of the force of gravity, on a universal scale.

As it is described in greater details in the chapter titled "What is Gravity?", the force of gravity at any given location in this universe is the drag force induced by the flow of aether which is accelerating through that particular location. Any given particle generates an aether flow towards itself from all of its surrounding volume of space. The rate at which aether flow accelerates towards a given particle and hence the drag force (gravitational force) it exerts on any other particle which happens to be in its path is inversely proportional to the square of the distance from that particle.

This relationship was given by Mr. Newton and it is known as the inverse square law,

$$F = Gm_1 m_2 / x^{2.000}$$

Where, F is the gravitational force that two objects of masses m_1 and m_2 separated by a distance of x exert on each other. And, G is the universal gravitational constant.

In the future, more accurate measurements will indicate that the force of gravity associated with aggregates of matter particles such as stars and planets (in relatively short ranges, such as within a given galaxy) changes slightly faster than it is predicted by the inverse square law. This is due to the gradient that exists in the density profile of aether as it approaches an aggregate of matter particles, since its density decreases by a minute amount. Hence, its speed increases to compensate, in order to allow for the same amount of aether to flow through, at all distances from the matter particle aggregate.

Therefore, since the same amount of aether is passing through but at slightly higher speeds than expected (as compared to if aether density were to stay constant), the induced drag force in close proximity of the gravitating body becomes slightly greater than it is predicted by the inverse square law of Mr. Newton. In other words,

On relatively smaller scales (within galaxies), <u>the dependence of the force of gravity on distance</u> is slightly stronger than it is dictated by Newton's inverse square law. For example,

$$F \approx Gm_1 m_2 / x^{2.001}$$

Also, the separations of galaxies (particularly between galaxies that are more distant from each other) are changing <u>due to their receding motions</u> with respect to each other, receding motions that are <u>in the aether medium</u>. This receding motion of galaxies is over and above their receding motion that is due to their being carried by the expanding aether medium (space), as a whole. The motions of galaxies <u>in the aether medium</u> expectedly lead to a stronger force of gravity than expected over long distances, particularly between distant galaxies.

This is due to the fact that, longer distances are observed after the passage of longer periods of time which include older times, when galaxies were closer to each other than the overall expansion rate of the aether medium would predict, and also <u>when aether was denser</u>. Hence, back then galaxies were justifiably exerting stronger gravitational forces on one another. In other words,

On intergalactic scales, <u>the dependence of the force of gravity on distance</u> is slightly weaker than it is dictated by Newton's inverse square law. For example,

$$F \approx Gm_1 m_2 / x^{1.999}$$

The force of gravity which is the drag force induced by the flow of aether accelerating through any given location in this universe, is directly dependent on aether's density and pressure at that particular location. Therefore, as the overall density and pressure of the aether medium are decreasing with time, due to aether's expansion and leakage, the flow of aether is gradually becoming less effective in inducing the force of gravity. In other words,

"The Universal Gravitational Constant is gradually decreasing in magnitude."

Based on the calculations performed by two independent groups of scientists in 1998, using data collected through direct observation of various supernovas in distant galaxies, the universe was found to be accelerating rather than deceleration in its expansion process. The overall history of the expansion of the universe is shown below, in a simple fashion.

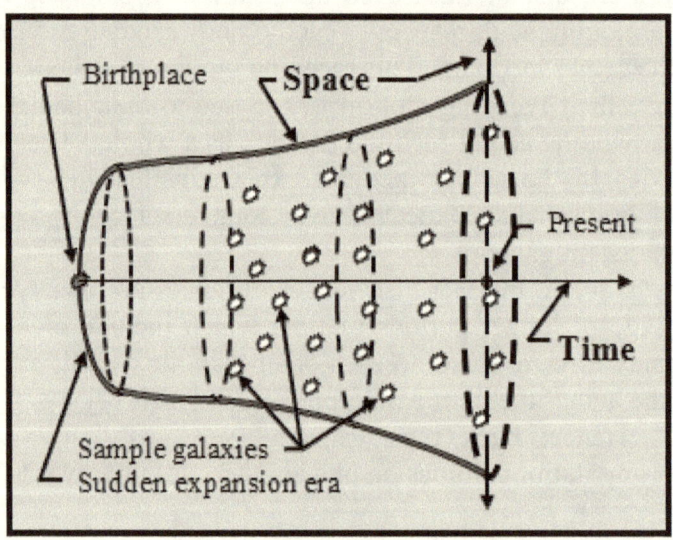

Even though, various theories have been put forward suggesting that some unknown form of energy, namely dark energy, is responsible for the current accelerated expansion process of this universe, there have not been any explanations on what can potentially be the source of this dark energy.

However, once the effects of the gradual decreases in both the density and the pressure of the aether medium in this universe on the overall expansion process of aether medium and also on the strength of the force of gravity are taken into account, on a universal scale, the reason for such accelerated expansion process can be readily understood and such findings can be well expected.

Notes:

- **<u>The internal pressure of the medium of aether is the driving force for its expansion,</u> and the expansion rate of the aether medium is directly proportional to its internal pressure.** This characteristic of the aether medium is analogous to that of a volume of pressurized gas which is allowed to expand uniformly in all spatial directions, into its surrounding environment.

- **The gravitational force that celestial bodies exert on one another is roughly proportional to the square of any variations in the pressure of aether that is in this universe**, because any decrease in the pressure of aether in this universe:

  1- Directly reduces the pressure difference that exists between aether that is in this universe and aether that is in the accompanying universe, the same pressure difference that is the driving force pushing aether towards and through any and all of the available drain holes, namely matter and anti-matter particles, as well as their aggregates (including neutron stars and black holes).

  As the pressure difference (of aether) between the two universes decreases, aether's flow rate towards matter and anti-matter particles drops. Hence, the drag force exerted by aether on any and all other particles that happen to be in its path gets reduced accordingly, as well.

2- Directly reduces the density of aether in this universe, because aether is a compressible fluid. This reduction in aether density in turn leads to even weaker drag force (weaker force of gravity) induced by the accelerated flow of aether, anywhere and everywhere in this universe.

Therefore, if the pressure of aether in this universe is reduced to one half, the force of gravity will be reduced to **about** one quarter. Since, as aether is transferred from this universe into the accompanying universe the overall pressure of aether that is in the accompanying universe is gradually rising.

- Other than due to aether's expansion in this universe, **aether's density and pressure are also decreasing further because of aether leaking from this universe**, through matter and anti-matter particles, as well as their aggregates (including neutron stars and black holes). Aether's leakage from this universe directly leads to a more pronounced effect due to expediting weakening both the expansion force and the force of gravity.

- **Due to the ongoing expansion of this universe the separation distances between the galaxies are increasing.** As a result, galaxies are gradually exerting weaker and weaker gravitational forces on each other, that is even if the pressure and the density of aether were to stay constant. It should be noted that, the receding motion of galaxies is due to two types of motions, namely:

1- The motion of galaxies IN the aether medium.
The **general** direction of this motion of galaxies is away from the birthplace of the universe.
This motion of the galaxies can be likened to the movements of individual ants running away from a

central point which is drawn on the surface of a balloon that is inflating.

2- The motion of galaxies due to being carried by the expanding aether medium.

The direction of this particular motion of galaxies is **exactly** away from the birthplace of the universe.

This motion of the galaxies can be likened to the motion of raisins in a raisin cake/dough relative to the center of that cake/dough, as it is baking.

- The aether medium in the accompanying universe is expanding at a much slower pace as compared the aether medium in this universe. Since, its pressure is much lower than the aether pressure in this universe. Therefore, the pressure of aether in the accompanying universe **is rising as it is receiving aether from this universe**.

In fact, due to aether's pressure and volume being much greater in this universe as compared to those of aether in the accompanying universe, aether's pressure in the accompanying universe is rising at a much faster rate as compared to the rate at which aether's pressure is dropping in this universe.

Therefore, as the overall volume of aether is doubled in this universe, even though the pressure of aether which is the driving force for the expansion of this universe, as a whole, is reduced to roughly one half, the gravitational force which is responsible for slowing down the existing expansion rate of this universe, is reduced to less than one quarter.

The following figure shows the variations in the relative magnitudes of the force of gravity and the force of expansion in this universe (both normalized to their current values), as the overall volume of this universe is gradually doubled in size.

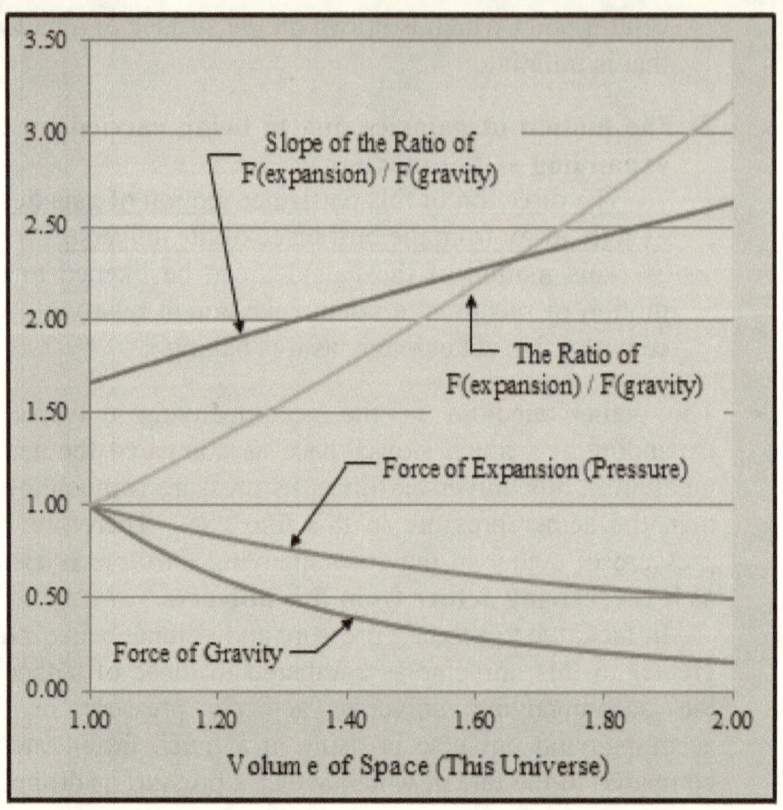

As it is obvious in the figure, the ratio of the force of expansion to the force of gravity has a positive slope. Also, the magnitude of this slope is gradually increasing as well. This is indicated by the curve marked as "Slope of the Ratio of F(expansion) / F(gravity)". In other words,

"Over time, the ratio between the force that is responsible for the expansion of this universe and the force that is trying to slow the rate of its expansion down is gradually increasing. Hence, <u>the expansion of the universe</u> (the matter medium) <u>is accelerating instead of decelerating</u>."

Note that, the universe (the matter medium) is accelerating in its expansion rate, due to:

- **The overall weakening of the gravitational force, and**

- **Being dragged at faster and faster paces by the local aether which is experiencing an overall expansion of its own.**

However,

"Due to the gradual decrease in aether's internal pressure which is the driving force for that medium's expansion, the aether medium in this universe, as a whole, is continuing to slow down in its overall expansion rate."

The overall expansion rate of the aether medium in this universe is still much faster than the speed of light in that medium. The current rate of expansion of the matter medium is only a fraction of the speed at which the fluid aether medium is expanding. However, as time goes on, the aether medium is dragging the matter contents of this universe at faster and faster paces. This is while the overall expansion rate of the aether medium is gradually slowing down, due to gradual decrease in aether's internal pressure.

Eventually, the expansion rate of the matter medium will become equal to that of the aether medium. Afterwards, both will gradually slow down towards a complete halt, as the overall aether pressure gradually approaches zero.

The following figure shows the accelerating expansion process of the matter medium inside the decelerating expansion of the fluid aether medium, as a whole, which defines 'space' in this universe.

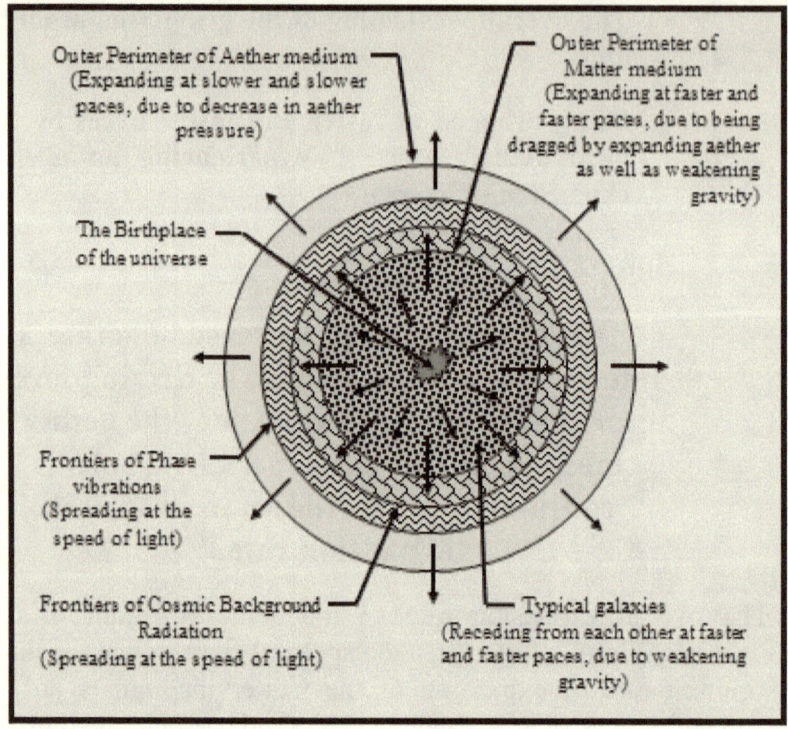

The following figure shows a general overview (NOT to scale) of **the relative expansion rates** of the fluid aether medium and the matter medium, as well as the expansion rates of the cosmic microwave background radiation due to the initial phase vibrations that were introduced in the aether medium and the cosmic microwave background radiation due to phase vibrations which were generated later on, as most of the matter and anti-matter particles annihilated each other.

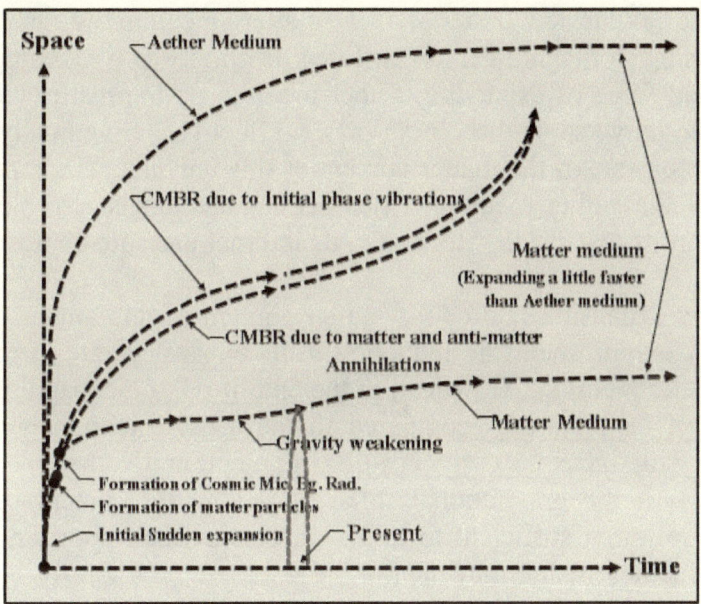

This figure shows how the initial rapid expansion of the aether medium (space) ended, as matter (and anti-matter) particles gave rise to the force of gravity with their very formation. By the time most of the matter and anti-matter particles had annihilated each other, the overall spreading rate of the leftover matter particles and the expansion rate of the whole aether medium were drastically reduced.

As it is shown, the expansion rate of the portion of the cosmic microwave background radiation (CMBR) which was due to the initial phase vibrations was changed according to the reduction in the expansion rate of the aether medium, as a whole. The portion of the cosmic microwave background radiation (CMBR) which was due to the matter and anti-matter annihilations experienced the same fate as the initial phase vibrations, since both were phase vibrations and both had the same speed of propagation in that medium.

Over time, both types of CMBR have gradually slowed down, due to the reduction in the expansion rate of the aether medium. But, they will gradually speed up, as the density of the fluid aether medium continues to decrease. Eventually, the propagation speed of all types of phase vibrations in the medium of aether will approach infinity, as the density of aether approaches zero.

The expansion rate of the matter medium is gradually increasing in magnitude, as the force of gravity is weakening and the drag force of expanding aether medium on the matter content of this universe is becoming more effective. Eventually, the expansion rate of the matter content of this universe will approach that of the aether medium, since aether medium is also slowing down in its expansion rate, due to its internal pressure approaching zero.

The expansion rate of the matter content of this universe will closely exhibit the same reduction in its expansion rate as that of the aether medium. However, <u>at the end, it will be expanding at a little bit faster rate as compared to the aether medium, since it would still have its own receding momentum IN that medium.</u> That is, if the gravitational force of the matter content of this universe is not sufficient to totally neutralize their spreading rate IN the aether medium, by then.

As aether's density and pressure are decreasing with time, aether's accelerated flow towards particles and their aggregates is becoming less effective in dragging other particles (objects) that happen to be in its path. Therefore, stars are gradually exerting less of a gravitational force on their respective planets. This is in turn leading to the widening of the orbital paths of planets around stars, on a universal scale. One can say that, stars are gradually losing their grips on their respective planets and consequently,

"The planetary orbits are gradually becoming wider."

This side effect of decrease in aether's density and pressure has already been detected in regards to earth's orbit around the sun. According to the collected data, the orbital path of earth around the sun is currently widening by about 7 meters (about 23 feet) per century. A small portion of this observed amount of expansion in the orbital path of earth is justifiably due to:

1- Sun's gravity is gradually becoming weaker since it is losing mass as it is continuously transforming some of its matter into energy. It is also literally throwing huge amounts of matter particles / ions into space, in the form of solar flares / solar storms.

2- Earth is being pushed away, as it is absorbing and/or deflecting particles that are thrown at it by the sun, in the form of solar storms.

Also, due to the very same effect, namely weakening of the force gravity, the orbital paths of all stars around the center of mass of their respective galaxies are widening. As a result, **not just the star systems but also all of the galaxies are gradually expanding in diameter.**

Since particles of matter (as well as anti-matter) are literally bubbles in the medium of aether, the physical size of any given particle is particularly dictated by the local aether's density and pressure. Consequently, as the density and the pressure of aether are gradually decreasing, due to its overall expansion as well as its leakage into the accompanying universe,

"In this universe, any and all of the matter and anti-matter particles are gradually becoming larger in size."

Note that, when the universe was much younger, due to the density and the pressure of the fluid aether being much, much higher than what they are at the present time, matter and anti-matter particles were comparatively much, much smaller in size, back then.

Also, as the density and the pressure of the fluid aether is decreasing with time, the resistance that the fluid aether can generate in response to changes in the movements of particles is becoming weaker, the same resistance that is manifested as the mass of the particles. Therefore,

"In this universe, the mass of any and all of the particles is gradually decreasing."

Note that, when the universe was much younger, due to the density and the pressure of the fluid aether being much, much higher than what they are at the present time, any and all of the matter and anti-matter particles were comparatively much, much more massive, back then.

As density and pressure of aether are gradually dropping, in this universe, the speed at which phase vibrations including electromagnetic waves such as light and so on, propagate through aether is increasing. In other words,

"The speed of light is gradually increasing."

Note that, the current of electricity is induced by certain type of phase vibrations in the medium of aether which is superimposed by a conducting matter medium. Therefore, as the propagation speed of phase vibrations in the medium of aether is gradually increasing, due to decreasing aether density in this universe,

"The speed at which phase vibrations associated with the generation of electrical currents propagate through any and all matter mediums is gradually increasing."

In short,

"The speed of electrical signals is gradually increasing, in this universe."

According to the information presented in the chapter titled "What is Time?", the rate at which time is experienced by any object (inanimate objects as well as living beings) is directly dependent on that object's speed relative to its local aether medium, as compared to the speed of phase vibrations in that medium. As this relative speed approaches the speed of phase vibrations in the local aether medium, the object experiences the passage of time at an ever slower pace.

The gradual decrease in the density of the aether in this universe implies that the speed of phase vibrations in that medium is gradually increasing. However, the speed of matter particles in aether is not changing by much, due to their mutual gravitational forces. Hence, the speed of matter particles relative to their local aether medium is gradually becoming a smaller fraction of the speed of phase vibrations in that medium. Therefore, the matter contents of this universe are gradually experiencing the passage of time at a faster pace. In other words,

"In this universe, 'Time' is gradually speeding up."

The ongoing expansion of the fluid aether that is in this universe is stretching the wavelengths of all types of phase vibrations that exist in it. Such phase vibrations include the cosmic microwave background radiations which are in fact composed of the remnants of the initial phase vibrations that were introduced into the aether medium, and the phase vibrations that were generated due to matter and anti-matter annihilations.

Therefore, the magnitude of the Planck constant (in this universe) is gradually decreasing, since Planck constant is proportional to the amplitude of the background phase vibrations in this universe. In other words,

"In this universe, the magnitude of the 'Planck constant' is gradually decreasing."

According to the current estimates, the universe is about 13.7 billion years old, and it is about 93 billion light years in diameter. The contents of this universe, on different scales, have fairly stabilized in their developments and abide by certain fairly well defined laws, which are the same everywhere. For example,

- All physical entities are experiencing the passage of time. Even though, due to either moving at incredibly high rates of speeds, or being in strong gravitational, magnetic or electric fields, they can experience the passage of time at slower paces, but they all are experiencing the passage of time in the positive direction (always towards the future).

- All physical and chemical laws are the same across this universe.

- The atoms belonging to any specific element have the very same structure and exhibit the very same physical and chemical properties, regardless of where in this vast universe they were/are formed.

Current Conditions in
the Accompanying Universe

Currently, the size of this universe is much larger than the size of the accompanying universe. The following figure shows the relation between the two universes, as well as the overall size of space which is currently occupied by aether in both universes.

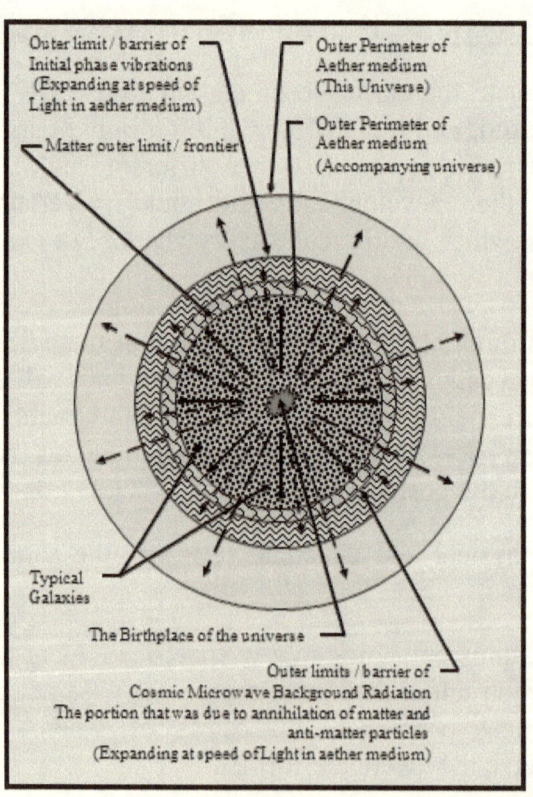

Outer limit / barrier of
Initial phase vibrations
(Expanding at speed of
Light in aether medium)

Outer Perimeter of
Aether medium
(This Universe)

Outer Perimeter of
Aether medium
(Accompanying universe)

Matter outer limit / frontier

Typical
Galaxies

The Birthplace of the universe

Outer limits / barrier of
Cosmic Microwave Background Radiation
The portion that was due to annihilation of matter and
anti-matter particles
(Expanding at speed of Light in aether medium)

The fluid aether that is in the accompanying universe is still under a much, much lower pressure as compared to aether that is in this universe. Consequently, its density is also much lower than that of aether that is in this universe. However, the internal pressure and the density of aether that is in the accompanying universe are gradually approaching those of aether that is in this universe.

As it is described in different parts of this book, a variety of phenomena are directly affected by the pressure and the density of the local fluid aether medium. Over time, such phenomena are bound to be affected by the variations in the internal pressure and the density of aether that is in the accompanying universe, as well. Therefore, the following are readily expected to be true in the accompanying universe,

- **Speed of light**

    The density of the fluid aether being much lower in the accompanying universe, as compared to what it is in this universe automatically implies that,

"The propagation speed of phase vibrations (such as light) is much higher in the accompanying universe, as compared to what it is in this universe."

Note that, higher propagation speed of electromagnetic waves in the accompanying universe as compared to what it is in this universe raises the possibility that the accompanying universe can be used to establish relatively instantaneous communication channels between various planets and even between star systems within a given galaxy.

- **Time**

    Due to much faster speed of phase vibrations (such as light) in the accompanying universe, the speed of anything and everything that exists in that universe is a smaller fraction of the speed of phase vibrations in that medium. Since, they are directly connected to their counterparts in this universe. Therefore,

"Currently, the rate at which the passage of 'Time' is experienced in the accompanying universe is at a much faster pace as compared to the rate it is experienced in this universe."

- **Gravity**

    According to the newly proposed theory of gravity presented in this book, the force of gravity at any given location in this universe is the drag force induced by the flow of aether which is accelerating through that particular location, as shown below. In this universe, the fluid aether is flowing towards matter and anti-matter particles (or their aggregates such as planets, stars, neutron stars and even black holes). Once the fluid aether reaches its destination in this universe, it gets transferred into the accompanying universe.

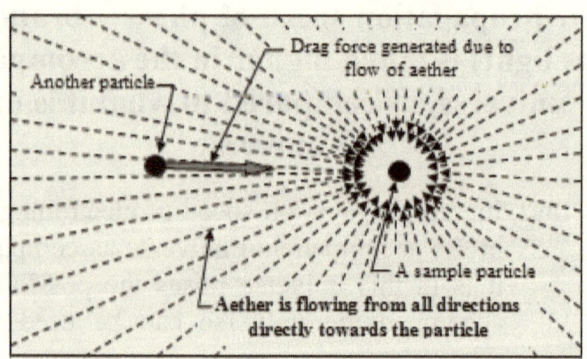

In this universe, as shown in the figure, the accelerated flow of aether towards matter and anti-matter particles gives meaning to the force of gravity. However, in the accompanying universe there is no such phenomenon as aether flowing **towards** matter and anti-matter particles. Therefore,

"There is no force of gravity in the accompanying universe."

In fact, as shown in the following figure, as the fluid aether is flowing away from matter and anti-matter particles in the accompanying universe, it gives rise to a phenomenon in that environment which can be truly referred to as **'anti-gravity'**.

Such anti-gravitating forces become stronger near larger aggregates of matter and anti-matter particles that correspond to planets, stars and particularly black holes in this universe.

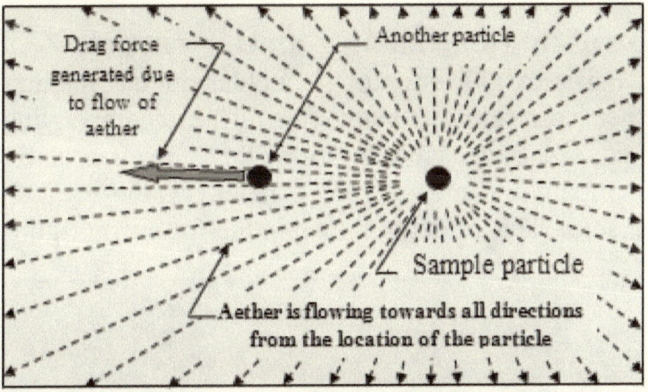

The following figure clearly shows the flow of aether towards particles in this universe, a flow that gives rise to the force of gravity.

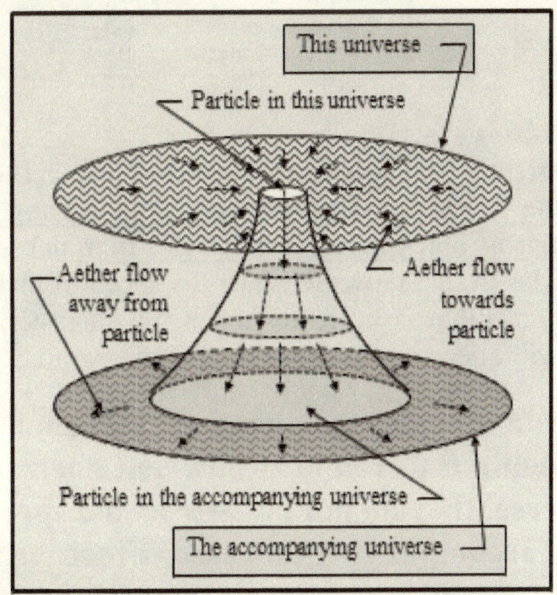

Correspondingly, the flow of aether is shown to be away from particles in the accompanying universe, a flow that gives rise to what can be truly referred to as the force of anti-gravity.

Note that, even though there is no force of gravity in the accompanying universe, aggregates of matter and anti-matter particles hold on to their structures due to the force of gravity that exists in this universe.

- **Size of matter and anti-matter particles**

  Particles are essentially bubbles in the medium of fluid aether, connecting this universe to the accompanying universe. Their sizes, as viewed from this universe and from the accompanying universe are dictated by the local fluid aether pressure. This effect is shown in the following figure.

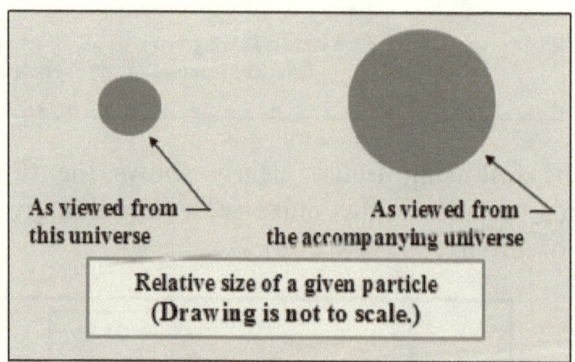

As viewed from → this universe

As viewed from → the accompanying universe

Relative size of a given particle
(Drawing is not to scale.)

- **Black holes**

  Currently, **the speed at which the fluid aether is flowing into the accompanying universe even through such huge openings as black holes, is much less than** the speed of phase vibrations in that universe, since the speed of phase vibrations in that medium is currently much, much faster than that of phase vibrations in this universe. In other words,

  "Even though black holes possess such a special domain referred to as an event horizon in this universe, they do not possess such a special zone in the accompanying universe. Since, as the fluid

aether that flows out of black holes, its speed is less than that of phase vibrations in that medium."

As black holes devour all of the phase vibrations (of all types and at all frequencies) that reach their event horizons, they transfer them into the accompanying universe. In so doing, they not only appear and act as fountains of fluid aether but also as strong sources of a variety of phase vibrations (including visible light), in the accompanying universe. In other words, as shown in the following figure,

"Contrary to what they pretend/appear to be in this universe, black holes act as brilliantly shining spherical light sources in the accompanying universe."

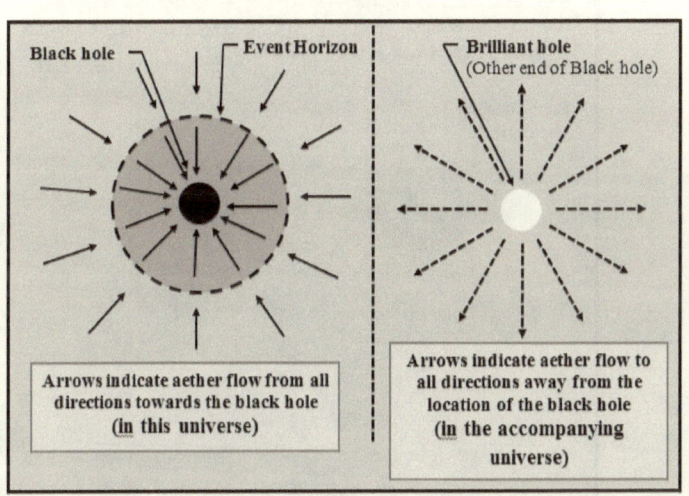

Also, stars (in this universe) are fair size collectives of passageways through which the fluid aether is transferred into the accompanying universe. In so doing, they allow the fluid aether to carry some of the phase vibrations along into that universe. In other words, one may state that,

"The visibility of stars in the accompanying universe is comparable to that of planets in this universe."

However, stars will be found to be orbiting the brilliant light sources located at the center of their corresponding galaxies. Those light sources are in fact giant black holes in this universe. In other words, as shown in the following figure,

"The accompanying universe is hosting a variety of light-giving sources of its own, just as this universe is."

The following figure shows how galaxies may appear to a hypothetical observer in the accompanying universe. The bright spots shown are the many black holes that exist in each galaxy. The sizes of the giant black holes, in the center of galaxies, are exaggerated to clearly indicate their existence.

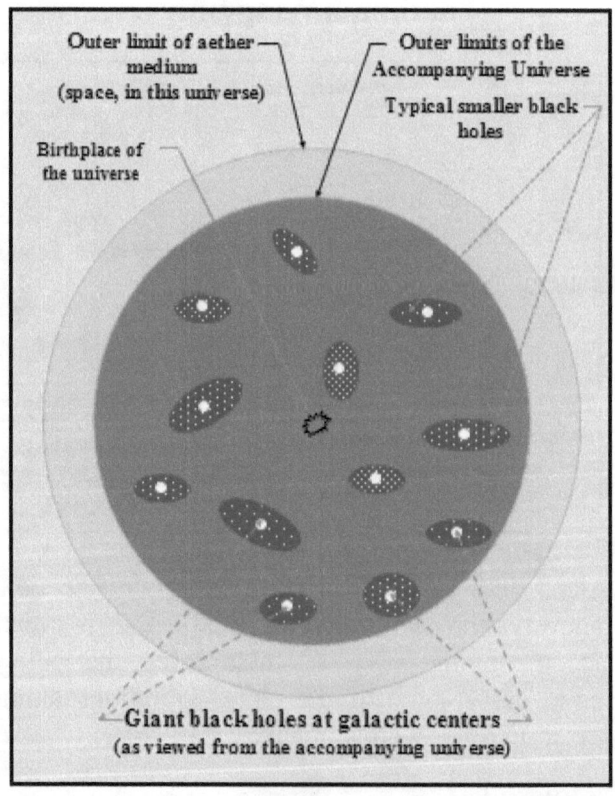

Giant black holes at galactic centers
(as viewed from the accompanying universe)

Note that, the general background in the accompanying universe is not pitch black, at all. Since, a small fraction of phase vibrations that reach freely wandering particles, at 90 degrees to their surface, in this universe, are transferred into the accompanying universe.

The larger the aggregate of matter particles in this universe, the brighter they appear to be in the accompanying universe. Hence, the giant particles, namely black holes in this universe, appear the brightest in that universe.

• Cosmic Microwave Background Radiation (CMBR) (Phase Vibrations)

As the fluid aether is transferred into the accompanying universe, particularly through black holes, it carries with it the varieties of phase vibrations that are propagating through aether medium in this universe. Such phase vibrations include the **cosmic microwave background radiations which are composed of the remnants of the initial phase vibrations and the phase vibrations that were generated due to matter and anti-matter annihilations.**

However, upon their entrance into the accompanying universe, the wavelengths of such background phase vibrations automatically become stretched, due to much lower aether pressure in that universe. Consequently, their amplitudes get reduced substantially.

As shown in the following figure, a simple setup can be used to demonstrate how the frequency and the amplitude of phase vibrations change as they are allowed to propagate from this universe into the accompanying universe.

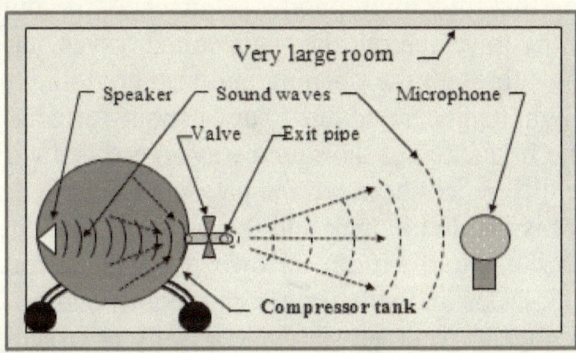

In order to perform such an experiment, one needs to use an air compressor, a speaker connected to a stereo system and a sound receiving system with frequency selection capability. The speaker should be installed inside the compressor's tank so that it is facing an opening which is equipped with a valve. The microphone which is connected to the receiving system should be positioned far away from the tank, but inside the very same large room wherein the compressor's tank is located.

Note that, the interior of the compressor's tank should be acoustically non-reflective, to reduce the amount of echo/noise as much as possible.

After the compressor's tank is filled to the highest possible pressure, the compressor should be turned off and the whole compressor should be left alone for a few minutes. This amount of time will allow the compressed air inside the tank to stop carrying any kind of noise that is related to the operation of the compressor, as well as stopping the physical internal commotion that is due to the initial momentum of air molecules, generated as air was pumped into the tank.

Next, while the microphone and the related sound system is prepared to receive and record the sound waves, the valve should be opened to allow the compressed air inside the tank to flow through and get mixed with the air that is in the room. Then, the stereo should be turned on to broadcast a monotonic sound of a certain frequency.

The microphone will pick up the vibrations corresponding to the monotonic sound waves that are generated by the speaker that is inside the air tank. Of course, the receiving system that the microphone is attached to, must have the capability to screen out the static sound waves generated as the air flows through the opening and also any other sound waves that might be generated due to any echoing effects.

The frequency of the sound waves received will be affected by the differences between the pressure (and the density) of the air that is inside the tank and the pressure (and the density) of the air that is in the room. However, as the air pressure inside the tank drops and equalizes with the air pressure in the room, the frequency received should gradually approach and exactly match the frequency of the monotonic sound that is broadcast by the speaker.

Note that, the very same type of experiment can be carried out for two different cases, namely **Subsonic flow of air through the opening** which will correspond to the flow of aether leaving this universe through individual particles, including their aggregates as large as planets, stars and even neutron stars, and **Supersonic flow of air through the opening** which will correspond to the flow of aether leaving this universe through black holes.

In other words,

"The accompanying universe hosts cosmic microwave background phase vibrations, also, but those phase vibrations are at different frequencies as compared to their counterparts in this universe."

The Future / Destiny
of the physical
Universe and its Contents
(This universe and the accompanying universe)

Over time, massive stars will experience a nova moment of their own, as many already have. During this process, while their outer layers get literally thrown into the surrounding space, their inner parts implode and give birth to neutron stars and black holes. Black holes can only grow in size. They will keep on devouring any and all of the planets and stars that get trapped in their one-way aether flow and are brought close enough.

Black holes will eventually devour all of the matter particles (and their aggregates) in their reach, as they follow their respective trajectories in galaxies. They will also form giant black holes, as they unite with each other.

Note that, the effectiveness of any black hole in devouring whatever that is nearby is directly dependent on the aether pressure difference between the two universes.

As time goes on and aether expands in this universe and also leaks into the accompanying universe, its pressure is continuously decreasing in this universe. Meanwhile, the pressure of aether is gradually increasing in the accompanying universe. In the future, the pressure difference between aether that is in this universe and aether that is in the accompanying universe will approach a critical value. After that, the existing pressure difference will no longer be able to encourage the flow of aether towards even the largest black holes to reach the speed of light (the speed of phase vibrations) in the local aether medium. This critical aether pressure difference can be referred to as,

"Esmailzadeh Critical Aether Pressure Difference"

As this critical pressure difference is approached, all of the black holes, starting with the smallest ones, will lose their status as being black holes, since light and other electromagnetic waves will be able to travel against the incoming flow of aether and escape. In other words,

"Eventually, all of the black holes will become visible."

The following figure shows a typical black hole's fate, as the pressure difference between aether in this universe and aether in the accompanying universe is decreased passed its critical value.

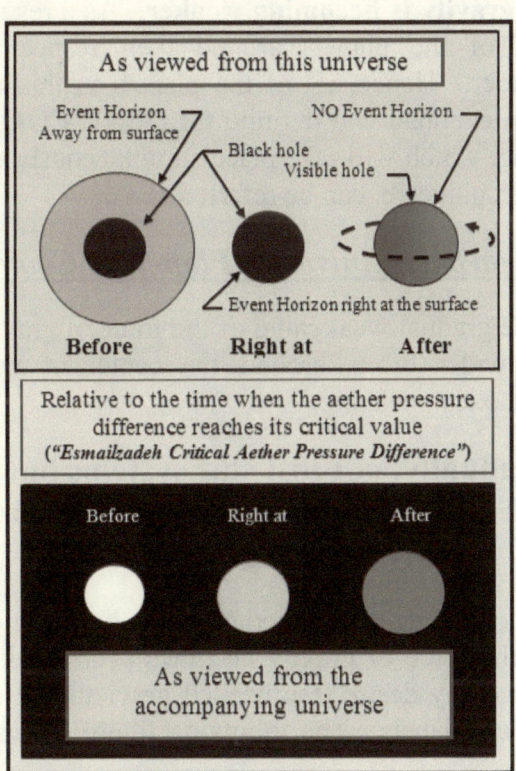

By **'visible hole'** it is meant that electromagnetic type of radiations of various frequencies can and will escape into the surrounding space. The colors chosen demonstrate the red shift and the blue shift generated due to the rotation of the 'then-visible' black hole.

Note that, as it is indicated in the figure, the physical size of black holes is gradually increasing, due to decreasing aether pressure in this universe. That is, even if they stop devouring their peaceful neighbors, they will still grow in size.

The very same effect will also be experienced by any and all of the particles in this universe, since they are bubbles in the medium of aether. Hence, as the internal pressure of aether is gradually decreasing the physical size of particles will be increasing.

As aether's density and pressure are decreasing with time, the force of gravity is becoming weaker. As a result, the orbital paths of all of the planets around their respective stars are becoming wider. Hence, all of the planets in this universe will experience an eventual and definite **GLOBAL COOLING**. This global cooling, which will take place simultaneously on all planets existing in this universe, can be referred to as,

"Scharback Universal Planetary Cooling"

Due to the gradual weakening of the force of gravity, stars will also follow wider orbits around the center of mass of their respective galaxies, meaning,

"Gradually, not just star systems but also each and every one of the galaxies in this universe will grow in size."

As pressure and density of aether that is in this universe reduce further, and the force of gravity becomes even weaker, stars will no longer have any desire (the needed gravitational attraction) to hold on to their planets. The following figure shows a couple of planets which have just started to experience their total freedom from their respective star.

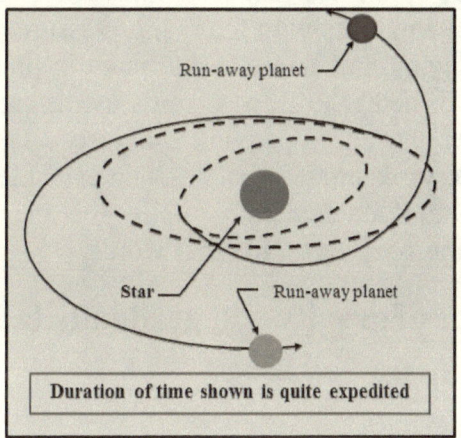

Stars also will no longer be motivated to stay in their own orbits around the center of mass of their respective galaxies, either. Therefore, gradually all of the star systems and galaxies will lose their structural integrities and all planets and stars will freely float in space. In other words,

"All of the celestial bodies will eventually become independent wanderers in this universe."

As the force of gravity becomes even weaker, all of the stars will expand, as their outer layers will no longer weigh as much. Consequently, the outer layers will not exert sufficient pressure on the inner layers to promote the ongoing nuclear fusion reactions. At that point, one by one, stars will literally turn off and start to cool down. They would basically become cold gas giants. Therefore,

"As all of the stars, one by one, turn off, the lights in the night sky will gradually become less in numbers and eventually all of them will disappear, forever."

Once all of the stars are turned off, the whole universe will start its eternal dark era. In other words,

"The universe will become pitch dark"

The pressure of aether that is in this universe will keep on dropping until it equalizes with the rising pressure of aether that is in the accompanying universe. At that point in time, there will be no motivation for aether to flow towards matter particles or black holes, and enter the accompanying universe. Therefore, as the flow of aether towards matter particles, and even black hole, comes to a complete halt, it automatically implies that the force of gravity will gradually approach zero. In other words,

"The force of gravity will gradually fade away."

Therefore,

"Just as the very formation of matter and anti-matter particles had given birth to the force of gravity, their continued existence will eventually cause it to fade away."

At that point in time, if the universe has any momentum left in its expansion, it will keep on expanding forever, since due to the lack of gravitational force, the expanding universe will literally lack its braking instincts. This era in the existence of this universe can be referred to as,

"Esmailzadeh Zero Gravity Era"

As the zero gravity era is approached and finally reached, the following will occur, since the density of aether will also gradually become identical in both universes:

- **Mass of matter and anti-matter particles**
  Due to decreasing aether density in this universe, the resistances that particles experience to changes in their motions will become weaker. In other words,

  "As the density of aether decreases in this universe, any given particle will gradually possess less 'Mass'."

• **Size of matter and anti-matter particles**

Sizes of various particles (which are literally bubbles in the fluid aether medium) are dictated by the internal pressure of their local fluid aether. Therefore, as shown in the figure below, due to the internal pressure of the fluid aether dropping in this universe, sizes of particles will gradually increase. Meanwhile, sizes of particles, as seen from the accompanying universe, will gradually decrease, due to the internal pressure of the fluid aether rising in that universe.

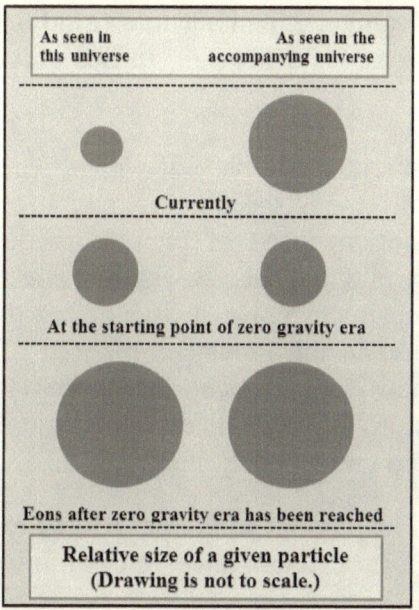

These opposite variations in the sizes of particles, in the two universes, will continue only until the pressure of aether in the accompanying universe becomes equal to the pressure of aether in this universe. At that point in time, sizes of the particles, as seen from this universe and as seen from the accompanying universe, will become equal to each other. Afterwards, as the mutual fluid aether medium continues to expand, its internal pressure will gradually decrease and consequently sizes of particles will gradually increase, regardless of from which universe they are observed.

- **Speed of light**

  The propagation speed of phase vibrations (including light) in this universe will gradually increase, due to gradual decrease in aether's density. In the meantime, the propagation speed of phase vibrations (including light) in the accompanying universe will gradually decrease, due to gradual increase in aether's density. Once, the density of aether in both universes become equal, which is when the zero gravity era has been reached,

  "The speed of phase vibrations in the accompanying universe will become equal to that of phase vibrations in this universe."

- **Time**

  Currently, the pace at which 'Time' is experienced in this universe is speeding up, due to gradual increase in the propagation speed of phase vibrations in this universe. Meanwhile, the pace at which 'Time' is experienced in the accompanying universe is slowing down, due to gradual decrease in the propagation speed of phase vibrations in that universe. Once, the propagation speed of phase vibrations in both universes become equal, which is when the zero gravity era has been reached,

  "The passage of 'Time' will be experienced at the same pace, in both universes."

- **Electric charge, electric field and magnetic field**

  According to the theories presented in this book, electric charge is due to certain type of aether flow into positively charged particles and out of negatively charged particles, in this universe. Hence, its effectiveness is directly dependent on (proportional to) the density and the pressure of the local aether medium. Therefore, as the density and the pressure of the aether medium are gradually decreasing in this universe,

"The strength of the electric charge of any given charged particle, as well as its directly dependent phenomena, namely its associated electric magnetic fields are gradually becoming weaker."

- **Black holes**

    As black holes are continually becoming weaker and weaker, gradually smaller fractions of the phase vibrations that are in this universe will be dragged into the accompanying universe through them. In other words,

"As black holes become weaker in this universe, less phase vibrations, including light waves, are being carried through them by the flow of the fluid aether. Hence, the spotlights in the accompanying universe are gradually becoming dimmer and dimmer."

Note that, once the zero-gravity era has been reached, only those phase vibrations that propagate directly towards black holes (from either universe) will be shared between the two universes.

- **Cosmic Microwave Background Radiation (CMBR) (Phase Vibrations)**

    The frequencies of the **cosmic microwave background phase vibrations** in both universes will gradually become the same, as the two universes will possess the same conditions. Afterwards, their frequencies will decrease, indefinitely, as the medium of aether expands.

- **Planck constant**

    As the amplitude of the cosmic microwave background phase vibrations keep on decreasing <u>in this universe, the magnitude of the Planck constant will be decreasing.</u> At the same time, the amplitude of the cosmic microwave background phase vibrations keep on increasing <u>in the</u>

accompanying universe. Hence, the magnitude of the Planck constant will also be increasing.

However, **when the zero-gravity era is reached**, the two universes will share the very same cosmic microwave background phase vibrations. At that point in time,

"The magnitude of the Planck constant will become the same in both universes."

After the zero gravity era has been reached, due to aether's ever decreasing internal pressure, the overall density of aether in both universes will simultaneously approach zero. Therefore, the following is expected to take place simultaneously in both universes:

- **The mass of particles will approach zero**, since the ever thinner fluid aether medium will exert an ever decreasing amount of resistance to any changes in their motions. In other words,

"All remaining matter as well as anti-matter particles will gradually become nearly massless, before they all dissolve in the fluid aether medium."

- **The physical sizes of particles in the two universes will gradually increase,** but only until density and pressure of aether reach a certain minimum, in both universes. Since at that point,

"All of the matter and anti-matter particles, including their aggregates, will literally dissolve in the aether medium."

Note that, how much the current size of various types of particles will increase, before they become dissolved and literally vanish in the aether medium, can only be speculated on.

However, at the present time, there is a way of performing an experiment with acoustically generated bubbles in a fluid medium such as water. Such an experiment can potentially shed some light on the processes involved, for bubbles to either continue to exist or to be dissolved.

Even such experiments will only provide a glimpse on what may be responsible for keeping bubbles intact or causing them to dissolve in the aether medium, since the aether medium and the water medium are alike in certain aspects, only.

Once, all of the matter and anti-matter particles as well as black holes dissolve in the fluid aether medium, there will be no connection between the medium of aether that is space in this universe and the medium of aether that is space in the accompanying universe. In other words,

"Eventually, the two aether mediums which form space in this universe and space in the accompanying universe, will become totally isolated from one another."

- **The propagation speed of phase vibrations in the medium of aether will increase indefinitely,** in both universes. In other words,

"The speed of light and other phase vibrations in the aether medium will gradually approach infinity."

- **The rate at which the passage of 'time' is experienced by all objects will increase indefinitely.** Since, their speeds (even if they be moving) with respect to their local aether medium will become an ever smaller fraction of the speed of phase vibrations in both universes. In other words,

"Gradually, as 'Time' speeds up, the duration of minutes, hours, days, years, even

millions of years will become equivalent to the duration of one 'SECOND, as it is experienced today."

- **Electric charge, electric field and magnetic field will gradually weaken, indefinitely,** due to the density and pressure of the aether medium gradually decreasing, that is, **until all particles are dissolved in the aether medium.**

- **Black holes will eventually dissolve and vanish**
  Over time, some of the black holes will gradually dissolve and vanish in the background fluid aether, while some others will remain as semi-permanent isolated holes in the fabric of space. However,

"Eventually, all of the black holes will dissolve in the aether medium, due to aether's density and pressure reaching certain low magnitudes."

- **Cosmic Microwave Background Radiation (Phase Vibrations) will be stretched, indefinitely**
  As the overall volume of the aether medium expands towards infinity, in both universes, the wavelengths of the cosmic microwave background phase vibrations will also be stretched towards infinity.

- **Planck constant will gradually decrease, indefinitely**
  As the overall volume of the aether medium expands towards infinity, in both universes, the wavelengths of the cosmic microwave background phase vibrations will be stretched towards infinity and their amplitudes will be gradually reduced towards zero. Therefore,

"The magnitude of the Planck's constant will gradually decrease towards zero."

Conclusion

Presented in this first chapter, has been a brief overview of the whole history, as well as the future of the universe, according to aether-based theories introduced in this book. Such theories clearly demonstrate that, once the existence of aether is accepted and its effects are properly taken into account, the formation and the development of this universe and its contents become readily comprehensible.

The following are only some of the phenomena that have been consistently explained in this book, using the variety of aether-based theories proposed. For the complete list, the reader is encouraged to refer to the table of contents, or the appendices.

- How did this universe come into existence?, and how can the birthplace of this universe be identified?,

- Why did the '**Initial Rapid Expansion**' of aether medium occur?,

- Why did the initial rapid expansion slow down?,

- What is '**Aether**'? and how does it relate to the variety of phenomena in this universe?,

- What is '**Space**'?, and why is it expanding?, and will its expansion ever come to a complete halt?,

- What is '**Time**'? and why is it gradually experienced at an increasingly faster pace?,

- What is '**Light**'? and why is its speed gradually increasing?,

- Formation of '**Matter**' and '**Anti-Matter**' particles,

- <u>Why do particles have specific **discrete physical sizes**?</u>,

- <u>Why are some particles stable while some others are not?</u>,

- <u>Why are some isotopes stable while some others are not?</u>,

- What is '**Dark Matter**'?,

- What is '**Dark Energy**'?,

- What is '**Vacuum Energy**'?,

- What is '**Electric charge**'?,

- Formation of '**Electric field**',

- Formation of '**Magnetic field**',

- What is '**Electricity**'?, and how is it formed?,

- What are '**Superconductors**?,

- Formation of the force of '**Gravity**', and why is its strength gradually decreasing?,

- Formation of the '**Cosmic Microwave Background Radiation**' (phase vibrations),

- What are '**Black Holes**'? and why are their properties related to their surface areas rather than their volumes,

- <u>Why are relativistic effects experienced as the speed of objects approach that of light in the aether medium?</u>,

- <u>What happens to the motion of atoms, nuclei and electrons at absolute zero degrees temperature?</u>

- Explanation for '**Casimir Effect**',

- Explanation for **'Einstein-Podolsky-Rosen Paradox'** of entangled particles,

- Explanation for **'Bose-Einstein condensates'**,

- Why electrons form an **'Electron cloud'** rather than definite orbital paths, as they revolve around atomic nuclei?

- Why do **'Lightning Sprites'** form?,

- Does **'Planck's Constant'** have a physical interpretation?, and why is it gradually decreasing in magnitude?,

- Explanation for different types of **'Double-Slit Experiment'**,

- Explanation for **'Photoelectric Effect'**, based on light being a wave and not a particle,

- How do **'Lasers'** function?, and how can their efficiencies be increased?,

- Explanation for **Mr. Galileo's Experiment**,

- Explanation for **Mr. Michelson and Mr. Morley's Experiment**,

- **'Precession of the Perihelion of Mercury's Orbit'**,

- The **'Equivalence Principle'**,

- How are **'Magnetic Hills'** formed?, and how can artificial magnetic hills be constructed?,

- Explanation for **'Near-Earth Probe Flyby Anomalies'**,

- The **'Science of Astrology'**, and why it is no longer as accurate as it used to be,

- Why are major **'Natural Disasters'** becoming more frequent, on a global scale?,

- Why is the '**Physical Size of Particles**' gradually increasing?,

- Why is the '**Mass of Particles**' gradually decreasing?,

- Why are the planetary orbits gradually becoming wider?,

- Why is the expansion process of this universe accelerating?,

- What events can be expected to occur in the future?,

And, many, many more.

The bottom line is that,

Aether IS Everything.

Even though the approach taken in this book is purely qualitative / conceptual, yet it clearly demonstrates the capabilities of aether-based theories. Therefore,

It is very crucial that, the scientific community accepts the existence of aether and takes its various effects into account.

What
is
Aether?

Introduction

In 1865, Mr. Maxwell proposed his theory on electromagnetism, the very same theory that is still valid to this day. Mr. Maxwell had based his theory on the existence of some kind of stationary medium through which light and other electromagnetic waves were assumed to propagate. At that time, this medium was referred to as 'Aether'. What aether was made of or what kinds of properties it had were not known. The aether medium to electromagnetic waves was thought to be as a medium such as air is to sound waves.

Over time, scientists became curious about aether and based on the assumption that it is stationary in space and all waves and even planets are moving through it, they tried to detect and measure the drift velocity of earth in that medium. The assumed motion of earth in the medium of aether is shown in the following figure.

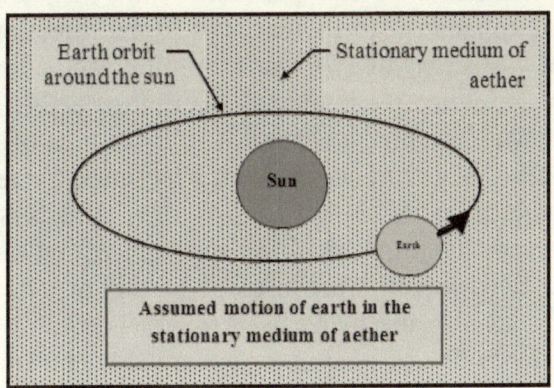

Amongst the multitude of experiments that were proposed and conducted was the one designed and performed by Mr. Michelson and Mr. Morley in 1887. As it is shown below, their experiment was based on reuniting the wave patterns of two light beams from the same monochromatic light source, after they were made to follow two paths which were orthogonal to each other, one in the direction of earth's motion around the sun and the other at 90 degrees to the said direction. By reflecting these two light branches back and superimposing them on each other, Mr. Michelson and Mr. Morley were expecting to see the formation of some sort of interference pattern.

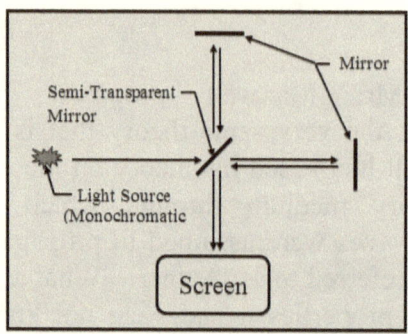

They conducted their experiment on the ground level. They also conducted their experiment at high altitudes, with the help of a balloon. However, they did not observe any interference patterns forming during any of their attempts.

In 1904, Mr. Fitzgerald and Mr. Lorentz, independently, provided reasoning for detecting no trace of any drift effect in all of the experiments. They proposed that the length of the experimental apparatus' arm which was in the direction of earth's motion was affected by the said motion and therefore had automatically annulled the intended effect of the aether drift.

In 1904, Mr. Lorentz published his "Principles of Relativity". With this theory Mr. Lorentz explained the effects due to moving at relatively high velocities. To this day, effects such as the shortening of the length of an object in the direction of its motion, as well as the slowing down of the rate at which time is experienced and also the increase in the mass of an object, as its speed approaches that of light in a vacuum, are calculated using his transformation equations.

It should be noted that, Mr. Lorentz and Mr. Fitzgerald had proposed their theories without rejecting the existence of aether which was commonly accepted by the then-current scientific community.

In 1905, Mr. Einstein used Mr. Lorentz's "Principles of Relativity" and added two postulates to it, namely:

- Light is made of particles, called photons, which do not need a medium to travel in, and

- The speed of light in a vacuum is the maximum possible speed and it is independent of the observer's state of motion relative to the light source.

Important notes:

1- The experiments performed by Mr. Michelson and Mr. Morley, by no means had indicated that there was no aether in space, or that the speed of light was the maximum speed possible.

A revised version of the experiment performed by Mr. Michelson and Mr. Morley is proposed and presented in detail in the chapter titled "What is Light?".

2- Since Mr. Zwicky published his report on gravitational effects of some kind of invisible mass (matter) in 1933, a new term, namely dark matter has entered the astrophysicists' vocabulary. Later on, due to the discovery of the fact that the expansion of this universe is accelerating, in 1998, another term, namely dark energy, also entered the scientific vocabulary. According to the current estimates, the contents of this universe consist of about %73 Dark Energy, %23 Dark Matter and only %4 normal forms of matter and energy.

In this chapter, first aether is introduced in sufficient details. Then, its effects on different phenomena are described. In other words, it is demonstrated that once the existence of a dynamic aether medium is accepted and its various effects are properly taken into account a variety of phenomena, that otherwise are inexplicable, become well understood and their occurrences and/or existences become readily expected.

Understanding Aether

Theories presented in this book explain the very essence of various phenomena such as "light", "gravity", "magnetic field", "electric field", "electricity", "black holes", "time", "space", "energy", "dark energy", "vacuum energy", "matter" and "dark matter". These phenomena directly rely on the existence of aether. They also rely on and affect each other in measureable ways. The following are brief descriptions of what aether is and how it relates to these phenomena.

- **Aether,** which is also referred to as **"fluid aether"** in this book, is a continuous, compressible, elastic, non-viscous, fluid-like medium which is under tremendous amount of pressure and occupies the whole physical universe **and more**. Aether is expanding, due to its internal pressure.

  Aether does not interact directly with matter or anti-matter particles. However, under certain conditions, it allows the formation of and hosts various matter and anti-matter particles. It also acts as the carrier of all forms of phase vibrations such as electromagnetic waves which may be generated in that medium. Different flow patterns of fluid aether give rise to such phenomena as gravity, magnetic field and electric field, amongst many others.

- **Space** is filled with aether and it is literally dependent on its presence. Space is divided into two major sections/regions, one section is this universe and the other section is another universe which is accompanying this universe.

- **Time** is experienced by an object (or a living being) only if its motion relative to its local aether is at a speed that is less than the speed of phase vibrations in that medium.

  As an object's speed relative to its local aether approaches the speed of light (the speed of phase vibrations) in the local aether medium, that object experiences the passage of time at progressively slower paces. If its speed becomes equivalent to that of phase vibrations in the local aether medium, that particular object (or living being, for that matter) will no longer experience the passage of time.

  > **Note that**, presence of aether and phase vibrations are necessary for experiencing "time'. Therefore, the very first phase vibrations that were introduced into the medium of aether literally gave meaning to "time" in this universe.

- **Regular Energy** is the energy associated with the variety of motions such as translational, rotational and vibrational motions of matter, anti-matter and dark matter particles, as well as the variety of phase vibrations such as electromagnetic waves which are hosted by the aether medium. In this universe, energy manifests itself in a variety of ways / forms, yet on the microscopic scale all of them are simply different types of motion in the medium of aether.

- **Dark Energy** is the potential energy associated with aether's internal pressure which is the driving force for its expansion.

- **Vacuum Energy** is the energy associated with the overall translational motion of aether which is manifested as its expansion process.

- **Matter and Anti-Matter Particles** are like bubbles, formed as a result of induced 'cavitation', in the very fabric of the fluid aether medium. They form as the local phase vibrations resonate into sufficiently high amplitudes.

Each particle (bubble) acts as a 3-dimensional gateway or entrance into a tiny narrow tube (tunnel) between this universe and the accompanying universe. Such tubes allow the fluid aether that is in this universe to flow into the accompanying universe, simply due to the pressure difference that exists between aether that is in the two universes.

- **Dark matter** particles are matter and anti-matter particles that are not readily visible. They can be in such forms as neutron stars, black holes, but mainly as individual particles, particularly a variety of particles which are either quite short lived due to their continuous production and destruction by resonances/spikes in the amplitudes of phase vibrations in the aether medium, or are quite small in size, among many others.

  Dark matter exists in vast quantities throughout this universe, but its existence can only be detected through its side effects such as its force of gravity. Since, each and every dark matter particle generates an accelerated aether flow towards itself. The induced aether flow, in turn, gives rise to a gravitational force which is in fact the drag force it exerts on any and all other particles/objects that happen to be in its path.

- **Force of Gravity** at any given location is the drag force induced by the flow of fluid aether which is **accelerating** through that particular location. The flow of aether towards subatomic particles (all of which act as drain holes) is an accelerated motion of aether which drags anything and everything that happens to be in its path, as shown in the figure below.

  The influence of this kind of motion of aether is long-range and reaches the end of space. The flow of aether towards particles is like the flow of air that is inside a compressor's tank as it flows towards an opening such as a valve. Since, once the valve is opened, an air flow is induced in the entire volume of the tank.

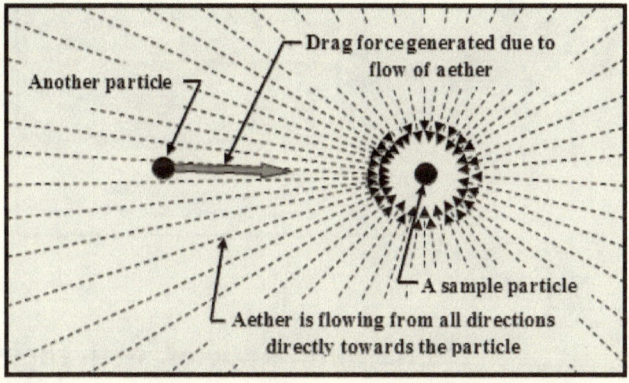

Note that, the drag force exerted on an object, by the accelerated flow of aether relative to that object, gives meaning to that object's "**Weight**". In other words,

> **"Weight of an object is induced by either that object <u>accelerating</u> in the medium of aether or the local aether <u>accelerating</u> through its position. And, the magnitude of the weight induced is dictated by the rate of the acceleration experienced, <u>regardless of that object's instantaneous speed relative to its local aether medium.</u>"**

In the case of gravity, the flow of aether is a one way trip towards individual subatomic particles. As fluid aether reaches any subatomic particle, it goes through it and enters the accompanying universe.

Note that, the gravitational effects of dark matter on regular matter is a clear indication that dark matter is made of matter and anti-matter particles, since only the flow of aether towards such particles can give rise to the drag force which is referred to as the force of gravity.

- **Magnetic Field** is a type of flow induced in the fluid aether medium which forms a complete loop within this universe. A sample is shown in the figure below.

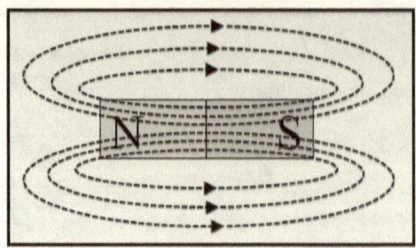

Note that,

"Charged particles, with their very motion in this universe, have given birth to Magnetic Field."

- **Electric Field** is a one-way type of flow induced in the aether medium which is from the negatively charged particles towards the positively charged particles. A sample is shown below.

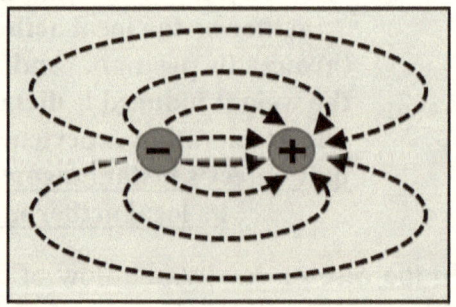

Note that,

"Charged particles, with their very presence in this universe, have given birth to Electric Field."

- **Electricity** is the motion of electrons in a conducting matter medium. Electrons' motion is induced by a certain type of phase vibrations in aether that is occupying the volume of that matter medium. Electrons are encouraged to let go of their atomic attachments and flow along the propagation path of such phase vibrations.

Such encouragements are exactly like the type of encouragements electrons receive in the case of photoelectric effect.

The containment of such phase vibrations, within the aether medium that is superimposed by the matter medium, is just like the containment of light waves in an optical cable. It can also be likened to the containment of sound waves within matter mediums which are surrounded by a vacuum.

Note that, "phase vibrations associated with electrical currents are generated by the variations in the local magnetic field, at some point along their path."

- **Light and other Electromagnetic Waves** are phase vibrations in the medium of aether. Such vibrations do not encourage aether to have any kind of translational motions, but simply make it vibrate locally, just like the waves formed on the surface of water in a pond, as a pebble is dropped into it.

  The speed of phase vibrations in a given medium of aether is constant and it is independent of the relative motions of the producer and the receiver. However, the density of the local aether and its pressure do affect the speed at which phase vibrations are allowed to propagate through.

- **Black Holes** are enormous drain holes made of countless number of matter particles (bubbles) that have literally joined (unified) together, forming a single giant gateway through which the fluid aether that is in this universe flows into the accompanying universe.

  As aether approaches a given black hole its flow speed increases exponentially and at some distance from the black hole it reaches the speed of the phase vibrations in the local aether medium. That particular distance to the center of the black hole, when considered in all possible spatial directions all around the black hole, forms a sphere. The surface of that sphere is referred to as the 'event horizon' of that black hole. The following figure shows the flow of aether from all spatial directions towards a black hole.

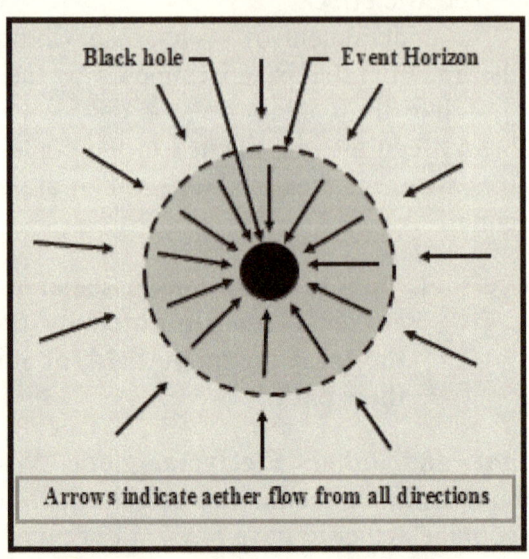

Arrows indicate aether flow from all directions

Note that, as matter particles that are dragged by the aether flow reach a black hole they merge with it and cause it to grow in size and become an even wider/larger gateway for aether to flow through.

The joining of the matter particles with a black hole can be better understood if they were visualized as bubbles in a fluid medium such as water (which in fact they are but in the medium of fluid aether). As smaller bubbles representing matter particles reach and touch a very large bubble which can represent a black hole, they simply unite/merge with it and form an even larger bubble. In fact,

"Black holes are the largest matter particles in this universe."

Aether versus Water or Air

Aether can best be likened to the water that is in an ocean (including all forms of ice and bubbles that may exist in it), while the ocean, as a whole, represents the space. Therefore,

- **Fluid Aether** can be likened to the fluid water itself which under different conditions hosts various forms and sizes of bubbles. Just as fluid aether acts as the carrier of various types of phase vibrations such as electromagnetic waves, water also acts as the carrier of many forms of phase vibrations such as sound waves.

- **Space,** is analogous to an ocean where aether has replaced the water. The essence of an ocean is its being filled with water. If all of the water that is inside an ocean is somehow taken out, what remains can no longer be called an ocean. The very same is also true about the direct relationship between space and aether, since for space to exist it must be occupied by aether. Therefore,

"The relation between aether and space is just like that of water and ocean."

In other words,

"Space without aether is literally meaningless."

- **Time** is experienced by any object (or any living being) based on its motion relative to its local aether medium.

There are two ways that an object can experience such a relative motion with respect to its local aether medium:

1- The object (or living being) moves in the aether medium,

In this scenario, as the object speeds up, it experiences time at slower and slower paces. And, if it manages to reach the speed of phase vibrations in the local medium of aether, it will literally catch up with the phase vibrations that it generates.

This effect can be likened to when an airplane, that is capable of supersonic flight, holds its speed right at Mach 1 (speed of sound in the local medium of air). In such a scenario, as it is shown in the following figure, all of the sound waves that were generated by the airplane engines from the instant the airplane had reached and locked its speed at Mach 1 up until the present time will be heard simultaneously. As time goes on, and the airplane keeps its speed locked right at Mach 1, this accumulated sound wave will only get stronger and stronger, just like a choir whose members one by one join in and sing along.

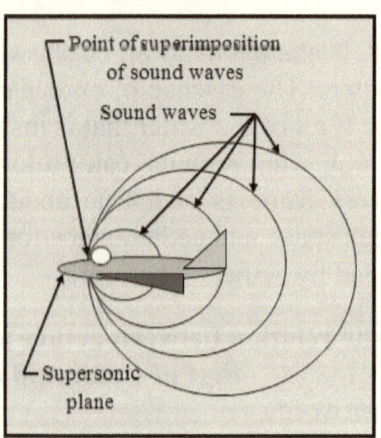

Note that, in the case of the supersonic airplane, the sound level will increase only until it reaches its saturation point. That saturation point corresponds to when the amount of added sound waves currently being generated becomes equal to the amount of sound waves that are fading away due to being generated in the distant past, as they weaken due to distance from their generation points.

Of course, in this scenario concerning an airplane, if the amplitudes of the sound waves detected are also taken into account, the timing of their generation can be isolated.

However, in the case of the singers in a choir, since they are all standing next to each other, there will be no variations in the amplitudes of the sound waves generated by its individual members. Hence, their voices cannot be isolated from each other.

2- Aether flows through the position of the object (or living being).

In this scenario, an object such as an airplane can be visualized as being mounted inside a wind tunnel. Then, the very same type of experiment as in the first case can be conducted by simply adjusting the wind tunnel power so that it generates a wind at a speed that is exactly equal to that of sound waves in the local air. The following figure shows such a scenario.

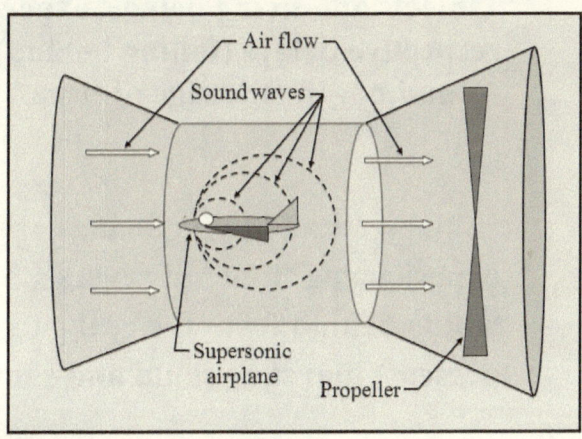

In other words, <u>once an object's (or a living being's) speed reaches that of phase vibrations in its local aether medium, it will no longer experience the passage of time,</u> **<u>since it will no longer be able to distinguish between the events that have occurred in the past and those happening at the present.</u>** Even though it sounds strange, one can actually state that,

"Time is progressing at the speed of phase vibrations in the aether medium."

Therefore,

"Any object that moves (relative to its local aether medium) at the same speed as the speed of phase vibrations in that medium will automatically be in sync with its own past and present, concurrently."

In short, one can state that, any object that is in motion relative to its local aether medium at a speed that is slower than the speed of phase vibrations in that medium, that object is literally falling behind from the wave fronts associated with its own past. Therefore, in normal circumstances,

"Objects and living beings, experience their respective delays (falling behind) in 'time' and not the passage of 'time' itself."

In other words,

"Objects and living beings are in fact experiencing 'time' in reverse, by literally falling behind from the real current 'time' (present) that they could and should be at."

- **Regular Energy** is like the energy associated with the variety of motions such as translational, rotational and vibrational motions which are exhibited by the individual objects / icicles / bubbles which are floating in water, as well as the energy associated with all forms/types of phase vibrations such as sound waves that may be propagating in the water medium.

- **Dark Energy** is the energy associated with the internal pressure of the aether medium. This energy of aether can be likened to the energy associated with the internal pressure of water in an ocean.

- **Vacuum Energy** is the energy associated with the overall motion of the aether medium, as it spreads towards its outer perimeters. This energy of aether can be likened to the overall kinetic energy associated with the motions of water molecules forming the ocean currents.

- **Matter and Anti-Matter particles** can be likened to bubbles which can form in a medium such as water due to cavitation, particularly when such cavitations are induced by ultrasound waves which propagate through that medium as phase vibrations.

  The following picture shows the entrance of a water-shoot inside the reservoir of a dam. Such water-shoots prevent the water that is behind the dam to rise beyond certain safe level. They perform their important task by simply allowing the excess water to fall in them and get transferred downstream, bypassing the dam structure altogether. The flow of aether towards and through matter and anti-matter particles is quite the same as the flow of water towards and through such openings.

One has to keep in mind that, the flow of aether towards a given particle is from the whole volume of space surrounding that particle, a volume of space which is three dimensional, while the flow of water towards the inlet of such a water-shoot is from a planar surface which is two dimensional.

Such water intakes clearly demonstrate how the flow of water changes direction, and literally disappears from the two dimensional plane in which it existed before. The very same process takes place with the fluid aether. Since, as it reaches a given matter or anti-matter particle it literally disappears from the familiar three dimensional volume of space that forms this universe and enters the physical dimensions corresponding to the accompanying universe.

Note that, in the two-dimensional surface of water, the middle part of such water-shoots is observed as a void. The very same is the case for the matter and anti-matter particles in the aether medium, since they too are bubbles, voids in the three-dimensional space which is this universe.

As shown below, in this universe the flow of aether is towards matter (as well as anti-matter) particles. While, in the accompanying universe the flow of aether is away from matter (as well as anti-matter) particles.

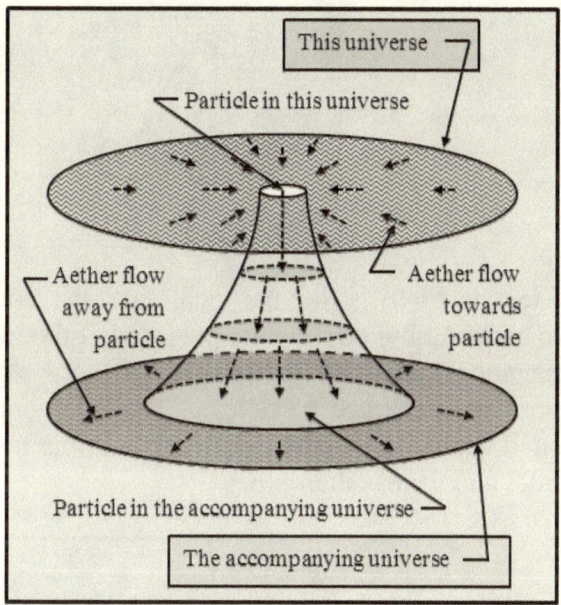

Note that, the difference in the physical size of particles in the two universes is due to the internal pressure of aether being different in those universes.

- **Dark Matter** is like the tiny bubbles that are always present in the ocean water and yet are too fine to be isolated. Dark matter that corresponds to black holes can be likened to the surface of the water in an ocean since all other bubbles are drawn towards it.

- **Force of Gravity** is like the drag force exerted on objects by the one way accelerated flow of water towards drain holes at the bottom of swimming pools which allow water to go through and enter a different environment / container.

- **Magnetic Field** is like activities such as a whirlpool which forms giant doughnut shaped flow patterns (a complete loop) in the medium of water. Magnetic field can also be likened to the overall motion of air molecules induced by a jet engine or even a household fan. The following figure shows the same type of flow (a complete loop) as it is induced by a household fan.

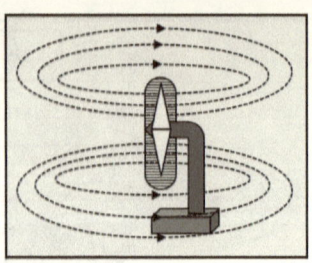

- **Electric Field** is like the local motions generated in water in a swimming pool by the presence of the water inlets (like negatively charged particles) and the water outlets (like positively charged particles). The following figure shows such water flows induced in a swimming pool by its water inlet and water outlet pipes.

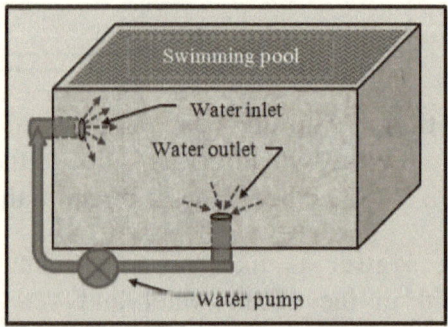

- **Electricity** is the motion of electrons (or ions) in a conducting matter medium, a motion that is due to encouragements by a certain type of phase vibrations in the medium of aether which is superimposed by that matter medium. Phase vibrations associated with an electrical current in a conductor can be likened to a tsunami which encourages the movements of objects floating on the surface of water. The wave form of a tsunami is the result of a sudden vibration introduced in the medium of water by an earthquake. Correspondingly, the phase vibrations associated with an electrical current are generated by the variations in the local magnetic field.

- **Light and other Electromagnetic Waves** being phase vibrations in the medium of aether, can be likened to sound waves that are also phase vibrations but in a matter medium such as water or air.

- **Black Holes** can be viewed as being giant drain holes at the bottom of an ocean, as shown in the following figure.

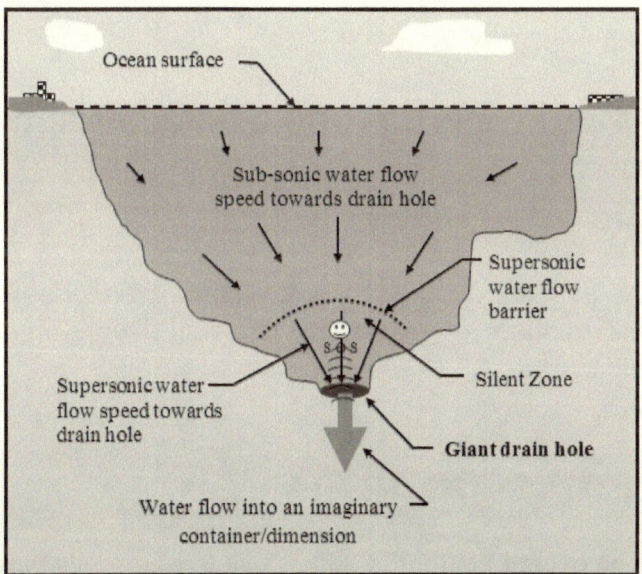

If the ocean is deep enough to cause the pressure to become sufficiently high, the free flow of water going through will be higher than the speed of sound waves (phase vibrations) in water.

In this case, if someone who is being dragged into such an opening attempts to broadcast an S.O.S. signal, regardless of the strength of the speakers used, his/her cry for help will not travel away from the drain hole. Because, the generated sound waves will not be able to propagate through water faster than the speed at which water is flowing in the opposite direction, approaching the drain hole. Therefore, all of the generated sound waves will get carried towards the drain hole and will mandatorily pass through it. Such a region in an ocean can be referred to as the **"Esmailzadeh Silent Zone"**. Because, no sound waves will ever be heard coming out of it.

Properties and Characteristics
of Aether

As they are described below, some of aether's properties are similar to those of regular matter, while some others are quite unique to aether.

1- Aether cannot be created or destroyed

The overall amount of aether that exists in this universe and in the accompanying universe is fixed. Aether is a unique medium that basically acts not only as the carrier of a variety of waves and forces, but also as the medium in which everything can exist, since it also gives meaning to space itself.

In the beginning, there was only aether in its fluid form in a very compressed state. As it experienced a rapid expansion in its overall size and therefore a reduction in its density and pressure, under different conditions, parts of it have given meaning to and have hosted matter, anti-matter, dark matter and so on.

In other words, aether is not created and will not be destroyed. However, under different circumstances, it manifests as being in different states/forms, just as water can manifest as being in different states such as solid (ice), gas (steam) and liquid (water).

Over time, aether that is in this universe is shared with the accompanying universe. But, its overall amount remains unchanged.

2- Aether is a continuous and fluid medium

Aether is a continuous medium that is freely flowing everywhere in this universe. It can be likened to a thick cloud that is encompassing everything. It possesses somewhat different local densities/pressures, but not by much since it is always freely flowing.

3- Aether is compressible and elastic

The ongoing expansion process of the aether medium, as a whole, and the gradual decrease in its overall internal pressure are clear indications that <u>aether is a compressible medium</u>.

The compressibility of the aether medium can be likened to that of rubber or a gaseous medium. However, aether's compressibility is not due to its being composed of individual particles which can be spaced differently, as its internal pressure is varied, but rather due to its elasticity.

4- Aether's viscosity is essentially nonexistent

Aether's flow between neighboring regions is always continuous since it has no viscosity. Therefore, it readily flows towards any and all openings that allow it to be transferred into the accompanying universe.

5- Variations in aether's density and pressure in different regions of space

Aether's density and pressure gradients, between neighboring regions, are quite gradual since aether readily flows from regions of higher pressure to regions of lower pressure, due to its non-existent viscosity. However, these gradual density and pressure gradients become noticeable on larger scales, when aether that is in the internal regions of galaxies are compared with aether that is in their respective surrounding regions.

6- Aether formed the initial container (space) for this universe

Aether was the only entity that existed in this universe, at the beginning of time. Even 'time' started its existence as phase vibrations were induced in the aether medium. Aether basically became the medium or the host for everything else such as matter, anti-matter and electromagnetic waves all of which are literally different manifestations and/or effects of aether itself. In other words,

"Aether can exist independently of any and all other apparent contents of this universe, while the very existence of aether is a prerequisite for the existence of anything and everything else."

7- Aether is a dynamic medium

Aether is not a stationary medium, as it was assumed to be by the 19th century physicists such as Mr. Lorentz and Mr. Maxwell, as well as others. Aether is quite a dynamic medium. The force of gravity associated with the individual particles, as well as celestial bodies both near and far, and the forces associated with the magnetic and the electric fields are due to different flow patterns in the medium of aether, from the most microscopic scales to the most macroscopic scales. Such flow patterns experienced by aether are just like the variety of flow patterns that are experienced by air molecules in the atmosphere or by water molecules in the oceans.

8- Aether can flow at sub-light and super-light speeds

Aether flows towards individual matter and anti-matter particles at sub-light speeds, since sizes of such drain holes are too small to allow unrestricted flow of aether through. However, on its path towards black holes, aether's speed reaches the speed of light in that medium as it reaches their event horizons. And, as it continues, its speed enters the super-light speed regime. The figure below shows aether flow speed regions near a typical black hole.

Note that, aether's speed at the event horizon of any black hole is exactly equivalent to the speed of phase vibrations (speed of light or other electromagnetic waves) in the local aether that exists at that particular black hole's event horizon.

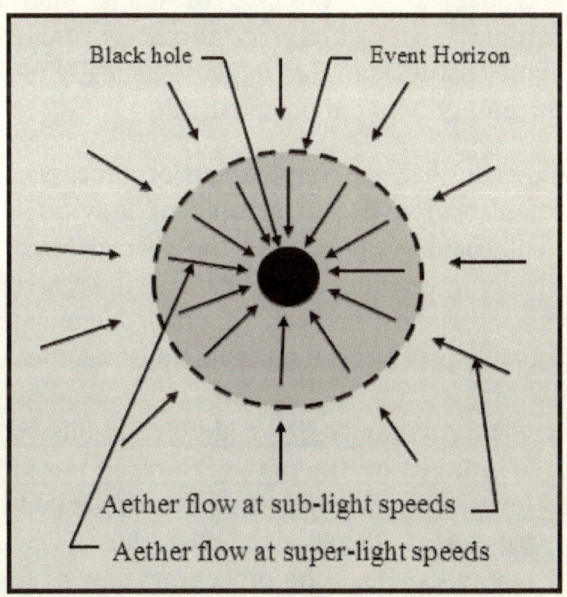

9- Aether in fluid state does not possess gravitational properties

Other than aether's overall outward flow that is due to its ongoing overall expansion process, aether only has the tendency to flow, in an accelerated fashion, towards matter and anti-matter particles or their aggregates, through which it leaks out of this universe. In so doing, it induces a drag force on other particles (or aggregates of particles, as well as phase vibrations) that happen to be in its path, the same drag force which is known as the force of gravity.

Aether in its fluid state is not made of matter or anti-matter particles or any other types of particles which function/serve as drain holes or gateways connecting this universe to the accompanying universe. Therefore, in the absence of such particles, no net accelerated flow of aether can exist. Hence, no drag force which is the force of gravity can be generated, either.

Note that, the initial rapid expansion of the aether medium (space) was due to the absence of matter and anti-matter particles. Since, the very absence of matter and anti-matter particles in that era automatically meant that the force of gravity did not exist, yet.

One could state that, the very expansion process of the aether medium must also give rise to a force of gravity of some sort which is actually acting opposite of the normal force of gravity. Since, by expanding the domain of space in this universe, instead of encouraging particles (as well as their aggregates such as galaxies) to come closer to each other it is pushing them apart from each other.

However, the term gravity as it is defined in this book refers to **the drag force induced by the net accelerated motion of aether at any given location in this universe**, an accelerated motion that is towards particles of matter and anti-matter. The motion of aether that is due to its expansion process, at any given location in this universe, does not contribute to any flow of aether at that location (in space), because space itself is literally expanding with the aether medium that is occupying it. Hence, no flow of aether is generated by the expansion of its medium.

The spreading action/motion of aether that is due to its medium's expansion can be readily likened to the spreading of the molecules in a piece of rubber as it is being stretched. Since, even though two dots drawn on the surface of such a rubber piece will recede from each other, as the rubber piece is stretched, no flow of rubber molecules is generated through the location of either one of those two dots.

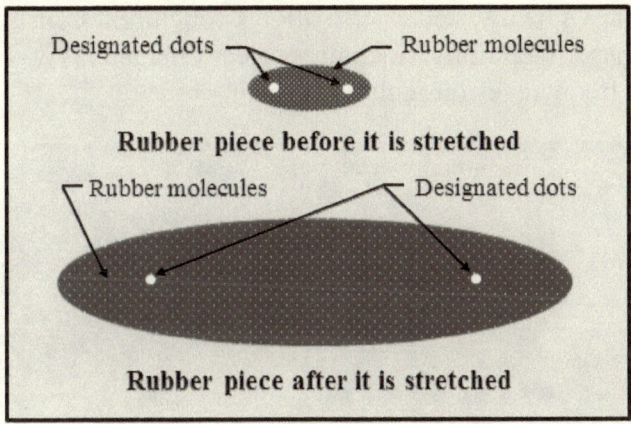

10- Temperature of the matter medium indirectly affects aether

As the temperature of a matter medium is raised that matter medium expands. Such an expansion directly implies that there are less particles (drain holes), per unit volume of space, available for aether to flow through. Hence, the flow of aether towards that particular region of space gets affected accordingly.

11- Aether can carry many different phase vibrations, concurrently

In this regard, the fluid aether behaves just like air. Air can concurrently host countless number of phase vibrations such as sound waves that are at a variety of frequencies. This is commonly experienced, as a variety of sound waves are received/heard concurrently, by anyone who has walked on sidewalks of city streets, particularly during busy hours, or as he/she may have listened to a live performance by a symphony orchestra. The sound waves generated by each and every instrument propagate in all directions, through the same medium of air.

As shown below, aether also allows concurrent transmission of many phase vibrations (including electromagnetic waves) with different frequencies through its fabric.

The very reception of various colors from different objects such as flowers, trees, buildings and so on, let alone the presence of millions of radio, TV and cell phone signals, all of which are transmitted as phase vibrations through the very same volume of space, clearly demonstrate aether's capability in handling such a task.

12- Aether does not interact directly with regular matter or anti-matter particles

Aether, in its fluid state, is a neutral environment for anything and everything that exists in this universe. It is literally a host for all of this universe's contents and a carrier for the variety of forces through which matter (as well as anti-matter) particles interact with each other. This is like air, since various objects such as airplanes use air to not only support themselves against the force of gravity but also propel themselves.

One can also say, by its very presence, aether literally enables the contents of this universe to exist, to move around and to interact with each other in a variety of ways which they could not do otherwise. In short,

"Aether is not only the host for the contents of this universe but also the carrier of the various forces that are exchanged between them."

13- Aether's internal pressure is quite high

Aether that is in this universe is under a tremendous amount of pressure, just like the compressed air that is contained inside a compressor's tank. However, its pressure is gradually decreasing due to aether leaking into the accompanying universe, as well as the overall expansion of its medium.

14- Contradictory effects of aether on the expansion rate of the universe

The force of gravity that all matter and anti-matter particles exert on each other is trying to slow down the expansion process of this universe. Yet, the internal pressure of the fluid aether is encouraging the expansion process of this universe to speed up. Therefore, one can say,

"Aether concurrently has contradictory influences on the expansion process of this universe, as a whole."

These contradictory effects can be better understood by visualizing two giant magnets which are freely floating in deep space with their North poles (or South poles) pointing at each other. In this case, while their gravitational forces will pull them towards each other, their repulsive magnetic forces will push them apart. Such a scenario is shown in the following figure.

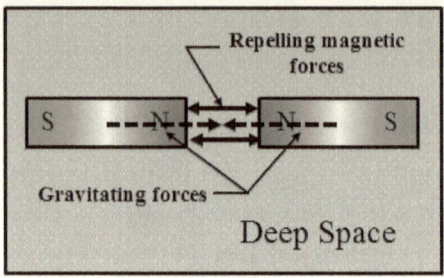

15- Aether in the accompanying universe

Formation of matter and anti-matter particles in this universe allowed the fluid aether to leak out of this universe and by accumulating in a different environment formed the accompanying universe. In other words, the density and the pressure of aether in the accompanying universe started from zero. They have been increasing ever since, due to the continuous leakage of aether from this universe through matter and anti-matter particles, as well as their aggregates of all sizes, particularly the black holes.

Currently, the fluid aether that is in the accompanying universe is still under a much, much lower pressure as compared to aether that is in this universe. Hence, its density is correspondingly much lower than that of aether that is in this universe.

16- Fluid aether cannot be detected directly

Wanting to detect 'aether' directly is exactly like wanting to detect 'space' directly. Aether can only be detected indirectly, through various tasks that it performs in this universe. For example:

- The very existence of matter and anti-matter particles is an indication of aether's presence since matter and anti-matter particles are literally bubbles that only form in the aether medium.

- If the force of gravity is felt, it can only be due to the flow of the local aether which is in an accelerated state of motion towards some matter or anti-matter particle (or their aggregates) that may be located anywhere in this universe.

- Any magnetic field is due to a flow that is generated in the local aether, a flow that is forming a complete loop.

- Any electric field is due to a one-way flow that is generated in the local aether, a flow that is only a local disturbance in that medium.

- If any electromagnetic type of wave such as light is transferred through anywhere, it is automatically an indication that aether exists in that volume of space.

17- Aether is the only content of this universe

Aether was the first content of this universe, as this universe came into existence. Over time, and due to different circumstances, aether has manifested its many faces which include various forms of matter and energy.

For instance, based on the information presented in the following section, matter (including anti-matter) particles are bubbles formed as a result of cavitation in the medium of aether, due to spikes forming in the phase vibrations present in that medium. Also, the electromagnetic waves in general are phase vibrations in the fluid aether medium. In other words,

"Everything in this universe including matter, anti-matter, dark matter, as well as all forms of energy, including dark energy and vacuum energy, are different manifestations of aether, the one and only content of this universe."

Relation between Aether and Matter and Anti-Matter Particles

This section provides sufficient details on how matter and anti-matter particles are formed and how their interactions with each other and their survival, as in their being stable or unstable, are affected by their local aether medium and the variety of phase vibrations which are continuously propagating through that medium.

1- Formation of matter and anti-matter particles

Phase vibrations in the medium of aether are just like regular sine waves. They have a positive half and a negative half, during every complete cycle.

Under certain conditions, phase vibrations in the aether medium form full-wave or even half-wave spikes (resonances / harmonics) in that medium. Simple presentations for both full-wave and half-wave spikes are shown below.

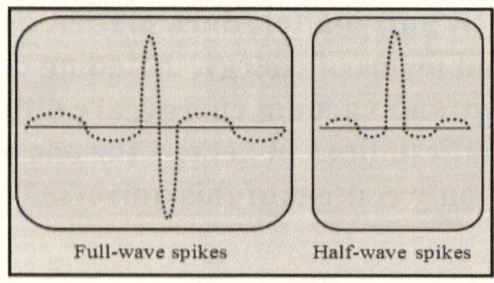

| Full-wave spikes | Half-wave spikes |

When spiked, the positive halves of the waves lead to the formation of particles that are of positive (+) charge. Correspondingly, the negative halves lead to the formation of particles that are of negative (-) charge.

According to the currently accepted definitions in the field of particle physics, the only difference between matter particles and their anti-matter counterparts is that their electric charges are opposite of each other. For example, a proton is a positively (+) charged particle, while an anti-proton which has the same amount of mass is a negatively (-) charged particle.

However, as far as the relations between aether and matter (and anti-matter) particles are concerned, all of the positively (+) charged particles belong to one group and all of the negatively (-) charged particles belong to another group, regardless of their belonging to the normally defined matter or anti-matter particle families. In other words,

According to this newly proposed matter particle categorization strategy, protons (+) and anti-electrons or positrons (+) are grouped together, just as anti-protons (-) and electrons (-) are grouped together.

Therefore, full-wave spikes result in the formation of particle pairs (such as proton and anti-proton, or electron and positron) that are of opposite electric charges. Correspondingly, half-wave spikes result in the formation of single particles that are of either positive or negative charge, depending on their respective half-wave spikes being positive or negative. In short,

"All of the matter and anti-matter particles are basically byproducts of the full-wave and half-wave spikes that were/are formed in the aether medium."

Note that, the types of particles forming during these processes solely depend on the amplitude of the spikes that are formed. For example, the more pronounced full-wave spikes give birth to proton and anti-proton pairs, while the weaker ones can only promote the formation of electron and positron pairs, and so on. In other words,

"Phase vibration spikes in the aether medium manifest as matter and/or anti-matter particles."

<u>To be more precise,</u>

"Matter and anti-matter particles are like bubbles that are formed as a result of induced 'cavitation' in the very fabric of fluid aether medium, as the local phase vibrations resonate into sufficiently high amplitudes."

The following picture shows the entrance of a water-shoot inside the reservoir of a dam. Such water-shoots prevent the water that is behind the dam to rise beyond certain safe level. They perform their important task by simply allowing the excess water to fall in them and get transferred downstream, bypassing the dam structure altogether.

In the two-dimensional planar surface of water, the middle part of such water-shoots is observed as a void. The very same is the case with the matter and anti-matter particles, as well as black holes, in the aether medium, since they too are bubbles, voids, in the three-dimensional space which is this universe.

In fact, the formation of matter and anti-matter particles (bubbles) in the aether medium by resonating phase vibrations is analogous to the formation of bubbles in a medium such as water when it is exposed to ultrasound waves which are also phase vibrations in the medium of water.

Bubbles (particles) formed by cavitation in the medium of fluid aether are basically openings into the accompanying universe. They automatically act as drain holes (openings / tubes) through which fluid aether can flow from this universe into the accompanying universe.

In the beginning, phase vibrations were distributed uniformly in the fluid aether medium, except for its outer limits which were (and still are) void of any kind of phase vibrations. Therefore, as the pressure and the density of the fluid aether medium gradually decreased, due to its expansion, the same circumstance was provided almost throughout its volume. Consequently, the phase vibrations present in the fluid aether resulted in the formation of matter and anti-matter particles everywhere, at about the same time. In other words, one can say,

"The whole universe literally bloomed with matter and anti-matter particles across its volume (the portion that was hosting phase vibrations), at about the same time."

Neutral particles, such as neutrons, where formed from the joining of electrons and protons (also anti-protons and positrons), as they attracted each other due to having unlike charges.

Note that, full-wave and half-wave spikes can also form randomly, as resonances or harmonics are formed, due to the ongoing interactions between the varieties of phase vibrations that readily exist almost anywhere and everywhere in the medium of fluid aether.

Such events lead to the **random production and hence random appearance of particles, either as in single particles or as in pair production.** Such events can also cause **the random disappearance of existing particles** (without any direct connection to other particles in this universe). Because, by forming resonances at the right place and at the right time, the existing phase vibrations can nullify the wave functions of locally existing particles.

Such interactions between phase vibrations in the fluid aether medium and matter (as well as anti-matter) particles are analogous to the destruction of bubbles in a medium such as water with the help of ultrasound waves.

Ultrasound waves (phase vibrations) of certain frequency and amplitude introduced in the medium of water can encourage the formation of bubbles by inducing cavitation. The very same type of phase vibrations can also cause the existing bubbles to collapse and literally disappear in that medium.

It must be mentioned and emphasized at this point that,

"The individual particles such as electrons, protons and neutrons, as well as their corresponding anti-particles, are simply bubbles in the medium of aether, bubbles which are not composed of any smaller constituents."

In other words,

"There are no such things as quarks in this universe."

Therefore,

"There is no such a force as the strong nuclear force in this universe."

As it is explained in the chapter titled "What is Gravity?", according to the aether based theory of gravity, it is the force of gravity that is holding the nucleons together in any and all complex nuclei. Hence, there is no such a force as the weak nuclear force in this universe, either.

2- Relation between aether and mass of particles

What is referred to as the 'mass' of a particle is actually the resistance that the local aether medium demonstrates in response to any change in the motion of that particle. Mass phenomenon manifests only as a particle is encouraged to move, stop or it is forced to change the magnitude of its speed in any particular direction, or any combination of different directions. Once a given particle is in motion, it continues to do so at a constant rate of speed forever, until it is acted upon by an external force of some kind. In other words,

"As the variety of forces act upon matter and anti-matter particles, the local aether medium generates a resistance in response to any changes in their motions, a resistance which is due to the formation of a wake-like wave in that medium. This aether resistance manifests itself as what is commonly referred to as the 'mass' of particles."

Note that, the ratio of masses of any two particles, in this universe, is directly proportional to their surface areas. Therefore,

"Due to being about 1836 times more massive than electrons, <u>the diameter of protons must be about 42.85 times larger than that of electrons</u>."

As the density of aether in this universe is gradually decreasing, due to its expansion and leakage, its resistance to the motion of particles (bubbles in that medium) is decreasing in magnitude. In other words,

"Mass of any and all particles, in this universe, is gradually decreasing."

In other words,

"In the earlier times, when this universe was much younger, any and all of the existing particles were more massive, due to the density of aether being much higher back then."

Note that,

"Fluid aether is not made of any kind of particles and therefore does not have any 'mass' associated with it."

3- Relation between aether and size of particles

Particles of matter are bubbles in the medium of fluid aether. Their sizes are dictated by the local fluid aether's density and pressure. Consequently, as the density and pressure of aether are gradually decreasing, due to its overall expansion as well as its leakage into the accompanying universe,

"Any and all matter and anti-matter particles are gradually becoming larger in size, in this universe."

In other words,

"In the earlier times, any and all existing particles were much smaller in size, since the aether medium was much denser and under much higher internal pressure back then."

4- Why particles are of specific discrete sizes?

The general size of particles is dictated by the pressure and density of the local fluid aether medium. However, their specific discrete sizes are dictated by the harmonics that can form by the phase vibrations present in that medium. The two main groups of phase vibrations present in the medium of aether are:

- The initial phase vibrations that were introduced in that medium, causing this universe to start its existence, and

- The phase vibrations that were generated due to the annihilation of matter and anti-matter particles nearly at the end of the rapid expansion era.

Of course, to a lesser extent there are also other phase vibrations, such as the variety of electromagnetic waves which were and still are produced due to a variety of atomic and subatomic interactions, as well as those artificially generated by various intelligent beings in this universe.

Note that, the sizes of particles which are known to exist are those that are the most stable (such as protons and electrons) and relatively stable particles (such as neutrons). Such particles are the more stable bubbles which are forming in the medium of aether due to harmonics of phase vibrations.

However, **particles of a variety of other sizes form as well but they are not long lasting due to their wave functions being in and out of phase with regularly and most importantly more frequently occurring harmonics of the phase vibrations in the medium of aether**.

As it is explained in the following subsections, particle pair productions and even single particle productions (as well as their respective annihilations) are quite common in nature and also in particle accelerators. But, only those that are more stable bubbles in the medium of aether are detected and/or verified to exist. Other particles form but literally momentarily and even though they really do come into existence they can be referred to as **virtual particles**, since they cannot be detected, directly.

5- Apparent random formation and destruction of matter and anti-matter particles in nature

Various phase vibrations existing concurrently in the medium of fluid aether form harmonics which, if strong enough, result in the formation of matter particles by literally inducing cavitation in that medium. The formation of matter and anti-matter particles in the fluid aether medium by resonating phase vibrations is analogous to the formation of bubbles due to cavitation in a medium such as water when it is exposed to ultrasound waves.

The newly formed particles can either manifest as matter and anti-matter particle pairs or as single particles, depending on the generation of full-wave and half-wave spikes, respectively.

Even though such processes took place on a massive scale, when the universe was going through its infancy period, over time its occurrence has become less frequent. This is expected, since phase vibrations in the medium of fluid aether are getting weaker as they are being literally stretched by the expansion of the aether medium, and also because aether is becoming less dense with time.

Note that, the very same type of phase vibrations in the medium of fluid aether can nullify the wave patterns of existing matter (as well as anti-matter) particles, hence causing the disappearance of particles, without any connection to other particles that may be nearby. This very effect also takes place in the case of bubbles in a medium such as water when they are exposed to ultrasound waves, since they simply collapse and vanish.

6- Apparent formation and destruction of various matter and anti-matter particles in particle accelerators

In particle accelerators, matter particles (which are in fact bubbles in the aether medium) are guided to collide with each other, at nearly the speed of light.

The interaction between accelerated particles can be likened to interaction of bubbles in a medium such as water, as they can be made to collide with each other at high speeds. During such processes a variety of other bubbles form, as the original bubbles lose their identities. Some of the newly formed bubbles are smaller and some others are larger in size, as compared to the original bubbles. Also, as newly formed bubbles continue to interact with each other, some of them get dissolved while some others grow in size.

This is the phenomenon that is observed in particle accelerators, as matter particles are smashed against each other at very high rates of speed.

Note that, <u>the artificial generation of more massive particles in various particle accelerators</u> is only indicative of the formation of ever more pronounced (higher amplitude) full-wave as well as half-wave spikes (bubbles) in the medium of aether. Such particles are manufactured either as the existing phase vibrations in the local fluid aether medium are amplified or as new phase vibrations are generated.

However, <u>by no means, such particles are representatives of the previous generations of particles that supposedly by splitting have led to the formation of the present-day particles.</u>

The formation of heavier particles in accelerators can be simply looked at as if someone is literally turning up the 'volume' on a stereo system or as if more speakers are used, in sync with each other, to generate a specific vibration but at a much higher amplitude.

The newly formed heavier particles (artificially amplified spikes in phase vibrations) have nothing to do with what have naturally existed in the past or will naturally exist in the future, in this universe. In short,

"The massive particles produced in particle accelerators do not represent what may have naturally existed in the past or may naturally exist in the future, in this universe."

7- Aether and formation of stable vs. unstable particles

If the matter particles (spikes or resonances formed by the phase vibrations in aether medium) are isolated and do not encounter other spikes of the opposite type (out of phase) they will be stable particles. However, if they encounter other harmonics, normally forming by phase vibrations in the local aether medium, that are of the opposite type (out of phase), they will face challenges, and will eventually lose their identities. These particles manifest as being unstable particles.

8- Why are isolated neutrons unstable?

Each and every matter (and anti-matter) particle type (such as electrons, neutrons and protons) are spikes with a certain frequency, and respond to vibrations that are of the same frequency. The varieties of phase vibrations which exist everywhere in the aether medium generate a variety of harmonics at fixed time intervals. These harmonics resonate at different frequencies.

If the frequency of a given particle does not match with the frequency of any of the harmonics generated by phase vibrations in the local aether medium, they will simply cross each other's path without affecting each other in any permanent way. Such particles manifest as being **stable particles**.

However, if matter particles encounter phase vibration harmonics that counter their very own frequencies (due to being of the same frequency and yet completely out of phase with them), those particles can and will lose their identities by instantaneously dissolving / disintegrating. Such particles manifest as being **unstable particles**.

Neutrons are among such unstable particles, whose vibration frequencies are so that they are matched with certain regularly occurring harmonics in the aether medium. As a result, if not interrupted by other vibrations (such as the presence of other particles in their respective immediate vicinities) they do go through a change of identity, as they unavoidably encounter such harmonics that are out of phase with respect to their (neutron's) vibrations, at fixed time intervals.

Note that, splitting of a particle such as a neutron due to encountering a phase vibration resonance of a certain frequency can be directly likened to bubbles dissolving in a medium such as water when exposed to certain ultrasound waves.

9- Relation between aether and unstable isotopes

Every matter particle is in fact a manifestation of a phase vibration spike in the medium of aether. As each of these matter particles chases its own respective path in the volume of a nucleus, it causes a disturbance / turbulence in aether that is in its immediate vicinity. If particles generate disturbances that are too far from each other or are mismatched in their frequencies, they will not usually interfere in each other's wellbeing.

However, if the generated disturbances are close enough to each other and/or their frequencies are nearly matched, they can form harmonics, as those particles repeatedly fly passed each other. In so doing, the harmonized disturbances can cause instabilities in the livelihood of one, two or even more matter particles (even as a group) that are of the same frequency and yet out of phase with the generated disturbances. In most cases particles will:

- Divide into smaller spikes (beta decay) or

- Lose part of their energy in the form of a newly formed phase vibration that is broadcast/spread in the surrounding fluid aether medium, as in electromagnetic waves (gamma rays) or

- Get knocked out of their path within the nucleus, either individually or as a collective (alpha decay or total splitting of the nucleus).

Either way, the overall process leads to a permanent change in the identity of the nucleus, as a whole. Such processes lead to the formation of new isotopes or even new elements.

Note that, the nucleus can be likened to an isolated star system in which as planets follow their respective orbital paths, after every so many revolutions, they will eventually line up in a relatively straight line or form other rare configurations with respect to each other. Such occurrences will trigger certain unique situations that can lead to one or more of the planets to get dragged out of their normal orbital paths. Such scenarios can lead to serious mishaps in the internal structure of that star system. Such events may have temporary or permanent side effects. The very same scenario applies to the nucleons in a nucleus.

10- Relation between aether and half-lives of unstable isotopes

The matter particles in the nucleus are in orbit around each other and they are literally continuously repeating a sequence of flyby routines that are apparently in random order. Every so often during their flybys they encounter a rare occasion when their generated disturbances become in sync and create harmonics that are strong enough to cause one or more matter particles to:

- Lose part of their energies in the form of a newly formed phase vibration that is broadcast/spread in the surrounding fluid aether medium, as in electromagnetic waves (gamma rays) or

- Divide into smaller spikes (beta decay) or

- Get knocked out of their path within the nucleus, either individually or as a collective (alpha decay or total splitting of the nucleus).

Such occurrences are not based on chance, since the motions associated with each and every matter particle is governed by certain forces that they encounter along the way. Some of these forces are caused by the other members of the same nucleus, while other forces originate from the outside. The external forces are due to the neighboring nuclei as well as the other particles that are freely wandering in the space between the nuclei.

In other words, the occurrences of various types of nuclei disintegrations can be viewed as if they are totally based on chance, yet they are not. Their behavior is exactly like flipping a coin. If it is done only once, the outcome is uncertain until the coin is actually thrown and has come to rest. This scenario corresponds to considering one specific nucleus since its disintegration cannot be predicted (according to commonly accepted rules/laws of probability) until it actually takes place.

However, if the very same coin is thrown many, many times, the outcomes will average out at around 50/50, since there are only two possible outcomes (heads or tails). Such an outcome is simply due to the existence of so many variables that affect the motion of the coin in its journey during each throw until it comes to rest. Therefore, the outcome evenly averages out between the two possible outcomes and the overall process seems to be based on chance, yet it is not. Because, if one were to repeat exactly the very same effects, during each and every throw, over and over, the results will always be identical. In such a case, all of the results can be either heads or tails, as desired.

Similarly, the nuclei present in any radioactive sample experience occasions when their own constituents (as well as external particles and the neighboring nuclei) generate disturbances that are strong enough to cause the formation of harmonics. There are literally millions of millions of millions of nuclei present even in a single gram of any kind of radioactive sample. Therefore, the occurrences of such harmonics among all nuclei that are present, average out.

"However,

When one specific nucleus is considered, if the state of every particle within that nucleus is known, one can calculate the exact timing of when that nucleus will encounter that very rare occasion when the disturbances caused by its own constituents will generate strong enough a harmonic that will change its identity."

Note that, shorter and longer half-lives are simply due to how often the needed circumstance occurs in a given nucleus that can lead to the formation of strong enough a harmonic.

If not influenced externally, the nucleus will experience the needed circumstance after a specific number of flybys of its own constituents and will disintegrate, regardless of how many other nuclei may be present in the same sample.

The nucleus can be likened to an isolated star system in which as planets follow their respective orbital paths, after every so many revolutions, they will eventually line up in a relatively straight line or form other rare configurations with respect to each other. Such occurrences will trigger certain unique situations that can potentially alter the normal orbital paths of one or more planets. If those planets receive sufficient energy from their neighbors they can even get out of their orbits and cause serious mishaps in the internal structure of that star system. Such events may have temporary or permanent side effects. The very same scenario applies to the nucleons in a nucleus.

Note that, as it is explained in greater detail in subsection "16- How do neutrons enable protons to bond together in nuclei of atoms?" below, **it is due to the lack of sufficient number of neutrons or excess number of neutrons that certain nuclei (isotopes) are unstable.**

The lack of sufficient number of neutrons in a given complex nucleus, necessary to properly bond the existing protons together, eventually causes that nucleus to lose its overall structural integrity and split in a variety of fashions, as the particle constituents of such a nucleus get maneuvered around by the local phase vibrations, as well as their harmonics.

Correspondingly, extra neutrons in a given nucleus also motivate instability in the overall structure of that nucleus. Since, the excess neutrons tend to cause protons to move in an abnormal / unstable fashion, as they try to uphold their bonds with too many neutrons, concurrently. Hence, they are literally encouraged to get out/off of their stable tracks, eventually leading to the overall instability of the nucleus, as a whole.

11- How does motion relative to aether affect the half-lives of unstable particles and unstable isotopes?

Various phase vibrations existing concurrently in the medium of fluid aether form harmonics, every so often. These harmonics are of a variety of frequencies and strengths. If strong enough, they can and do affect the wellbeing of particles and their aggregates in many different ways.

A particle or an aggregate of particles such as the nucleus of an atom that happens to be stationary in the medium of aether encounters such harmonics in relatively speaking fixed intervals. However, a particle (or an aggregate of particles) that is in motion in that medium encounters such harmonics at different intervals, depending on its rate of speed in that medium, due to missing its encounter appointments at certain set times for any given specific location in the medium of aether.

Also, as the rate of speed of such a particle or an aggregate of particles increases in the medium of aether and approaches that of light (phase vibrations) in that medium, that particle or that aggregate of particles literally not only skips more and more encounters with such harmonics but also alters how such harmonics can affect its wellbeing. Since, <u>any particle (let alone an aggregate of particles) that is in motion in the medium of aether automatically generates a wake-like wave in its immediate frontal surroundings. Such wake-like waves affect the shape of the harmonic waves reaching that particle, hence automatically altering the final effect of such harmonics on the wellbeing of that particle.</u>

These effects can be likened to the effects forced on a boat's haul by the waves that continuously exist on the surface of water, on which it may be floating. The following are three of many different scenarios that such a boat can experience while floating on the surface of a given lake:

- **In the first case**, the <u>boat is anchored in the middle of the lake.</u>

  In this case, the boat may show a gentle translational motion due to changing local wind direction, as it will be made to rotate around its anchor point. However, it will readily demonstrate very strong random wavy motions that are due to being exposed to the waves that exist on the surface of the water at its location.

- **In the second case**, the same <u>boat is moving at a slow speed on the surface of that same lake.</u>

  In this case, the boat clearly possesses a steady translational motion as it traverses the lake surface at a gentle speed in a particular direction. It will also demonstrate rather weaker random wavy motions that are due to being exposed to the variety of waves that exist on the surface of the water at any of its instantaneous locations.

- **In the third case**, the same <u>boat is moving at its fastest speed on the surface of that same lake.</u>

  In this case, the boat clearly possesses a steady translational motion as it traverses the lake surface at a high rate of speed in a particular direction. However, it will barely demonstrate any kind of random wavy motions that can be traced to its being exposed to the variety of waves that exist on the surface of the water at any of its instantaneous locations.

It is due to such effects that the rate of speed of already known unstable particles and unstable isotopes affects their respective half-lives. And, the faster they move in the medium of aether, the slower becomes their respective decay rates. In other words,

"The slowing down of decay rates of particles and/or radioactive isotopes which happen to be in motion at a high rate of speed relative to their local aether medium are due to their lessened and altered encounters with harmonics generated by the various phase vibrations existing concurrently in their respective immediate aether surroundings."

12- Effects of magnetic field and electric field on the half-lives of unstable isotopes and unstable particles

According to this book, magnetic field and electric field are different types of motions induced in the local aether medium. Therefore, the very presence of a strong magnetic field or a strong electric field will have the same effect on the half-lives of unstable isotopes and unstable particles as their motions relative to their respective aether mediums. Such effects are introduced in chapters "What is Magnetic Field?" and "What is Electric Field?" by proposing specific experiments. In short,

1- As it is stated in experiment "(Magnetically induced 'Time Dilation' on laboratory scale)", in chapter "What is Magnetic Field?",

"The half-life of any unstable isotope, or unstable particle, can be extended by exposing it to a strong and steady magnetic field."

2- Also, as it is stated in experiment "(Using electric field to induce 'Time Dilation')", in chapter "What is Electric Field?",

"The half-life of any unstable isotope, or unstable particle, can be extended by exposing it to a strong and steady electric field."

13- Effect of temperature on the half-lives of unstable isotopes and unstable particles

It is explained in great detail in the previous section "How does motion relative to aether affect the half-life of particles and radioactive isotopes?" that, the rate of speed of particles and their aggregates in aether affects their half-lives. Such an effect is due to how often and in what fashion these particles (or their aggregates) encounter and interact with the various harmonics that are generated by the phase vibrations that are continuously and concurrently propagating in the local medium of aether.

As the temperature of a matter medium is raised, its atomic and molecular constituents vibrate more vigorously. As far as the motion relative to the local aether is concerned, such motions of atoms and molecules are quite the same as those of the constituents of the same material if that material, as a whole, were propelled to a higher rate of speed in any given direction in this universe. Therefore, the very same effects are expected to be experienced by such isotopes that are known to be unstable. In other words,

"As the temperature of a radioactive isotope is raised, its half-life is expected to lengthen, accordingly."

14- Matter particles and their wave properties

Each and every particle (as well as aggregates of particles) has a vibration associated with it. This vibration is due to that individual particle's response, or the response of that aggregate of particles, to the phase vibrations that exist in the aether medium, in its immediate vicinity. The amplitude and the frequency of the vibrations manifested by any given particle (or an aggregate of particles) are directly dependent on its size and mass in comparison with the size (amplitude and wavelength) of the vibrations that are present.

The lighter and smaller the particles are, the more responsive they are to the local vibrations. For example, because of being much lighter than protons and neutrons, electrons respond more easily to the existing vibrations in their local aether medium. That is why their wave-like behavior is readily demonstrated and detected.

Note that,

> **"It is due to being readily responsive to the three dimensional waves (phase vibrations), present in their immediate vicinities, that,**
>
> **"Electrons form electron clouds, rather than following well-defined orbital paths, as they go around their respective nuclei."**

One can compare the lighter and heavier matter particles and even their aggregates to ping pong balls, basketballs, small boats and large ships such as oil tankers which react differently to the vibrations that exist on the surface of the same body of water on which they all may be freely floating. The following figure shows a few different size particles (or their aggregates) floating on a wavy medium such as the surface of water in a lake.

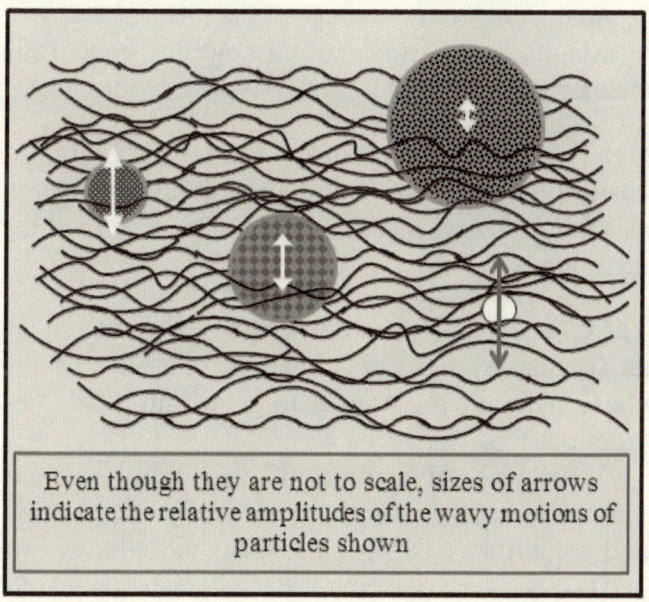

Even though they are not to scale, sizes of arrows indicate the relative amplitudes of the wavy motions of particles shown

Note that, nuclei, atoms, molecules, objects, planets, stars, solar systems and even galaxies and their clusters, demonstrate their own individualized reactions (vibrations) to the forces that are exerted on them, as a whole, by the variety of phase vibrations that exist in the medium of aether in their respective immediate vicinities.

15- What happens to various motions of atoms, nuclei and electrons at absolute zero degrees temperature?

Individual particles are literally bubbles in the medium of aether. As they form aggregates of any size or form, as in atoms or even molecules, they still are literally in the medium of aether. Currently, this overall medium of aether is also hosting a variety of phase vibrations, some of which were generated as this universe was born and some were generated as matter and anti-matter particle merged together and annihilated each other, nearly at the end of the rapid expansion era of space, in this universe.

At normal temperature ranges, each and every atom or molecule existing as a part of a solid object or even a liquid or a gaseous matter medium possesses two types of vibrations.

- One type is due to their individual kinetic energies which is basically representative of their individual temperatures and as a whole is demonstrated as the temperature of that matter medium.

- The other type of vibration which is very fine in amplitude, as compared to the first type of vibration, is due to the phase vibrations propagating within their respective vicinities.

  This type of motion of the individual atoms and molecules can be likened to the motion of individual boats and ships that is due to the local vibrations / waves that exist on the surface of the water on which they are floating.

The following figure shows the atomic structure of regular materials, at room temperature. The top section shows the atoms in liquids, gasses and even some of the solids (as their temperatures are raised). Their atoms / molecules as a whole can have certain amount of vibrational motions as well as translational motions associated with them. The bottom section shows the atomic structure of a solid material that is very organized in its lattice format. In this case, the atoms / molecules only possess local vibrational motions, but no translational motions. The shaded areas around individual atoms indicate the spatial range of their vibrations which is due to their kinetic energies manifesting as the overall temperature of that matter medium.

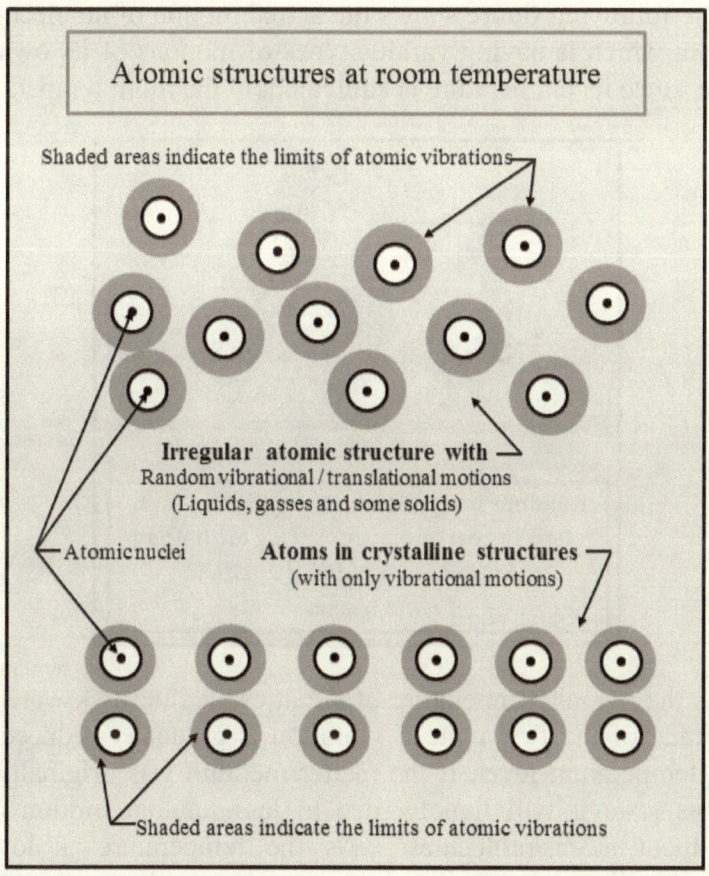

At normal temperature ranges, all of the electrons which are attached to the atomic nuclei are literally carried by their respective nuclei, as the whole atoms or molecules vibrates. Therefore, all of such electrons not only experience their own direct interactions with the phase vibrations existing in the local aether medium, but also are made to have motions due to being carried by their respective nuclei.

Note that, even the trajectories of the individual nucleons inside any given nucleus, as they follow their orbital paths around each other, vary due to the general motion of the nucleus, as a whole, as well as due to the presence of phase vibrations in the local aether medium.

The following figure shows the actual motion of an electron in an atom which is having various types of motions of its own as a whole, since its temperature is equivalent to the room temperature.

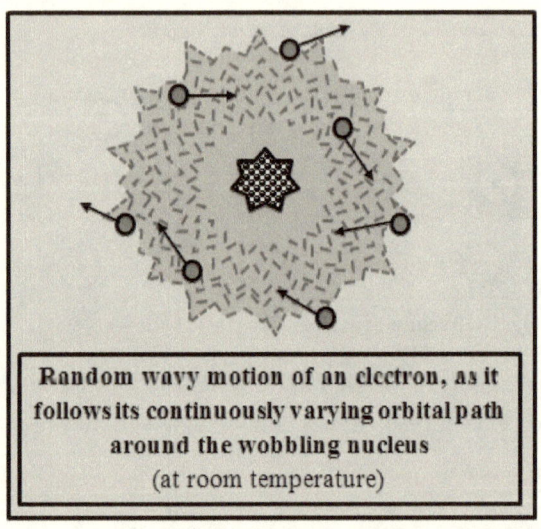

Random wavy motion of an electron, as it follows its continuously varying orbital path around the wobbling nucleus
(at room temperature)

As the overall temperature of a matter medium is lowered, the amplitude of the first type of vibration is gradually reduced. At some temperature level, if the matter medium was originally in a gaseous state it will liquefy, due to less random motion of its constituent atoms/molecules. As the temperature is lowered further, the liquid form will solidify. And, once the temperature of the matter medium reaches the absolute zero degrees Kelvin, the kinetic energy of the individual atoms and molecules will also reach zero. Hence, they become relatively frozen in their respective positions, just like boats or ships that have been tied with ropes to the peer next to them.

However, the individual atoms and molecules would still possess some fine motions / vibrations that are due to the phase vibrations propagating in the local aether medium. Such fine motions of atoms and molecules are analogous to the fine motions of boats or ships that are tied with ropes to the peer next to them, a motion that is simply due to the existence of waves on the surface of water in their respective immediate vicinities.

The following figure shows the atomic structures of materials which have been cooled to the absolute zero degrees Kelvin. As it is shown, in this case, atoms/molecules are nearly fixed in their positions, whether it be in a lattice type of structure or not.

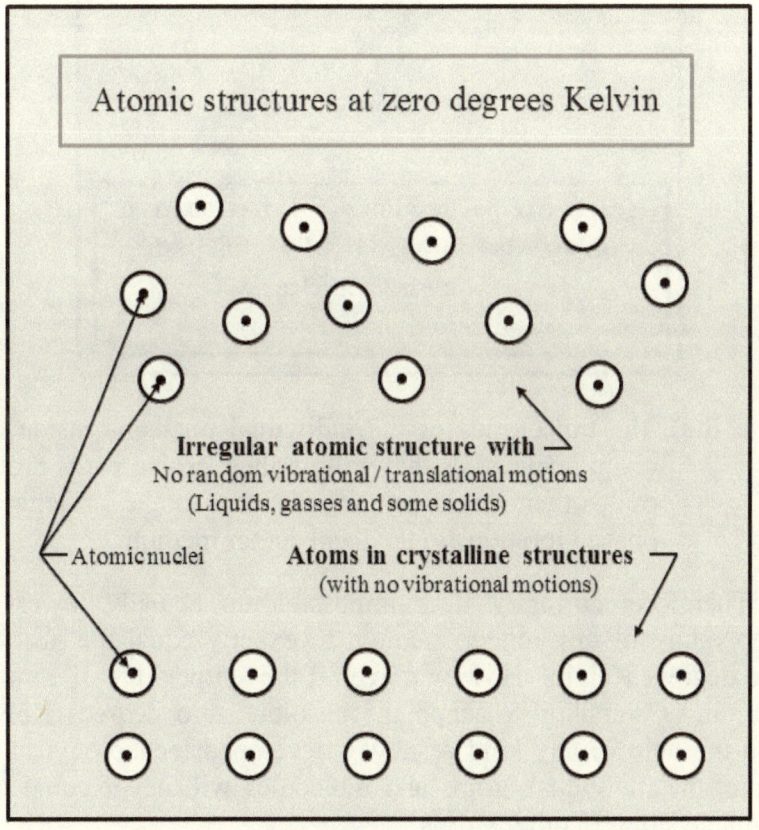

At zero degrees Kelvin, all of the electrons which are attached to the atomic nuclei follow nearly circular paths around their respective nuclei, since at zero degrees Kelvin, all of the atoms and hence their nuclei are relatively fixed in their positions. Therefore, all of such electrons only experience their own direct interactions with the phase vibrations existing in the local aether medium.

The following figure shows the actual motion of an electron in an atom which is not having any type of motions of its own as a whole, except for its minimal vibration which is due to the local phase vibrations in the medium of aether, since its temperature has been lowered to zero degrees Kelvin.

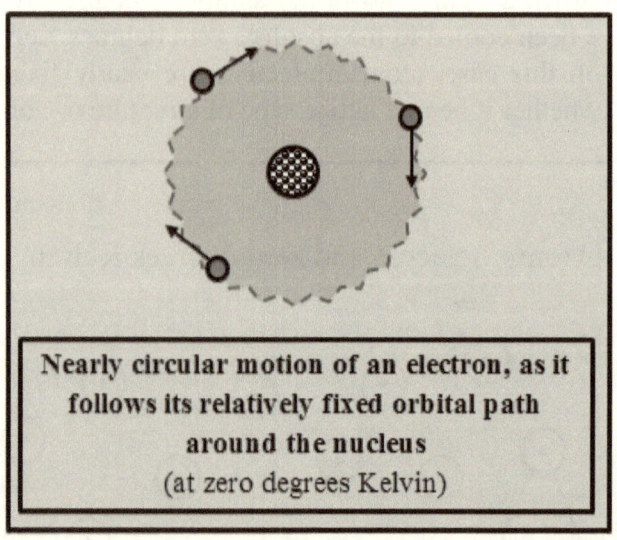

Nearly circular motion of an electron, as it follows its relatively fixed orbital path around the nucleus
(at zero degrees Kelvin)

Note that, the trajectories of the individual nucleons inside any given nucleus, as they follow their orbital paths around each other, also vary slightly due to the presence of phase vibrations in the local aether medium.

Therefore, currently it is impossible to actually lower the temperature of any matter medium to exactly equal the absolute zero degrees Kelvin. However, even if the temperature of a matter medium is somehow reached the absolute zero degrees Kelvin, with the help of any kind of newly developed technology, in the future, the individual atoms and molecules will never come to a complete stop. In other words,

"Even if the temperature of a matter medium could be lowered to zero degrees Kelvin, its individual atoms / molecules would still possess a certain amount of vibrations, and all of the nucleons and electrons will also possess certain amount of wobbliness in their orbital motions. All of these vibrations are due to the phase vibrations existing in their local aether medium."

16- How do neutrons enable protons to bond together in nuclei of atoms?

Protons have positive electrical charge and so when they happen to get close to each other, they repel one another. In the case of complex nuclei, where two or more protons are necessarily coexisting within relatively speaking short distances of each other, there is a need for an outside help to encourage protons to get along. Neutrons are the particles that basically do just that in any and all complex nuclei.

The way protons resist being in close proximity of each other and how the very presence of neutrons persuades them to get along can readily be demonstrated by two strong magnets and a chunk of magnetic but not magnetized material. This is shown in the following figure.

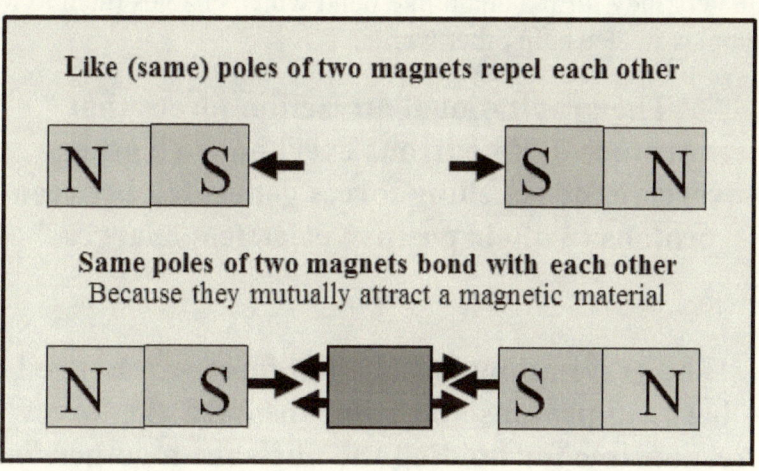

~ 161 ~

Shown in the figure are two different circumstances. In one case, the like poles of the two magnets are made to directly face each other. In such a scenario, the like poles of the two magnets definitely repel one another. While in the other case, a chunk of magnetic material is placed in the space between the two magnets. In such a scenario, the two like poles of the two magnets literally get distracted by the presence of the magnetic material and both attract that same magnetic material towards themselves. In so doing, they actually bond with each other through the magnetic material. In other words,

The attraction forces that like poles of two magnets exert on a magnetic material that is placed in between them overrides their mutually repelling force towards each other.

The very same effect takes place as two protons are made to get close to each other. If there are no neutrons in between them, the two protons definitely repel one another from far away distances. Yet, when there are one or more neutrons present in the space between the two protons, those two protons literally ignore each other's electrical charges and simply pay attention to the one or more neutrons that happen to be present between them. One could say, **protons concentrate their efforts in exchanging gravitational forces with those neutrons.** In so doing, each and every proton gets attracted towards the existing neutrons and ultimately they form a chain like bond which enables them to form a complex nucleus. In other words,

"The gravitational attraction forces that protons and neutrons exert on each other override the repelling forces generated between protons by their positive electrical charges."

Therefore,

"The gravitational attraction forces exchanged between protons and neutrons ARE the forces responsible for holding the nucleons together, in complex atomic nuclei.

Hence, according to aether-based theory of gravity,

"There is no such a force as the weak nuclear force in this universe."

The following figure shows three different scenarios. In the first one, there are only two protons that resist getting too close to each other. In the second one, the two protons notice that there is a single neutron between them and try to exchange gravitational forces with that neutron, and in so doing, they form a semi bond between themselves, as well. In the third case, the two protons notice that there are two neutrons in between them, and as each proton tries to attract the two neutrons and also by being attracted by them both, the four particles form a much better and more stable bond together, as a whole.

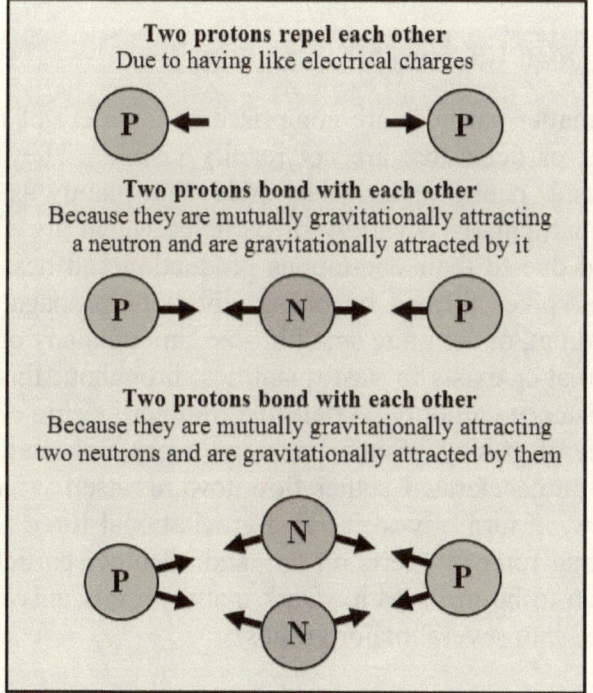

It must be emphasized that, it is due to the lack of sufficient number of neutrons or excess number of neutrons that certain nuclei (isotopes) are unstable.

The lack of sufficient number of neutrons in a given complex nucleus, necessary to properly bond the existing protons together, eventually causes that nucleus to lose its overall structural integrity and split in a variety of fashions, as the particle constituents of such a nucleus get maneuvered around by the local phase vibrations, as well as their harmonics.

Correspondingly, extra neutrons in a given nucleus also motivate instability in the overall structure of that nucleus. Since, the excess neutrons tend to cause protons to move in an abnormal / unstable fashion, as they try to uphold their bonds with too many neutrons, concurrently. Hence, they are literally encouraged to get out/off of their stable tracks, eventually leading to the overall instability of the nucleus, as a whole.

17- Formation of dark matter in this universe

Dark matter particles are comprised of a variety of matter and anti-matter particles that are not readily visible. They can be in such forms as neutron stars, black holes, but mainly as individual particles, particularly a variety of particles which are either quite short lived due to their continuous production and destruction by resonances/spikes formed by phase vibrations propagating in the aether medium, or are quite small in size, among many others.

Dark matter exists in vast quantities throughout this universe, but its existence can only be detected through its side effects such as its force of gravity. Since, each and every dark matter particle generates an accelerated aether flow towards itself. The induced aether flow, in turn, gives rise to a gravitational force which is in fact the drag force it exerts on any and all other particles/objects that happen to be in its path. Dark matter in this universe can be categorized into several major groups:

- The existence of some forms of dark matter, namely black holes, particularly within older galaxies, is due to the explosion of previous generations of large stars in such galaxies. As stars reach the end of their respective energetic life-cycles, if large enough, they go supernova.

During such a process, the inner layers of each star implode into a compressed state and form a black hole.

- In a given galaxy, as large stars experience a supernova moment the contents of their outer layers are literally thrown outwards, in all spatial directions. Some of the matter particles are thrown towards the outer regions of the galaxy. Eventually, due to the mutual force of gravity between such particles and the rest of the matter particles in the galaxy, they follow an orbital path around the center of mass of that galaxy. Over time, these matter particles form a significant portion of the overall mass of that galaxy.

    Initially, such particles follow orbits that as a whole occupy a huge volume that engulfs the whole galaxy. However, over time, as the galaxy evolves into its planar shape such particles also share/compromise in their orbital inclinations with respect to the overall plane of the galaxy and basically become more or less superimposed on that galaxy's plane.

    Matter particles forming such a collective within a typical older galaxy, by allowing the local aether to flow through them, and into the accompanying universe, they generate an accelerated flow of aether and hence a drag force (force of gravity) that is oriented towards the whole volume of galaxies. Hence, their very existence affects the orbital velocities of particularly those star systems that are closer to the outer edges of such well-developed, planar galaxies.

    The following figure shows the development / formation of such a mist throughout the volume of a galaxy, as that galaxy evolves from its spherical/elliptical shape into its planar shape.

    The density of such a mist-like gathering of particles in a given galaxy gradually tends to be higher towards the outer edges of that galaxy, due to lower number density of stars and their associated planets in those outer regions that would otherwise capture them.

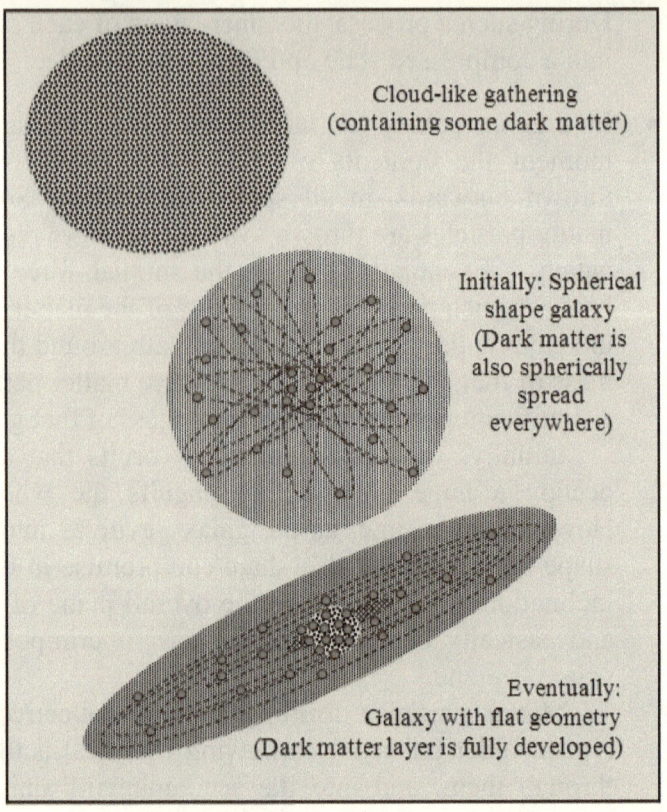

Cloud-like gathering
(containing some dark matter)

Initially: Spherical
shape galaxy
(Dark matter is
also spherically
spread
everywhere)

Eventually:
Galaxy with flat geometry
(Dark matter layer is fully developed)

Even the individual star systems possess such localized regions which are composed of matter particles that are literally thrown outwards by the stars, as particle storms. Over the life-cycle of a star, a significant amount of matter particles is thrown outwards. Those particles follow orbital paths in all directions, far away from the central star.

However, planets exert much weaker gravitational forces on such particles orbiting stars, as compared to the gravitational forces exerted by stars on such particles within galaxies. Hence, as shown in the following figure, such layers of dark matter forming within and around star systems, unlike in the case of galaxies, more or less tend to hold on to their random orbital paths around stars, instead of becoming nearly superimposed on the overall plane of that star system.

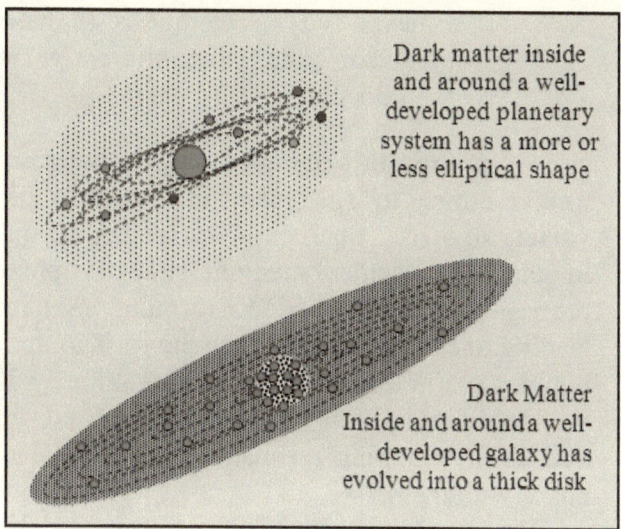

Dark matter inside and around a well-developed planetary system has a more or less elliptical shape

Dark Matter Inside and around a well-developed galaxy has evolved into a thick disk

Therefore, their gravitational effects on the orbital speeds of the outer planets are much, much weaker, as compared to the gravitational effects of such layers around galaxies on stars that are far from those galaxies center of mass.

Also, star systems hosting older stars possess higher overall concentrations of matter particles (in the form of dark matter). This is simply due to the fact that, matter particles thrown into the outer regions or outside the main rotational planes of star systems stay in those regions, indefinitely. Therefore, over time, their concentration can only grow.

Note that, The extra gravitational forces associated with the dark matter buildup within star systems, even though are relatively quite weak, they cause faster than expected deceleration of probes that are travelling from the interior regions towards the outer regions of star systems such as our solar system. <u>**This is exactly what has been experienced by the two probes, namely Pioneer 10 and Pioneer 11, which were sent towards the outer regions of our solar system.**</u> The extra (unexpected) decelerations of these two probes were confirmed since the early 1980s, and are well documented.

- However, **the majority of dark matter particles in this universe are in the form of unstable particles of a variety of sizes.** Such particles are continuously produced in the fluid aether medium, due to phase vibrations propagating in that medium forming resonances / spikes, leading to cavitation and hence the formation of a variety of bubbles (particles) in that medium. Shortly after forming, such particles become dissolved in the fluid aether medium, due to encountering resonances / spikes which are out of phase with them.

    The temporary appearance of such particles in the fluid aether medium can be likened to the temporary formation of bubbles inside a fluid medium such as water, when such a medium is exposed to ultrasonic waves of variable frequencies. Since, certain ultrasonic frequencies encourage the formation of bubbles by inducing cavitation in such a medium, while some other ultrasonic frequencies cause the bubbles existing in that medium to dissolve and vanish.

    Note that, the very existence of regular, more stable matter particles, as well as the variety of radiations they generate, within star systems and galaxies, encourage the formation of the needed resonances which would then form cavitation in the local aether medium. Since, they literally interrupt the propagation of the background phase vibrations, which otherwise would resonate less frequently.

- **Some of dark matter particles in this universe are in the form of very small particles which may be either stable or unstable, particles such as neutrinos.** Such particles are continuously forming in the fluid aether medium, due to a variety of interactions which are continuously occurring between other particles, particularly within stars due to their ongoing nuclear fusion reactions.

Such tiny particles can be readily likened to the miniature air bubbles in the medium of water that is forming the oceans, bubbles which are literally everywhere in that medium but are not readily detectable. Yet, the very livelihood of nearly all types of fish in such mediums depends on the very existence of such tiny air bubbles.

According to the above mentioned sources of dark matter within galaxies, the dark matter content (percentage of the overall mass) of the older galaxies must be greater than that of the younger ones. To confirm this overall proposal, one can perform the following experiment.

- Experiment:
(Relation between age of galaxies and the percentage of their overall mass that is in the form of dark matter)

To demonstrate the relation between the age of galaxies and the percentage of their overall mass that is in the form of dark matter within them one must study the orbital velocities of stars in many galaxies which are of a variety of shapes and forms. Particularly, the galaxies selected must include younger ones which are still spherical / elliptical in their overall geometries, as well as those that are much older, well developed, demonstrating well-defined flat geometries.

The data needed is the magnitude of the redshifts associated with many stars that are following their orbital paths around their respective galaxy's center of mass, at different distances. The following figure shows the relative motions of several typical stars in two different types of galaxies, namely a spherical/elliptical galaxy and a spiral/flat galaxy.

Note that, such sets of data can be extracted from the existing data which have been collected for stars of many galaxies, already.

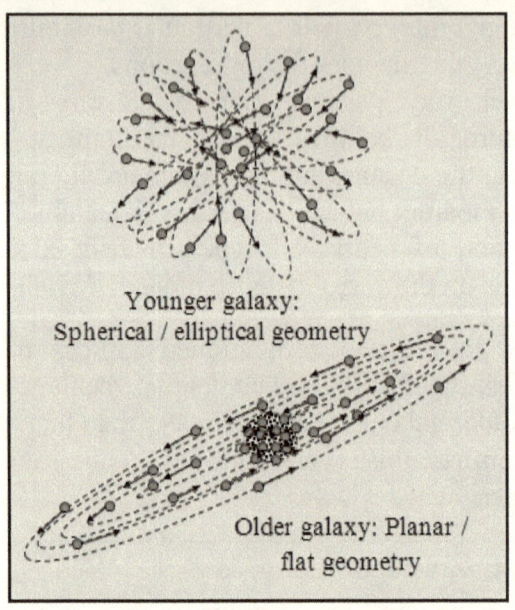

Younger galaxy:
Spherical / elliptical geometry

Older galaxy: Planar /
flat geometry

18- Aether and the cosmic microwave background radiation

Shortly after coming into existence, matter and anti-matter particles slowed down due to their mutual force of gravity. Soon after, most of the matter and anti-matter particles joined together and annihilated each other. Their annihilation processes generated phase vibrations which then spread in the surrounding aether medium just like shock waves. The following figure shows the end result of such a process in a very simple fashion.

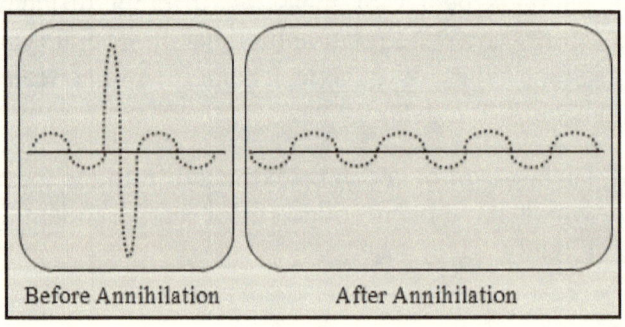

Before Annihilation   After Annihilation

The phase vibrations generated due to the annihilation of the matter and anti-matter particles were further spread and flattened by the ongoing expansion of the aether medium, as a whole.

Over time, these phase vibrations not only blended together by crossing each other's paths but also blended with the existing phase vibrations in that medium, phase vibrations (the original ripples) which were introduced into the medium of aether, and had caused this universe to come into existence. Eventually, this mixture of phase vibrations has formed a uniform background noise in the medium of fluid aether.

This very effect can be likened to the formation of many waves on the surface of a pond as multiple pebbles are thrown in the middle of it, at about the same time. As the newly formed waves spread outwards, they not only blend together but also blend with the waves that already exist on the surface of the water. Eventually, they become a uniform disturbance throughout the surface of the pond.

This uniform background noise (the mixture of phase vibrations in the medium of fluid aether) has been detected, already. It is known as the **Cosmic Microwave Background Radiation**.

Since the cosmic microwave background radiation is a mixture of phase vibrations in the medium of aether, its propagation speed in that medium is equivalent to the propagation speed of the phase vibrations which were introduced / generated when this universe had come into existence. Therefore,

"The cosmic microwave background radiation's outer limits or frontiers were and in fact still are basically chasing the frontiers of the phase vibrations that were introduced / generated at the time of this universe's birth."

It should be emphasized that,

"The phase vibrations in the outer regions of this universe and those in its inner regions have two totally different textures."

The variations in their textures can be likened to the differences that can exist between the textures of waves on the surface of the outer regions of a vast lake as compared to the textures of waves formed on its inner regions in which many small pebbles are dropped, at about the same time. As the newly generated waves propagate and spread towards the outer boundaries of the lake, they mix/blend with the existing waves which are already covering the entire surface of the lake.

In the case of a lake, the surface of the lake has fixed boundaries. Therefore, after a certain amount of time (which will depend on the overall size of the lake surface area) nearly the whole surface of the lake will exhibit the very same wavy texture. Only the outermost regions will be slightly different, due to waves being reflected off of the outer edges.

However, in the case of the phase vibrations in the fluid aether medium and their intermixing in this universe, **the outer regions will always be void of any kind of phase vibrations that are due to matter and anti-matter annihilations.** This is due to the fact that, both frontiers, namely that of the original phase vibrations and the one corresponding to the matter and anti-matter annihilations, are expanding / spreading at the very same rate of speed which is equivalent to the speed of phase vibrations in the medium of fluid aether.

This is analogous to the complex sound waves generated by a symphony orchestra that propagate through the medium of air. Such complex sound waves propagate at the speed of sound (speed of phase vibrations) in the medium of air. Therefore, they cannot catch up with other sound waves that were generated before the conductor had prompted the musicians to play their respective pieces.

In the case of the phase vibrations in the aether medium, in this universe, the overall modifications imposed on the texture of the existing phase vibrations in the inner regions depended on the frequencies of the two groups of phase vibrations. Since, their frequencies dictated how the two groups of phase vibrations could affect each other, as they strived to reach a compromise in their energies, as well as wave forms. The following figure shows the frontiers of the cosmic microwave background radiation in relation to other major frontiers in this universe.

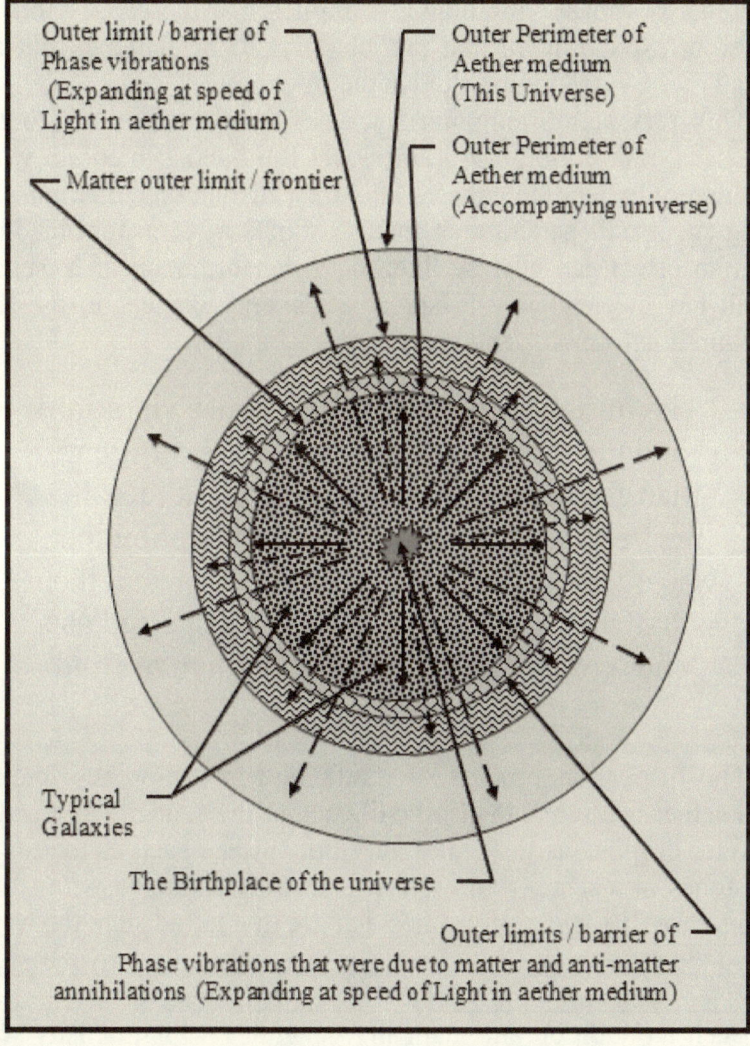

It should be emphasized that, all of the outer limits / frontiers shown in the figure are spherical in their overall geometries.

19- Matter-Energy equivalence

Any matter particle is in fact a phase vibration spike (a bubble) in the fluid aether medium. If for any reason a given matter and anti-matter particle pair or a single matter particle (a full-wave spike or a half-wave spike, respectively) is to be transformed into pure energy (phase vibrations), it implies that the phase vibration spike corresponding to that particle(s) is to be flattened out and spread in the surrounding fluid aether medium.

This very action temporarily causes the formation of a higher than usual phase vibration in the local fluid aether medium which will naturally tend to spread into the surrounding medium. In doing so, it will generate an intense shock wave in that medium. Such an effect can best be likened to the formation of a tsunami which has the tendency to spread its energy content in the fluid medium it is formed. Therefore,

"The initial matter (the spiked phase vibration) and the resulting energy wave are different manifestations, as well as effects, of literally the very same amount of phase vibration that was once condensed into a spike form and then is flattened out and is eventually dispersed as a 'shock wave' in the general medium of fluid aether."

The dispersion process of matter/anti-matter spikes into the fluid aether medium can also be likened to the sudden vaporization of water droplets as they hit a very hot surface such as that of the hot stones in a steam room which is already saturated with steam. In this case, the water molecules that are in a liquid state (as water) are suddenly vaporized and are consequently added to the steam molecules that are in the surrounding environment.

Such an addition automatically generates a high density wave front that will spread in all accessible spatial directions. Basically, what is taking place is a rapid dispersion of gaseous water molecules in the immediate surrounding medium which is also composed of gaseous water molecules.

If many particles go through such a process simultaneously, they will cause a very strong expanding shock wave in the local aether medium. This shock wave would literally carry with it anything and everything in its path. The following figure shows such a process for the case involving the annihilation of a pair of matter and anti-matter particles.

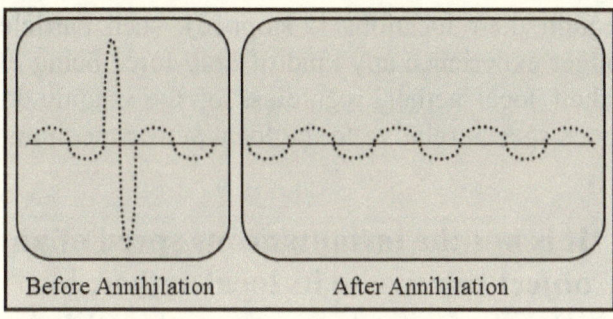

Before Annihilation   After Annihilation

20- How does an object's acceleration (not its speed), relative to its local aether medium, affect its physical dimensions?

As it is explained in the chapter titled "What is Gravity?", the drag force experienced by particles / objects at any given location in this universe, is due to the accelerated flow of aether through that location. And, uniform motion of aether flowing at constant speed through the location of particles / objects does not exert any kind of drag force on those particles / objects, regardless of the magnitude of the instantaneous speed of aether relative to the position of those particles / objects.

The very same applies when the local aether medium is relatively stationary and particles / objects are in motion in that medium. Since, it is only when such particles / objects are accelerating in their local aether medium that the local aether exerts a drag force on them. However, while they move at a constant rate of speed relative to their local aether medium they do not experience any kind of drag force being exerted on them by their local aether, regardless of the magnitude of their instantaneous speed relative to that medium.

Therefore, as individual nuclei, atoms, molecules, and objects as a whole, accelerate in their local aether medium (or as the local aether is encouraged to accelerate through their positions), they experience a drag force being exerted on them by their local aether. The magnitude of this drag force depends on the rate at which they are accelerating relative to their local aether medium. Once they stop accelerating in that medium (or the acceleration of the local aether through their locations is stopped), such particles/objects will no longer experience any kind of drag force being exerted on them by their local aether, regardless of the magnitude of their instantaneous speeds relative to the local aether medium. In other words,

"It is not the instantaneous speed of an object relative to its local aether, but rather its <u>instantaneous acceleration</u> in that medium that causes its dimension along its direction of acceleration to contract."

In order to understand the explanations provided by aether-based theories for why an object that happens to be accelerating relative to its local aether at a very high rate contracts along its direction of acceleration, one has to examine that object on the atomic and subatomic scales.

In some objects, atoms are arranged in an organized fashion and form a lattice structure of some sort. In others atoms do not form any particular pattern in their spacing. Each individual atom is also a 3-D entity which includes a central nucleus that is surrounded by an electron cloud. Even each nucleus (except for the case of Hydrogen) is composed of different number of nucleons, namely protons and neutrons.

The following figure shows the changes that would take place in the internal structure of an object, its atoms as well as the nuclei of its atoms, as the rate at which such an object accelerates relative to its local aether medium increases.

As shown, the object as a whole, its atoms and even the nuclei of its atoms become flattened only in one dimension which is along the direction the object is accelerating relative to its local aether medium. However, the other dimensions of the object, along the other two directions that are perpendicular to that object's direction of acceleration, are not affected by the acceleration of the object, since there is no change in the drag/push that is experienced from the sides.

In other words, the greater amounts of energy used to increase an object's rate of acceleration relative to its local aether medium are in fact stored in that object as different forms of potential energies. These stored potential energies are shared amongst various components of the object, as they are literally compressed towards each other and are flattened out of their normal shapes/positions. Such stored energies are like the energy that is stored in a spring, as it is compressed, or like the energy that is stored in air inside a cylinder, as it is squeezed in one particular direction, by pushing on the piston.

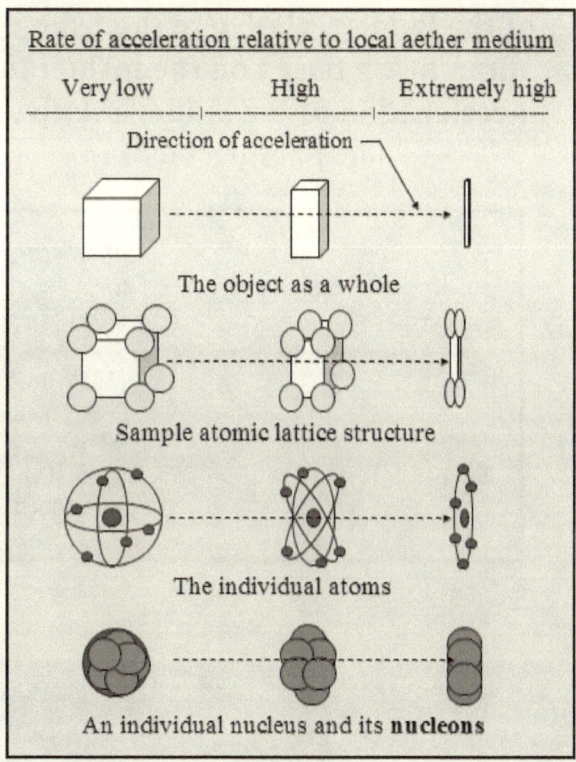

Note that, the individual subatomic particles do not get squeezed in the direction of acceleration.

Even though the above figure clearly demonstrates the compliance of the individual constituents, on different levels, as an object with a crystalline lattice structure accelerates at faster and faster rates, relative to its local aether medium, the very same effects can be visualized for an object consisting of atoms that are not well organized.

Note that, all examples in regards to an extra-long ladder moving lengthwise, at a very high rate of speed, that is nearly equivalent to that of light in a vacuum, contracting so much that it would fit inside a small barn, as shown below,

<u>**are fundamentally flawed**</u>. **Since,**

they are based on the instantaneous speed of the ladder relative to the barn, rather than being based on the instantaneous acceleration of the ladder relative to its local aether medium.

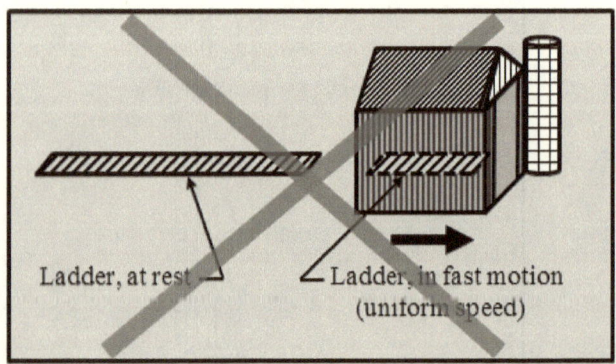

Ladder, at rest

Ladder, in fast motion (uniform speed)

21- Relation between density, weight and mass of an object and that object's speed and acceleration relative to its local aether medium

As an object accelerates at higher and higher rates relative to its local aether medium, it contracts along the direction it is accelerating. However, as shown in the following figure, the other two dimensions, that are perpendicular to its direction of acceleration, remain unchanged.

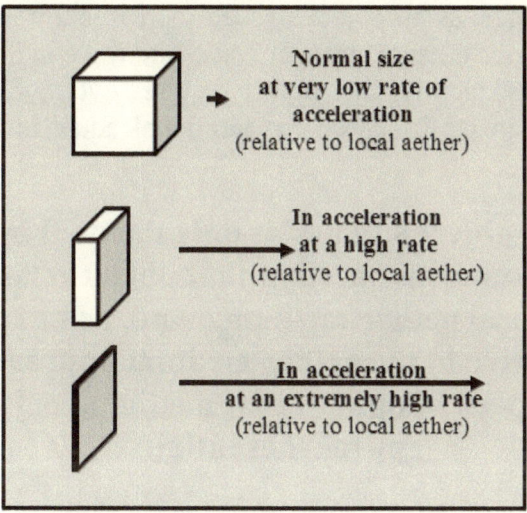

Therefore, as an object accelerates at faster and faster rates its volume decreases linearly with its length (dimension that is in the direction of its acceleration). In the meantime, its matter particle content remains unchanged. In other words,

"As an object's acceleration relative to its local aether medium increases, its 'density' increases proportional to the decrease in its dimension that is along the direction it is accelerating."

Also, regardless of its instantaneous speed relative to its local aether medium, the drag force experienced by an object is directly dependent on its instantaneous rate of acceleration relative to its local aether medium, the very same drag force that gives meaning to the weight of that object. In other words,

"An object's 'weight', regardless of its speed relative to its local aether medium, is only dictated by the magnitude of its instantaneous acceleration relative to its local aether medium."

Mass of an object is the resistance that the local aether medium demonstrates to any changes in the motion of that object, a resistance that is due to the formation of a 'wake-like wave' in that medium. Since, as an object's speed increases towards the speed of light, it forms an ever stronger wake-like wave in the medium of aether. Consequently, that object requires evermore amounts of energy to increase its speed by additional equal increments. In other words,

"An object's 'mass' is only dictated by the instantaneous speed of that object relative to its local aether medium. And, as its speed relative to the aether medium approaches the speed of light in that medium, its 'mass' approaches infinity."

Note that,

"If an object manages to reach the speed of light (the speed of phase vibrations) in its local aether medium, it will literally experience the very same effects as if it were at rest right at the event horizon of a black hole, where the speed of aether rushing by is exactly equal to the speed of light in that medium."

Explanations for Various Effects and Paradoxes

This section provides explanations for why certain effects such as Casimir effect are observed. It also explains what the uncertainty principle and the Planck's constant represent, in physical terms, as they relate to aether and phase vibrations which are constantly propagating through that medium. The entanglement paradox proposed by Einstein-Podolsky-Rosen is also explained with simple reasoning. Finally, the results of various types of double slit experiments are explained using the theories presented in this book.

Note that, these few subsections are included in this chapter, just to demonstrate the capability of aether-based theories in explaining what is truly behind the mechanism by which this universe operates.

1- Casimir effect

Casimir effect in brief can be described as a mysterious force that manifests itself when two electrically neutral flat plates are brought very close to each other, so that they are parallel but not touching one another. The following figure shows a simple presentation of such a set up.

The medium of aether anywhere in this universe is concurrently hosting multitude of phase vibrations, including electromagnetic waves, with a variety of frequencies. Since phase vibrations in the medium of aether are localized motions of that medium, it is the medium of aether itself that is literally vibrating when such vibrations propagate through it. In a sense, this type of motion can even be referred to as the temperature of that medium. Such vibrations (localized cyclic dynamic motions) in the medium of aether give rise to a dynamic pressure in that medium. Such a pressure is automatically exerted on any and all of the particles (bubbles) present in the vicinity.

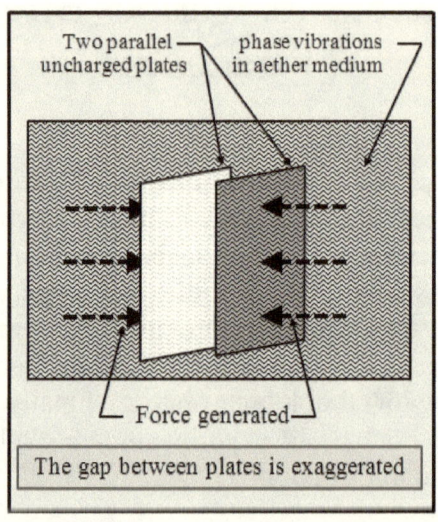

Therefore, if less of such vibrations are literally allowed to exist in one region of space, as compared to its neighboring regions, a force directed towards that particular region will manifest itself. This force is induced by the pressure difference that is generated between that particular region of space and its neighboring regions.

In the case of the Casimir effect, as the two flat plates are brought into close proximity of each other, phase vibrations face a challenge as they try to propagate in the small gap that is formed between the two plates. However, such waves have no difficulty, whatsoever, propagating outside that gap.

Smaller gaps between the two parallel plates allow fewer of the phase vibrations to propagate through, without bumping into the local particles forming the plates. Consequently, waves that are propagating outside the gap, by exerting a force on those two plates from outside, tend to push them towards each other.

Therefore, as the gap between the two plates is decreased, the magnitude of this force becomes greater.

Note that, at room temperature the atoms and molecules forming the two plates have their own vibrations. Therefore, the magnitude of the force manifesting as Casimir effect, in between the two plates, due to phase vibrations in the local aether medium is cushioned by the vibrations of the atomic constituents of the two plates. Since, at room temperature, there is no strict division in the range of frequencies that can pass in between the two plates. And, certain range of frequencies can partially penetrate the gap in between the two plates.

However, by housing the two plates in a vacuumed chamber and lowering their temperatures to near the absolute zero degrees Kelvin, one can eliminate such cushioning effects. Hence, the effect observed, the magnitude of the force measured, will be more pronounced, for any give size of a gap in between the two plates.

The difference between the magnitudes of the force generated between the two plates in these two cases, when the two plates are at room temperature and when they are cooled down to near absolute zero degrees Kelvin, can be compared to the magnitudes of the force exerted on a nail by a rubber mallet and a solid steel hammer.

2- The uncertainty principle

According to quantum mechanics, the uncertainty principle arises due to the matter-wave nature of objects. As it was explained in the section 'Matter particles and their wave properties', each and every particle, as well as aggregates of particles, exhibits a specific vibration which is due to the phase vibrations present in the immediate vicinity of that particle (or aggregate of particles). There are two types/groups of fundamental background phase vibrations in this universe.

1- The initial phase vibrations that were introduced into the fluid aether medium, the same phase vibrations that encouraged the aether medium to start its rapid expansion process.

2- The phase vibrations that were generated as most of the matter and anti-matter particles united and annihilated each other, nearly at the end of the rapid expansion era, when universe was still quite young.

The very existence of such background vibrations which are in fact phase vibrations in the fluid aether medium, automatically introduces uncertainties in the instantaneous position and speed of any and all objects, regardless of their sizes and masses. In fact, all other phenomena such as momentum are also directly affected by such vibrations.

As shown in the section 'Matter particles and their wave properties', the effect of local phase vibrations on various objects can be likened to the effect of waves present on the surface of a lake, on different size objects that may be floating on that body of water.

The phase vibrations in the aether medium affect larger objects only weakly as compared to smaller objects, just as the waves on the surface of a lake barely affect a large ship, yet they quite readily affect smaller objects such as Ping-Pong balls.

Note that, any and all of the phase vibrations present in the aether medium are being stretched by the ongoing expansion process of that medium. Therefore, the averaged amplitude of the overall background phase vibrations in that medium, which limits value of the uncertainty in measurements of position and speed of particles, is decreasing in magnitude. In other words,

"The limiting value of the uncertainty principle is gradually decreasing in magnitude."

3- Planck's constant

The background phase vibrations in the fluid aether medium can be defined by their frequencies and their amplitudes. This is analogous to defining waves on the surface of a lake or defining sound waves in any medium such as air or water using such properties of those waves.

"Planck's constant is proportional to the 'Amplitude' of the background phase vibrations which exist in the aether medium, everywhere in this universe."

Note that, the background phase vibrations in the region of space that is hosting matter as well as anti-matter particles are a mixture of two different types of phase vibrations, namely the initial phase vibrations that were introduced into the fluid aether medium and the phase vibrations that were generated as most of the matter and anti-matter particles united and annihilated each other.

The value/magnitude of the Planck's constant is almost exactly the same everywhere in this universe, since the variations in the amplitude of the background phase vibrations in different regions of this universe are minimal. Therefore,

"Planck's constant can be considered as a constant on a universal scale."

Note that, as time goes on, **amplitudes of all of the phase vibrations present in the fluid aether medium are gradually decreasing since any and all of the phase vibrations present are being stretched by the very expansion process of that medium.**

Also, **due to evermore mixing and blending, the existing phase vibrations are averaging out by partially neutralizing each other.**

Therefore, the averaged amplitude of the background phase vibrations in the aether medium, to which the Planck's constant is proportional, is decreasing in magnitude. In other words,

"The magnitude of the Planck's constant is gradually decreasing."

4- Einstein-Podolsky-Rosen entangled particles (Explained)

Two scenarios of particle entanglements have made lots of headlines over many decades, since Einstein-Podolsky-Rosen first proposed such issues. These two cases are dealt with separately, below.

1- **According to Einstein-Podolsky-Rosen paradox of entangled particles**, as it is described on Wikipedia on the internet, once the state of one of the two entangled particles is known, at any time after their separation, automatically and instantly the state of the other particle becomes accurately known, regardless of how far they may have travelled away from each other

This case of entanglement of particles can be explained readily due to the very nature of such entanglements. The specific physical states, in which the entangled particles are, at any given moment after their separation, are just like the identities of the two sides of a single coin.

Suppose a coin is thrown and has come to a complete rest on the surface of a table top which is made of clear glass. One person looks at its upper surface and takes a picture of it from above, while another person looks at its lower surface and takes a picture of it from below. At any time later, once one of these individuals reveals the photograph he/she has taken, hence identifies which side of the coin he/she has seen and photographed, automatically and instantly the surface of the coin seen and photographed by the other individual becomes known. Such a setup is shown below.

In such cases, the connectedness of the outcomes shown by the photographs of the two sides of a given coin does not imply that somehow the second photograph has become aware of what side of the coin was viewed in the revealed photograph and so has instantly changed/selected to represent the other possible face of the coin. The top and bottom sides of the coin, as shown in the photographs, have had their identities set since the very instant the coin had come to a complete rest on the table top.

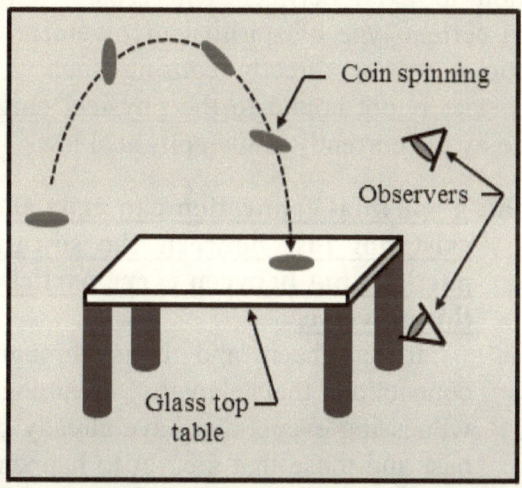

The face of the coin revealed by looking at one or the other photograph at any time later, is only a matter of clarifying for a third individual that, which face of the coin was on which side. The observation of the photograph has nothing to do with any decision making by the second photograph (or second side of the coin) to decide which face of the coin it should demonstrate/represent.

The very same applies to the so-called entangled particles. Since, the action of the observer becoming aware of the state of the first particle has nothing to do with what the state of the second particle is. Since, the state of the second particle was already set from the instant it had separated from the first particle.

Therefore the observer's act of identifying the state of the first particle does not imply that the second particle has received a telepathic call/message to be what it has already been, all along.

2- In other cases, it is claimed that if any changes are imposed onto the state of one particle, the state of the separated particle instantly changes in a predicted manner, as if there is a direct line of communication between the two particles.

It is the opinion of this author (based on his book titled "the Evolution of Spirits", 2012) that, the only way for such a connection to exist between two separated particles is by having a certain type of spiritual involvement accompanying the particles and/or directly causing such effects. Since, spiritual state is not bound to the physical universe and does not abide by the currently known physical laws.

Note that, **a spiritual connection can exist and in fact does exist not just between the so-called entangled particles but between every particle that exists in this universe.**

It has been and it is through such direct connections that telepathy, dreaming, premonition, witnessing events that have already occurred in the past and those that are yet to happen in the future, even the all common imagination experienced by all life forms that possess both spirit and soul, as well as the very 'thought', amongst many other functions, have been possible and are possible.

It should be emphasized that, over the last few decades, there have been many occasions that bogus, misleading information (fictitiously) presented as scientific breakthroughs by the so-called front-line scientists. Also, after certain real breakthroughs were announced they were later on retracted as having been mistakenly interpreted earlier. For examples,

- About two years ago, the scientists at CERN had officially announced that they had detected neutrinos travelling faster than light. Thousands of scientists were involved in these experiments and they had repeated their experiments literally 3,000 times before they had the courage to come forward with their findings.

  Yet, in less than a year later, they retracted their report and stated that their calculations were not quite right, since one of the fiber optic cable plugs was not inserted into its socket properly. One can only wonder if these scientists should even be allowed to be on welfare, let alone have access to the most expensive scientific equipment on this planet.

- **Scientists have recently claimed that they have managed to isolate and directly and visually study individual particles such as single electrons and even single photons, without destroying or disrupting their normal states.**

  While they have made such claims, yet to this day, it is quite amazing that **none of these scientists have been able to measure the true diameter of a proton or that of a neutron, let alone the diameter of an electron or the shape and size of the imaginary photon, let alone measure and identify different types of particles by their shapes, colors, spin and so on.**

- As other examples one can mention **General Theory of Relativity** and the **Quantum Theory**, both of which have been worked on by literally millions of scientists and graduate students globally since their introduction, nearly ONE hundred years ago, and yet, neither one is completely explored or fully understood and examined. Our current front-line scientists are still trying to figure out how to introduce more fudge factors in both theories so that they become compatible with each other, since they both are supposed to be applicable to/describing the one and only universe in which we live.

5- Double-slit experiments (Explained)

The double-slit experiment has been tried in a variety of ways. Such experiments may be categorized into three general types.

- **Type I**: The formation of interference patterns has been demonstrated using light waves shined at such double slits.

The results of this type of double-slit experiment, which are already well explored and documented, are clearly due to the wave nature of light.

- **Type II**: The formation of interference patterns has been demonstrated using particles such as electrons, atoms and even molecules. The formations of such interference patterns are vaguely contributed to the wave properties of matter particles. An example of such an experimental setup is shown below.

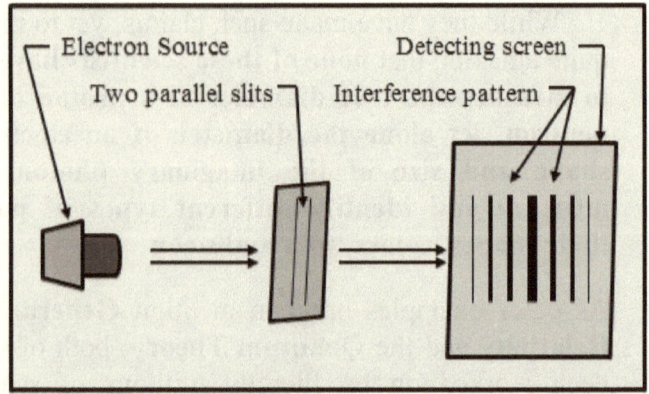

It must be emphasized that, in the case of electrons and other matter particles, an interference pattern forms only when the very same process of throwing individual electrons for example, at a screen, is repeated many, many times. If only one electron is thrown, it cannot be even predicted through which slit it will pass. The result will become known only after the individual electron is actually thrown and it is allowed to complete its journey towards the detecting screen. However, every time such an individualized throwing is repeated, the outcome (the highlighted spot on the screen) will be at a different position. In other words,

No two particles will ever hit the exact same spot on the detecting screen.

The individualized trajectory followed by each of the thrown particles is affected by different forces exerted on that particle, both along its trajectory and as it passes through one slit or the other. These forces, which cause minute variations in the path of the thrown particles (or aggregates of particles), are associated with the background phase vibrations that exist in the local aether medium, through which each and every particle must necessarily pass, as they travel from the source to the detecting screen.

<u>Since the background phase vibrations in the medium of aether are totally random in their positions, each particle starts and finishes its journey by following a totally unique route between the source and the detecting screen. Therefore, the destination spot on the detecting screen will understandably be a unique one for each particle, as well.</u>

When, such a process of throwing individual particles towards a designated target screen is repeated with a great number of particles, the formation of an interference pattern is unavoidable, unless other forces affect the particles motions along their paths.

To clarify this process of individuality that is experienced by each and every thrown particle, one can propose the following experiment.

• Experiment:
 (Throwing tennis balls at a lake)

A standard tennis ball throwing machine is needed to perform this experiment. Such a machine must be positioned, on a tall stand which is directly resting on the ground (the bottom of the lake), so that the machine is just above the surface of the water that is in the lake. This way, the thrown balls will repeatedly touch and bounce off the surface of water, before they come to rest. The machine used can be accurately fine-tuned so that every ball that is thrown will leave the machine with nearly identical speed and direction of motion.

All of the tennis balls thrown will follow the very same trajectory but only up until they hit the water surface.

Since, as it can be expected even before performing such an experiment, regardless of how accurately the machine is adjusted in its delivery of momentum to the balls, NO two balls will ever follow the very same exact route, once they touch the surface of the water. The following figure shows such a setup.

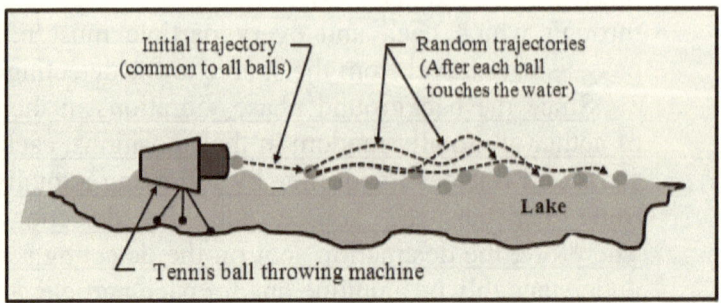

Each tennis ball, during its repeated touch and bounce off scenarios, will have encounters with the water surface at different angles as compared to the other tennis balls. Even the number of times each ball will hit and bounce off of the water surface will be different. Therefore, each tennis ball will come to rest at a different spot on the surface of the lake.

Note that, during such an experiment, the only variable that **<u>cannot</u>** be adjusted or compensated for is the randomness of the waves (vibrations) that are constantly present on the surface of the lake. The exact contact angle between each ball's initial trajectory and the portion of the wave form on the surface of water directly affects how it will be bounced off.

Therefore, not only the magnitude of the forces exerted by the water onto each and every ball during their first touch and bounce off experience will be different, but also during each and every one of their following touch and bounce off scenarios each ball will have its own unique experience, as well. <u>Such differences will include changes in their directions towards</u>

the sides, due to the vibrations on the surface of water being 2-dimensional.

If many, many balls are thrown and are allowed to come to rest on the water surface, a random pattern will be demonstrated by the rest positions of the balls on the lake.

To complete this experiment, one needs to place a narrow wall with two side-by-side, narrow vertical openings just after where the balls are expected to have already bounced a couple of times off of the water surface. The width of the openings can be about twice the diameter of the balls used. Other ranges of opening sizes can also be considered and experimented with.

In such a case, as the tennis balls are bounced off at different angles they may go through one or the other opening and due to their entry angles they will be either bounced off of the sides of the opening as they go through, or will go right through without touching the sides.

Once many, many balls are thrown and are allowed to follow their desired paths, a pattern will emerge by the resting positions of the balls that have actually passed through either of the two openings. This pattern is exactly analogous to the interference pattern that is observed to form as electrons and other particles or their aggregates are thrown at a detecting screen, through a double-slit membrane.

Note that, the waves on the surface of water in this experiment quite closely resemble the phase vibrations that exist in the fluid aether medium, the same phase vibrations that directly affect the routes taken by individual electrons or other matter particles, as they are thrown towards a detecting screen.

Here is an excellent spot to mention that, such an experiment can be repeated with appropriate equipment, at the very same location and with the very same water surface conditions, even at the very same time, but with a variety of balls such as ping pong balls, tennis balls, baseballs and even basketballs, or even with several of such balls attached together resembling a molecule of any desired size.

During such experiments, it will become clear how the size and mass of such balls, or those of their aggregates, affect the amount of variations induced in their paths, as they traverse from their respective sources to their respective rest spots on the lake surface.

The overall results obtained will quite closely, if not identically, match the behavior observed when particles such as electrons, atoms and complex molecules are used, during such experiments.

Note that, a smaller similar setup can be designed and built to perform such experiments indoors. In such setups, the needed random waves on the water surface can be generated by the cyclic motions of several paddles near the outer walls of the water container which can be a swimming pool.

- **Type III**: In some cases of double slit experiments, apparently a strange phenomenon takes place. If the experiment (the throwing of electrons at a detecting screen, for example) is not recorded (watched), an interference pattern forms on the detecting screen. However, if the process is recorded (watched), no interference pattern forms.

  In such cases, it appears as if the electrons (for example) are aware of being watched and refuse to play along and do not form an interference pattern. Yet, when the experiment is conducted without being recorded (watched), the electrons willingly form an interference pattern. In other words,

It seems like as if the very action of watching the process affects the results of such double-slit experiments.

The following experiment is proposed to expose the affecting phenomenon, in such cases.

- Experiment:
  (Effect of camera position)

First, the process of throwing electrons should be conducted without being recorded (watched). In such a case, the electrons will form a well-defined interference pattern on the detecting screen. Then, the very same process should be repeated, but this time it must be recorded. In such a case, as it has been reported by various researchers, most likely no interference pattern will form.

Next, the very same process must be repeated with the camera rolling but not recording. In this case, either an interference pattern will form or will not form. The very same must be repeated many times (with the camera rolling but not recording) to explore the effect associated with the position of the camera with respect to the path taken by the electrons.

For example, the camera can be mounted on the sides, above or below, at different angles and/or distances from the path taken by the electrons, as they travel between their source and the detecting screen. Such a setup is shown below.

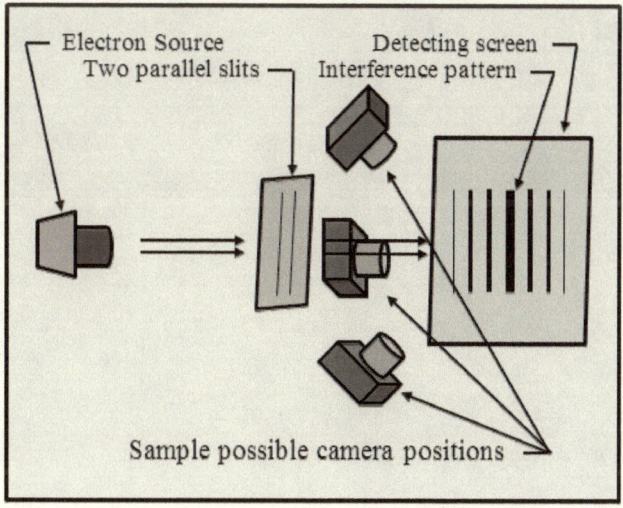

Sample possible camera positions

The same must be repeated with the camera being on and actually recording what electrons are doing.

Upon the completion of such an experiment, it will become clear that the position of the camera will have a definite effect on electrons reactions. Because, an active electrical / electronic device such as a camera generates an electric field and/or a magnetic field which directly affects the background phase vibrations in their immediate vicinities.

The electric / magnetic fields generated by the camera are much stronger than the background phase vibrations in the local aether medium. Hence, they override the background waves and in so doing affect and alter the conditions of the experiment. In other words,

"The very presence of an <u>active</u> electric / electronic device such as a recording device which naturally generates an electric field and/or a magnetic field in its immediate vicinity, directly affects the background phase vibrations which are present in the local medium of aether through which the particles pass, during double slit experiments."

Explanations for Gradual Variations in Different Phenomena

This section provides explanations for gradual variations in various phenomena on a universal scale.

1- Force of gravity is gradually weakening

As space in this universe is expanding and as aether is leaking into the accompanying universe by going through matter and anti-matter particles and their aggregates, the overall aether content of this universe is gradually decreasing. As a result, the overall density and pressure of aether that is in this universe are gradually decreasing.

The force of gravity, at any given location in this universe, which is the drag force induced by the flow of aether which is accelerating through that location is directly dependent on aether's density and pressure at that particular location. Also, the 'mass' of individual particles is dependent on the density of the aether, as well. Therefore, as the overall density and pressure of the aether medium are decreasing with time, due to aether's expansion and leakage, the flow of aether is gradually becoming less effective in inducing the force of gravity. In other words,

"The Universal Gravitational Constant is gradually decreasing in magnitude."

2- Gradual widening of the planetary orbits

As aether's density and pressure are decreasing with time, due to its expansion and leakage, the flow of aether is becoming less effective in dragging objects that happen to be in its path. Therefore, stars are gradually exerting weaker gravitational forces on their respective planets. This simply means that stars are gradually losing their grips on their respective planets and consequently

"The planetary orbits are gradually becoming wider."

This side effect of decrease in aether's density has already been detected in regards to the orbit of planet earth around the sun. According to the collected data, the orbital path of earth is widening by about 7 meters (about 23 feet) per century. Of course, a small portion of this observed amount of expansion in the orbital path of earth is justifiably due to:

1- Sun's gravity is becoming weaker due to losing mass, as it is continuously transforming some matter into energy. It is also literally throwing huge amounts of matter into space as solar flares / solar storms.

2- Earth is being pushed away, as it is absorbing and/or deflecting particles that are thrown at it by the sun, in the form of solar storms.

Predicted Eventual Outcome of Various Phenomena

This section provides certain predictions about what can be expected to eventually become of certain phenomena on a universal scale.

1- Speed of light and other phase vibrations in the aether medium is gradually increasing

The speed of light and other types of phase vibrations in the aether medium is dependent on aether's density. Higher aether density limits the speed at which phase vibrations can travel in that medium. Therefore, in the very beginning, when the density of aether was literally next to infinite, the speed of phase vibrations (the speed of light) in that medium must have been next to nil. In other words,

> **"In the beginning, during the rapid expansion of the aether medium, the existing phase vibrations were spreading more effectively as they were carried by the expansion of the aether medium rather than by propagating in that medium."**

During the rapid expansion era, as aether itself was expanding, it stretched the wavelengths of the phase vibrations that were present in it.

As the density and pressure of aether were reduced, due to its expansion, the speed of phase vibrations in that medium was increased. The very same process is still taking place. In other words,

"As the density and the pressure of aether are gradually decreasing, in this universe, the speed of light (the speed of phase vibrations) in that medium is gradually increasing."

Since aether's density is approaching zero, with time, the speed of phase vibrations in that medium will correspondingly approach infinity. In other words,

"The speed of light and other phase vibrations in the aether medium will gradually approach infinity."

2- Time is gradually experienced at a faster pace

As it is explained in the chapter titled "What is Time?", the rate at which the passage of time is experienced by any object or any living being is directly dependent on the speed of that object or living being relative to its local aether, as compared to the speed of phase vibrations in that medium. As the speed of phase vibrations in aether is gradually increasing due to reduction in aether's density, the pace at which time is experienced by any and all objects, universally, is speeding up. In other words,

"Time is gradually speeding up, in this universe."

- **In the distant past**

   To calculate how slow time was actually progressing in the beginning, one has to know how fast phase vibrations were propagating through the aether medium of the early universe. The propagation speed of phase vibrations in turn depended on what the density of aether was back then.

The absolute value of aether's density during the first moments of this universe's life cannot be calculated. However, its value can be estimated relative to what it is at the present time. Currently, the universe is estimated to be about 93 billion light years across. In other words, at the present time the diameter of the universe in meters is,

$$(9.3 \times 10^{10}) \times (365) \times (24) \times (3600) \times (300{,}000) \times (1{,}000) = 8.80 \times 10^{26} \text{ m}$$

If, in the beginning, the diameter of the universe was ONE meter, the volume of the universe has grown by the cube of the above number which is (6.81×10^{80}). As a result, the density of aether at the beginning has been higher than what it is at the present time, at least by the same ratio. Based on that ratio alone, one can readily estimate that the rate at which 'time' was experienced back then must have been literally billions of billions of times slower as compared to the rate it is experienced, at the present time. Therefore, one can confidently make the following statement,

"The duration of the very first 'Second', when this universe started its existence, was possibly as long as millions of years based on today's time scale."

- **In the distant future**

As the density of aether is continuously decreasing and it is headed towards zero, the speed of phase vibrations in that medium is correspondingly approaching infinity. This automatically implies that the speed of objects will become smaller and smaller fractions of the speed of phase vibrations in the aether medium. Therefore, the passage of time will be experienced at progressively faster paces. As the speed of phase vibrations in the aether medium approaches infinity, the rate at which time is experienced by all objects in this universe will also approach infinity. In other words,

"Gradually, as 'Time' speeds up, the duration of minutes, hours, days, years, even millions of years will become equivalent to the duration of one 'Second', as it is experienced today."

3- Esmailzadeh critical aether pressure difference

At a certain point in the expansion process of this universe, the pressure difference between aether that is in this universe and aether that is in the accompanying universe will approach a critical value. That value of the pressure difference corresponds to the minimum required pressure difference that would encourage the flow of aether through the largest possible black holes to barely reach the speed of light (speed of phase vibrations) in the aether medium. This critical aether pressure difference can be referred to as:

"Esmailzadeh Critical Aether Pressure Difference"

As this critical value of the aether pressure difference between this universe and its accompanying universe is approached and finally reached, all of the black holes, starting with the smallest ones, will lose their status as being black holes. This is due to the fact that, the speed of the aether reaching their surfaces will be just under the speed of light. This in turn implies that even though black holes will keep on devouring more planets and stars, light and other electromagnetic waves will be able to travel against the incoming flow of aether and escape.

Note that, three factors directly affect how fast/soon this critical aether pressure difference will be experienced:

1- The rate at which the overall pressure of aether is decreasing in this universe, due to aether's expansion and leakage,

2- The rate at which the overall pressure of aether is increasing in the accompanying universe (while that universe is expanding), due to receiving aether from this universe, and

3- The rate at which the speed of phase vibrations in aether (the speed of light and other electromagnetic waves) is increasing in this universe, due to decreasing aether density.

Therefore, as time goes on, the aether pressure difference between this universe and the accompanying universe is decreasing while the speed of light and other phase vibrations in aether that is in this universe is increasing. At some point in time, it would not be possible for the existing aether pressure difference between the two universes to cause the flow of aether through the biggest openings (the largest black holes) to reach the speed of light. Encountering such a circumstance will serve as the clear indication that *'Esmailzadeh Critical Aether Pressure Difference'* has been reached.

4- All of the black holes will gradually lose their status as black holes

As the Esmailzadeh critical aether pressure difference is approached, all of the black holes, starting with the smallest ones, will lose their status as being black holes. At that point, the local aether pressure difference will no longer be sufficient to cause the flow speed of aether towards the surface of even the largest black holes to reach the speed of light in that medium.

As shown in the following figure, eventually none of the black holes will possess an event horizon, because light and other electromagnetic waves will be able to escape their grips.

In other words,

"Eventually, all of the black holes will become visible."

By 'visible hole' (in the figure) it is meant that electromagnetic type of radiations of various frequencies will be able to escape into the surrounding space. The colors chosen demonstrate the blue-shift and the red-shift generated due to the rotation of the 'then-visible' black hole.

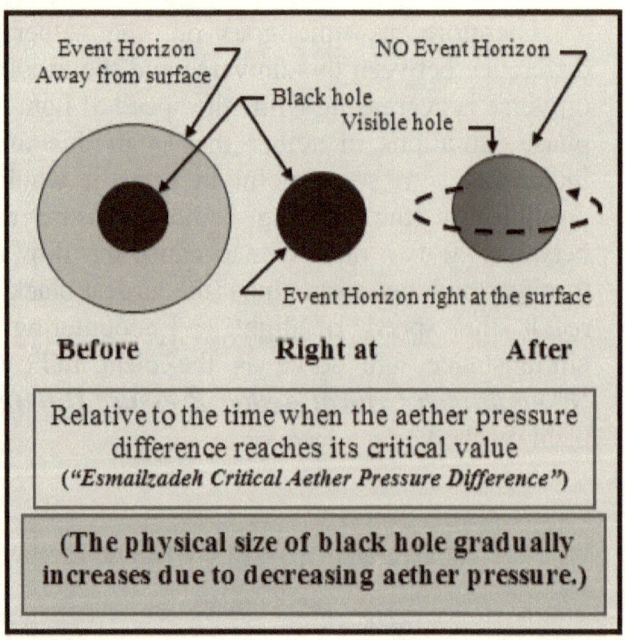

Event Horizon
Away from surface

NO Event Horizon

Black hole
Visible hole

Event Horizon right at the surface

Before　　　　**Right at**　　　　**After**

Relative to the time when the aether pressure difference reaches its critical value
(*"Esmailzadeh Critical Aether Pressure Difference"*)

(The physical size of black hole gradually increases due to decreasing aether pressure.)

Note that, as it is indicated, in the figure, the physical size of black holes will gradually increase due to decreasing aether pressure in this universe. That is, even if they do not devour any of their peaceful neighbors, they will still grow in size.

The very same effect will also be experienced by any and all of the particles in this universe, since they are bubbles in the medium of aether. Hence, as the internal pressure of aether is gradually decreasing the physical size of particles and that of black holes is gradually increasing.

5- All of the galaxies, star systems, stars and even planets will lose their structural integrities

As the force of gravity becomes weaker and weaker, gradually, the following events will take place on the universal scale:

1- Stars will gradually let their respective planets widen their orbital paths. Consequently, all of the planets will experience a definite **GLOBAL COOLING**. This global cooling, which will take place simultaneously on all planets existing in this universe, can be referred to as,

"Scharback Universal Planetary Cooling"

As planets follow ever wider orbits, all of the star systems will gradually grow in size. Galaxies will also gradually grow in size, due to their stars following wider orbits.

2- Long before the force of gravity reaches its total ineffectiveness, stars and planets that used to form star systems, will no longer have any desire (the needed gravitational attraction) to stay together. The following figure shows a couple of planets which have just started to experience their total freedom from their respective star.

At that stage, not only star systems but also all of the galaxies will lose their structural integrities. As a result, all of the celestial bodies will act as independent atoms do in a gaseous medium. In so doing, they will literally float everywhere without having any preference or desire to go towards any particular direction in this universe. They will simply follow their momentums.

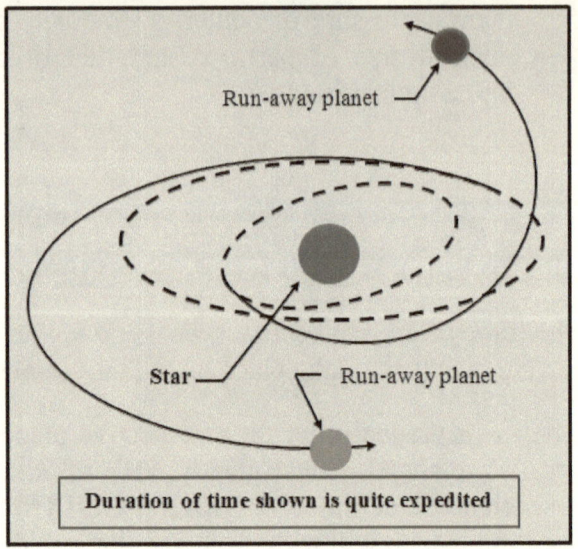

Run-away planet

Star

Run-away planet

Duration of time shown is quite expedited

3- As the force of gravity becomes even weaker, all of the stars will expand in their physical sizes, as their outer layers will no longer weigh as much. Consequently, the outer layers will not exert sufficient pressure on the inner layers to promote the ongoing nuclear fusion reactions. At that point, stars will literally turn off and start to cool down. They would basically become cold gas giants

As all of the stars, one by one, turn off, the lights in the night sky will gradually become less in numbers and eventually all of them will disappear, forever.

Once all of the stars are turned off, the whole universe will start its eternal dark era. In other words,

"The universe will become pitch dark"

4- During later stages, even the individual stars and planets will lose their structural integrities, since their gravitational forces will no longer be strong enough to hold them together.

6- Eventually, the force of gravity will be neutralized, altogether

Eventually, the pressure of aether that is in this universe will become equal to the pressure of aether that is in the accompanying universe. Once this pressure is reached, the flow of aether towards matter particles will slow down to a complete stop. In other words, as the pressure difference between this universe and the accompanying universe approaches zero,

"The force of gravity will gradually fade away."

Therefore,

"Just as the very formation of matter and anti-matter particles had given birth to the force of gravity, their continued existence will eventually cause it to fade away."

At that point in time, if the universe has any momentum left in its expansion, it will keep on expanding forever, since it would literally lack its braking instincts. This era in the existence of the universe can be referred to as,

"Esmailzadeh Zero Gravity Era"

7- Universe will never stop expanding

The volume of aether will always keep on expanding, even after its pressures in this universe and in the accompanying universe become equal. This expansion process will continue forever, since the universal values of aether's density and pressure can only approach zero, but will never reach zero, as aether's volume which is the size of the universe approaches infinity.

Conclusion

As it is briefly presented here, and in greater details throughout this book, this universe is simply made of fluid aether which under different conditions manifests in a variety of states / forms such as matter, anti-matter, dark matter. The varieties of motions that exist in the aether medium give meaning to different forces such as gravity, magnetic field and electric field, among many others. The fluid aether is also carrying a variety of phase vibrations such as the electromagnetic waves.

The following are some of the phenomena that have been reasonably and consistently explained in this book, using aether-based theories which allow aether to play its role in the overall operation of this vast universe:

- The location of the **Birthplace of this universe,**

- The '**Initial Rapid Expansion**' of the aether medium,

- The slowing down of the initial rapid expansion of the aether medium,

- What is '**Aether**'? and how does it relate to various phenomena in this universe?,

- What is '**Space**'?, and why is it expanding?, and will its expansion ever come to a complete halt?,

- What is '**Time**'? and why is it gradually experienced at an increasingly faster pace?,

- What is '**Light**'? and why is its speed gradually increasing?,

- Formation of '**Matter**' and '**Anti-Matter**' particles,

- Why particles have specific discrete physical sizes?,

- Why some particles are stable while some others are not?,

- Why some isotopes are stable while some others are not?,

- What is '**Dark Matter**'?,

- What is '**Dark Energy**'?,

- What is '**Vacuum Energy**'?,

- What is '**Electric Charge'?**,

- Formation of '**Electric Field**',

- Formation of '**Magnetic field**',

- What is '**Electricity**'?, and how is it formed?,

- Formation of the force of '**Gravity**', and why is its strength gradually decreasing?,

- Formation of the '**Cosmic Microwave Background Radiation**' (phase vibrations),

- What are '**Black Holes**'? and why are their properties related to their surface areas rather than their volumes?,

- Why relativistic effects are experienced as objects approach the speed of light in a vacuum?,

- What happens to the motion of atoms, nuclei and electrons at absolute zero degrees?

- Why are '**Lightning Sprites**' forming?,

- Explanation for **Casimir Effect**,

- The 'Equivalence Principle',

- Explanation for Einstein-Podolsky-Rosen Paradox of entangled particles,

- Does 'Planck's Constant' have a physical interpretation?, and why is it gradually decreasing in magnitude?,

- Explanation for different types of 'Double-Slit Experiments',

- Explanation for 'Photoelectric Effect' based on light being a wave and not a particle,

- What are 'Lasers'?, and how can their efficiencies be increased?,

- Mr. Galileo's Experiment,

- Mr. Michelson and Mr. Morley's Experiment,

- 'Precession of the Perihelion of Mercury's Orbit,

- How are 'Magnetic Hills' formed, and how can artificial magnetic hills be built?,

- Explanation for 'Near-Earth Probe Flyby Anomalies',

- Why has the 'Science of Astrology' lost its accuracy?

- Why are major 'Natural Disasters' becoming more frequent, on a global scale?,

- Why is the 'Physical Size of Particles' gradually increasing?,

- Why is the 'Mass of Particles' gradually decreasing?,

- Why are the planetary orbits gradually becoming wider?,

- Explanation for 'Pioneer (10 and 11) anomaly',

- Why is the expansion process of this universe accelerating?,

- **What events can be expected to occur in the future?**,

And, many, many more. The bottom line is that,

Aether Exists and it is Everything.

Therefore, it is very crucial that, the scientific community accepts the existence of aether and takes its various effects into account.

What
is
Space?

Introduction

This universe owes its existence to two fundamental phenomena, namely 'Time' and 'Space'. Even though, in every aspect of his life, work and essentially his existence man has always been dependent on these two phenomena, to this day he has not been able to identify their true nature or essence. He has only come up with descriptions of them which are dependent on each other. In other words, man has only managed to understand 'Time' and 'Space' just enough to indicate that he is aware of their existences.

At the present time, space is measured and analyzed using units that are dependent on each other. For example, one kilometer is one thousand meters, or one light year, by definition, is equivalent to the distance that light travels in a vacuum in one year, and so on. According to the international standards, since 1960, one meter was accepted to be equivalent to 1,650,763.73 wavelengths of Krypton 86. However, in 1983 one meter was redefined based on the distance travelled by light in $(1/299,792,458^{th})$ of one Second.

In other words, even though man is capable of measuring space intervals very precisely, yet none of the internationally offered and accepted definitions for units of length have provided or will provide any understanding or insight into what 'Space' really is. Currently, the only direct definition that man can propose for 'Space' is that,

"Space is the gap between objects."

Therefore, the question still stands: **What is Space?**

In order to comprehend what space really is, one needs to understand how aether and space relate to each other, since space owes its existence to aether.

The relationship between space and aether is just like that of an ocean and water. The existence of water gives meaning to an ocean. If the amount of water in an ocean is somehow increased, it will directly lead to an expansion of that ocean's volume. However, if all of the water that is inside an ocean is somehow taken out, what remains can no longer be called an ocean.

The very same relationship exists between aether and space. For space to exist, it must be occupied by aether. Space is basically the container in which everything that exists in this universe is floating. The very existence of aether has given meaning to space, and the expansion of the overall medium of aether has led to the current size of space in this universe.

The internal structure of space in this universe can be analyzed on two extreme scales.

- **On the macroscopic scale**
  Space in this universe was formed as aether which was initially in quite a compressed state started to expand. Later on, as matter and anti-matter particles (drain holes in space) were formed, aether was allowed to leak out of this universe. During this process, aether started to accumulate in an adjacent environment and hence gave meaning to 'space' in the accompanying universe. In other words,

"The very formation of matter and anti-matter particles in this universe initiated the existence of the accompanying universe. "

Currently, space is divided into two distinct regions. One region encompasses this universe, and the other region encompasses the accompanying universe. In other words,

"Space (aether medium) is a continuous medium that encompasses this universe and its accompanying universe."

It should be mentioned that, the existence of the accompanying universe enables this universe to function properly. In fact, this universe and its accompanying universe are complementary parts of each other.

Once matter and anti-matter particles were formed across the inner region of the aether medium (the inner region of space in this universe), due to their abundance in this universe, the rate at which fluid aether was flowing into the accompanying universe was quite high. However, as most of the matter and anti-matter particles annihilated each other shortly after their formations, the flow rate of aether into the accompanying universe was drastically reduced.

The internal pressure and the density of the fluid aether in the accompanying universe literally started from zero, and their magnitudes have been rising continuously ever since, due to the ongoing leakage of aether from this universe, through matter and anti-matter particles.

Note that, the fluid aether that is in the accompanying universe is expanding at an ever increasing rate, due to its internal pressure continuously rising. However, the overall size of the portion of space that is the accompanying universe is closely related only to the size of the region of space in this universe that is hosting matter and anti-matter particles.

The following figure shows the two universes, before and after matter and anti-matter particles were formed in this universe.

Aether that is in this universe has been and still is under a much, much higher pressure as compared to aether that is in the accompanying universe. Therefore, space in this universe has been and still is expanding at a much faster pace as compared to space in the accompanying universe, because the internal pressure of aether in each universe has been and still is the driving force for its expansion rate. Hence, the portion of space that hosts this universe is much, much larger than the portion of space that is hosting the accompanying universe.

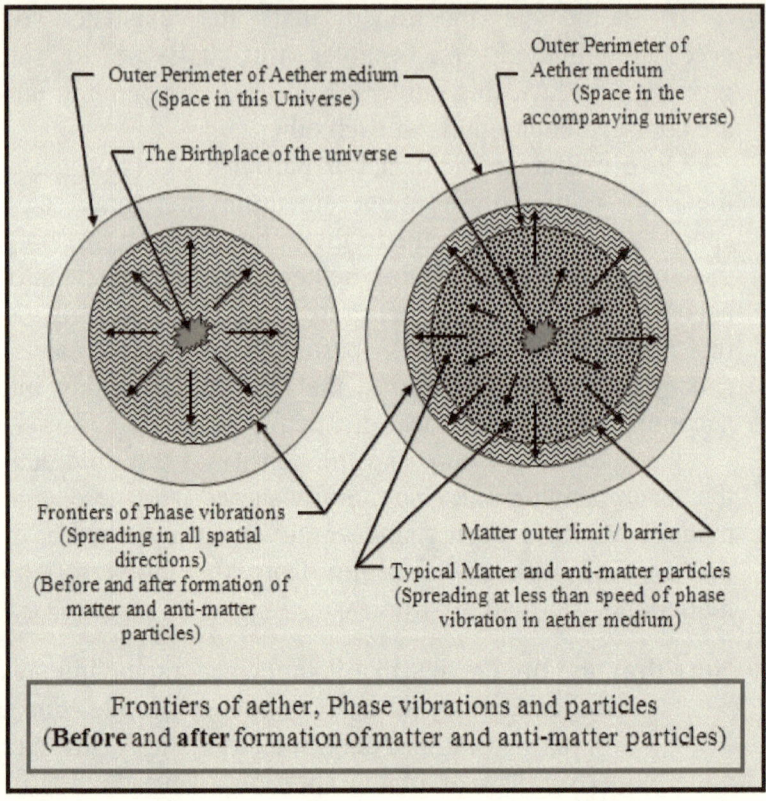

Outer Perimeter of Aether medium
(Space in this Universe)

Outer Perimeter of
Aether medium
(Space in the
accompanying universe)

The Birthplace of the universe

Frontiers of Phase vibrations
(Spreading in all spatial
directions)
(Before and after formation of
matter and anti-matter
particles)

Matter outer limit / barrier

Typical Matter and anti-matter particles
(Spreading at less than speed of phase
vibration in aether medium)

Frontiers of aether, Phase vibrations and particles
(**Before** and **after** formation of matter and anti-matter particles)

- **On the microscopic scale**

    The internal structure of space in this universe, in the central region of its volume, as shown in the following figure, is quite like a sponge which has countless number of tiny holes. These holes are the matter and anti-matter particles which are formed due to the formation of spikes in the phase vibrations in the fluid aether medium. Matter and anti-matter particles act as drain holes through which fluid aether is escaping from this universe and entering into the accompanying universe.

Note that, as shown in the following figure, aether medium that is near the outer limits of this universe does not host any kind phase vibrations, let alone any kind of matter and/or anti-matter particles or their aggregates.

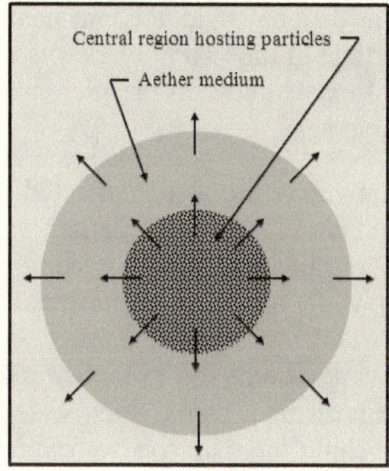

A sample of such a connection point between the two universes, namely a matter particle, is shown below.   As shown in the figure, when viewed from this universe, matter (as well as anti-matter) particles are like drain holes at the bottom of a swimming pool through which aether is leaking out.   While, when viewed from the accompanying universe, matter (as well as anti-matter) particles are like water spring jets in a swimming pool through which aether is flowing into and is joining the aether that is already there.

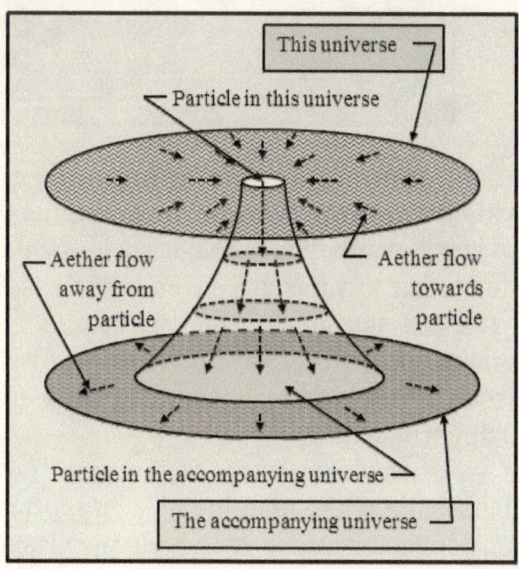

Note that, the difference in the physical size of any given particle in the two universes is due to the difference that exists between the pressures of aether in those universes.

The two universes are also connected through black holes which are in fact much, much larger aggregates of matter particles that by literally merging (unifying) together have formed much wider openings (gateways) for aether to flow through.

The following picture shows the entrance of a water-shoot inside the reservoir of a dam. Such water-shoots prevent the water that is behind the dam to rise beyond certain safe level. They perform their important task by simply allowing the excess water to fall in them and get transferred downstream, bypassing the dam structure altogether.

The flow of aether towards and through matter and anti-matter particles, as well as black holes, is quite the same as the flow of water towards and through such openings. However, the flow of aether towards any given particle is from the whole volume of space surrounding that particle, a volume of space which is three dimensional, while the flow of water towards the inlet of such a water-shoot is from a planar surface which is only two dimensional.

Such water intakes clearly demonstrate how the flow of water changes direction, as it literally disappears from the two dimensional planar surface in which it existed before. The very same type of a process takes place with the fluid aether.

Since, as it reaches a given matter or anti-matter particle (or a black hole) it literally disappears from the familiar three dimensional volume of space that forms this universe and enters the physical dimensions corresponding to the accompanying universe.

Note that, in the two-dimensional surface of water, the inlet part of such water-shoots is observed as a void.

The very same is the case with the matter and anti-matter particles, as well as black holes, in the aether medium, since they too are bubbles, voids, in the three-dimensional space which is this universe.

Space is Finite and it is Spherical in its Overall Geometry

The overall geometry of space is spherical, since aether's expansion has been and still is symmetrical in all spatial directions. Aether's outer limits or its overall frontline is expanding just like a balloon, and is not encountering any kind of variations in resistance against its advancement towards different directions. This automatically implies that,

"The overall size of space is finite."

One may ask:

What is aether expanding into? Isn't that the true space?

Such questions are philosophical questions. They are just like the questions that one can ask regarding the source of aether and the source of that source, and so on.

Any answers to such questions can only be based on pure speculation. This is due to the simple fact that, in order to even know what the outside of this universe (or even a house, for that matter) looks like, let alone where this universe is located, or into what it is expanding, one needs to see it from the outside. Because, even if the whole of the interior of a house (for example) can be surveyed and precisely mapped out, still no guess can be made about what the outside of the house truly looks like, due to the lack of knowledge about the thicknesses of different exterior walls.

Questions regarding the source of aether as well as the overall picture of the void into which aether (along with this universe and the accompanying universe) is expanding must be put on hold, at least for the time being. Since, the human race has just begun to have a glimpse at the structure of this universe and how its contents interact with each other. Therefore, its members have a long journey ahead of them before they can expect to have reasonable answers for such questions.

Space Will Continue to Expand, Forever

In the beginning, space was very small in its volume, since the fluid aether was in a super compressed state. Due to phase vibrations that were somehow introduced in it, aether medium experienced a rapid expansion. As aether expanded, so did the vastness of the space which was occupied (hence, formed) by aether, and over time, space has become the size it is observed to be, today.

Space (and not just the physical universe) is still expanding and its expansion process will never come to a complete stop, since the internal pressure of the aether medium can only approach zero but it will never actually reach zero. In other words,

"The radius of space that is hosting this universe is expanding and will keep on expanding at an ever slower rate, forever."

All Spatial Dimensions are Straight

Progressively stronger gravitational forces that are experienced near massive stars and especially near black holes (shown in the following figure) have been interpreted by the General Theory of Relativity as if they represent curvatures in the spatial dimensions (or space-time dimensions, as it is referred to), while they are not. The curvatures that are calculated in fact indicate the gradients that exist in the flow speeds of the local aether which is headed towards the star, the galaxy or the black hole.

Stars, galaxies and black holes, each on their own scales, are relatively small regions of space that are densely populated with matter particles. Denser matter concentrations simply imply that there are more available drain holes, in a given volume of space, for aether to flow through. Hence, the fluid aether is encouraged to flow towards such a region of space at a faster pace.

The strength of the gravitational force of any celestial body, particularly that of a black hole, as shown in the following figure, can be represented by a funnel type of a shape because the strength of the gravitational force is directly proportional to the rate at which the flow of aether is accelerating at any given location (distance from such celestial bodies).

In other words, the progressively steeper slopes of the smooth surface of the funnel shape shown in the figure actually represent the gradual increases in the speed of aether that is approaching the black hole. Since, the closer aether gets to a black hole, the faster it accelerates.

The very same is also true about the gravitational forces of other celestial bodies such as stars, galaxies and even planets.

However, in their cases, the slope of the central portion does not become as pronounced.

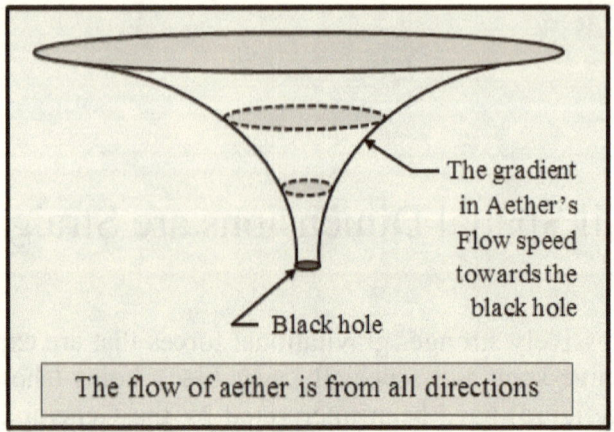

The gradient in Aether's Flow speed towards the black hole

Black hole

The flow of aether is from all directions

The experiment that was performed in 1919, during a total solar eclipse, had apparently confirmed the prediction made by the General Theory of Relativity. However, such experiments did not and do not indicate the existence of any curvature in the fabric of space (or space-time), but rather simply indicate how light waves were and are dragged by aether that is flowing in an accelerated fashion towards gravitating bodies such as stars. The following figure shows such an effect as it is induced by the accelerated aether flow, in a space that consists of straight dimensions.

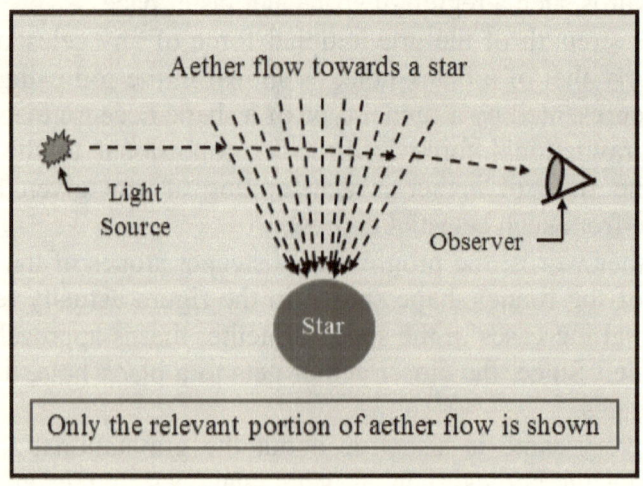

Aether flow towards a star

Light Source

Observer

Star

Only the relevant portion of aether flow is shown

Note that, the prediction made by the General Theory of Relativity agrees with the observed amount that light waves bend as they pass near a star, simply because the very same equations derived in the General Theory of Relativity apply to the newly proposed theory of gravity presented in this book, a theory that introduces the force of gravity, at any given location in this universe, as the drag force induced by the flow of aether which is accelerating through that particular location.

What
is
Time?

Introduction

The universe owes its existence to two fundamental phenomena, namely 'Time' and 'Space'. Even though, in every aspect of his life, work and essentially his existence, man has always been dependent on these two phenomena, to this day he has not been able to identify their true nature or essence. He has only come up with descriptions of them which are dependent on each other. In other words, man has only managed to understand 'Space' and 'Time' just enough to indicate that he is aware of their existences.

Currently, in physics 'Time' is defined as the separation of events. For example, having breakfast and having lunch can take place 4 hours apart, or two consecutive sun rises are said to take place 24 hours apart. Definitions provided for units of time, such as seconds, hours, years and so on, are all interdependent, and none of them offers any insight into what 'Time' really is. According to the internationally accepted standards, one second is defined to be equal to the duration of 9,192,631,770 vibrations of the Cesium 133.

In other words, Even though man is capable of measuring time intervals quite accurately, yet none of the internationally offered and accepted definitions for units of time have provided or will provide any understanding or insight into what 'Time' really is. Currently, the only direct definition that man can propose for 'Time' is that,

"Time is the gap between events."

Therefore, the question still stands: **What is Time?**

The universe has existed for a long time in the past. It does exist at the present time. And, it will exist for some time into the future. Therefore, to correctly understand the structure of this universe, one must have a proper and concise understanding of 'time' itself.

To truly comprehend the essence of 'time' one needs to understand how aether and time relate to each other.

So far, mankind has identified and even proven some of the characteristics of time. For example:

- Even though objects and living beings may experience the passage of time at different rates, but all of them always experience it in the positive direction, only towards the future.

- According to Mr. Lorentz's Relativity (transformation) equations, the passage of time experienced by any object directly depends on the speed of that object in the physical space. As the speed of the object approaches that of light waves in a vacuum, the object experiences the passage of time at slower and slower paces.

- According to Mr. Einstein's General Theory of Relativity, the rate at which time is experienced anywhere in this universe also depends on the strength of the local gravitational field. Time is experienced at a slower pace where the force of gravity is stronger.

According to this book,

"'Time' is one of the by-products of the presence of phase vibrations in the aether medium."

The whole volume of the physical space is occupied by aether. The existence of aether in the physical space is exactly like the existence of air in the earth's atmosphere, with only one difference and that is, aether covers the whole of the physical space, and not just the space between the planets and stars.

It is proposed here that,

"The rate at which 'time' is experienced by an object depends on that object's instantaneous speed relative to its local aether medium, as compared to the propagation speed of phase vibrations in that medium."

As the speed of an object, relative to its local aether, approaches that of phase vibrations in that medium, the object experiences the passage of time at progressively slower paces. In case the speed of the object reaches the speed of phase vibrations in its local aether medium, the passage of time will no longer be experienced by that particular object.

Even if the object stands relatively still in one place, anywhere in this vast universe, and somehow the local aether be encouraged to pass by it, the resulting effect will be the very same and identical. That is, as the speed of aether is increased, relative to the object, the passage of time experienced by that object will be at slower and slower paces. Again, if the speed of aether relative to the object reaches the speed of phase vibrations in the local aether medium, the object will stop experiencing the passage of time, altogether.

Note that, <u>even though 'Time' can be measured using a variety of scales, it is a continuous phenomenon. It is not made of small discrete pieces, as objects are made of molecules / atoms / particles.</u>

It should also be mentioned that, light and all other electromagnetic waves are transmitted as phase vibrations through the medium of aether, exactly as sound waves are transmitted as phase vibrations through various mediums such as air. The propagation speed of phase vibrations in a given medium is independent of the speeds of the source and the receiver of the waves. However, it is dependent on the density of aether and the internal pressure of that medium. For example, phase vibrations such as light propagate at slower speeds through denser aether.

Various Phenomena Affecting
the Flow of 'Time'
(According to this book)

It is important to identify various natural phenomena and/or artificial methods, which may result in the desired effect, namely encouraging the motion of a given object relative to its local aether medium to approach the speed of phase vibrations in that medium.

- **Motion of the object**
  If the local aether is relatively stationary, any increase in the speed of an object such as a spaceship directly leads to an increase in its speed relative to its local aether. This is analogous to a boat moving on the surface of a lake, where water is relatively stationary. The figure below shows the motion of a spaceship relative to the local aether medium which is relatively stationary.

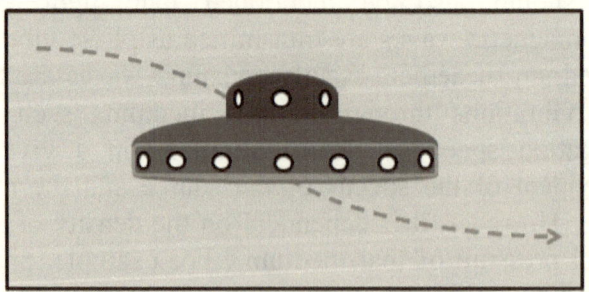

However, if the local aether is in motion, only the speed of the object relative to aether that is in its immediate vicinity is important and must be used. Such a scenario can be likened to a boat's motion on the surface of a running river.

In both cases, the passage of time experienced by the object will be affected by its proper speed relative to aether in its immediate vicinity. Therefore, the speed used in the transformation equations provided by Mr. Lorentz must be that of the object relative to its local aether medium.

Note that, as it is explained later on in this chapter, along with experiments to verify the information presented, **acceleration of an object does not necessarily induce 'Time Dilation'.**

- **Gravity**

The force of gravity is the drag force induced by the accelerated flow of aether towards matter (as well as anti-matter) particles, see figure below. (Please, refer to the chapter titled "What is Gravity?")

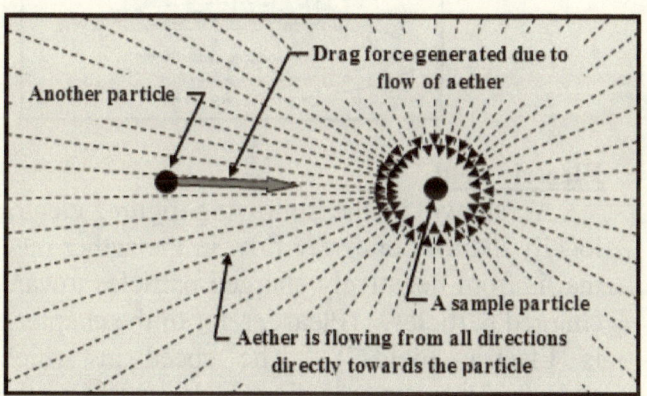

The flow speed of aether towards a certain region of space is directly dependent on the amount and number density of matter particles that are present in that particular region. Higher concentration of matter particles in a given region of space induces faster flow of aether towards that locality. Hence, the local aether will be flowing faster relative to any object that may be present, nearby.

Note that, the fastest aether flow takes place close to black holes, since aether's speed literally reaches the speed of light (the speed of phase vibrations in the local aether) at the event horizon of black holes.

- **Magnetic field**

    Magnetic field is a type of round trip motion that is induced in the local aether, just like the flow of air that is generated by a jet engine or an open-ended wind tunnel, or even a regular household fan. (Please, refer to the chapter titled "What is Magnetic Field?")

    The speed at which aether is encouraged to move is directly dependent on the strength of the magnetic field present. The following figure shows the aether flow induced by a magnet, manifesting as its magnetic field.

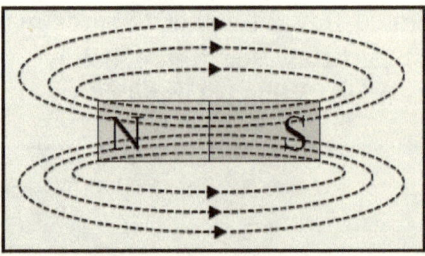

- **Electric field**

    As shown in the following figure, electric field is a locally induced one-way flow in the aether medium, a flow that is from negatively charged particles towards positively charged particles. (Please, refer to the chapter titled "What is Electric Field?") The speed at which aether is encouraged to move is directly dependent on the strength of the electric field present.

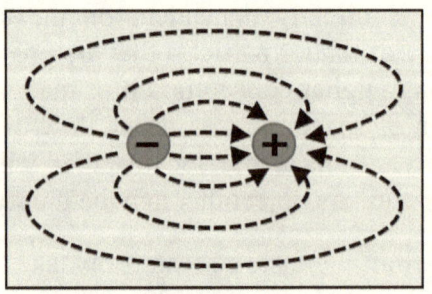

Therefore, stronger gravitational fields, stronger magnetic fields and stronger electric fields, each encourage aether to move faster relative to objects that are nearby. Higher relative speeds of aether in turn cause the more pronounced effects on how time is experienced by those objects. The very same effect is experienced by such objects when they move at high rates of speed relative to their local aether medium.

Important note:

To obtain accurate results from Mr. Lorentz's transformation equations, one must use the speed of the object relative to its local aether and not relative to space, and definitely not relative to any other object.

The effects associated with the movement of the object, the local gravitational field, the local magnetic field and the local electric field are accumulative. That is, if the object is moving in the very same direction as the aether is encouraged to flow due to the presence of a strong gravitational and/or magnetic and/or electric fields, they will counteract each other's effects. Since, the object and the aether will be moving in the same direction. Therefore, their relative motion is actually reduced.

However, if the object is moving in the opposite direction of the induced flow in aether, the effects will be complementary to each other and will enhance the overall effects.

It must be emphasized that,

"Aether has never been and will never be at a complete rest at anywhere in this universe."

Aether is always in motion. The continuous motion of aether in space may be likened to the continuous motion of the air molecules in the atmosphere or the continuous motion of the water molecules in the oceans.

The rate at which 'time' is experienced by any object (or any living being) is directly dependent on the motion of that object relative to its local aether medium. As either the object speeds up or the local aether is made to pass by faster, the object experiences time at progressively slower paces. If an object's speed relative to its local aether reaches the speed of phase vibrations in that medium, <u>that object literally catches up with the phase vibrations that it generates in that medium.</u> At that point, the object will not experience the passage of time, because it can no longer distinguish between current events and those that have taken place in the past. Even though it sounds strange, one can say,

"Time progresses at the speed of phase vibrations in the local medium of aether."

Therefore,

"Any object that moves at the same speed relative to its local aether medium as the speed of phase vibrations in that medium, will automatically be in sync with its own past and present, concurrently. Therefore, it cannot experience the passage of time."

Explaining the Passage of 'Time' Using Sound Waves

Consider a supersonic airplane flying at the speed of sound in air. While the airplane is travelling at exactly the speed of sound, using a very strong speaker, the pilot can broadcast his/her own voice counting 1, 2, 3, …., 100,….

Since the airplane is moving at the speed of sound in air, the portion of the sound waves, corresponding to each and every number stated, that are propagating towards the front end of the airplane will be literally superimposed on each other, as if they were/are all stated at the very same time. The following figure demonstrates the sound waves generated during such a scenario.

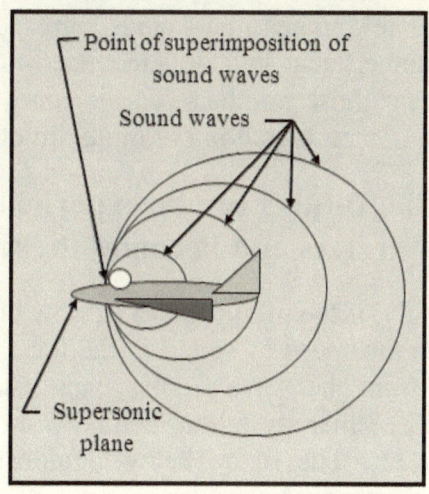

In other words, all of the numbers that were announced through the speaker up to and including the one announced at the present will be heard simultaneously. Basically, the 'past' starting from the instant the airplane had reached and had stabilized its speed at exactly the speed of sound in air, and the 'present' will be experienced at the very same time.

If the pilot decides to fly his airplane faster than the speed of sound in the local air, the sound waves generated due to numbers stated in the past will fall behind and the only number heard will be the one that is currently stated. The following figure demonstrates the sound waves generated during such a scenario.

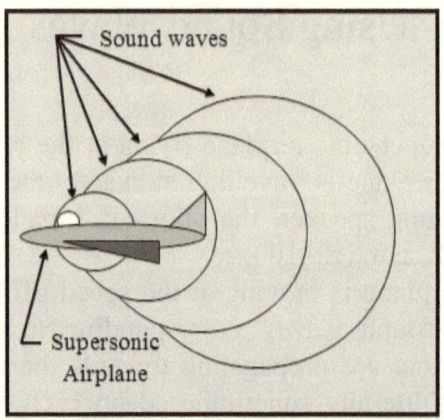

Note that, even though the airplane is moving faster than the speed of sound in the local air, it does not imply that the pilot will be able to hear the sounds corresponding to any of the numbers that will be announced in the future. He will hear those numbers, one by one, as they will be stated, in their own proper timing. In other words,

"The 'future' can be experienced, only as it arrives and becomes the 'present'."

Correspondingly, if the pilot decides to slow his airplane speed down to less than the speed of sound in the local air, the airplane will fall behind from the wave fronts generated by the sounds. Therefore, he will be literally falling behind from the real present that he could be at. The figure below demonstrates the sound waves generated during such a scenario.

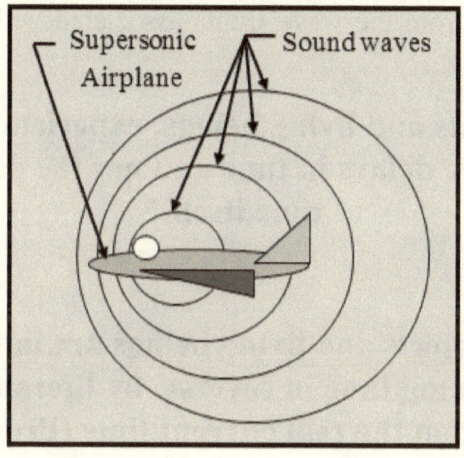

In cases where the object is relatively stationary and the local aether is made to pass by it, either due to gravity, magnetic or electric phenomena, it would be as if the airplane is mounted inside a wind tunnel and the medium of air is made to pass by it, at various speeds. Such a setup is shown below. The effects experienced due to the speed of the wind that is generated by the wind tunnel increasing up to and beyond the speed of sound will be exactly the same as they were stated in the above scenarios.

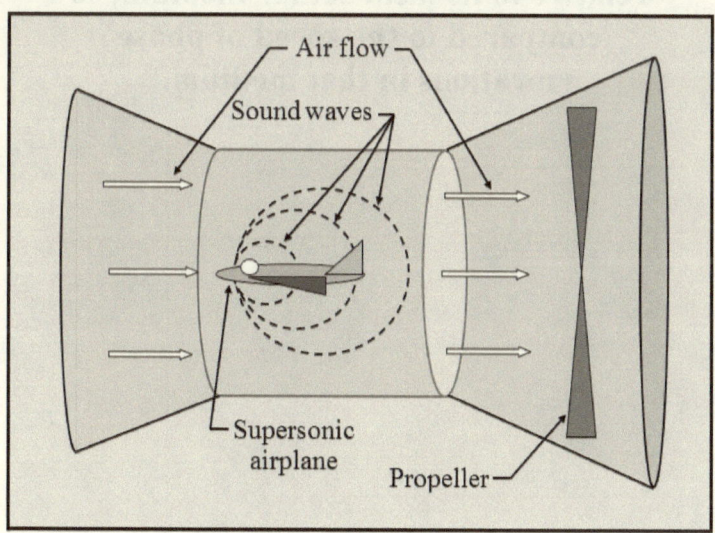

In short, any object that is moving at a speed that is slower than the speed of phase vibrations in its local aether medium, is literally falling behind from the wave fronts associated with its own past. Therefore,

"Objects and living beings, experience their respective delays in time and not the passage of time itself."

In other words,

"Objects and living beings are in fact experiencing time in reverse, by literally falling behind from the real current time (Present) that they can and should be at."

It must be emphasized that,

"The rate at which 'time' is experienced has nothing to do with 'space'. 'Time' experienced by any object is dictated by that object's instantaneous speed relative to its local aether medium, as compared to the speed of phase vibrations in that medium."

The Beginning of 'Time'

Time is experienced only by objects that are moving at speeds that are slower than the propagation speed of phase vibrations in the local aether medium. In other words, as shown in the following figure,

"'Time' started as the first ripples, phase vibrations, were formed in aether and started to propagate in that medium."

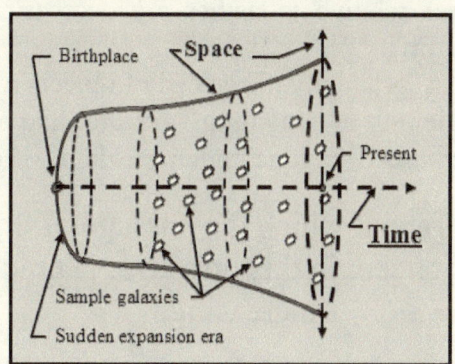

However,

"The meaningful startup point for experiencing 'Time' was when matter and anti-matter particles were formed. Since, by falling behind relative to the ripples that had formed them, they were the very first entities that experienced the passage of time."

Speed of "Time"
Along the History/Future of the Universe

Just as the density of air affects the propagation of sound waves in that medium, the density of aether also affects the propagation of light and other electromagnetic waves (phase vibrations) in that medium. <u>Phase vibrations propagate slower in denser aether medium</u>. As aether is expanding, its density is gradually decreasing. This reduction in aether's density leads to higher phase vibration speeds in that medium. Also, the speeds of objects (galaxies) relative to aether are decreasing due to their mutual force of gravity.

In other words, gradually, the speed of objects relative to aether that is in their vicinity is becoming a smaller and smaller fraction of the speed of phase vibrations in that medium. Therefore,

> "The rate at which the passage of time is experienced, in this universe, is gradually increasing."

In other words,

"Time is gradually speeding up."

- **'Time' in the distant past**
  To calculate how slow time was actually progressing in the beginning, one has to know how fast phase vibrations were propagating through the aether medium of the early

universe. The propagation speed of phase vibrations in turn depended on what the density of aether was back then.

The absolute value of aether's density during the first moments of this universe's life cannot be calculated. However, its value can be estimated relative to what it is at the present time. Currently, the universe is estimated to be about 93 billion light years across. In other words, currently the diameter of the universe in meters is,

$$(9.3 \times 10^{10}) \times (365) \times (24) \times (3600) \times (300,000) \times (1,000) = 8.80 \times 10^{26} \text{ m}$$

If, in the beginning, the diameter of the universe was ONE meter, the volume of the universe has grown by the cube of the above number which is (6.81×10^{80}). As a result, the density of aether at the beginning has been higher than what it is at the present time, at least by the same ratio.

Based on that density ratio alone, one can readily estimate that the rate at which 'time' was experienced back then must have been literally billions of billions of times slower as compared to the rate it is experienced, at the present time. Therefore, one can confidently make the following statement,

"The duration of the very first 'SECOND, when this universe started its existence, was possibly as long as millions of years based on today's time scale."

- **'Time' in the distant future**

  As the density of aether is continuously decreasing and it is headed towards zero, the speed of phase vibrations in that medium is correspondingly approaching infinity. This automatically implies that the speed of objects is becoming smaller and smaller fractions of the speed of phase vibrations in the aether medium. Therefore, the passage of time is progressively experienced at faster paces.

  As the speed of phase vibrations in the aether medium approaches infinity, the rate at which time is experienced by all objects in this universe will approach infinity. In other words,

"As 'Time' is gradually speeding up, the duration of minutes, hours, days, years, even millions of years will become equivalent to the duration of a 'SECOND' as it is experienced today."

- **Estimating the rate at which 'Time' is currently speeding up**

    The rate at which time is speeding up cannot be measured directly, since any kind of time measuring device will be automatically affected, as well. However, it can be estimated based on the variations in the speed at which phase vibrations such as light propagate through aether.

    The speed of phase vibrations in turn is dependent on the density of aether. Therefore, in order to measure the rate at which the speed of light, as well as that of all other phase vibrations in the medium of aether, is currently increasing, the following must be known:

    - **The relationship between the magnitude of the speed of light and the density of aether, and**

        As a first approximation, the relationship between the speed of light and the density of aether can be assumed to be the same as that between the speed of sound waves in air and the density of air. Later on, this relationship can be refined, upon the availability of more specific data and experimental capabilities.

    - **The rate at which the density of aether is currently decreasing, in this universe.**

        At the present time, the variations in the density of aether in this universe can be roughly estimated using Hubble's constant. Hubble's constant is a direct indication of the rate at which the physical contents of this universe (not the overall aether medium) are spreading by receding from each other. The spreading of the physical contents of this universe is due to the expansion of the aether medium, as a whole, as well as the motion of galaxies IN that medium.

Therefore, the speed at which matter contents of this universe are dragged at any given location is proportional to (but not equal to) the speed at which aether itself is moving due to its ongoing expansion process.

The relative motion of matter contents of this universe with respect to aether in the overall picture of this universe is literally like the relative motion of objects that are allowed to free fall towards a gravitating celestial body. Such objects will be dragged by the flow of aether, yet they will not be moving at the same rate of speed as the local aether itself. However, if those same objects were allowed to start their free fall at infinite distance from the gravitating body, during their entire free fall journey, they will be moving at just about the same rate of speed as that of their local aether.

As the matter contents of this universe were formed, they fell behind from the phase vibrations that had formed them. Even more so, they fell behind with respect to aether that was expanding. Yet, they all were moving IN that medium. The general direction of their motion / momentum was away from the birthplace of this universe, since the phase vibrations that caused the generation of the matter and anti-matter particles were propagating away from where this universe had started its existence.

Gradually, all of the matter contents of this universe are being dragged by their local aether at faster and faster paces, due to the weakening of the force of gravity which is directly opposing their receding motions from each other.

According to the most recent measurements, Hubble's constant is **about** 70 km/sec/Mega parsec. Since one Mega parsec is equal to the distance traversed by light in free space in 3.26 years, the current value of Hubble's constant can be rewritten as,

$$(HC = 7.157 \times 10^{-11} \text{ km/year//km}).$$

Therefore,

The current annual rate of <u>increase</u> in the volume of aether that is in this universe <u>is greater than</u>

$$((R_2)^3 - (R_1)^3) / (R_1)^3 = 2.1477 \times 10^{-10}$$

Where, R_1 and R_2 are the radii of a given amount of aether, as of last year and as of this year, respectively.

The annual variations in the density of aether that is in this universe are numerically the same but in reverse of the variations in its aether's volume. In other words,

The current annual rate of <u>decrease</u> in density and pressure of aether in this universe <u>is greater than</u>

$$(2.1477 \times 10^{-10})$$

Note that, this rate of variation in aether density (calculated using Hubble's constant) is **only** based on the receding motion of galaxies from each other. It is not based on the expansion of the aether medium, as a whole. It does not take into account the rate at which aether is gradually leaking through matter and anti-matter particles (including neutron stars or black holes), either.

Using the above estimated magnitude in the equations governing the speed of sound waves in a matter medium such as air will provide an initial estimate of the rate at which the speed of light and other electromagnetic waves is increasing in this universe.

'Time' in the Accompanying Universe

Currently, the fluid aether that is in the accompanying universe is under a much, much lower pressure as compared to aether that is in this universe. Hence, its density is correspondingly much lower than that of aether that is in this universe. Therefore, the speed of phase vibrations (such as light) is much higher in the accompanying universe, as compared to what it is in this universe.

Due to much faster speed of phase vibrations (such as light) in the accompanying universe, the speed of anything and everything that exists in that universe is a smaller fraction of the speed of phase vibrations in that medium. Since, they are directly connected to their counterparts in this universe. Therefore,

"At the present time, the rate at which 'time' is experienced in the accompanying universe is at a much faster pace as compared to the rate it is experienced in this universe."

In the future, the pressure and the density of aether that is in the accompanying universe will gradually become equal to those of aether that is in this universe. Consequently, the speed of phase vibrations will become the same in both universes. Therefore, the passage of 'time' will be experienced at the same rate in both universes, as well.

Is 'Time Travel' Possible?

'Time Travel' can be defined in a variety of ways. For example, it can be defined as **experiencing the passage of 'Time' at a different pace, relative to the other entities.** Such a 'Time Travel' scenario is in fact experienced by anything and everything in this universe. Since, by being exposed to different gravitational forces, different strengths of local magnetic and/or electric fields and/or moving in different directions at different rates of speeds, everything in this universe is moving at a different rate of speed relative to its respective local aether medium. Therefore, they are experiencing the passage of 'Time' at different paces relative to each other.

However, true 'Time Travel' can only be referring to **experiencing the same 'Time' period in a multiple fashion.** Such a 'Time Travel' scenario implies the physical coexistence of an object (or living being) with its own past self (or its own future self). In other words, the object would experience multiple existences in this universe, concurrently. Such 'Time' Travel scenario must be examined for two totally different cases:

- **Physical 'Time Travel'**

    If physical 'Time Travel' of this type were possible to take place, a variety of paradoxes will arise due to alterations in the chain of events taking place in this universe. For example, an individual can go back in 'Time' and prevent his/her own parents to even get to know each other in the general sense, let alone in the biblical sense. Also, an individual who has access to much more advanced weapons could travel back in 'Time' and assist one side or another in any of the major wars and literally change the history.

If such a scenario were truly possible, and even if it were only permitted to be used for peaceful applications, the members of the societies of millions of years into the future would have at least eliminated all of the serious unpleasant experiences of the past. Therefore, the very existence of such major unpleasant events in our history books alone proves otherwise. In other words, the recorded accounts of the occurrences of such events in our own past can serve as proof that at least <u>physical intervention in past events (of any kind) is not possible.</u> The very same argument applies to 'Time Travel' into the future, as in skipping even minutes, let alone years or millennia. In short,

"Physical 'Time Travel' is not possible."

- ## Spiritual 'Time Travel'
    Spiritual 'Time Travel' is quite possible, since spirits do not experience 'Time'. In fact, there have been individuals in the past who could perform such tasks, individuals who were capable of seeing the events which had occurred in their past as well as events that were going to take place in their future.

Notes:

- The distance in time, that any individual can possibly travel to spiritually, to directly witness events, depends on his/her concentration and ability to stay connected with his/her soul which remains in his/her physical body and continues to operate its various internal organs.

- The variety of prophesies mentioned in various religious books, as well as by individuals such as Mr. Nostradamus are clear indications of spiritual travel in 'Time'.

- The variety of books written by individuals such as Mr. Joules Vern, movies such as star trek, in which equipment used are directly related to what have become possible technologically only in the future, are clear indications of such writers' abilities of seeing into the future. Such writers had chosen to dazzle their readers / viewers by expressing their visual findings in an entertaining fashion.

- Currently, there are individuals in different countries who are using their 'spiritual time travel' abilities to help others by solving various otherwise unsolvable issues such as robberies, accidents or by locating various misplaced items.

- North American Indians used such spiritual travels to locate the herds of buffalos or other needed necessities.

It must be emphasized that,

"Any event that is seen, as in spiritually viewed (not to be mistaken as being imagined), to take place in the future, will take place at the time and place it is seen to take place."

Since, even the very knowledge of the future events becomes part of the information needed for certain actions to be taken so that they eventually lead to the occurrence of such foreseen future events. In other words,

"The future that has been seen cannot be altered."

The reason for 'Time Travel' being possible spiritually and not physically is that, in this universe, the path of 'Time' is like a train track which is quite straight, with no bends whatsoever. On that track, there is a single passenger wagon which is moving along at a gradually faster pace. The occupants of that wagon represent the current inhabitants / contents of this universe. The instantaneous speed of the wagon represents the rate at which the passage of 'Time' is experienced on a universal scale. The instantaneous position of the wagon along the 'Time' track represents the instantaneous universal 'Time'. In such a scenario,

- The physical bodies of the occupants are contained by the wagon's walls / ceiling / floor. Hence, they are limited to their common universal 'Time'. The individual contents / living beings in the wagon can experience the passage of 'Time' at slightly different paces as compared to each other, by moving at different speeds relative to their local aether mediums. That is, they can only move quite gently along the length of the wagon's interior.

- Spirits of the occupants of the wagon are not directly bound to the 'Time' track, as their corresponding physical bodies are. However, they are temporarily bound to their respective physical bodies, since parts of them that are serving as the souls of those bodies, must stay superimposed on their respective physical bodies and continue performing their tasks which are the internal operations of those bodies, otherwise those being will die.

  If the spirit of a living being can develop the ability to get out of the body and yet in a controlled fashion stay directly connected to its respective soul part, it can get out of the limited volume of the wagon (its physical body's current 'time') and observe events that either have already taken place in the past (where the wagon used to be) or will happen in the future (where the wagon is headed).

Effect of Gravity on the Passage of 'Time'

A variety of experiments can be designed and conducted that will demonstrate how the speed of objects (or living beings) relative to their respective local aether mediums affects the rate at which they experience the passage of time. The following four experiments are proposed to confirm the validity of the aether-based theory of 'Time' which states,

> **"The rate at which 'time' is experienced by an object (or a living being) depends on its instantaneous speed relative to its local aether medium, as compared to the propagation speed of phase vibrations in that medium."**

These experiments will also confirm the validity of the aether-based theory of 'Gravity' which states,

> **"The force of gravity, at any given location, is the drag force induced by the flow of aether which is accelerating through that particular location."**

- First experiment:
  (Orbital free fall vs. direct free fall)

Suppose two atomic clocks, "A" and "B", are onboard a space rocket and are placed in a circular orbit around earth, at an altitude of say 7,500 Km from the center of mass of earth. Once in orbit, clock "A" is gently released so that it keeps on orbiting the earth. As a certain coordinates are reached, the secondary rocket fires up and carries clock "B" to a much higher altitude in such a way that it is no longer orbiting the earth. As shown below, to accomplish this required task, two possible trajectories can be followed by the second stage of the rocket:

1- It can follow a trajectory as shown below in the left figure, so that after all of its fuel is spent, it coasts and eventually comes to a complete stop relative to earth in such a way that it is no longer orbiting the earth.

2- It can follow an orbital transfer trajectory to reach the desired higher altitude, as shown below in the right figure. However, in this case, after reaching the destination altitude, the rocket must fire in reverse and bring its orbital motion to a complete stop at the right time and at the right coordinates.

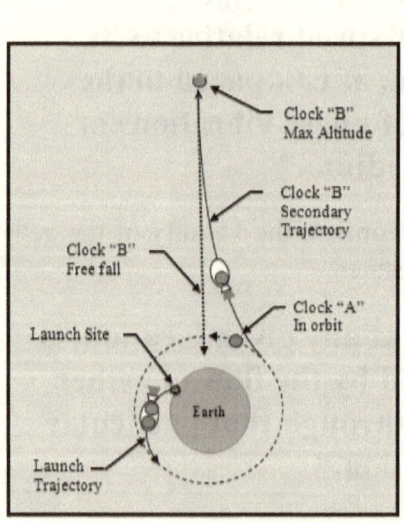
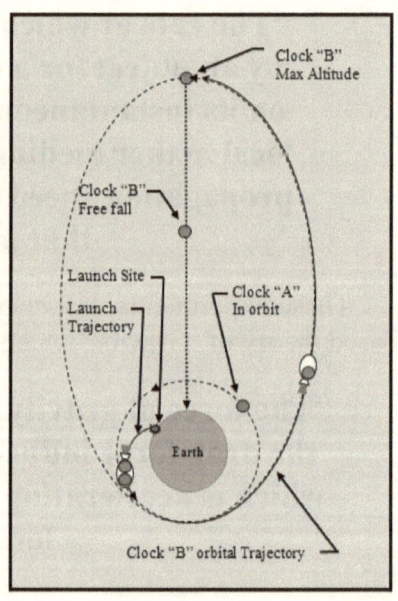

In either case, from the instant clock "B" comes to a complete stop at its highest altitude and starts its free fall journey directly towards the center of mass of earth, the rates at which the two clocks experience the passage of time must be recorded.

The results will indicate that, during such a scenario, the two clocks will experience the passage of time at different rates due to their motions relative to their respective local aether mediums being different.

- **Clock "A" will experience the passage of time at a fixed rate**, as it follows its circular orbital path around earth. This is <u>because its speed relative to its local aether will be constant.</u>

    As shown below, the true speed of clock "A" relative to its local aether will be equal to the vector sum of its orbital speed and the flow speed of the local aether which is moving directly towards the center of mass of earth.

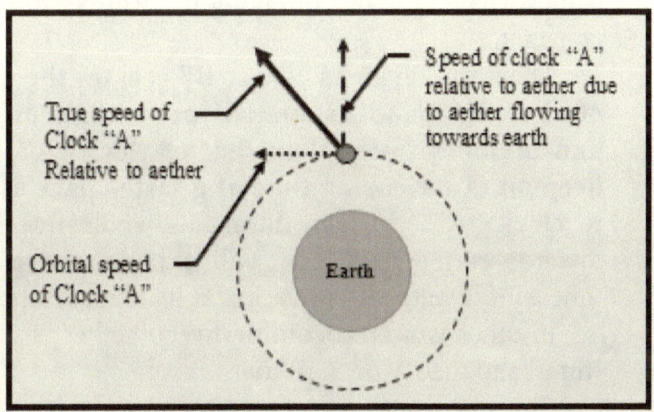

Note that, the flow speed of aether towards any gravitating body (giving rise to its force of gravity) at any given altitude, distance from the center of mass of that gravitating body, is equivalent to the escape velocity corresponding to that particular altitude. This is due to the fact that, aether that flows passed any given elevation / altitude starts its journey from a state of relative rest at an infinite distance and gradually speeds up as it approaches the gravitating body, which in this case happens to be the planet earth.

- **Once clock "B" comes to a complete stop at its highest altitude, it will experience the passage of time at a faster pace as compared to clock "A".** This is because the speed of clock "B" relative to its local aether will be simply that of aether flowing towards earth at that altitude (equivalent to the escape velocity at that particular altitude) which is far less than that of clock "A" relative to its local aether medium.

During the whole period of time that clock "B" free falls towards earth it will experience the passage of time at the same rate as it did while it was at its highest altitude. This is because during its entire free fall journey (until it encounters the atmosphere) its speed relative to its local aether which is in fact dragging it towards earth will remain constant.

Therefore, **even as clock "B" crosses the orbital path of clock "A" and its spatial speed keeps on increasing and becomes faster than that of clock "A", it will still keep on experiencing time at a faster pace as compared to clock "A".** In fact, during its whole free fall journey towards earth, clock "B" will experience the passage of time as if it were still hovering at its highest altitude.

In other words, according to the aether based theory of 'time' and theory of 'gravity':

**The results of such an
experiment will reveal that,
for the whole duration of its free
fall directly towards earth,
regardless of its instantaneous
elevation and speed relative to earth,
clock "B" will continuously experience
the passage of time at a CONSTANT rate
which will be exactly equal to the rate it
had experienced the passage of time when
it was at its highest altitude.**

Notes:

1- The maximum height reached by clock "B" can be calculated / selected so that as clock "B" crosses clock "A's" orbital altitude, while freely falling towards earth, its speed relative to earth will be equal / comparable to that of clock "A's" orbital speed.

2- This experiment can also be conducted so that the two clocks pass very close to each other.

In such a scenario, even though the two clocks will temporarily be:

- **at the very same altitude,**
- **be in a state of free fall towards earth, and**
- **be moving at the same rate of speed relative to space,**

they will be experiencing time at two different rates, simply due to their speeds relative to their temporarily common local aether medium being different.

The following figure shows a different version of such an experiment. In this case, the two clocks are sent out to two different heights and both are allowed to free fall so that they will pass by a certain altitude at about the same time. In such a case, **even though the two clocks will be in a state of free fall directly towards earth and also be at the very same altitude, they will experience the passage of time at two different rates.**

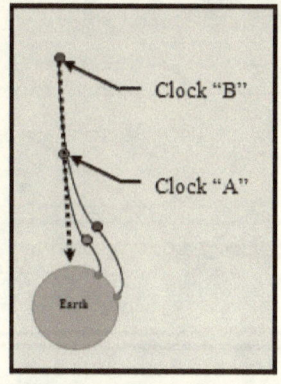

Each clock will experience the passage of time as if it is still hovering at its respective maximum height, from where it starts its free fall directly towards the center of mass of earth. In the case shown in the figure, clock "B" will experience 'time' at a faster pace as compared to clock "A", due its speed relative to the local aether being less than that of clock "A".

- Second experiment:
(How does an object experience the passage of 'Time' as it <u>directly</u> approaches a black hole, in a state of free fall?)

Suppose two atomic clocks, "A" and "B", as shown below, are onboard two space rockets.

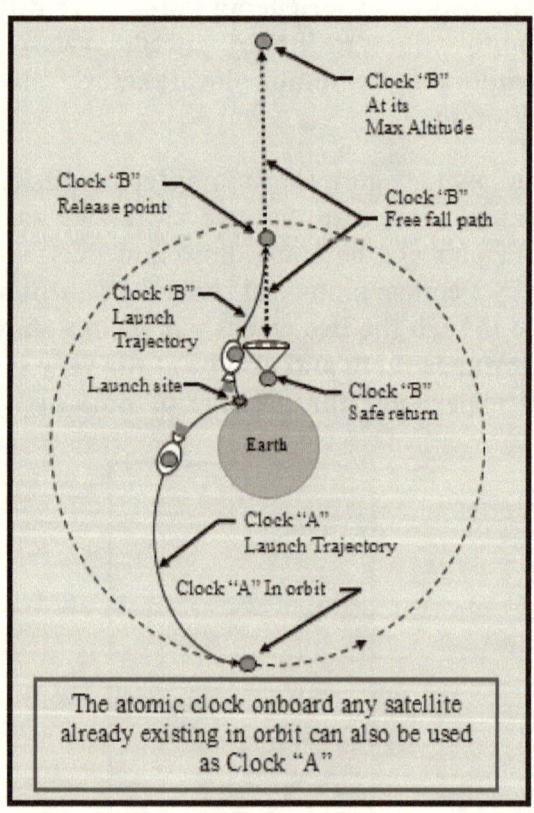

The atomic clock onboard any satellite already existing in orbit can also be used as Clock "A"

First, the rocket carrying clock "A" is launched in the same direction as earth's rotation and follows a trajectory that takes it into a nearly circular orbit around earth, at an altitude of say 7,500 Km from the center of mass of earth. Then, the other rocket carrying clock "B" is launched in the opposite direction of earth's rotation and follows a trajectory straight up with no orbital motion around earth whatsoever.

Clock "B" must be speeded up so that as it is released prior to reaching clock "A's" orbital altitude, its speed is comparable to the orbital speed of clock "A", but at 90 degrees to it, going straight away from the center of mass of earth. Clock "B" should continue on with its upward free fall journey until it gradually slows down and comes to a complete stop, at its maximum altitude. Then, it should start its downward free fall motion towards earth. Clock "B" will cross the orbital path of clock "A" at the very same speed as it did while going upward.

Notes:

- The launch time of the second rocket can be selected so that the upward free fall trajectory of clock "B" takes it just passed clock "A", as it crosses its orbital path.

- The orbital altitude of clock "A" and the size and the engine power of the second rocket carrying clock "B" can be selected so that as clock "B" falls back towards earth it passes by clock "A", again.

During such an experiment:

While in a nearly circular orbit, clock "A" will experience the passage of time at a constant pace, since it will be moving at a constant speed relative to its local aether which is flowing towards earth (the same aether flow that is giving rise to earth's force of gravity).

Clock "B", however, will experience the passage of time at different rates, as it follows its free fall trajectory upward and then downward:

1- As clock "B" crosses clock "A's" orbital path, on its way up, it will be:

- in a state of free fall,
- at the very same altitude as clock "A", and
- be moving at the same speed as clock "A".

Yet, it is expected that, clock "B" will experience the passage of time at a slower pace as compared to clock "A". Since, clock "B" will be moving in the opposite direction of the local aether flow which is directly towards the center of mass of earth. Therefore, the speed of clock "B" relative to its local aether will be faster than the speed of clock "A" relative to its local aether.

Note that, in this part of the experiment, if the two clocks were to move at the very same speed relative to their respective, temporarily common, aether medium they would experience the passage of time at exactly the very same pace, regardless of their directions of motion being at 90 degrees to each other, and having a relative speed of a fair magnitude with respect to each other.

2- As the upward speed of clock "B" decreases, while it is approaching its maximum altitude, it will gradually experience the passage of time at faster and faster paces.

This is expected, since it will be moving at slower speeds relative to its local aether, at higher altitudes. Therefore, its speed relative to its local aether will gradually become smaller fractions of the speed of phase vibrations in that medium. Hence, gradually, it will experience the passage of time at faster paces.

3- While temporarily at rest at its highest altitude, clock "B" will experience the passage of time the fastest.

At that height, while it will temporarily be at a complete rest relative to earth's center of mass, without any orbital motion around earth, the speed of clock "B" relative to its local aether will be exactly equal to aether's flow speed towards earth at that particular altitude (its elevation with respect to the center of mass of earth).

4- As clock "B" speeds up during the downward part of its free fall journey, it will be continuously experiencing the passage of time at a CONSTANT pace which will be EXACTLY equal to the pace at which it was experiencing the passage of time when it was at its maximum altitude.

This effect is expected since during its downward journey, clock "B" will be constantly moving at the very same speed relative to its local aether, which will in fact be dragging it towards earth. Hence, during the entire trip downwards, clock "B" will experience the passage of time as if it were still hovering at its maximum altitude.

Notes:

- During its free fall journey, clock "B" will cross the same altitudes twice, once on its way up and again on its way down. However, **even though at any given altitude it will be moving at the very same speeds relative to earth (while going up and while coming down), it will experience the passage of time at two different rates.**

- **During the upward part of its free fall journey, clock "B" will gradually experience the passage of time at faster and faster paces (until it reaches its maximum altitude). However, it will experience the passage of time at a CONSTANT fixed rate during its entire trip downwards** (until it hits the atmosphere).

The results of such experiments will clearly confirm the following answer to such an important question / issue in regards to how the passage of time is experienced by objects that directly approach black holes, in a state of free fall and cross their event horizons:

"As an object allows itself to free fall directly towards a black hole, it will continuously experience the passage of time at a CONSTANT FIXED RATE, as

if it were still hovering at the same distance to the black hole from where it starts its free fall. Such an experience will last for the entire duration of that object's free fall, not just until it crosses the event horizon, but rather until it literally hits the surface of the black hole."

- Third experiment:
  (Esmailzadeh paradox)

Suppose three identical atomic clocks, "A", "B" and "C", as shown below, are onboard three satellites in orbit around earth.

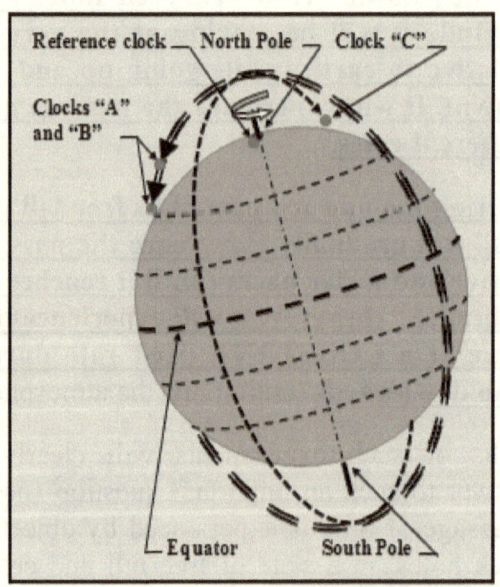

The three satellites are following orbital paths that are:

- Polar orbits, crossing over the polar regions,
- At the same altitude,
- Two satellites corresponding to clocks "A" and "B" share the same orbital path, as they move in-line (one after the

other). However, their common orbital path is at 90 degrees to the orbital path of the third satellite which is carrying clock "C".

In such a case, as the third satellite repeatedly passes by the other two satellites, twice during every full orbit around earth, their time readings can be cross checked against a reference clock that is stationary on the surface of earth.

The polar orbits are selected for this experiment because they would ensure that all of the satellites (all three clocks) will experience:

- The very same exposure to earth's magnetic field, as they orbit around earth.

- They will experience the very same gravitational effect, since their altitudes are nearly identical to each other.

- Their speeds will also be identical, due to their altitudes being the same, relative to the center of mass of earth.

- The three clocks will be in an identical state of free fall.

Satellites carrying clocks "A" and "B" will be moving at about 40,000 km/hr relative to the satellite carrying clock "C", as they pass by each other, twice per every revolution around earth (over the poles).

Therefore, according to Mr. Einstein's special theory of relativity and even Mr. Lorentz's relativity:

- Clocks "A" and "B" should not indicate any time dilation with respect to each other, since they have no apparent in-line motion relative to each other.

- Clocks "A" and "B" should indicate a different rate of passage of time as compared to the rate it is experienced by clock "C". Since, according to an astronaut who might be onboard satellite carrying clock "A" or "B", as clocks "A" and "B" pass by clock "C", due to their relative speed of 40,000 km/hr, he/she can calculate that clock "C" should experience the passage of time at a little slower pace.

While, another astronaut onboard satellite carrying clock "C" sees clocks "A" and "B" to be moving at 40,000 km/hr and therefore according to his/her calculations the passage of time experienced / indicated by them is expected to be slower.

Such a scenario leads to a paradox. Since, both astronauts' expectations are based on applying the very same theory and mathematical equations provided by Mr. Lorentz and Mr. Einstein, equations which are supposed to be the foundations of today's high speed physics. Yet, as each astronaut is expecting the other to experience time at a slower pace, they both cannot be correct. This paradox can be referred to as:

"Esmailzadeh Paradox".

It can be easily verified that, in such a scenario no relative time dilation will be registered by any of the clocks, as they may be compared to each other. Since, once the three clocks onboard the three satellites are synchronized, they will stay synchronized regardless of how long they stay in orbit and/or how many times they pass by each other. All three clocks can even be checked against any other atomic clock (as a reference) that might be stationary on the surface of earth.

It can be confirmed that the three clocks that are in orbit will experience and indicate the passage of time to be at EXACTLY very same pace as they cross each other's path.

The reasoning for such a discrepancy in their expectations (calculation of expected amounts of time dilations) derived when using equations of relativity is that, in relativity the speeds used are the speeds of the clocks with respect to each other. However,

"To arrive at the correct answer, the speeds of clocks relative to their local aether must be used, regardless of their relative speed with respect to each other."

As it becomes clear during such experiments, the three clocks will experience the passage of time at exactly the same pace. Since, all three satellites (hence, all three clocks "A", "B" and "C") will be moving at the very same rate of speed relative to the local aether which is flowing towards earth. Therefore, once they are synchronized, the three clocks will continue to stay in sync, indefinitely. In other words,

"Once the speed of objects relative to their respective local aether is used in calculating how fast they experience the passage of time, the "Esmailzadeh Paradox" becomes resolved, automatically."

- Fourth experiment:
  (Esmailzadeh Orbital Altitude Range)

Depending on the altitude of the orbit selected, the clock in the orbit can experience the passage of time at a pace that is slower than, equal to or even faster than the pace it is experienced by a reference clock that is stationary on the surface of earth.

The speed of the reference clock on the surface of earth relative to its local aether is equivalent to the vector sum of:

- **The inflow speed of aether towards earth (giving rise to its force of gravity) at that particular altitude (location on earth surface)**

   This aether flow speed is equivalent to the escape velocity corresponding to that particular altitude. Note that, the sea-level elevation at the equator is about 11 kilometers higher than the sea-level elevations at the two poles.

- **The local rotational speed of earth surface around its axis**

   The magnitude of this eastward speed depends on the latitude of the location on earth surface. It varies from zero (at the two geographic poles) to a maximum of about 1,650 km/hr, corresponding to the equator.

- **The speed of aether flow due to the local magnetic field**

   The direction of this aether flow can be vertically downwards (at the South Pole), vertically upwards (at the North Pole) or horizontally southward at the equator. At other locations in between, the direction of the aether flow will depend on the direction of the local magnetic field lines.

Therefore,

"The speed of an object on the surface of earth relative to its local aether is just over the escape velocity corresponding to the altitude of that particular location on earth surface."

However, <u>the speed of satellites in low altitude orbits</u> relative to their local aether medium is equivalent to the vector sum of:

- **The inflow speed of aether towards earth (giving rise to its force of gravity) at that particular altitude**
  This speed is directly dependent on the altitude of the orbit. The higher the altitude the slower will be this downward aether flow speed. Since, this aether flow speed is equivalent to the escape velocity corresponding to that particular altitude.

- **The orbital speed of the satellite**
  This speed varies with the distance from the center of mass of earth and it is gradually reduced as the distance is increased.

Therefore,

"Satellites in low altitude orbits, due to moving at faster speeds relative to their respective aether medium, as compared to objects on the surface of earth, experience the passage of time at a slower pace as compared to a reference clock which is positioned on earth surface."

As the orbital altitude is increased, not only aether is flowing slower towards earth, but also satellites require slower speeds to stay in orbit. At certain mid-range altitudes the pace at which time is experienced by a clock onboard the satellite will match the rate at which it is experienced by clocks positioned at certain locations on earth surface.

At the minimum and the maximum limits of these mid-range altitudes the speed of the satellite relative to its local aether will be exactly equal to the speed at which aether is flowing past objects, on the surface of earth, which are positioned at the intersections of the geographic equator and the magnetic equator and at the Magnetic North Pole, respectively. Since, objects positioned at the equator experience the highest speed relative to their local aether. Whereas, the objects positioned at the Magnetic North Pole experience the lowest speed relative to their local aether, due to the intensity and direction of the local magnetic field.

This particular range in the orbital altitudes which allow satellites to experience 'time' at the same pace as it is experienced by objects at particular locations on the surface of earth can be referred to as,

"Esmailzadeh Orbital Altitude Range"

The following figure shows samples of a variety of orbital altitudes which can be used by satellites.

As the orbital altitude is increased further, due to having lower speeds relative to the local aether, such satellites experience the passage of time at a faster pace as compared to clocks positioned anywhere on the surface of earth. The orbital path which is followed by earth's moon is one such high altitude orbit which is far enough from earth's center of mass and any clock stationed in such an orbit will experience the passage of 'time' at a faster pace as compared to the rate it is experienced by clocks on the surface of earth.

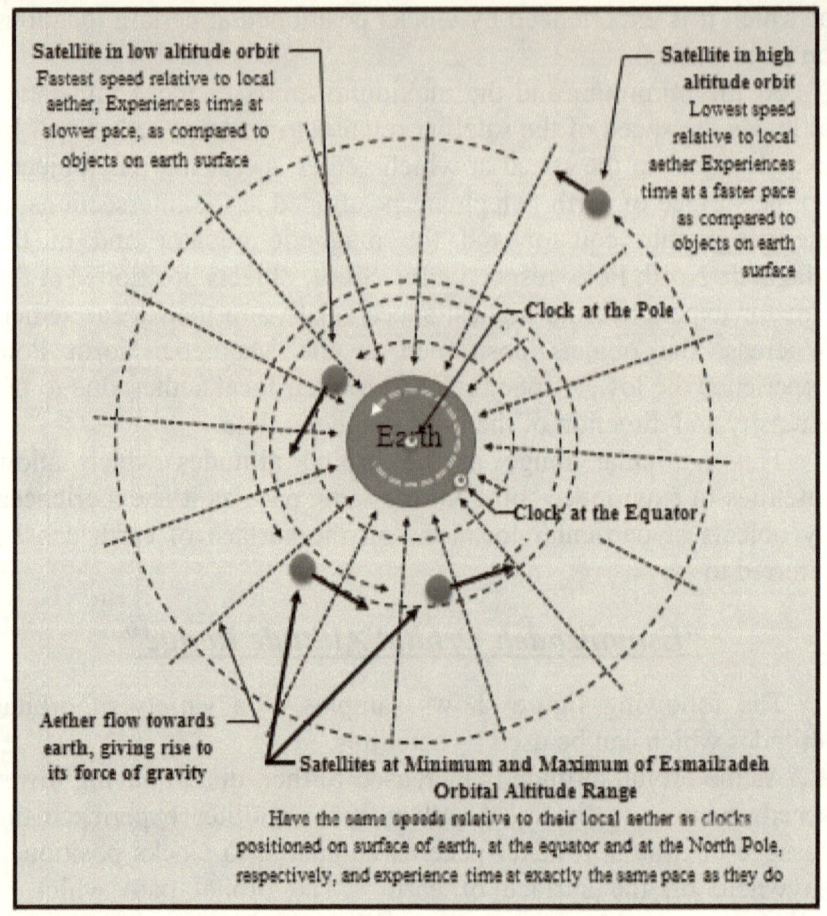

Satellite in low altitude orbit
Fastest speed relative to local
aether, Experiences time at
slower pace, as compared to
objects on earth surface

Satellite in high
altitude orbit
Lowest speed
relative to local
aether Experiences
time at a faster pace
as compared to
objects on earth
surface

Clock at the Pole

Earth

Clock at the Equator

Aether flow towards
earth, giving rise to
its force of gravity

Satellites at Minimum and Maximum of Esmailzadeh
Orbital Altitude Range
Have the same speeds relative to their local aether as clocks
positioned on surface of earth, at the equator and at the North Pole,
respectively, and experience time at exactly the same pace as they do

Does Acceleration Induce 'Time Dilation'?

According to the theory of "Time", presented in this chapter, along with various experiments provided to confirm its validity,

"The rate at which 'Time' is experienced by an object depends on that object's instantaneous speed relative to its local aether medium, as compared to the propagation speed of phase vibrations in that medium."

As the speed of an object, relative to its local aether, approaches that of phase vibrations in that medium, the object experiences the passage of time at progressively slower paces. In case the speed of the object reaches the speed of phase vibrations in its local aether medium, the passage of time will no longer be experienced by that particular object. In other words,

"It is the instantaneous speed of an object relative to its local aether that dictates how fast that particular object experiences the passage of time."

It must be emphasized that,

The speed of an object relative to anything else that exists in this universe is totally irrelevant, as far as experiencing 'Time' by that object is concerned.

Also, the medium of aether in this universe is a very dynamic medium and manifests its various types of motions as different phenomena such as the local force of gravity, the magnetic field and the electric field (amongst many others). Therefore,

"As an object accelerates, its speed relative to its local aether can vary in any surprisingly different ways."

For example, as an object accelerates in free space that is not near any gravitating body or any charged particle, its acceleration, regardless of the direction of motion chosen, will have the very same effect which is an increase in its rate of speed relative to its local aether. Since, in such a case, the local aether is relatively stationary.

However, if the very same object is placed close to a gravitating body such as a planet, let alone a star or a black hole, or when the same object is placed in a magnetic field or in an electric field, the direction of its acceleration is very decisive in how that particular object's speed relative to its local aether changes. Since,

- **The object may be accelerating in the very exact same direction as that of the existing flow of aether in its immediate vicinity**

    In this case, at least for some time, that object's acceleration will actually cause its instantaneous speed relative to its local aether to decrease. Therefore, <u>the object will experience the passage of 'Time' at an increasingly faster pace.</u>

    Once the object catches up with its local aether, it will have no relative motion with respect to it. <u>At that instant, the object will experience the passage of 'Time' the fastest.</u>

    After that instant, however, as it continues to accelerate, it will actually be moving faster than its local aether. Therefore, <u>as the object moves faster and faster relative to its local aether, it will experience the passage of 'Time' at a gradually slower pace. Since,</u>

It is the instantaneous speed of an object relative to its local aether that is affecting how fast that object experiences the passage of 'Time'. It does not

matter that the object is moving faster or the local aether is moving faster.

- **The object may be accelerating in the opposite direction of the existing flow of aether in its immediate vicinity**

   In this case, that object's acceleration will cause its instantaneous speed relative to its local aether to increase. Therefore, the object will experience the passage of 'Time' at a gradually slower pace.

- **The object may be accelerating in any other direction**

   In such cases, to calculate the true instantaneous speed of that object relative to its local aether, the angle between that object's direction of motion and the flow direction of its local aether must be taken into account.

The following experiments are specifically designed to explore the validity of the above statements.

> - First Experiment:
>   (Three probes accelerating in different directions)

As shown below, to perform such an experiment one needs to place three probes in an elongated elliptical orbit around earth. The altitude of the orbital aphelion can be equivalent to twice that of a geosynchronous orbit.

Each of these probes must be equipped with its own adequate propulsion system and must be carrying an atomic clock onboard. There also needs to be an atomic clock left on the surface of earth to serve as a reference.

When the rocket carrying the three probes reaches its highest altitude, it must bring itself to a complete stop with respect to earth (no orbital motion) and start hovering at that altitude. Then, as probe "A" keeps on hovering, the other two probes must be propelled / accelerated in opposite directions. Probe "B" must be accelerated directly away from earth and probe "C" must be accelerated directly towards earth.

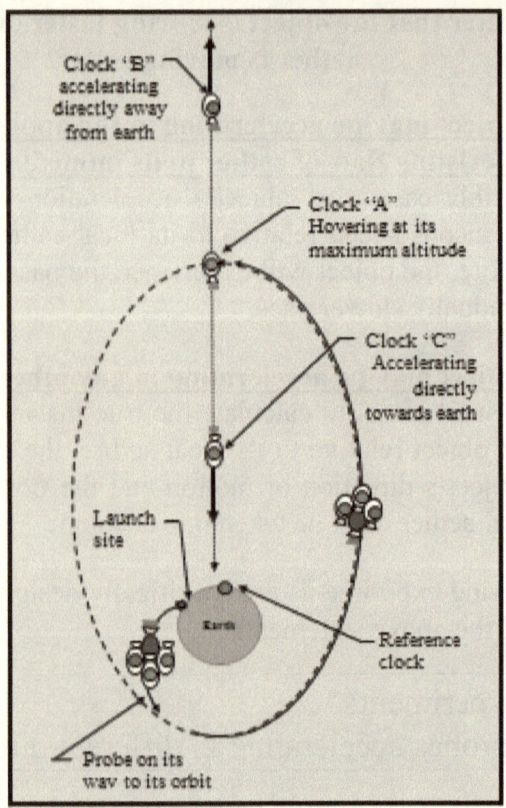

Probes "B" and "C" do not necessarily have to be accelerating at the very same rate, but it is recommended that their accelerations (which do not include the acceleration of the local aether towards earth) to be constant uniform accelerations, each in their own respective directions. It must be emphasized that probe "C" is not freely falling towards earth, but it is propelled directly towards earth, just as probe "B" is propelled directly away from it.

Once the three probes start their journeys along their respective separate paths, their clocks will start to experience the passage of 'Time' at different paces.

Clock "A"

- The speed of clock "A" relative to its local aether will be constant (being equal to the escape velocity from earth's gravity at that altitude). Since, **even after** it runs out of fuel and starts to free fall towards earth, its speed relative to its local aether stays relatively constant. Therefore, **Clock "A" will experience the passage of 'Time' at a fixed pace**, until it hits earth's atmosphere.

Clock "B"

- While the rate of acceleration of probe "B" (carrying clock "B") going away from earth is constant, depending on the magnitude of its rate of acceleration, clock "B" will experience 'Time' in two different ways:

  1- If <u>(at the aphelion of the initial elliptical orbit)</u> the magnitude of the constant acceleration of clock "B", relative to the center of mass of earth, be larger than the acceleration of the local aether flowing towards earth, its instantaneous speed relative to its local aether will only gradually increase, as it gets farther and farther away from earth's center of mass. That is, until it runs out of fuel. Therefore, **during the powered portion of its flight it will experience the passage of 'Time' at an ever decreasing pace.**

  2- If <u>(at the aphelion of the initial elliptical orbit)</u> the magnitude of the constant acceleration of clock "B", relative to the center of mass of earth, be smaller than the acceleration of the local aether flowing towards earth, its instantaneous speed relative to its local aether will decrease, but only until its rate of acceleration becomes equal to that of its local aether which is accelerating towards earth.

      Therefore, **<u>contrary to the expectations based on the principle of equivalence and the General Theory of Relativity,</u>**

<u>During this portion of its journey, even though clock "B" will be accelerating away from earth (over and above its being dragged by earth's gravity), its instantaneous speed relative to its local aether will be gradually decreasing, hence</u>

it will be experiencing the passage of 'Time' at an ever faster pace.

At the instant clock "B" experiences the slowest speed relative to its local aether, it will experience the passage of 'Time' the fastest.

After that instant, because the rate of its acceleration going away from earth will be greater than that of the local aether accelerating towards earth, the speed of clock "B" relative to its local aether will start to increase, but only until it runs out of fuel. **During this portion of its powered flight, clock "B" will experience the passage of 'Time' at an ever slower pace.**

Note that, clock "B" will experience the passage of 'Time' at a fixed pace, as if it were still hovering at the same altitude as clock "A", **if and only if** its rate of acceleration going away from earth is made to decrease exponentially so that the increase in its speed going away from earth exactly matches and compensates for the decrease in the local aether's flow speed towards earth, due to distance from earth.

In other words, in such a case, the instantaneous speed of clock "B" relative to its local aether will remain constant, until it runs out of fuel. **In such a scenario, during its powered flight, clock "B" will experience the passage of 'Time' at a constant pace.**

Clock "C"

- Even though the rate of acceleration of probe "C" (carrying clock "C") towards earth due to its own rocket engines is constant, it is in addition to the acceleration it receives from its local aether due to being dragged by it towards earth. In other words, the instantaneous speed of clock "C" relative to its local aether will gradually **decrease**, until it catches up with its local aether.

In other words, **contrary to the expectations based on the principle of equivalence and the General Theory of Relativity,**

<u>During this portion of its journey, even though clock "C" will be accelerating towards earth (over and above its acceleration which is due to earth's gravity), its instantaneous speed relative to its local aether will be gradually decreasing, hence it will be experiencing the passage of 'Time' at an ever faster pace.</u>

At the instant clock "C" catches up with its local aether flow, it will experience the passage of 'Time' the fastest.

After that instant, because the rate of its constant acceleration towards earth will continuously be greater than that of its local aether flowing towards earth, clock "C" will move fasters and faster towards earth as compared to its local aether. In other words, it will gradually build up a speed relative to its local aether. **During this portion of its journey, clock "C" will experience the passage of 'Time' at an ever slower pace.**

It should be emphasized that,

It is the instantaneous speed of an object, such as clock "C" in this case, relative to its local aether that is affecting how fast that object experiences the passage of 'Time'. It does not matter if the object is moving faster or the local aether is moving faster.

- Second Experiment:
  (Two probes falling at different speeds)

A different way of performing such an experiment is to send two probes into two different highly elongated elliptical orbits around earth. Both elliptical orbits must be in one plane and their aphelions must be nearly along the very same line from earth's center of mass. The aphelion of one orbit must be at a much higher altitude than that of the other orbit. The following figure shows the general layout of such orbital paths around earth.

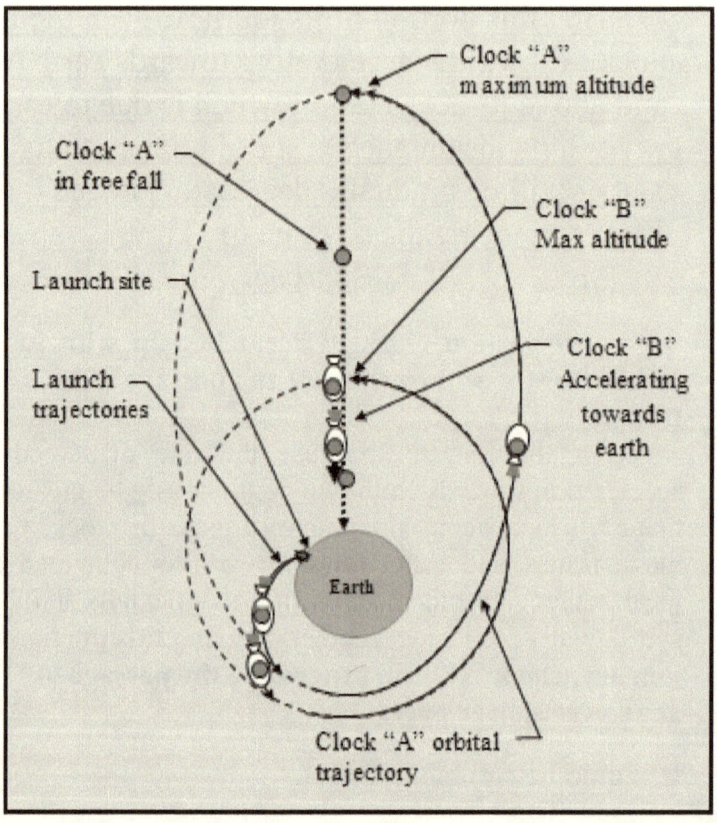

Once the two orbital paths are selected, the two probes "A" and "B" can be launched so that the following can be executed in their proper sequence as stated:

- As probe "A" which is in the longer elliptical orbit reaches the aphelion of its orbit, its rocket must fire up and bring it to a complete stop (no orbital motion around earth) so that it starts a free fall journey directly towards the center of mass of earth with zero initial velocity.

- Probe "B" which is in the lower orbit must be timed so that it reaches the aphelion of its orbital path as probe "A" passes by it. At that instant the rocket engine onboard probe "B" must not only bring probe "B" to a complete stop (no orbital motion around earth) but also keep on accelerating it at a constant rate, directly towards earth's center of mass. The acceleration of probe "B" will be over and above its acceleration towards earth which is due to local earth gravity. In other words, it should be approaching earth along the very same path as that taken by probe "A" and as time goes on, while still accelerating it should catch up with probe "A".

- As probe "B" passes by probe "A", its speed relative to earth's center of mass must be equivalent to that of escape velocity from earth's gravity at that particular altitude.

- At the instant probe "B" passes by probe "A", its rocket engine must be turned off so that it starts its own unique free fall journey towards earth's center of mass, from that particular altitude.

- The maximum altitudes of the two elliptical paths selected, the orbital timing of the two probes and the rate of acceleration of probe "B" must be so that it catches up with and passes probe "A", while still outside earth's atmosphere, with sufficient time left to transmit the rate at which it is experiencing the passage of 'Time', while free falling towards earth.

Note that, the two probes must be transmitting their respective time experiences during the whole experiment.

Long after such an experiment is conducted and the collected data are analyzed, they will indicate that, the two probes had experienced the passage of 'Time' in two totally different fashions, as follows:

- **The transmitted data from probe "A" will clearly indicate that,** once it had started its free fall towards earth from the aphelion of its elliptical orbit, **it had experienced the passage of 'Time' at a constant pace, as if it were still hovering at the altitude corresponding to the aphelion of its elliptical orbit.**

This result is quite expected, since once probe "A" had started its free fall from the highest altitude of its orbit, it was actually dragged by the aether that was flowing passed its position towards earth. Also, due to not resisting, its speed relative to its local aether had stayed constant. Therefore, it had experienced the passage of 'Time' at a constant pace, until it had hit the atmosphere of earth.

However, as probe "A" had gradually slowed down due to atmospheric friction, its speed relative to its local aether had increased and it had experienced the passage of 'Time' at a gradually decreasing pace. That is, until it had vaporized due to excess heat.

- **The transmitted data from probe "B" will clearly indicate that,** once it had started its accelerated fall towards earth from the aphelion of its lower elliptical orbit around earth, **it had experienced the passage of 'Time' at an ever increasing pace, but only until it had caught up with the flow speed of its local aether towards earth, and had turned its rocket engine off.**

Note that, at the instant probe "B" had caught up with its local aether and had no speed relative to it, it had experienced the passage of 'Time' the fastest, <u>as if it were hovering in the medium of aether at an infinite distance from earth.</u>

Once probe "B" had started its unique free fall towards earth center of mass, **it had started experiencing 'Time' at a constant pace, as if it were hovering at an infinite distance from earth's center of mass.**

These results are quite expected, since as probe "B" had started its acceleration towards earth at a constant rate, it was actually catching up with aether that was already flowing towards earth, along its path. And, consequently its speed relative to its local aether was gradually decreasing. Hence, it was experiencing 'Time' at an ever faster pace.

However, once it had caught up with probe "A" and had also caught up with the flow of its local aether, and had turned its rocket engine off, and had started its unique free fall towards earth, its speed relative to its local aether had stayed nearly equal to zero. Consequently, it had kept on experiencing 'Time' at a constant pace, until it had hit earth's atmosphere.

However, as probe "B" had gradually slowed down due to atmospheric friction, its speed relative to its local aether had increased and it had experienced the passage of 'Time' at a gradually slower pace. That is, until it had vaporized due to excess heat.

During such an experiment it will become evident that, even though the two probes will experience a free fall state of motion towards earth, at a common altitude from earth's center of mass, each will experience the passage of 'Time' at a totally different pace. **Probe "A" will experience the passage of 'Time' as if it were still hovering at the aphelion of its elliptical orbit, while probe "B" will experience 'Time' as if it were hovering at an infinite distance from the center of mass of earth.**

In other words, <u>**contrary to the expectations based on the principle of equivalence and the General Theory of Relativity,**</u>

<u>"Even though probe "B" is accelerated to a much higher speed as compared to probe "A", it will be experiencing the passage of 'Time' at a faster pace as compared to probe "A".</u>"

- Third Experiment:
  (Using regular satellites launched into orbit)

A different way of performing such an experiment is to use the atomic clocks that are placed onboard any new satellite scheduled to be launched in the near future.

The vectors in the following figure show the instantaneous speeds of the clock used relative to its local aether, along its trajectory from earth surface up until it reaches the entry point of its orbital destination.

As it is shown in the figure, as the rocket carrying the clock starts its journey and picks up speed, it changes its direction of motion along the way. Its overall speed relative to its instantaneous local aether is the vector sum of two speeds. One speed is that of the local aether flowing towards earth (giving rise to its force of gravity) and the other speed is due to clock's motion, due to its rocket engines propelling it forward.

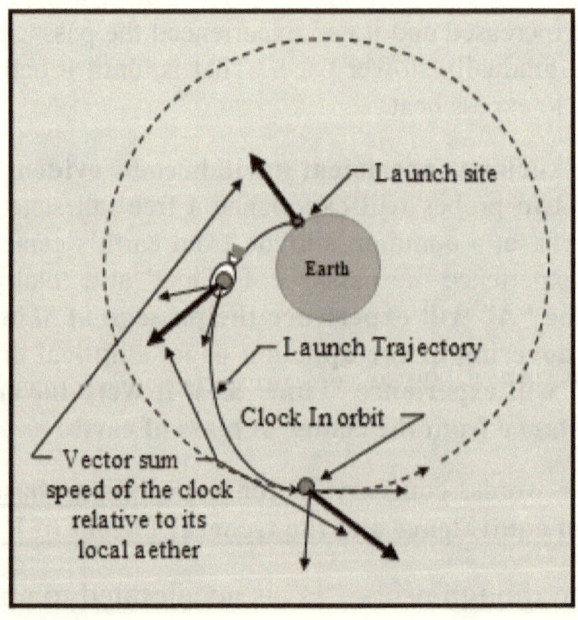

In the beginning, at the launch site, the speed of the satellite relative to its local aether is only that of the aether flowing towards earth (giving rise to its force of gravity). However, as the rocket engines start operating, they propel the clock upward, against the local flow of aether, and then gradually propel the clock along a path that is increasingly diagonal to the direction of the local aether flow that is towards earth. And, as the rocket speeds up and approaches its orbital destination, the direction of its motion becomes nearly perpendicular to that of the local aether that is flowing towards earth. The magnitude of the speed of the clock and that of the local aether, at the orbital destination, are dependent on the altitude of the orbit reached.

In order to properly calculate the amount of 'Time Dilation' experienced during such a journey, one needs to calculate the instantaneous vector sum of the speeds of the satellite (the atomic clock onboard) with respect to its local aether, along the path taken, from launch pad to the entry point of its orbital destination.

Once such calculations are made, the result obtained will exactly match the amount of 'Time dilation' experienced and indicated by the atomic clock onboard the satellite. In other words, contrary to the expectations of the **General theory of Relativity** and the **Equivalence Principle,**

There will be no whatsoever effect on the rate the passage of 'Time' experienced, that is due to the extra g-forces generated by the acceleration of the rocket as it speeds up to its required orbital velocity.

Note that, if the precise amount of 'Time dilation' experienced by any of the previously launched satellites and the data regarding their respective trajectories which they had followed to reach their orbital destinations are recorded, they can be used for such calculations, as well.

Such experiments will readily demonstrated that the clock onboard any particular satellite will abide by the predictions made by the new theories of 'Gravity' and 'Time' which are presented in this book, theories which are based on the existence of aether in this universe. In other words,

"There will be no 'Time dilation' effect recorded that will be due to the acceleration the rocket towards its orbital destination."

Therefore, it must be particularly emphasized that,

According to the information presented in this book and the results of the experiments proposed to verify their validities,

"Even though force of gravity and uniform rate of acceleration give rise to certain identical effects, such as the weight of objects, they have nothing in common as far as inducing 'Time Dilation' is concerned."

Therefore,

"The paradoxes / twin stories about accelerating objects / living beings that experience time at a slower pace, as compared to those left relatively stationary on their home planets, are fundamentally flawed. Since, such paradox stories totally ignore the actual instantaneous speeds at which such objects / living beings are moving relative to their respective local aether mediums."

It is the instantaneous speed of an object relative to its local aether that gives rise to 'Time Dilation' effect and not its instantaneous accelerations.

Can Acceleration, Gravity and High Speed Lead to Experiencing 'Time' at a Faster Pace?

The two experiments described below explore the possibility of combining the effects of acceleration, gravity and high speed on how "Time" is experienced, as indicated by three identical atomic clocks.

- **First experiment:**
  (The balloon version)

To perform this experiment, three identical atomic clocks ("A", "B" and "C") should be synchronized, while all three are positioned side by side, on the ground level. The three clocks should be equipped with transmitters so that they can independently transmit their respective "Time" experiences to a computer located at a receiving station, fixed on the ground level, nearby. The computer should be programmed to graphically show the differences in the rates at which "Time" is experienced by the three clocks, simultaneously, while taking into account the effects due to its instantaneous distance to each one of the clocks. In other words, it should simultaneously show three curves,

- One corresponding to the difference in the rates at which "Time" is experienced by clocks "A" and "B"

- One corresponding to the difference in the rates at which "Time" is experienced by clocks "A" and "C"

- One corresponding to the difference in the rates at which "Time" is experienced by clocks "B" and "C"

Clocks "A" and "B" should be lifted up into a fairly high altitude, by separate balloons. Clock "C" should be kept near the receiving station on the ground level to act as a reference. Clock "A" should be lifted up while it is housed in a rocket. Therefore, due to the weight of the rocket, the balloon carrying clock "A" must be considerably larger in size, as compared to the balloon carrying clock "B".

Once the two balloons reach the desired altitude, the balloon carrying clock "B" should be kept hovering at that altitude, while clock "A" (along with its rocket) should be released from its balloon and with the help of its rocket engine should be propelled directly towards earth, at a high rate of acceleration.

The three clocks should be continuously transmitting the rate at which they experience the passage of "Time", to the computer, from the instant they are synchronized up until clock "A" either hits the ground level or melts / explodes due to generation of excessive heat as it travels through the lower layers of the atmospheric.

The following figure shows the overall concept of this experiment in a simple fashion.

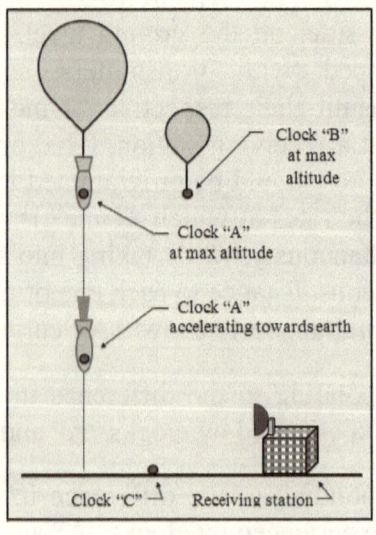

Note that, the acceleration of the rocket carrying clock "A" back towards earth can be of any rate, yet it is preferable that it be as high and over the longest period of time as possible.

According to aether-based theories, presented in this book, the following results are expected to be shown on the computer screen, during such an experiment:

Clock "C" will experience the passage of 'Time' at a constant pace, during the whole period of this particular experiment, due to its speed relative to its local aether being constant. Since, clock "C" will be in the path of an aether flow which is going through its position at a constant rate of speed equivalent to the vector sum of earth's rotational speed at its fixed latitude and altitude and the flow speed of aether which is moving directly towards earth's center of mass and is giving rise to earth's force of gravity, at that particular altitude measured from earth's center of mass.

Clocks "A" and "B", at first, will mutually experience the passage of 'Time' at gradually faster paces, as compared to clock "C", as their altitude from the center of mass of earth is gradually increased, that is until they reach the desired altitude. Since, as clocks "A" and "B" get farther away from earth's center of mass they will be in an aether flow which is moving towards earth at slower speeds.

However, from the instant that clocks "A" and "B" reach their highest altitude and clock "B" hovers there, while clock "A" is accelerated towards earth by its rocket, they will experience the passage of "Time" differently, since,

Clock "B" will experience the passage of "Time" at a constant rate. During the whole time that clock "B" can be kept hovering at that altitude, it will experience the passage of 'Time' at a faster pace as compared to the rate clock "C" will experience its passage. Since, the flow speed of aether through its position will be less than that of aether that is going through the position of clock "C", the same aether flow that is giving rise to earth's force of gravity.

For this part, the prediction made by the General Theory of Relativity is in agreement with that of the aether-based theories, but due to different reasoning. Since, according to the General Theory of Relativity clock "B" will experience the passage of 'Time' at a faster pace because of being exposed to a weaker gravitational force, as compared to clock "C", due to being farther away from the center of mass of earth.

Clock "A" will experience the passage of "Time" at gradually faster paces, as compared to clock "B" (and clock "C"). Since, its acceleration towards earth due to the rocket engine will be over and above its acceleration which is due to being dragged by the accelerated flow of aether towards earth. Therefore, the speed of the rocket (clock "A") relative to its local aether will gradually decrease. In short,

The rocket, along with clock "A", will be gradually catching up with the local aether flow which is accelerating towards earth's center of mass.

Consequently, as the speed of clock "A" relative to its local aether gradually decreases, it will experience the passage of 'Time' at faster and faster paces. Since, it is the speed of that clock relative to its local aether that dictates how fast it will experience the passage of 'Time'. In other words,

As clock "A" is accelerated towards earth, even though,

- **Its speed relative to earth and clock "B" will be continuously increasing, and**
- **It will be in an accelerated state of motion, going away from clock "B", and**

- **It will be experiencing a stronger gravitational force dragging it towards earth, than what is experienced by clock "B",**

yet, it will gradually experience the passage of 'Time' at faster and faster paces, as compared to clock "B", let alone compared to clock "C".

The following figure is a very simple and yet exaggerated presentation of the chart that will be shown on the monitor of the computer receiving and calculating the differences between the rates at which 'Time' is experienced by the three atomic clocks used. The timing of the chart starts from the instant clock "A" starts its accelerated (powered) fall directly towards earth.

As shown in the figure, the amount of 'Time' difference between the two clocks "B" and "C" will increase at a constant rate, since those two clocks will experience the passage of 'Time' at slightly different but fixed rates.

However, as clock "A" starts its powered acceleration towards earth it will gradually experience the passage of 'Time' at faster and faster rates, as compared to clock "B" which is already experiencing 'Time' at a slightly faster pace as compared to clock "C".

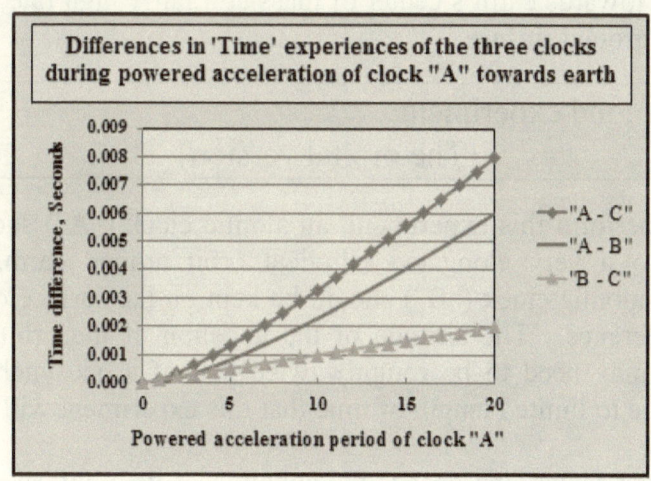

Such an outcome is clearly in contradiction with what is expected / predicted by the Special

<u>Theory of Relativity, the Equivalence Principle and the General Theory of Relativity. Since, they all predict that in such a scenario, clock "A" must be experiencing the passage of 'Time' at gradually slower paces, as compared to the rate clock "B" or clock "C" experiencing its passage.</u>

If the rocket used to accelerate clock "A" be powerful enough so that it can catch up with the local aether that is flowing towards earth, clock "A" will experience the passage of 'Time' the fastest. In fact, at that instant, clock 'A" will experience the passage of 'Time' as if it were positioned at an infinite distance from earth.

Depending on the precision / accuracy of the atomic clocks used, as well as those of the timings of the signals received / processed by the computer, such an experiment can be performed on a much smaller scale, since only the difference in the rates at which the two clocks "A" and "B" experience the passage of 'Time' is of prime interest.

Even a small rocket launched from a helicopter would be quite adequate to perform such an experiment, accelerating clock "A" towards earth, while clock "B" is kept onboard the helicopter, so long as the rocket used has sufficient propellant to accelerate directly towards earth's center of mass at a fairly high rate, until it hits the ground surface.

> • Second experiment:
>   (The orbital version)

To perform this experiment, an atomic clock ("A") should be sent into a very elongated elliptical orbit around earth, while another atomic clock ("B") should be kept on the surface of earth as a reference. The altitude of the aphelion of the orbital path chosen may need to be roughly twice that of a geosynchronous orbit, due to limited length of time that this experiment will take to perform.

As the rocket approaches the aphelion of its elliptical orbit, it must reverse its thrust and slow its orbital speed down, change direction and **accelerate directly towards earth**. The powered acceleration of the rocket can be of any rate, yet it is preferable that it be at such a rate that, as it runs out of propellant, its speed towards earth becomes equivalent to that of scape velocity from earth's gravity (at its then current altitude), before it reaches earth's atmosphere. The following figure shows such a scenario.

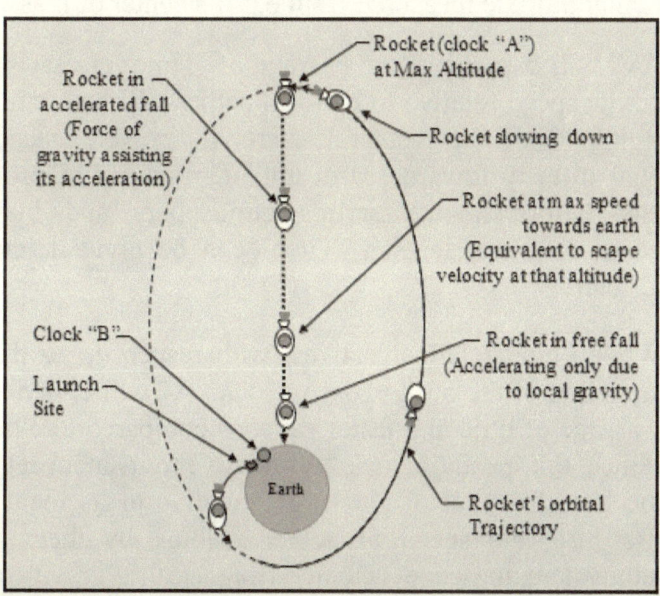

As clock "A" reaches the aphelion of its orbit and starts its powered accelerated descent directly towards earth, the two clocks must transmit their current 'Times' to a computer which should be positioned next to clock "B". The computer needs to calculate the true difference between the rates at which the two clocks experience the passage of 'Time', by taking into account the instantaneous distance of clock "A" as it accelerates towards earth (as well as earth's rotational speed at the altitude and latitude of the location where the computer and clock "B" are positioned), and show the result as a graph on its screen.

According to aether-based theories, presented in this book, the following results are expected to be graphically shown on the computer screen, during such an experiment:

Clock "B" <u>will experience the passage of 'Time' at a constant pace,</u> due to its speed relative to its local aether being constant.

Since, clock "B" is in the path of an aether flow which is going through its position at a speed equivalent to the vector sum of earth's rotational speed at its fixed latitude and altitude and the flow speed of aether which is directly towards earth's center of mass and is giving rise to earth's force of gravity at that particular altitude measured from earth's center of mass.

Clock "A" will experience the passage of 'Time' at varying rates. Since, its speed relative to its local aether will be continuously changing, from the instant it starts its powered accelerated descent directly towards earth until it runs out of propellant, shortly before it hits earth's atmosphere and burns up. However, its experiencing 'Time' can be divided into three major parts.

1- While being relatively at rest with respect to earth, at the aphelion of the orbit chosen, clock "A" will experience the passage of time at a faster pace, as compared to clock "B". Since, the speed of clock "A" relative to its local aether will be less than that of clock "B" relative to its local aether. Because, the speed of aether rushing by them will be equivalent to scape velocity from earth's gravity at their respective altitudes measured from earth's center of mass.

For this part, the prediction made by the General Theory of Relativity is in agreement with that of the aether-based theories, but due to different reasoning. Since, according to the General Theory of Relativity clock "A" will experience the passage of 'Time' at a faster pace because of being exposed to a weaker gravitational force, as compared to clock "B", due to being farther away from the center of mass of earth.

2- Once the rocket starts to accelerate towards earth, due to being pulled by earth's gravity and also being pushed by its own engines, clock "A" will experience the passage of time at gradually faster paces as compared to clock "B". Since, the acceleration of the rocket due to its engines will cause it to speed up over and above its acceleration which is due to being dragged by the accelerated flow of aether towards earth. Therefore, the speed of the rocket (clock "A") relative to its local aether will gradually decrease. In short,

The rocket, along with clock "A", will gradually catch up with the local aether which is accelerating towards earth's center of mass.

Consequently, as the speed of clock "A" relative to its local aether gradually decreases, it will experience the passage of 'Time' at faster and faster paces. Since, it is the speed of that clock relative to its local aether that dictates how fast it will experience the passage of 'Time'. In other words,

As clock "A" is accelerated towards earth and its speed is increased relative to earth, and while is in an accelerated state of motion, it will experience the passage of 'Time' at faster and faster paces, as compared to clock "B", which is stationary on the surface of earth.

The following figure is a very simple presentation of the chart that will be shown on the monitor of the computer.

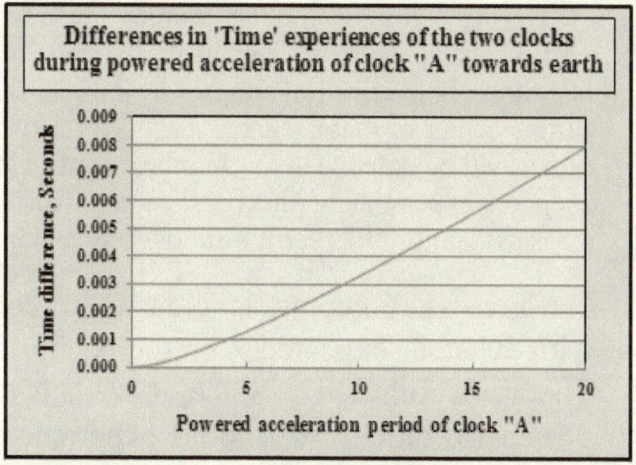

Such an outcome is clearly in contradiction with what is expected / predicted by the Special Theory of Relativity, the Equivalence Principle and the General Theory of Relativity. Since, they all predict that in such a scenario, clock "A" will experience the passage of 'Time' at gradually slower paces, to the point that it will be even slower than the rate clock "B" experiences its passage. Because, speed of clock "A" in space will be much faster than that of clock "B". Clock "A" will also be in a state of acceleration, over and above that of the local force of gravity.

3- Once the speed of the rocket towards earth becomes exactly equal to that of aether flowing towards earth, clock "A" will experience the passage of 'Time' the fastest. In fact, at that instant, clock 'A' will experience the passage of 'Time' as if it were positioned at an infinite distance from earth.

Since rocket engines will run out of propellant, as its speed becomes equivalent to that of its local aether flow towards earth, clock "A" will continue with its accelerated descent towards earth but in a state of free fall, as it will be dragged along by the accelerated flow of aether towards earth's center of mass.

During this portion of its journey towards earth, clock "A" will keep on experiencing the passage of 'Time' the fastest, as if it were hovering at an infinite distance from earth's center of mass. Since, its speed relative to its local aether will be equal to zero. In other words, clock "A" will keep on experiencing the passage of 'Time' faster than clock "B" until it hits earth's atmosphere and burns up.

Such an outcome is clearly in contradiction with what is expected / predicted by the Special Theory of Relativity. Since, it predicts that in such a scenario clock "A" should experience 'Time' at gradually slower paces, as compared to clock "B" which is stationary on the surface of earth, that is until it hits the atmosphere and burns up. Because, speed of clock "A" in space will be much faster than that of clock "B".

Mr. Einstein is quoted as having said,

"No amount of experimentation can ever prove me right; a single experiment can prove me wrong."

From the above quotation it is clear that, even though performing such experiments were not quite feasible during the first half of the twentieth century, Mr. Einstein was already fully aware of such potential scenarios.

Effects of Magnetic Field
on
the Passage of 'Time'

> ● First experiment
> (Magnetically induced 'Time Dilation', on a planetary scale)

Suppose three atomic clocks, "A", "B" and "C", are positioned at the magnetic equator, the magnetic North Pole and the magnetic South Pole, respectively, exactly as they are shown in the figure below.

The clock at the magnetic equator can be positioned anywhere along the magnetic equator, however for ease of reference with respect to the geographic equator, it can be positioned either in Kenya or in the Kiribati Islands, since those are the locations where the geographic equator and the magnetic equator intersect each other, located nearly exactly opposite of each other on the globe. Once the three clocks are in position, they can be synchronized and left alone to transmit the rate at which they experience the passage of time.

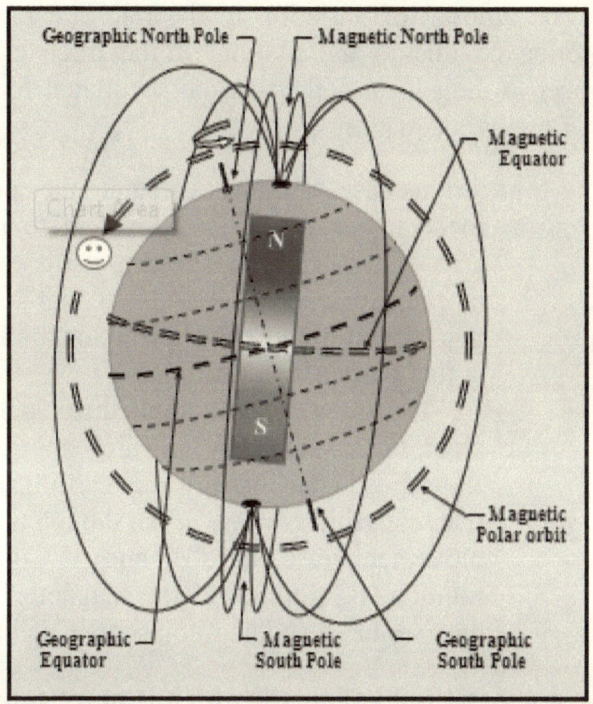

The duration of such an experiment can be a few days, a few months or even a few years, depending on the affecting phenomenon of interest, as well as the accuracy desired. Since, over time, these three clocks will demonstrate the effects associated with different phenomena such as:

- **Earth gravity, which is slightly less at the equator as compared to what it is at the two Poles**, due to slight elevation difference of about 11 Kilometers.

  The variation in the rate at which time is experienced due to this phenomenon is relatively easy to isolate. Since, clock "B" and clock "C" located at the magnetic poles experience slightly stronger force of gravity, which is due to faster local aether flow. Therefore, they will register the same amount of difference, running slower, as compared to clock "A" which is located at the magnetic equator.

- **The motion of clock "A" located at the equator and those of clocks "B" and "C" located at the magnetic Poles, due to earth's spin.**

The amount of variation in the rate at which time is experienced due to the motion of the three clocks with respect to their local aether can be accounted for by using Mr. Lorentz's equation.

Note that, in this case, one must calculate the speed of the local aether which is equivalent to the vector sum of aether's flow speed towards earth which is equal to escape velocity at that location (at that altitude) and earth's rotational speed (also, at that particular location).

The same must be applied to clock "B" and clock "C", since they too are experiencing the flow speed of aether towards earth that is giving rise to earth's gravity. Even though the physical movements of clocks "B" and "C", due to earth rotation, is minimal at the magnetic poles, yet they should be taken into account, as well.

- **The magnetic field of earth, which is more concentrated at the two magnetic poles,** as compared to all other places on the surface of earth, especially near and at the magnetic equator.

This effect is of prime concern in this particular experiment. Because, even though both magnetic poles experience the very same magnetic strength, yet they do differ in their effects on the atomic clocks used. This is due to the fact that, magnetic field of earth which is another type of aether flow, exits earth at one pole and enters earth at the other pole.

The expected flow speed of aether at the poles, due to the strength of earth's magnetic field, can be calculated using the results obtained in the experiments proposed in the chapter titled "What is Magnetic Field?". It involves measuring variations induced in the speed of light as it propagates in the same direction and in the opposite direction of an artificially generated magnetic field of known strength.

Once the effects of earth's rotation and gravitation are taken into account, it will become clear that the passage of time experienced by clock "B" and clock "C" will be off as compared to clock "A" by nearly very same amount but in opposite directions. The results obtained from such an experiment will confirm that the local aether flows at the poles are different with respect to each other due to the partial contribution attributed to the magnetic field of earth.

Note that, such an experiment will also clarify the direction of aether flow <u>outside a magnet</u> as being from the North Pole towards the South Pole, or the other way around.

- Second experiment:
  (Time anomaly experienced by clocks onboard satellites in polar orbits)

The aim of this experiment is to explain a specific type of time anomaly which is experienced particularly by satellites on orbital paths that take them nearly over the magnetic poles. This anomaly can be described as follows:

"As a satellite's orbital path takes it over both magnetic Poles, regardless of its altitude, on the southbound portion of its orbital journey it experiences time at a bit faster pace, as compared to the northbound portion of its orbital journey."

This anomaly can be referred to as,

"Esmailzadeh Magnetic Polar Orbit Time Anomaly"

For this experiment two atomic clocks are required, one should be onboard a satellite in an orbital path that takes it over the magnetic poles of earth and the other should be stationed on the surface of earth.

Note that, this experiment can be readily performed using any of the already existing satellites which are in Polar orbits around earth. Since, every so often, such satellites pass directly over earth's magnetic poles, as earth literally spins under them.

During the northbound portion of its orbital journey around earth, such a satellite (and the clock onboard) will be moving faster relative to its local aether medium, as compared to when it tracks the southbound portion of its journey. This is due to the fact that,

> **As a satellite in a polar orbit tracks the northbound part of its orbital path it will be moving against the aether flow associated with earth's magnetic field. However, during the southbound portion of its orbital path the satellite will be moving with this aether flow.**

The following figure shows one such orbital path, during which the satellite passes nearly over the magnetic poles of earth.

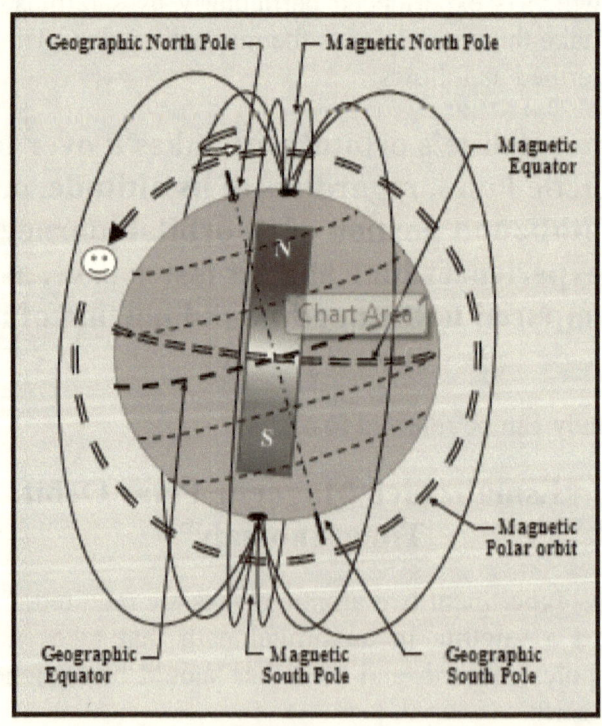

Therefore, the clock onboard the satellite will experience and indicate the passage of 'time' to be slightly slower as it travels towards the northern magnetic pole as compared to when it travels towards the southern magnetic pole.

Note that, nearly all of the other satellites in orbit around earth (except for those orbiting earth right at the magnetic equator), also experience such an effect, but to lesser degrees. The severity of such an experience by any given satellite depends on how close its orbital path takes it to the magnetic poles. The closer it passes over the magnetic poles the stronger it will experience such an effect.

> • Third experiment:
>   (Magnetically induced 'Time Dilation', on a
>   Laboratory scale)

Suppose an atomic clock "A" is placed in between the two poles of a very strong horseshoe-shaped electromagnet. Also, another atomic clock "B" is placed at far enough a distance so that it is not affected by the local magnetic field and hence can serve as a reference. Such a setup is shown in the following figure.

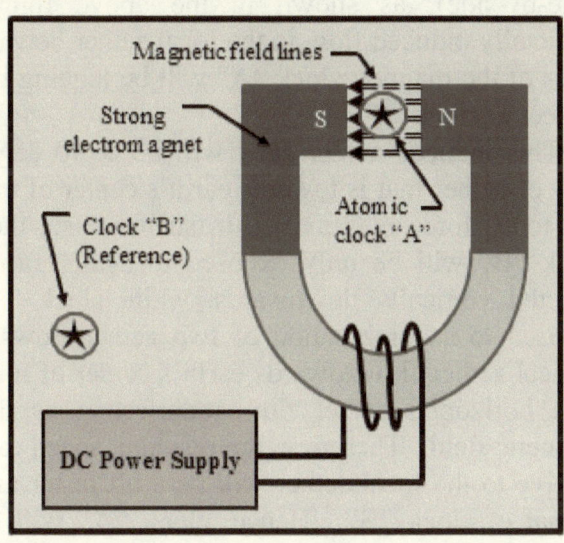

Note that, if need be, identical amounts of radioactive materials can be used instead of the two atomic clocks.

Once the two clocks are synchronized, the electromagnet must be energized. Over time, clock "A" which is placed in the magnetic field will reveal that it is experiencing the passage of 'Time' at a different rate as it is experienced by clock "B". Also, by activating the electromagnet with different voltages, it can be demonstrated that the rate at which 'Time' is experienced by clock "A" will depend on the strength of the magnetic field to which it is exposed. In other words,

"The very presence of a magnetic field gives rise to what is known as 'Time dilation'."

The time dilation induced in such an experiment is due to the induced aether flow which manifests as the magnetic field present, a flow which gives rise to a relative motion between the local aether and the clock.

However, it must be emphasized that, such an experimental setup can give quite different results, depending on the orientation of the magnet relative to the vertical line.

- **If the electromagnet poles are oriented horizontally** (side-by-side), as shown in the above figure, due to artificially induced flow in the local aether between the two poles of the magnet, clock "A" will be lagging behind with respect to clock "B".

  This induced aether flow will be at 90 degrees to the flow of aether that is towards earth's center of mass, giving rise to its force of gravity. In such a case, the reference clock "B" will be only exposed to aether flow which is directed vertically downwards, while clock "A" will be exposed to a combination of two aether flows, namely a vertical aether flow towards earth's center of mass, as well as a horizontal aether flow manifesting as the induced magnetic field. Therefore, the resultant speed of clock "A" relative to the local aether will be a bit faster as compared to that of clock "B". Hence, **clock "A" will experience 'time' at a slower pace as compared to clock "B".**

- **If the electromagnet poles are oriented vertically** (one above the other), two different scenarios must be considered:

  1- If the South Pole of the electromagnet is at the top and its North Pole is at the bottom, the aether flow induced in between the poles will be upwards, against the flow of aether that is headed directly towards earth and is giving rise to its force of gravity. The overall speed of clock "A" relative to its local aether will be equal to the difference between the two flow speeds (the vertical aether flow towards earth's center of mass, and the vertical aether flow away from earth, manifesting as the induced magnetic field). These two aether flows will partly cancel each other out.

  Therefore, the overall speed of clock "A" relative to the local aether will be less than that of clock "B". Hence, **clock "A" will experience the passage of 'time' at a bit faster pace as compared to clock "B".**

  2- If the North Pole of the electromagnet is at the top and its South Pole is at the bottom, the aether flow induced in between the poles will be downwards, in the very same direction as the flow of aether that is headed directly towards earth and is giving rise to its force of gravity. Therefore, the overall speed of clock "A" relative to its local aether will be equal to the sum of the two flow speeds (the vertical aether flow towards earth's center of mass, and the vertical aether flow manifesting as the induced magnetic field). These two aether flows will complement each other. Therefore, the overall speed of clock "A" relative to its local aether will be faster than that of clock "B". Hence, **clock "A" will experience the passage of 'time' at a bit slower pace as compared to clock "B".**

- **If the electromagnet is oriented at any other angle**, the direction of the aether flow manifesting as the magnetic field with respect to the flow of aether that is directly headed towards earth's center of mass and is giving rise to its force of gravity must be taken into account.

Note that, such an experiment will also clarify the direction of aether flow manifesting as the magnetic field as being from the North Pole towards the South Pole, or the other way around.

During such an experiment, the relationship between the strength of the magnetic field and the magnitude of the time dilation induced can be determined. Then, an experimental setup can be designed that would be able to handle the required high voltage, generated by a very large capacitor bank, to slow down the rate at which time is experienced by any desired amount, or **to literally stop the flow of time**, even if momentarily.

Of course, during an experiment testing such an extreme case, the very structural integrity of the atomic clock can be jeopardized. Such an experience can also have a range of unknown consequences, including **the disappearance of the clock from this universe**. Also, all of the charged particles, namely electrons and protons, in the atoms of anything put in such an environment can be readily ripped off of their atomic bindings, as well.

Note that, **if such an experiment is performed using identical amounts of radioactive materials** (instead of the atomic clocks), the materials used should have sufficiently short half-lives so that the effects can be readily measured after a short period of exposure to such an environment. The decay rates of the materials used will clearly indicate the effects of a magnetic field on the rate at which the passage of time is experienced by such sensitive materials. In other words,

"Strong and steady magnetic fields can be used to extend the half-lives of radioactive isotopes and those of unstable particles."

Of course, as demonstrated in this experiment, the direction of the induced magnetic field in relation to the local aether flow must be taken into account.

- Fourth experiment:
  (Magnetic field powered 'Time Dilation' machine)

To slow down the aging process of objects and/or living beings, there is no need to build a starship capable of flying at near the speed of light to go for an interstellar journey and/or orbit a black hole and risk being pulled in by its force of gravity. One can accomplish this task of experiencing the time dilation right here on earth.

By applying the information provided in this chapter, one can build a sort of time machine which would accomplish such a task. The following drawing shows a simple presentation of such a device.

He/she needs to generate a **strong and constant magnetic field** in a closed space such as a room (which can serve as the bedroom, office, living room or even as the whole house) in which the objects of interest including living beings can spend most of their times, and so benefit from its induced effects. The longer time an individual spends in the space equipped with such special effects, the more of an age difference he/she will experience as compared to those who do not enter such a device/space.

It must be mentioned that,

If the magnetic field generated be a variable one, there will be a major issue/concern with living beings using such a 'time dilation inducing machine'. Since, the accelerated flow of aether which is manifesting as the variable magnetic field will also generate a drag force along its alternating flow directions. This drag force will be thousands of times stronger than the regular g-force experienced on the surface of earth.

Therefore, objects and living beings placed inside such chambers and exposed to such a super strong variable magnetic field will expectedly cyclically weigh thousands of times heavier than their normal weights under normal conditions on the surface of earth. Consequently, they will be literally squeezed alternately against two opposing sides of the chamber they are in, the same sides towards which the induced aether flow is alternately headed.

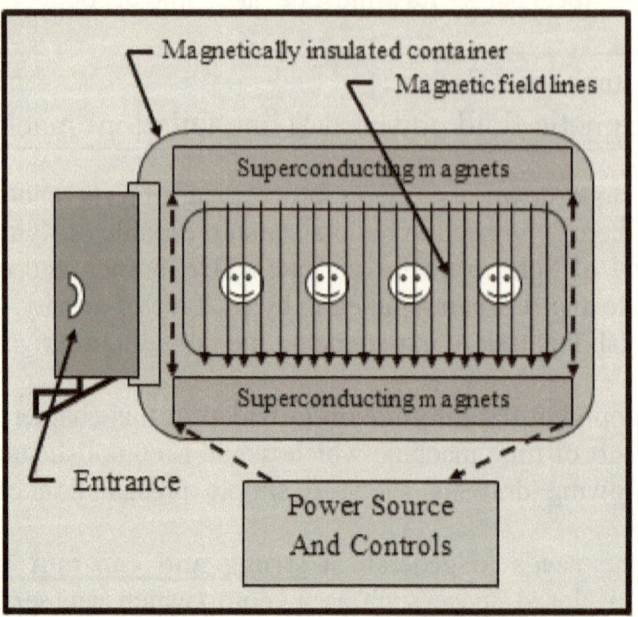

Effects of Electric Field
on
the Passage of 'Time'

- First experiment:
  (Using electric field to induce 'Time Dilation')

Suppose an atomic clock "A" is placed in an electric field generated between two plates which are connected to the two leads of a DC power supply. And, another atomic clock "B" is placed at far enough a distance so that it is not affected by the local electric field and hence can serve as a reference. Such a setup is shown in the following figure.

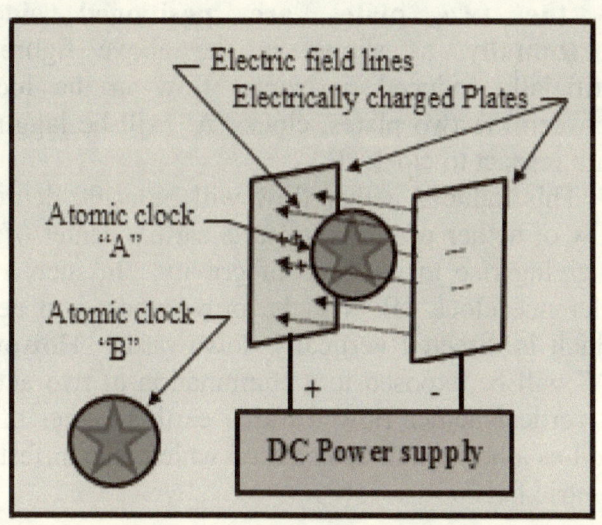

Note that, if need be, identical amounts of radioactive materials can be used instead of the two atomic clocks.

First, the two clocks must be synchronized, while the electric current is disconnected from the two plates. Then, the electric current of an arbitrary voltage setting must be supplied to the plates.

Over time, clock "A" which is exposed to the electric field will reveal that it is experiencing the passage of 'Time' at a different rate as compared to clock "B". By supplying the two plates with different voltages, it can be demonstrated that the rate at which 'Time' is experienced by clock "A" will depend on the strength of the electric field to which it is exposed. In other words,

"The very presence of an electric field can give rise to what is known as 'Time dilation'."

The time dilation induced during such an experiment is due to the induced aether flow which is manifesting as the electric field present, a flow which gives rise to a relative motion between the local aether and the clock.

However, it must be emphasized that, such an experiment can give rise to different results, depending on the orientation of the two plates relative to the vertical line.

- **If the two plates are positioned side-by-side, horizontally,** as shown in the above figure, due to artificially induced horizontal flow in the local aether between the two plates, clock "A" will be lagging behind with respect to clock "B".

  This induced aether flow will be at 90 degrees to the flow of aether which is towards earth's center of mass and is giving rise to its force of gravity. In such a case, the reference clock "B" will be only exposed to aether flow which is directed vertically downwards. However, clock "A" will be exposed to a combination of two aether flows (a vertical aether flow towards earth's center of mass, as well as a horizontal aether flow which is manifesting as the electric field).

Therefore, the resultant speed of clock "A" relative to its local aether will be a bit faster as compared to that of clock "B" relative to its local aether. Hence, **clock "A" will experience the passage of 'time' at a slower pace, as compared to clock "B".**

- **If the two plates are oriented vertically** (one above the other), two different scenarios must be considered:

1- **If the positively charged plate is at the top and the negatively charged plate is at the bottom**, the aether flow induced in between the two plates will be upward and against the flow of aether which is headed directly towards earth, giving rise to its force of gravity. The overall speed of clock "A" relative to its local aether will be equal to the difference between the two flow speeds (vertically downward flow of aether towards earth's center of mass, and vertically upward flow of aether manifesting as the induced electric field). These two aether flows will partly cancel each other out.

Therefore, the overall speed of clock "A" relative to its local aether will be less than that of clock "B" relative to its local aether. Hence, **clock "A" will be expected to experience the passage of 'time' at a bit faster pace, as compared to clock "B".**

2- **If the negatively charged plate is at the top and the positively charged plate is at the bottom**, the aether flow induced in between the two plates will be in the very same direction as the flow of aether which is headed directly towards earth, giving rise to its force of gravity. Therefore, the overall speed of clock "A" relative to its local aether will be equal to the sum of the two flow speeds (vertically downward flow of aether towards earth's center of mass, and vertically downward flow of aether manifesting as the induced electric field). These two aether flows will complement each other.

Therefore, the overall speed of clock "A" relative to its local aether will be more than that of clock "B" relative to its local aether. Hence, **clock "A" will be expected to experience the passage of 'time' at a bit slower pace, as compared to clock "B".**

- **If the two plates are oriented at any other angle**, their angle relative to the vertical, as well as the direction of the induced aether flow with respect to the flow of aether which is directly headed towards earth's center of mass must be taken into account.

Note that, such an experiment will also clarify the direction of aether flow as being from the positive plate towards the negative plate, or the other way around.

During such an experiment, the relationship between the strength of the electric field and the magnitude of the time dilation induced can be determined. Then, an experimental setup can be designed that would be able to handle the required high voltage, generated by a very large capacitor bank, to slow down the rate at which time is experienced to any desired amount or **to literally stop the flow of time**, even if momentarily.

Of course, during an experiment testing such an extreme case, the very structural integrity of the atomic clock can be jeopardized. Such an experience can also have a range of unknown consequences, including **the disappearance of the clock from this universe**. Also, all of the charged particles, namely electrons and protons, in the atoms of anything put in such an environment can be readily ripped off of their atomic bindings, as well.

Note that, if such an experiment is performed using identical amounts of radioactive materials (instead of the atomic clocks), the radioactive materials used should have sufficiently short half-lives so that the effects can be readily measured after a short period of exposure to such an environment. The decay rates of the materials used will clearly indicate the effects of the electric field on the rate at which the passage of time is experienced by such sensitive materials. In other words,

"Strong and steady electric fields can be used to extend the half-lives of radioactive isotopes and those of unstable particles."

Of course, as demonstrated in this experiment, the direction of the induced electric field in relation to the local aether flow must be taken into account.

- Second experiment:
  (Electric field powered 'Time Dilation' machine)

To slow down the aging process of objects and/or living beings, there is no need to build a starship capable of flying at near the speed of light to go for an interstellar journey and/or orbit a black hole and risk being pulled in by its force of gravity. One can accomplish this task of experiencing the time dilation right here on earth.

By applying the information provided in this chapter, one can build a sort of time machine which would accomplish such a task.

He/she needs to generate a **strong and constant electric field** <u>(using DC power source)</u> in a closed space such as a room (which can serve as the bedroom, office, living room or even the whole house) in which the objects of interest including living beings can spend most of their times, and so benefit from its induced effects. The following drawing shows a simple presentation of such a device.

The longer time an individual spends in the space equipped with such special effects, the more of an age difference he/she will experience as compared to those who do not enter such a device/space.

It must be mentioned that,

If the electric field generated be a variable one, there will be a major issue concerning living beings using such a 'time dilation inducing machine'. Since, the accelerated flow of aether which is manifesting as the variable electric field will also generate a drag force in its alternating flow directions. This drag force will be thousands of times stronger than the regular g-force experienced on the surface of earth.

Therefore, objects and living beings placed inside such chambers and exposed to such a super strong variable electric field will expectedly cyclically weigh thousands of times heavier than their normal weights under normal conditions on the surface of earth. Consequently, they will be literally squeezed alternately against two opposing sides of the chamber they are in, the same sides towards which the induced aether flow is alternately headed.

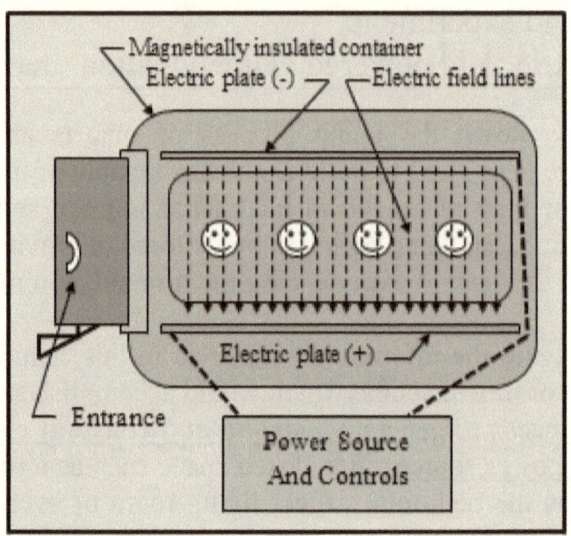

Conclusion

According to the information, as well as the experiments presented in this chapter,

"The rate at which 'time' is experienced by an object (or a living being) depends on its instantaneous speed relative to its local aether medium, as compared to the propagation speed of phase vibrations in that medium."

As the speed of an object, relative to its local aether approaches that of phase vibrations in that medium, that object experiences the passage of time at progressively slower paces. In case the speed of the object reaches the speed of phase vibrations in its local aether medium, the passage of time will no longer be experienced by that particular object.

Even if the object stands relatively still in one place, anywhere in this vast universe, and due to affecting phenomena such as gravitational, electric or magnetic fields which may be present, the local aether is encouraged to move passed it, the resulting effect will be the very same and identical. That is, as the induced speed in the aether is increased, relative to the object in mind, the passage of time experienced by that object will be at slower and slower paces. Again, as the speed of aether relative to the object reaches the speed of phase vibrations in the local aether medium, the object will stop experiencing the passage of time, altogether.

Also, the speed of phase vibrations is dependent on the density of the aether medium. Denser aether mediums limit the speed at which phase vibrations propagate in that medium. Therefore, as the aether medium, as a whole, is expanding and its density is decreasing on a universal scale, the speed of phase vibrations in that medium is gradually increasing. Hence, the speed of objects relative to their local aether environment is gradually becoming a lesser fraction of the speed of the phase vibrations in that medium. In other words,

"Time is speeding up, on a universal scale."

Note that, the density of aether is not only different at locations that are far from each other, but also is gradually decreasing due to aether's ongoing expansion, as well as its leakage from this universe. Therefore,

"A Constant ABSOLUTE 'Universal Time' CANNOT be defined as being the 'Time' experienced by any and all objects that are at rest with respect to the aether that is in their immediate vicinities."

Currently, due to lower aether density and pressure in the accompanying universe, as compared to those of aether that is in this universe, the rate at which 'Time' is experienced in the accompanying universe is faster than it is experienced in this universe. However, as time goes on, the pressure of aether that is in this universe is decreasing and the pressure of aether that is in the accompanying universe is increasing.

Eventually, the two universes will be occupied by aether that is at the same pressure and has the same density. At that point in time, 'Time' will be experienced at the same pace in both universes. The equal pace of experiencing 'Time' in the two universes will continue into the eternal future, but at an ever faster pace. Since, aether's density will continue to decrease indefinitely, due to the ongoing expansion of that medium.

It is concluded in this chapter that,

"Physical 'Time Travel' is not possible."

However, **spiritual 'Time travel' is quite possible.** In fact, it has been performed by many individuals in the past, individuals who were capable of seeing the events which had occurred in their past as well as events that were going to take place in their future. Currently, there are individuals who are capable of performing such tasks, as well.

It must be emphasized that,

"Any event that is seen, as in spiritually viewed (not to be mistaken as being imagined), to take place in the future, will take place at the time and place it is seen to take place."

Because, even the very knowledge of the future events becomes part of the information needed for certain actions to be taken so that they eventually lead to the occurrence of such foreseen future events. In other words,

"The future that has been seen cannot be altered."

Important note:

To obtain accurate results, the object's speed in the Lorentz transformation equations, used to calculate the time dilation, must be relative to its local aether, and not relative to space, and definitely not relative to another object such as a planet, a star, a galaxy or even the birthplace of this universe.

Also, it must be particularly emphasized that,

According to the information presented in this book and the results of the experiments proposed to verify their validities,

"Even though force of gravity and uniform rate of acceleration give rise to certain identical effects, such as the weight of objects, they have nothing in common as far as inducing 'Time Dilation' is concerned."

Therefore,

"The paradoxes / twin stories about accelerating objects or living beings that experience time at a slower pace, as compared to those left relatively stationary on their home planets,

are fundamentally flawed.

Since, such paradox stories totally ignore the actual speeds at which such objects or living beings are moving relative to their respective local aether mediums."

It is the instantaneous speed of an object relative to its local aether that gives rise to 'Time Dilation' effect and not its instantaneous accelerations.

Note that, even though 'Time' can be measured using a variety of scales, it is a continuous phenomenon. It is not made of small discrete pieces, as objects are made of molecules / atoms / particles.

What
is
Light?

Introduction

Light waves that are recognizable by human eyes make up only a very narrow section of the electromagnetic waves spectrum. Electromagnetic waves also include radio waves, microwaves, X-rays, Gamma rays and so on. Electromagnetic waves in a variety of ways testify to the existence of the contents of this universe and make them visible to or detectable by each other.

Over the centuries, a variety of theories have been proposed to describe how light and other electromagnetic waves behave in different situations. Mr. Newton believed that light was made of particles and different colors of light were due to the variety that existed in the sizes of these particles. While, the nineteenth century scientists believed that light was a kind of wave, since it behaves and demonstrates properties that are similar to those of sound waves.

In 1865, Mr. Maxwell proposed his theory on electromagnetism in which he introduced light and other electromagnetic waves as a kind of phase vibration and formulated their behavior, the very same formulas that are still valid. Mr. Maxwell had based his theory on the existence of some kind of a medium through which light and other electromagnetic waves propagate. At that time, this medium was referred to as aether. What aether was made of or what kinds of properties it had were not known. Aether was considered as a necessary medium through which electromagnetic waves travel. To electromagnetic waves, the medium of aether was assumed to be just as a medium such as air is to sound waves.

Over time, scientists became curious about aether and based on the assumption that it is stationary in space, and planets are moving through it, by conducting a variety of experiments, they tried to detect its drift effect on the surface of earth. Since, they knew that earth is orbiting the sun at about 107,000 kilometers per hour. Over decades, a variety of attempts were made by different scientists to detect and measure this drift velocity.

The assumed motion of earth in the stationary medium of aether is shown in the figure below.

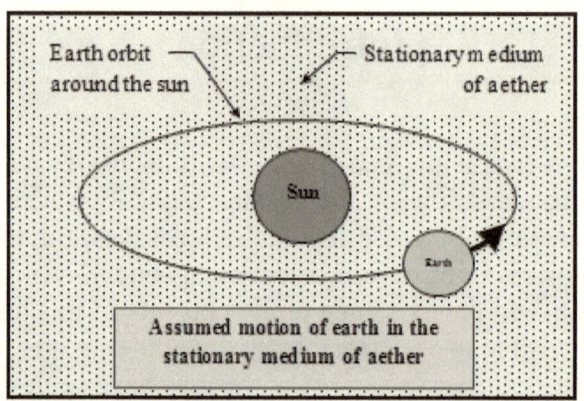

The most promising of all the proposed experiments were designed and performed by Mr. Michelson and Mr. Morley in 1887. As shown below, their experiment was based on light waves that were split into two branches which were then made to complete a round trip journey in different directions, one in the direction of the earth's motion around the sun and the other at 90 degrees to the said direction.

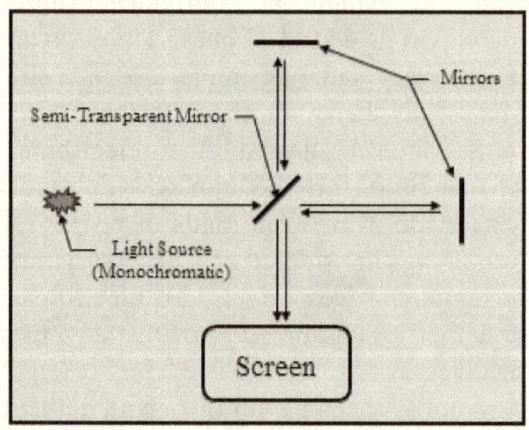

By reflecting these two light branches back and superimposing them on each other, Mr. Michelson and Mr. Morley were expecting to see the formation of some sort of an interference pattern. They conducted their experiment several times. They even performed their experiment up at high altitudes, with the help of a balloon. However, every time the result was the very same and no interference pattern was formed in any of their attempts.

In 1904, Mr. Fitzgerald and Mr. Lorentz, independently, provided reasoning for detecting no trace of any drift effect in all of the experiments. They proposed that the length of the apparatus's arm which was in the direction of earth's motion was affected by the motion of earth and therefore had resulted in annulling the intended effect of the aether drift. It must be emphasized that, Mr. Lorentz and Mr. Fitzgerald had proposed their theories without rejecting the existence of the aether medium, which was commonly accepted by the then-current scientific community.

In 1904, Mr. Lorentz proposed a theory based on which the effect of motion on the apparatus was clearly explained. Particularly, the effects such as the shortening of the length of the object in the direction of motion, as well as the slowing down of the passage of time, as one approaches the speed of light in a vacuum, were first proposed by Mr. Lorentz' theory and his transformation equations. These transformation equations are also known as Mr. Lorentz' Relativity Theory or "Principles of Relativity".

In 1905, Mr. Einstein introduced his special theory of relativity which was basically the same as the theory of relativity proposed by Mr. Lorentz, a year earlier. However, Mr. Einstein had added two extra postulates:

1- Light is made of particles, namely photons, and therefore does not require a medium such as aether to propagate.

2- The speed of light in a vacuum is the maximum possible speed and it is a constant which is independent of the observer's state of motion relative to the light source.

Note that, the experiments performed by Mr. Michelson and Mr. Morley, by no means had indicated that there was no aether in space, or that the speed of light was the maximum speed possible.

A revised version of the experiment conducted by Mr. Michelson and Mr. Morley is described in great details later on in this chapter.

Since Einstein, some scientists have accepted that light is made of particles, namely photons which by definition have no mass. In other words, a definition that in itself is contradictory. Photons are supposedly some kind of **particles** (a real physical entity) that supposedly actually exist, yet as they get stopped, there is nothing there. That is, no mass is gained by the absorbing target, regardless of how intense the light beam may be or for how long the absorption process may last. Of course, the target may heat up indicating the absorption of the energy received.

The majority of the present-day scientists believe that light is both, namely particle and wave. Their view is due to their attempts in trying to quantize the light phenomenon so that they can fit it in the equations of the quantum theory. They treat light as a wave when IT suits better, and treat it as particles when IT suits better.

However, light cannot have two identities that are totally different, particularly two identities that in fact oppose each other. Since, for a particle to exhibit a wave type of motion it has to either lose energy or be continuously acted upon by an outside force. Neither one of these scenarios applies to the propagation of light in space.

Basically, today's physicists are not sure what light and other electromagnetic waves really are.

The Proposed Theory

In order to explain what light really is, one needs to propose a theory that can provide consistent explanations for all of the known and proven characteristics of light. Also, if possible, it should predict specific verifiable new findings about light and its characteristics. The theory proposed in these pages is essentially based on the theory that was accepted by the nineteenth century scientists. In other words,

"Light is a form of wave (phase vibration) in the medium of aether."

Light and other electromagnetic waves travel as phase vibrations through aether, just as sound waves travel in a medium such as air.

The aether that is referred to in this presentation is the very same aether that was believed to exist by the 19th century scientists. However, contrary to what the nineteenth century scientists had come to accept, the medium of aether is not a stationary medium in space. Aether medium is quite an active and dynamic environment. The variety of movements that exist in that medium can be likened to the movements of the air molecules that form the atmosphere or the variety of movements exhibited by the water molecules that form the oceans.

Even though light and other electromagnetic waves motivate aether to vibrate, they do not cause it to have a translational motion. Of course, due to a variety of other reasons, aether can have different types of motions in the same direction or against the direction that certain waves are propagating. Hence, it will automatically carry the waves and as a result will cause their apparent speeds to become faster or slower.

According to the theories introduced in this book, aether is also responsible for various physical and energetic phenomena such as the force of gravity, magnetic field and electric field, among many others. As an example, the force of gravity is due to the drag force induced by the accelerated flow of aether towards individual subatomic particles. The accelerated flow of aether automatically carries any particles and/or waves that may be in its path. The flow of aether and the dragging effect that it exerts on other objects as well as waves that happen to be in its path can be likened to the flow of air (wind) and how it drags sail boats or sound waves that may exist in its path.

These effects can be readily observed in the case of the circular waves generated on the surface of water that is in a lake, as compared to the circular waves generated on the surface of water that is in a river stream, when a pebble is dropped in. The following figure, on the left, shows the generation of such waves on the surface of a lake.

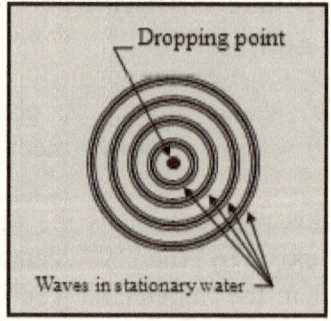 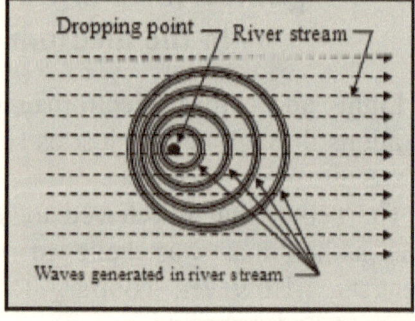

However, in the case of a river, as shown in the figure, on the right, the generated waves are carried by the flow of water. Therefore, they propagate faster in the downstream direction and propagate slower in the upstream direction.

According to the Proposed Theory

1- Light, being a phase vibration in the medium of aether can only propagate where aether exists.

2- The **actual speed** of light in aether is only dependent on the local aether density. Lower aether densities allow light and other electromagnetic waves to propagate faster.

3- The **actual speed** of light in the local aether is always the same and equivalent to the speed of phase vibrations in that medium, regardless of the medium being void of matter particles and other phenomena such as gravitational, magnetic or electric fields, or not.

4- The **apparent speed** of light, between its source and its receiver, is the fastest when it follows a straight path. Such a condition exists only at places that are far away from galaxies.

5- The **apparent speed** of light, between its source and its receiver, is reduced if the medium of aether is host to matter particles. The presence of matter particles, in the immediate vicinity, automatically induces a variety of cross flows in the local aether, causing light to follow a zigzag path. This type of situation is also experienced by waves that are formed on the surface of water flowing downstream in a river that has many rocks present at the water surface level, as shown below.

    Since, as water flow becomes deflected towards different directions due to encountering various obstacles such as rocks, waves mandatorily follow a zigzag path as they propagate in different directions. Hence, their apparent speed gets reduced.

6- The **apparent speed** of light, in any medium such as a gas, a liquid or a solid directly depends on:

- How densely the matter particles are packed, since they generate cross flows in the local aether which in turn cause the waves to follow a zigzag pattern,

- How the matter particles are arranged / organized, and

- Whether the matter particles are relatively rigid in their respective positions, as in a solid crystalline lattice structure or they are free to move around, as in a gas or a liquid.

7- As light is encouraged to follow a path through a crystalline structure, its **apparent speed** through that matter medium also depends on the direction of propagation of light, as well as the frequency of the light wave. Since, by entering a matter medium that has a fixed atomic (molecular) structure, particularly if it is of a crystalline type, as light waves are directed in different directions, they encounter different number of matter particles simply because the number density of matter particles in any given direction is different and it is fixed.

Light waves that enter a crystalline matter medium at a given angle encounter matter particles in **roughly** specific distances. It is stated as roughly specific distances because of the localized atomic vibrations which are in turn dictated by that medium's temperature. Such distances are an integer multiples of certain range of wavelengths (corresponding to certain range of frequencies). Therefore, light waves of those particular frequencies will be deflected / distracted more often or less often than the other frequencies, which means they will require longer time or shorter time to pass through the matter medium, respectively.

This type of effect becomes drastically more pronounced as the temperature of the matter medium is lowered to near absolute zero degrees Kelvin. At such low temperatures, the atoms have minimal vibrational movements and therefore cause a more specific frequency of light to experience the most zigzag deflections while passing through and therefore slow down to a crawling speed. Such a medium is referred to as the **Bose-Einstein Condensate**.

Note that, light waves at double or half of that particular frequency will also experience such slowing down of their respective propagation speeds as well, but to a lesser degree. While, light at all other frequencies will pass through normally.

8- The coherency of the light wave passing through any matter medium directly depends on the organization, as well as the uniformity in the distribution of matter particles that make up that medium. For example, a piece of glass that is made of a single and uniform crystalline structure (with flat, parallel surfaces on both sides) will allow the passage of light waves without causing it to disperse much. However, if a piece of glass is made of many crystalline pieces that are put together or it consists of many inhomogeneous regions, due to the manufacturing process used, every time light waves enter a different crystalline region, they get dispersed in different directions.

This is why images of objects seen through some crystalline materials can be quite crisp and clear, while when seen through others they may appear in duplicates. Other materials may barely allow the passage of light and only show a blurred image of the objects or simply cause the total blockage of the view.

9- Light, being a phase vibration, does not get absorbed by the matter particles, as it encounters them. Only the aether is absorbed due to flowing right into the matter particles. Of course, a very small fraction of the light waves, which are propagating perpendicular to the surface of the particle, are carried into the matter particles by the aether flow, as it enters them.

If matter particles are too densely populated and are not organized in a crystalline lattice structure, they will induce complicated deflections in the path of light waves. Therefore, as light waves pass through such a medium they spread in all directions, and blend together to the point that they weaken and eventually become totally unrecognizable.

10- Always some (even if it is reduced to a miniscule amount) light waves (electromagnetic waves, in general) cross a matter medium. The percentages of light waves' intensity that passes through a given matter medium depends on the frequency of the light wave and the internal atomic (molecular) structure of the medium, as well as its thickness.

It is due to this very reason that, to reduce the intensity of Gamma Rays and X-Rays a thick layer of Lead is used, while a thin layer of Aluminum or even wood is sufficient to relatively-speaking block the visible light waves.

11- The vibrations due to any and all light waves continue to spread and weaken, yet there will always be a reminiscence of them in this universe, just as the sound waves generated in the atmosphere anywhere on the planet spread in all directions and can eventually reach even the opposite side of the planet.

Light waves generated in aether and carried by that medium can even be likened to the waves generated in and carried by water in a river stream. Along their path, such waves encounter many rocks and obstacles of different sizes (as well as directional detours) and eventually their original wave pattern literally disperses into countless other waves.

Those waves are still there but they become so weak that their detection will only dependent on the sensitivity of the equipment used.

It is due to their sensitive wave detection (reception) abilities that humpback whales can communicate over distances that are in excess of 1,000 km and sing the very same songs.

Therefore, by using a sensitive enough a receiver for sound waves and a corresponding one for light waves, one should be able to not only hear the sounds that were generated, but also literally witness events that have occurred years ago. Since,

"Sound waves and light waves corresponding to the past events which have occurred on this planet of ours are literally still floating in earth's atmosphere and in space, respectively."

Effects Observed on the Universal Scale

The following are three of the most important effects experienced by light waves in this universe.

1- As light passes close enough to a galaxy (or even a star), its path bends towards that galaxy (or that star).

2- If light crosses the event horizon of a black hole, it will be pulled into the black hole.

3- The speeds at which galaxies are receding from earth are calculated using the amounts of red shifts that are detected in the light spectrum received from them.

As shown below, the above effects are quite similar to the effects experienced by sound waves in a medium such as air. For example, using two parabolic dishes, positioned so that the sound waves generated at the focal point of one can be heard at the focal point of the other, one can perform the following experiments.

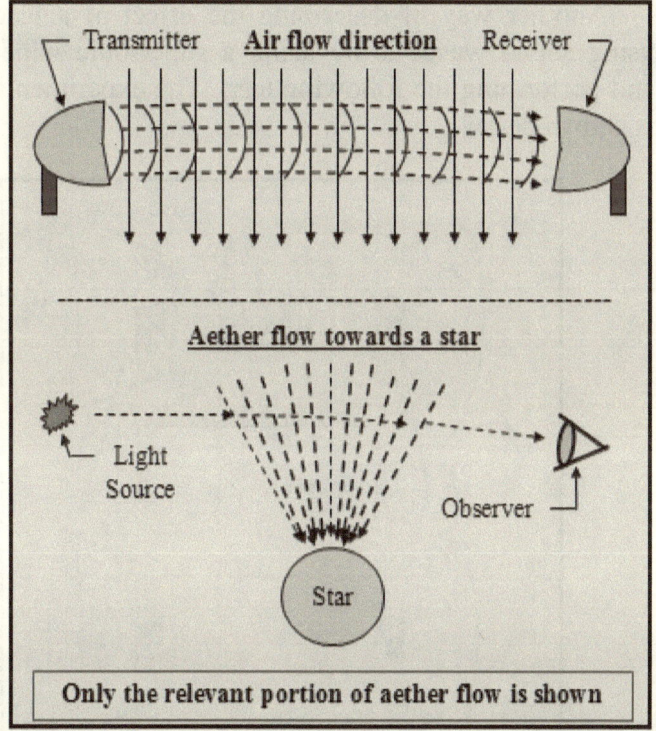

Only the relevant portion of aether flow is shown

- The effect of a gentle crosswind on sound waves traveling in air is the same as the effect associated with the existence of a star (or galaxy) near the path of light waves. Just as a breeze can literally drag and carry the sound waves downstream, the accelerated flow of aether towards a star also can drag the light waves towards that star.

- In case the wind is blowing at or faster than the speed of sound and it is in the direction that is in-line with the two parabolic dishes, from the receiver towards the source of the sound, the transmitted sound waves would never reach the receiver.

  This effect is analogous to the effect experienced by light waves that get too close to a black hole, since at the distance equivalent to the radius of the event horizon, the speed of the aether flowing inwards reaches the speed of phase vibrations in that medium. Therefore, light cannot propagate in the opposite direction, upstream, and gets dragged in.

Another way of describing the effect of a black hole using sound waves is by using a supersonic wind tunnel and performing the following test. The experimental setup is shown below.

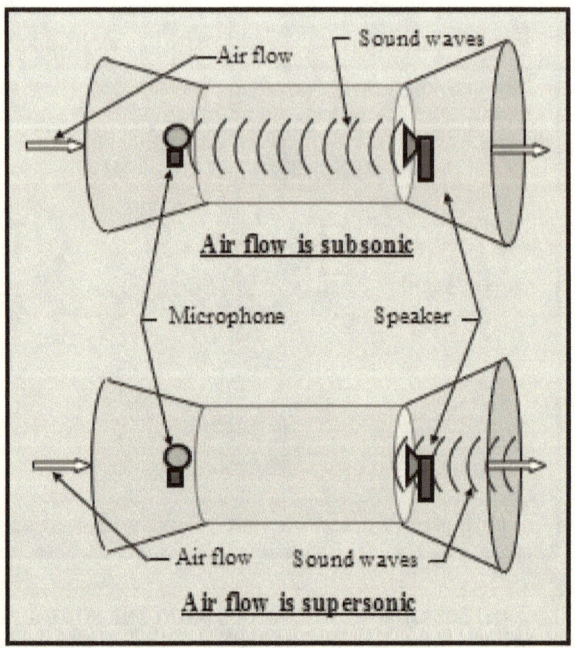

In this case, one needs to install a speaker (connected to a stereo system) near the outlet and a microphone (connected to a set of head phones) near the inlet of the supersonic portion of the wind tunnel.

By starting the wind tunnel and gradually bringing it up to its supersonic speed, while listening to his/her favorite music, he/she will notice that the sound heard through the headphone set gradually becomes weaker and weaker, as if it is broadcast from farther and farther away. So long as the speed of the wind generated by the wind tunnel is less than the speed of sound waves in that air medium, the sound waves generated by the speaker can propagate against the flow of air and reach the microphone.

However, once the sonic barrier is crossed, he/she will no longer hear the music, because the sound waves can no longer compete against the speed at which air is flowing in the opposite direction. In fact, they will get dragged downstream, even farther away from the microphone.

- Black holes can also be viewed as being giant drain holes at the bottom of an ocean. In this case, as shown below, if the diameter of the hole is big enough and the ocean is sufficiently deep, due to the enormous pressure that will be present, the speed of the flow of water towards the drain hole at a certain distance, which can be readily calculated, will reach and exceed the speed of sound waves in water.

  If someone who is being pulled into such a drain hole and is already passed that certain distance, tries to broadcast a sound wave type of an S.O.S. message towards outside, his cry for help will never be able to propagate outwards and it will be dragged into the drain hole. This is due to the fact that, the sound waves will not be propagating fast enough to counter the speed at which water is flowing towards the drain hole. Such a region in any ocean can be referred to as **"Esmailzadeh Silent Zone"** since no sound waves will ever be heard coming out of it.

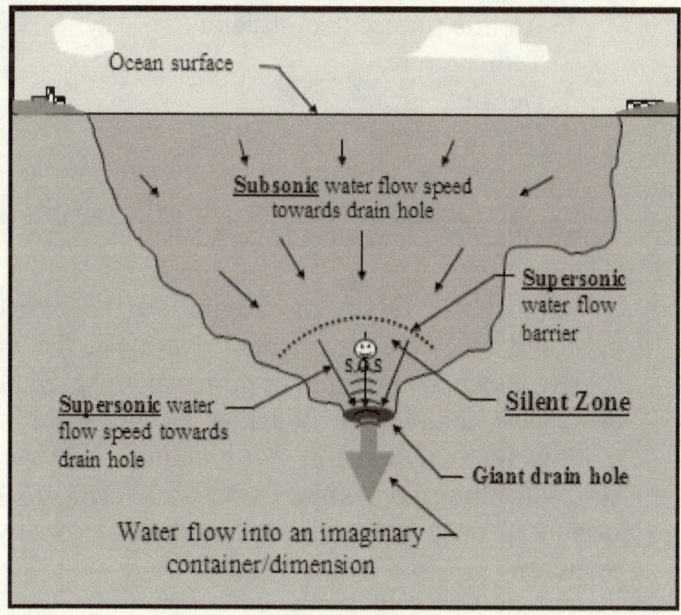

This is exactly what is taking place in the immediate vicinity of a black hole. A black hole is made up of an enormous amount of particles (drain holes) which have joined together and have formed a <u>single unified spherical gateway</u> for aether to pass through. For more detailed information in this regard, please refer to chapters titled "What is Gravity?" and "Black Holes and Their Properties".

The induced inward flow of aether at close proximity of any black hole is so great that the local aether is moving faster than phase vibrations such as light or other kinds of electromagnetic waves in that medium. Therefore, light waves simply cannot propagate upstream fast enough to counter the speed at which aether is flowing in the opposite direction, hence they get dragged into the black hole.

- The sound waves from any source to any receiver are transferred at exactly the very same rate of speed as that of the phase vibrations in the local medium which may be air. The propagation speed of phase vibrations in any medium is dependent on that medium's:

  1- Atomic (molecular) structure,
  2- Temperature,
  3- Density,
  4- Internal pressure, as well as
  5- Existing overall translational motions of its constituents, as in flow currents, relative to both the transmitter and the receiver.

Almost the very same applies to the propagation of light and other electromagnetic waves in the medium of aether. Since, their common propagation speed depends on the density and pressure of aether, as well as the varieties of motions that may exist in that medium. However, there is no such thing as temperature of aether or its atomic structure, since aether is not composed of any kind of particles.

Light Demonstrates Different Apparent Speeds as it Traverses Through Different Matter Mediums

Light and other electromagnetic waves, travel the fastest in outer space and their apparent speeds in different matter mediums are dependent on:

- The density of those matter mediums,
- The atomic / molecular structures of those matter mediums,
- The frequency of the light waves, and
- The angle/direction of propagation through those matter mediums.

As it is explained in details in the chapter titled "What is Gravity?", matter particles act as drain holes for aether. Therefore, their existence automatically encourages an inward flow of aether towards them. In doing so, they form a complex cross flow pattern in the local aether medium. As light waves pass through, they get dragged by these local cross flows and mandatorily follow a zigzag pattern. Presence of more matter particles near/along the path of light waves means more detours light waves will have to follow, hence they need longer time to cross through the medium.

The dependence of the speed of light in a given matter medium on the light's frequency and/or its angle of propagation, particularly in crystalline materials, is due to the number density and spacing between the matter particles that are encountered in any given direction through that medium.

Uniform crystalline structures result in more dependence of the speed of light on its frequency. The more compact the particles are organized in the matter medium in any given direction as compared to the other directions, the slower will be the speed of light in that particular direction. Since, light waves get dragged more often and have to travel longer actual distances (in a zigzag pattern) as they cross the same matter medium.

Note that, <u>the frequency dependence of the speed of light in a given matter medium is dependent on the temperature of that medium.</u> Since, as the temperature of a matter medium is changed, its atoms and molecules vibrate at a different frequency. Changes in the temperature of a matter medium also affect the spacing between atoms and/or molecules in that matter medium, due to expansion/contraction of that matter medium, as a whole.

For example, as the temperature of a cube made of Copper is raised, its atoms not only vibrate faster but also become spaced farther apart.

Light is Transferred Through Some Materials, While it is Partially or Even Totally Absorbed in Others

If the particles (atoms or even molecules) are organized in a crystalline structure, throughout the thickness of the medium (as in a single piece of glass that is perfectly homogenous), the light path propagating through will be composed of more or less symmetric zigzag detours. Such detours cancel each other in the amount they cause the light waves to be repeatedly dragged/deflected towards one side and then towards the other side. Therefore, light waves follow a general straight path as they go through such a medium. This scenario is shown below.

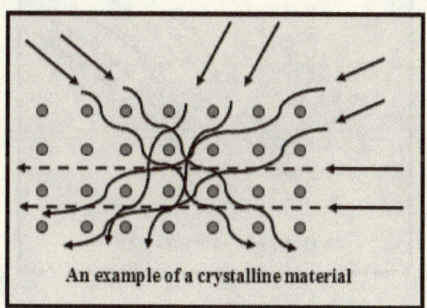

An example of a crystalline material

The same is also almost true when the medium consists of particles (atoms and/or molecules) that are in total random motion, as in gasses or liquids. In these cases, light waves get deflected in random as well and get dispersed. But, in such cases, only a fraction of the waves are affected. Because, as atoms and/or molecules move in and out of the pathway, most waves are allowed to continue, particularly when the particle density of the medium is not high enough to cause the total dispersion of the incoming light waves before they get the chance to reach the other side of the matter medium. This scenario is shown in the following figure.

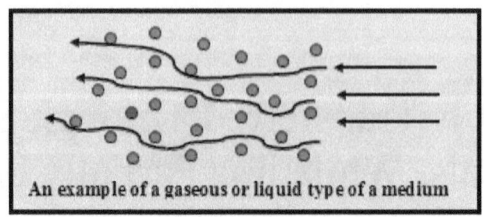

An example of a gaseous or liquid type of a medium

However, if the particles in the medium are not organized but they are densely packed and are relatively rigid in their respective locations, light waves get literally stranded by becoming dragged/deflected in complicated patterns by the cross flows that are induced in the aether medium due to aether flowing towards individual particles. Denser and more unorganized matter mediums lead to more severe zigzag patterns in the path of light waves passing through, hence causing them to disperse more and get lost in the matter medium. This scenario is shown in the following figure.

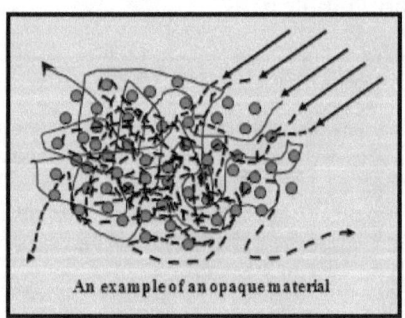

An example of an opaque material

Note that, always some of the light (electromagnetic) waves (even if their intensities are weakened by orders of magnitude) eventually cross any given matter medium, regardless of that medium's internal structure and/or its thickness.

Is Light a Wave or a Particle?

Currently, physicists believe that light is both, namely particle and wave. Their view is due to their attempts in trying to quantize light so that they can fit it in quantum theory equations. They treat light as a wave when IT suits better, and treat it as particle when IT suits better. **<u>Even Mr. Einstein stated in 1951:</u>**

<u>"Fifty years of pondering have not brought me any closer to answering the question, what are light quanta?"</u>

It is about time that, this matter is conclusively resolved. To answer this very fundamental question, one can propose a variety of experiments. The following are two versions of such an experiment.

> • Experiment:
> (Is light a wave or made of photons?)

Suppose an individual electron and an individual positron (anti-electron) are made to gently collide head-on. According to current understanding, they will annihilate each other and produce TWO photons which would then leave the scene of the collision, in opposite directions. Two versions of this experiment are described below.

1- Using only one detector

Suppose, as shown below, a SINGLE detector is placed nearby so that it has a clear view of the predetermined location of the collision. This way, if one of the TWO photons happens to come in its direction the detector would register / confirm its arrival.

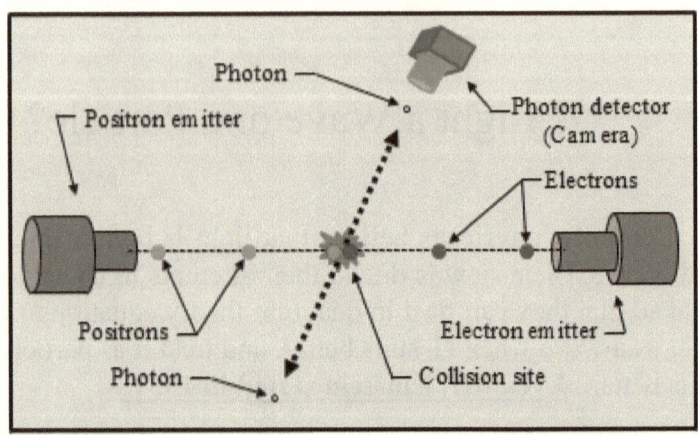

In such a scenario, regardless of its position being above, below or to the sides of the collision site, the single detector used will register the arrival of a "so-called" photon of light, every single time the experiment is performed.

However, neither of the two newly produced photons can be expected to be aware of the detector's position, and certainly cannot be reasoned that as such an experiment is performed one or the other photon instantly decides to willingly sacrifice itself so that its twin photon can escape unharmed.

2- Using too many detectors

Suppose, as shown below, TEN detectors are placed randomly nearby, all around the collision site, so that every single one of the detectors has a clear view of the collision site. This way, if by any chance, either one of the TWO photons happens to come towards any of the detectors that detector would record / report its arrival.

The detectors can be staggered so that no two detectors are placed exactly opposite of each other. Therefore, only one of the two photons can be captured by the detectors, hence doubling the chance of detecting / confirming the production of such photons.

In such a scenario, every single detector used will concurrently register the arrival of a "so-called" photon of light, every single time the experiment is performed. In other words, in this particular setup, TWENTY photons must be produced as a result of every electron / positron annihilation. Since, ten other photons must be flying in the directions that are opposite of the directions pointing towards the ten detectors used, where there are no detectors.

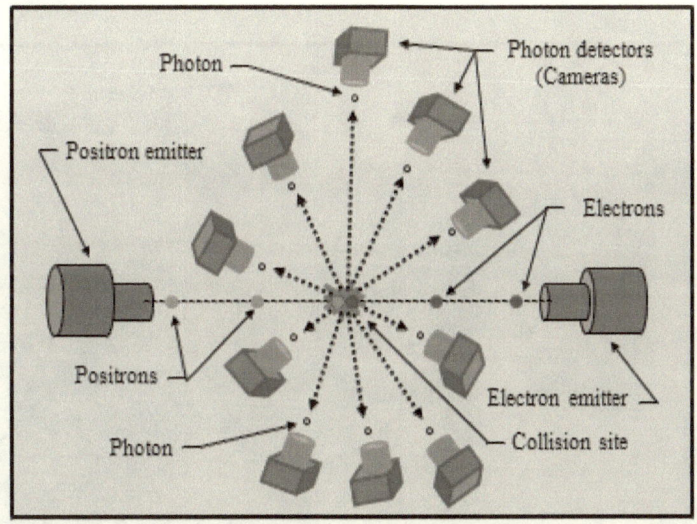

But, <u>only TWO photons are supposed to be produced during such a collision, and not twenty photons. Then, if light is made of photons, where do the other (18) photons come from?</u>

There can only be one possible explanation for obtaining such results, each and every single time the above experiments are performed.

"The collision between an electron and a positron results in the generation of a wave (an

electromagnetic wave) which then propagates spherically into all spatial directions."

The spreading action of such a wave from the electron/positron collision site can be readily likened to the spreading of the sound wave generated by a single clap of two hands into all possible directions. And, as shown below, each of the detectors used (either it be only one or as many as can be properly positioned), regardless of its position with respect to the collision site, will receive a portion of that wave, every single time such a collision takes place. Since, the generated wave will spread / expand spherically in all directions and will eventually pass through the position of each and every detector used.

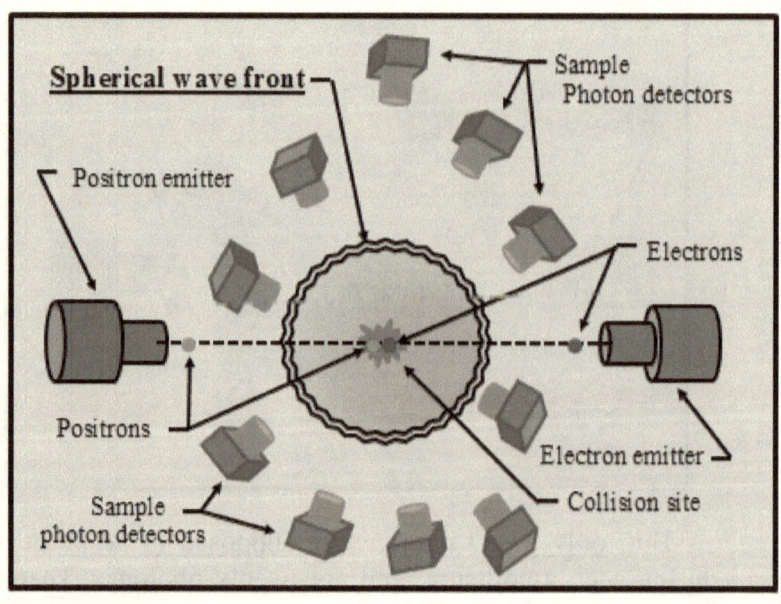

In other words,

"Light is a Wave."

Photoelectric Effect

Light waves of certain frequency, or higher, can cause electrons that are in the outer layers of certain atoms to let go of their atomic bonds and freely wander away. This process is referred to as the photoelectric effect and it is shown below.

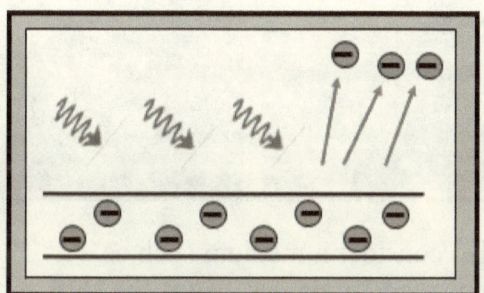

According to Mr. Einstein, photoelectric effect is due to light being made of particles, namely photons. According to his photoelectric theory, when photons with certain minimum amount of energy collide with electrons, they transfer sufficient amount of momentum to them and cause those electrons to break away from their respective nuclei.

It must be mentioned that, physicists consider effects such as photoelectric effect as strong enough of a proof and hence motivation for accepting light as being made of particles. However, as it is explained below, such effects can be readily explained by considering light as being a wave, as well.

The waves that are formed on the surface of water in a pond, demonstrate very gentle vertical movements. Such waves can have different amplitudes but their vertical motions are always quite gentle and slow. The vertical speeds of the waves generated on the surface of water or any other medium is directly dependent on the density and viscosity of that medium. For example, waves generated on the surface of water demonstrate easier movements as compared to the waves that may be introduced in other mediums such as honey or molten asphalt.

As shown below, different objects such as a piece of paper or a piece of wood or even a small boat demonstrate the very same alternating up and down motions while floating on the surface of water. Of course, various objects move by different amounts due to their own overall weights as well as their respective shapes and sizes as compared to the overall shapes and sizes of the waves (ripples) encountered on the surface of water.

One can also consider a medium such as air. The sound waves (vibrations) generated in air can cause the dust particles or even small objects resting on nearby surfaces to vibrate.

Sound waves with fairly high amplitudes and fairly low frequencies, as in low bass at high volume, promote visible motions in tiny objects they encounter, as they encourage them to repeatedly separate themselves from the surfaces they are resting on.

If the frequency is high enough and the amplitude is also fairly high, the vibrations can cause serious internal effects in the surrounding materials, such as breaking glasses and so on.

The photoelectric effect induced by **light waves,** in certain materials, can be likened to the effects described to take place in mediums such as water and air, due to the presence of waves.

In the case of photoelectric effect, electrons, due to being about 1,836 and 1838 times lighter than protons and neutrons, respectively, are readily affected by the presence of light waves.

If the frequency of the light waves encountered is low, electrons simply follow a roller coaster kind of a motion. Such motions are similar to the motions exhibited by small boats when they encounter waves generated by large ships, as they pass by. In this case, electrons do not gain sufficient momentum to break free from their atomic bonds. They just wobble as they follow their orbital paths around the nuclei.

However, if the frequency of the incoming light waves is above a certain minimum the electrons will be literally shaken out of their stable orbital paths and consequently break free from their respective nuclei.

Note that, if the frequency of the incoming light waves is not sufficiently high the electrons will not let go of their bonds with their respective nuclei, regardless of the intensity of the light waves encountered. In such cases, electrons will only experience more frequent wobbles, but will manage to hold on to their respective nuclei.

This effect is exactly like shaking dust and debris off of a carpet (or a rug or even a floor matt) by violently shaking the carpet in such a way that its surface exhibits a wavy motion. As it is shown in the following picture, if the generated wave motion (the ripple) is abrupt enough the particles of dust and debris will be shaken off. Otherwise, they will hold on to their respective positions and simply enjoy the ride.

Note that, even in the case of shaking dust particles off of a carpet or a rug, different particles let go of their bondages with the carpet (or the rug) as wave motions of different frequencies are encountered. Higher frequencies cause more abrupt (therefore more instantaneous) transfer of the needed momentum to particles that are more tightly stuck in their places.

It is due to the very same effect, namely the wave motion induced by the light waves of certain minimum frequency in the local medium of aether which is occupied by the conducting material, that electrons are encouraged to let go of their atomic bondages.

Laser

A Laser is simply an amplifier of light waves that are of a specific frequency. As it is shown in the drawing below, light waves with sufficient energy can cause electrons, in certain elements, to jump from lower orbits around the nuclei into higher orbits. But, due to the lower orbits being empty the electrons have the tendency to drop back down, as soon as possible. As electrons jump back to the lower orbits which are more stable and of a lower energy state, they generate waves in the local aether medium.

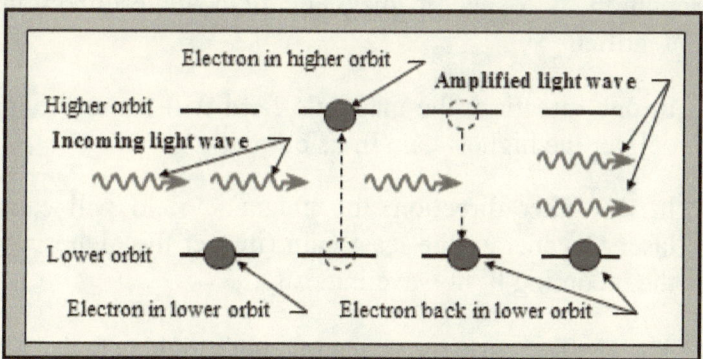

The waves generated are identical to the waves that had raised the electrons to their higher orbits. In case the generated waves and the incoming wave forms happen to be in sync, the outgoing waves will be an amplified version of the incoming waves. In other words, electrons temporarily store part of the energy associated with the incoming light waves, as they move to higher energy levels. Subsequently, as they jump back to the lower energy levels and release their excess energies they intensify /

complement / amplify the incoming waves, as they join them before leaving the instrument.

A laser can be likened to a hydroelectric dam which has pumps as well as generators built into it. By using grid's excess power, during part of the day, the pumps can be used to pump water up into the reservoir behind the dam. Later on, as the need for power necessitates, the stored energy in the pumped water can be used to drive the extra generators which will assist the grid. That is, if the power generated is in sync with the power that is already flowing in the local grid.

In fact, the hydroelectric power station at Niagara Falls (located between the United States and Canada) by using such a scheme is assisting the local power grid, by generating extra power when the demand for electricity outweighs the overall capacities of the nuclear and the coal power stations that are running continuously.

It should be noted that,

As shown below, by placing a laser in a uniform, unidirectional magnetic field which can be rotated in its orientation, two specific magnetic field angles/directions can be identified.

- In one direction, the magnetic field will cause the laser to deliver the highest gain in its output,

- In the other direction, the magnetic field will cause the laser to generate the least gain (in fact the highest loss) in the incoming light wave intensity.

These effects are due to the magnetic field actually pushing the electrons along the way and encouraging them to move away from the nuclei in one direction and to come closer to the nuclei in the opposite direction.

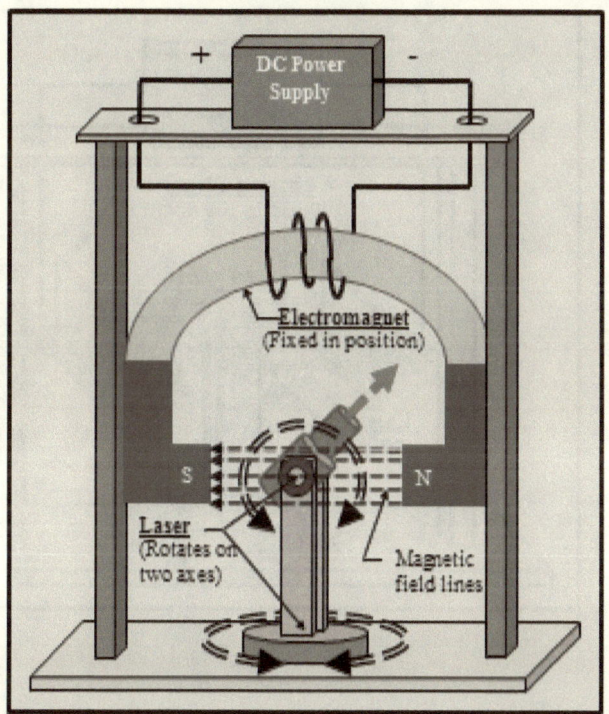

Therefore, if the electrons are encouraged to drop to the lower orbit where they will generate waves that are in sync with the incoming light waves, the generated waves will result in the amplification of the incoming light wave's intensity. However, if the electrons are made to drop to the lower energy level while their generated waves are 180 degrees out of phase with respect to the incoming light waves, they will actually neutralize part of the incoming light waves and hence lead to the most loss in its intensity.

Note that, as shown below, a uniform, unidirectional electric field would also cause the very same effects on the operation of laser equipment, as a magnetic field would. However, the angles / directions of the electric field that would generate the maximum and the minimum gains would differ from those of the magnetic field.

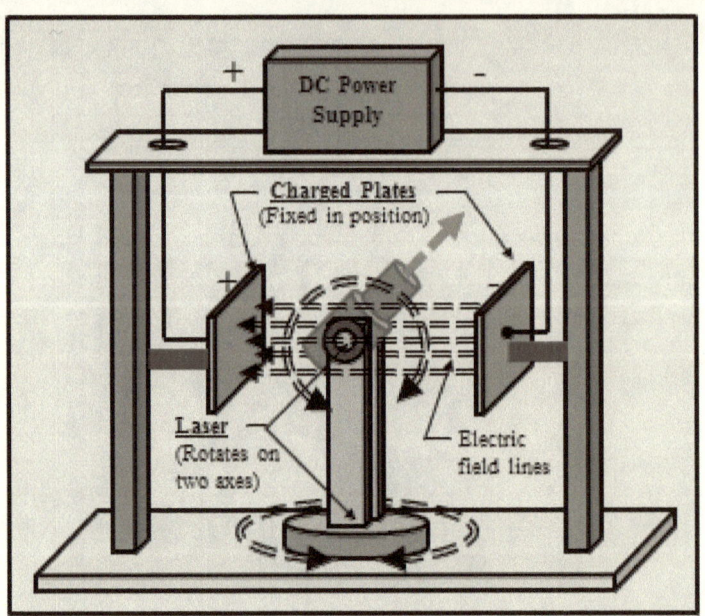

What is a Spark?

The phenomenon of spark is the manifestation of phase vibrations induced in the local aether medium by electrons or other charged particles that are transferred from one region of space to another at a high rate of speed, particularly as they are transferred between objects with electrical charge imbalance. Such phase vibrations can cover a wide range of frequencies, since individual electrons (particles) move at different rates of speed and follow quite a random path, as they are transferred between objects.

For example, as two objects with opposite electrical charges come in direct contact, they have the tendency to share their charges until they reach a mutually neutral state. As shown in the following figure, during such a process some of the electrons in the negatively charged object are literally pushed away by the other electrons in that material, and unto the other object which is starving for electrons and hence is attracting them.

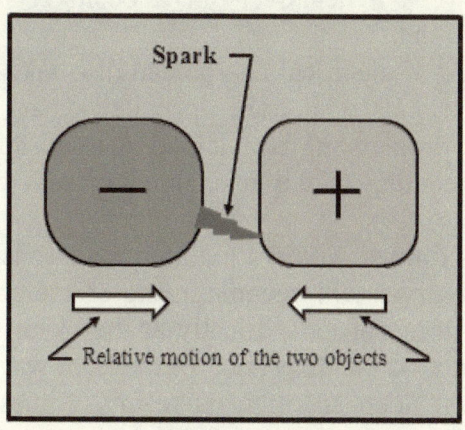

As a result, when the physical contact is **nearly** established, a spark forms between the two oppositely charged objects, due to the sudden transfer of multitude of electrons. The size of the gap between the two objects across which a spark can form depends on the relative amount of charge imbalance between those two objects, as well as the presence of a medium in between which can potentially become ionized and allow electrons (or other charged particles) to cross the gap with multiple hops.

Note that, sparks are electromagnetic waves in nature, whether they are visible or not. Therefore, they can form in a pure aether medium (so-called vacuum), as well. However, an observer can see such sparks only and only if he/she is looking directly at the exact loaction of the sparks, as they form.

Also, a spark formed in a pure aether medium will not have any kind of sound waves associated with it, since in order to propagate, sound waves require a matter medium.

Sparks generated due to the buildup of static electricity can be very small and relatively quite weak in nature, such as the ones formed by walking on carpets and then touching a metal doorknob. They can also be quite enourmus in their sizes and strengths. Lightnings are among such sparks. Lightnings are formed as other matter particles, by becoming ionized and hence acting as conductors, assist in providing a bridge or path for electrons to be transferred between a negatively or a positively charged cloud mass and the ground.

An exchange of electrical charge can also take place between two masses of clouds which are oppositely charged. In such a case, it is the droplets in both cloud masses that provide the necessary connection, as the two cloud masses come in direct contact.

If a spark is formed within a liquid or a gaseous type of matter medium, the observer will become aware of the spark happening even if he/she does not look directly at the exact location of the spark. This is due to the dispersion of the light waves by the local matter medium. Such an effect is readily witnessed by almost everyone, because at one time or another they must have noticed the flash of light in the air, as a lightning was occurring at a distance in any direction, including behind them.

Also, the very presence of objects nearby causes the light waves to become reflected towards the observer. That is how an observer can acknowledge the occurrence of a lightning just by receiving its reflected light waves off of the nearby objects, such as buildings.

Thunder is due to the sudden expansion of the immediate air medium, as the local atoms / molecules along the length of the spark receive considerable amounts of kinetic energy from the stampede of electrons. The sudden increase in their kinetic energies generates a compression wave-front in the surrounding air medium which in turn carries the generated wave (phase vibration) in all directions. The duration of the sound or thunder heard by an observer at a distance depends on the distance between the observer and different parts of the lightning, from one end to other end.

The following figure shows different cases, as the water droplets in the atmosphere, a lightning rod, or even a tall live tree assist in the formation of such a connection between highly charged clouds and ground, as well as between two oppositely charged clouds.

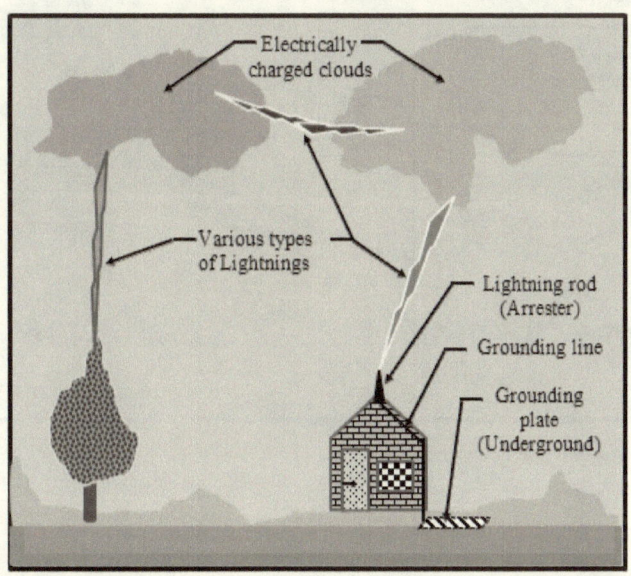

There are even larger lightnings than the usual lightnings which are referred to as **"Lightning Sprites"**. These lightnings form only between the top of supermassive, super-active clouds and the lower boundaries of the ionosphere. In other words, they can be as much as 40 miles or so in height. The cloud masses engaged in such activities have to be quite actively engaged with ground surface through multiple and nearly continuously occurring strong lightnings. The reason for such a requirement is because,

"As a huge cloud mass is directly and nearly continuously engaged in generating multiple strong lightnings with earth surface, the cloud mass as a whole in fact literally acts as a supersized lightning rod which is directly feeding off of the charge that is stored in earth itself."

The following figure shows such a massive lightning sprite in a simple fashion.

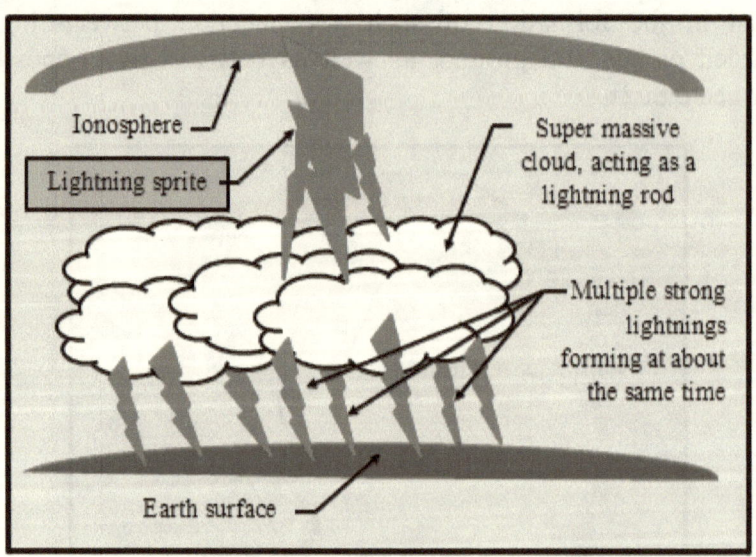

Are Light Waves Visible?

Light waves make their presence known to an observer only as they reach and enter his/her eyes. They do so as they get reflected by objects that surround him/her, since objects reflect light waves in literally all spatial directions. Even mirrors reflect part of their received light waves in other directions than it is dictated by the law of reflection. However, light waves will go unnoticed if they just pass by an observer without entering that observer's eyes. In other words,

"Light waves become visible only when they enter an observer's eyes."

To demonstrate such a characteristic of light waves, one needs to point his/her flashlight towards the sky and also shine it at nearby objects, both at night when the sky is also moonless and dark and the air is clean.

An observer, who may be positioned in front of the flashligh so that he/she is facing away from it, by receiving light waves that are reflected back towards his/her eyes by the nearby objects, will be able to testify that light waves are present (flashlight is turned on), due to the very visualization of those objects.

Yet, in the case of aiming the flashlight towards the night sky, the observvber would not be able to notice any difference between when the flashlight is turned on and when it is not, because the clear night sky will never be iluminated by light waves that are generated by the flashlight. Since, there is relatively nothing nearby in the clear night sky that will reflect those light waves back towards the observer's eyes.

Is the Speed of Light in Outer Space the Maximum Possible Speed for Objects in This Universe?

The speed of light in the medium of aether is equivalent to the speed of phase vibrations in that medium, since light and other electromagnetic waves are one type of phase vibration in the medium of aether.

Even though the speed of phase vibrations in any medium tends to act as a barrier for objects which try to travel faster in that medium, but it has nothing to do with the actual maximum speed attainable by objects in that medium. This fact has already been proven by supersonic airplanes such as the Concord, the Topolov 144 and many fighter planes, as well as a variety of rockets which are capable of reaching supersonic speeds in the medium of air.

Of course, objects need to either generate or receive the amount of energy that is required to increase their speeds passed that of phase vibrations in the medium in which they are travelling.

Therefore, just as the propagation speed of sound waves in the medium of air is proven not to be the maximum speed attainable by objects in that medium, in the future it will be demonstrated that the propagation speed of light waves in the medium of aether is not the maximum speed attainable by objects in that medium, either. In other words,

"In every medium, objects can travel at speeds that are slower or faster than the propagation speed of phase vibrations in that medium."

Consequences of the Dependence of the Speed of Light on the Density of Aether

The speed of light and other electromagnetic waves is dependent on the density of aether. Any decrease in aether density results in an increase in the speed at which light and other electromagnetic waves propagate in that medium.

According to the information presented in this book, the density and pressure of aether in this universe are gradually decreasing, due to both the overall expansion of the aether medium and the leakage of aether through matter and anti-matter particles, as well as through their aggregates such as stars, neutron stars and particularly black holes. Therefore, the speed of phase vibrations in that medium is gradually increasing. In other words,

"The speed of light is gradually increasing."

The consequences of this gradual increase in the speed of light are quite important in two regards, namely in the distant past and in the distant future, because the rate at which time is experienced by all objects and living beings is directly dependent on the speed of light which is the speed of phase vibrations in the local medium of aether.

As it is explained in details in the chapter titled "What is Time?", the rate at which the passage of time is experienced by any object or any living being is directly dependent on the speed of that object or living being relative to its local aether, as compared to the speed of phase vibrations in that medium. As the speed of phase vibrations in aether is gradually increasing due to reduction in aether's density, the pace at which time is experienced by any and all objects, universally, is speeding up. In other words,

"'Time' is gradually speeding up, on a universal scale."

- **In the distant past**

  To calculate the speed of light back in the beginning, when universe was just born, one has to know what the density of aether was back then. The absolute value of aether's density during the infancy of this universe cannot be calculated. However, its value can be estimated relative to what it is at the present time. Currently, the universe is estimated to be about 93 billion light years across. In other words, currently the diameter of the universe in meters is equivalent to,

$$(9.3 \times 10^{10}) \times (365) \times (24) \times (3600) \times (300{,}000) \times (1{,}000) = 8.80 \times 10^{26} \, m$$

  If, in the beginning, the diameter of the universe was only ONE meter then the volume of the universe has grown by the cube of the above number (6.81×10^{80}). Therefore, in the beginning, the density of aether was at least (6.81×10^{80}) times higher than what it is at the present time. Also, knowing that the speed of light and other phase vibrations in aether are dependent on the density of aether, one can confidently make the following statement,

  "During the very first "SECOND" that this universe had started its existence, the speed of light must have literally been billions of billions of times slower than what it is at the present time."

- **In the distant future**

  As time goes on, the density of aether is continuously decreasing and it is headed towards zero. The very gradual decrease in aether density towards zero implies that the speed of light and other phase vibrations in that medium are correspondingly approaching infinity. In other words,

"The speed of light will ultimately approach infinity."

- ## Speed of light in the accompanying universe
  Currently, the fluid aether that is in the accompanying universe is under a much, much lower pressure as compared to aether that is in this universe. Hence, its density is correspondingly much lower than that of aether that is in this universe. Therefore,

"Currently, the speed of phase vibrations (such as light) is much higher in the accompanying universe, as compared to what it is in this universe."

In the future, the speeds of phase vibrations in the two universes will approach each other and equalize. Then, it will increase and approach infinity, in both universes, simultaneously.

What is the Current Rate of Increase in the Speed of Light?

Light, as well as other electromagnetic waves are phase vibrations in the medium of aether, just as sound waves are phase vibrations in a matter medium such as air. The speed of phase vibrations in any medium is dependent on the density of that medium. Therefore, speed of light in this universe is expected to change with time since the density of aether in this universe is gradually decreasing, due to its overall expansion and also due to its leakage into the accompanying universe.

In order to measure the rate at which the speed of light, as well as all other phase vibrations in the medium of aether, is currently increasing, the following must be known:

- **The relationship between the magnitude of the speed of light and the density of aether, and**

   As a first approximation, the relationship between the speed of light and the density of aether can be assumed to be the same as that between the speed of sound waves in air and the density of air. Later on, this relationship can be refined, upon the availability of more specific data and experimental capabilities.

- **The rate at which the density of aether is currently decreasing, in this universe.**

   At the present time, the variations in the density of aether in this universe can be roughly estimated using Hubble's constant. Hubble's constant is a direct indication of the rate at which the physical contents of this universe (not the overall aether medium) are spreading by receding from each other. The spreading of the physical contents of this universe is due to the expansion of the aether medium, as a whole, as well as the motion of galaxies IN that medium.

Therefore, the speed at which matter contents of this universe are dragged at any given location is proportional to (but not equal to) the speed at which aether itself is moving due to its ongoing expansion process.

The relative motion of matter contents of this universe with respect to aether in the overall picture of this universe is literally like the relative motion of objects that are allowed to free fall towards a gravitating celestial body. Such objects will be dragged by the flow of aether, yet they will not be moving at the same rate of speed as the local aether itself. However, if those same objects were allowed to start their free fall at infinite distance from the gravitating body, during their entire free fall journey, they will be moving at just about the same rate of speed as that of their local aether.

As the matter contents of this universe were formed, they fell behind from the phase vibrations that had formed them. Even more so, they fell behind with respect to aether that was expanding. Yet, they all were moving IN that medium. The general direction of their motion / momentum was away from the birthplace of this universe, since the phase vibrations that caused the generation of the matter and anti-matter particles were propagating away from where this universe had started its existence.

Gradually, all of the matter contents of this universe are being dragged by their local aether at faster and faster paces, due to the weakening of the force of gravity which is directly opposing their receding motions from each other.

According to the most recent measurements, Hubble's constant is **about** 70 km/sec/Mega parsec. Since one Mega parsec is equal to the distance traversed by light in free space in 3.26 years, the current value of Hubble's constant can be rewritten as,

$$(HC = 7.157 \times 10^{-11} \text{ km/year//km}).$$

Therefore,

The current annual rate of <u>increase</u> in the volume of aether that is in this universe <u>is greater than</u>

$$((R_2)^3 - (R_1)^3) / (R_1)^3 = 2.1477 \times 10^{-10}$$

Where, R_1 and R_2 are the radii of a given amount of aether, as of last year and as of this year, respectively.

The annual variations in the density of aether that is in this universe are numerically the same but in reverse of the variations in its aether's volume. In other words,

The current annual rate of <u>decrease</u> in density and pressure of aether in this universe <u>is greater than</u>

$$(2.1477 \times 10^{-10})$$

Note that, this rate of variation in aether density (calculated using Hubble's constant) is **only** based on the receding motion of galaxies from each other. It is not based on the expansion of the aether medium, as a whole. It does not take into account the rate at which aether is gradually leaking through matter and anti-matter particles, and their aggregates such as neutron stars or black holes, either.

Using the above estimated magnitude (as the rate at which the density of the fluid aether in this universe is decreasing) in the equations governing the speed of sound waves in a matter medium such as air will provide an initial estimate of the rate at which the speed of light and other electromagnetic waves is increasing in this universe.

Subsequently, various types of experimental setups can be constructed to verify the results obtained. Such experiments will also lead to a better understanding of how the variations in the density of the fluid aether affect the speed of light in that medium.

Effects of Gravity on the Speed of Light and its Direction of Propagation

The force of gravity is the drag force that the accelerated flow of aether (towards matter and anti-matter particles) exerts on other particles that happen to be in its path. The very same flow of aether also carries any and all kinds of phase vibrations such as light and other electromagnetic waves that may be in its path.

The effect of gravity on the direction of propagation of light coming towards earth from distant galaxies has already been demonstrated in a variety of experiments. However, the effect of gravity on the speed at which light propagates was never attempted. The following experiment is particularly designed to explore the effect of gravity (earth's gravity) on the propagation speed of light waves.

> • Experiment:
> (Effect of gravity on the speed of light)

To begin with, one has to perform an experiment with sound waves. As shown in the figure below, he/she needs to install a speaker and a microphone in a long tube (with sound proofed interior surfaces) so that the sound generated by the speaker can be clearly picked up by the microphone. The speaker should generate a monotonic sound at a specific frequency.

The output of the microphone should be delivered to a multichannel analyzer, so that the sine wave form of the sound wave received can be seen on the screen.

Also, as shown in the figure, two fans must be mounted behind the speaker and the microphone. Their purpose is to alternately generate a flow in the air medium through which sound waves are going to propagate. The fan behind the microphone will generate an air flow which is in the same direction as the sound waves, while the fan behind the speaker will generate an air flow which is in the opposite direction.

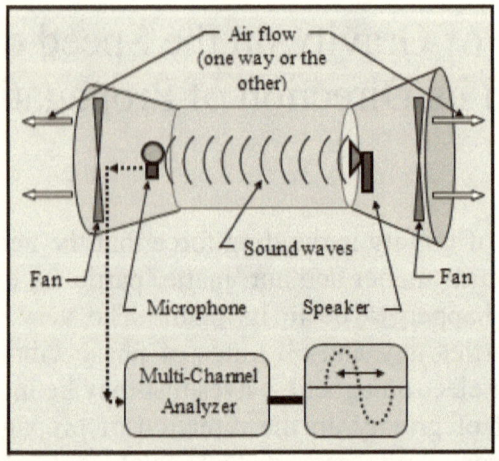

First, while both fans are inoperative, one needs to allow the speaker to generate a continuous monotonic sound wave which after it is picked up by the microphone, its sine wave form should be made visible on the screen of the multichannel analyzer. The position of the sign wave on the screen should be recorded.

Then, one of the fans should be turned on to generate a flow in the air medium, either from the speaker towards the microphone or from the microphone towards the speaker. The flow of air will affect the position of the sine wave on the screen of the multichannel analyzer. The amount of the shift in the position of the sine wave on the screen will directly depend on:

- The speed of the air flow generated,
- The distance between the speaker and the microphone,
- The speed of sound in that air medium.

The same procedures should be repeated while the first fan is turned off and the other fan is turned on to generate an air flow in the opposite direction. In this case, the sine wave shown on the screen of the multichannel analyzer will shift in the opposite direction as compared to the previous part of the experiment. This is due to the fact that, during the first and the second portions of the experiment, the sound waves are carried / dragged by the air flow in two opposite directions.

The very same effects can be demonstrated to take place with light beams due to the flow of aether which is giving rise to earth's force of gravity. The apparatus which can be used for such an experiment is shown in the following figure. The main components used in this apparatus include two monochromatic light sources and two light sensors which are housed inside two parallel vacuumed tubes, as well as a multichannel analyzer.

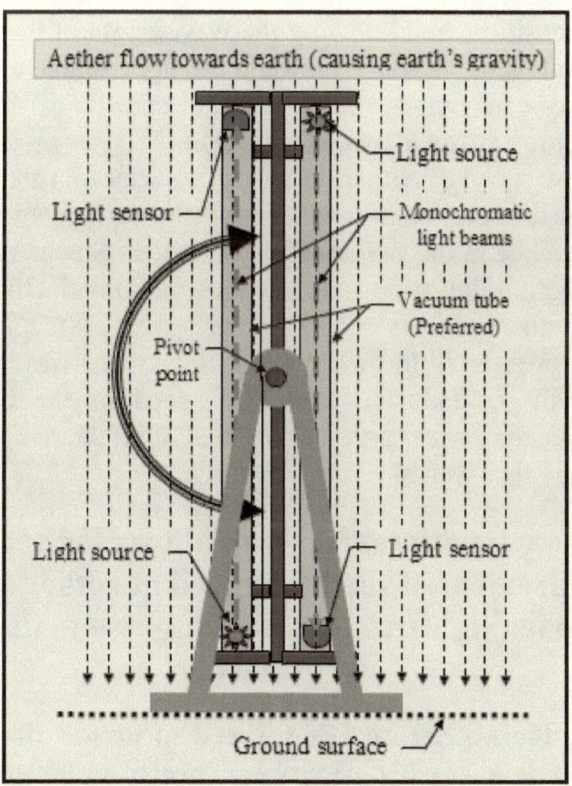

For the first part of this experiment, one needs to pay attention to only one of the two tubes, either the one on the right side or the one on the left side, as shown in the figure.

First, while the tube that is housing the light source and the light sensor is in the horizontal position, the position of the sine wave form corresponding to the light received by the light sensor chosen, and observed on the screen of the multichannel analyzer, should be recorded.

Next, after turning the housing tube so that the light source chosen is at the top end and its corresponding light sensor is at the bottom end of the tube, the location of the sine wave observed on the screen of the multichannel analyzer should be recorded. Then, the same should be repeated after the housing tube is turned by 180 degrees so that the light source chosen is at the bottom end and its corresponding light sensor is at the top end of the tube.

By comparing the two recorded locations for the sine waves with the location of the sine wave when the housing tube was in its horizontal position, and knowing the wavelength of the light wave used, as well as the distance between the light source and the light sensor, one can readily calculate the speeds of light waves propagating towards and away from earth.

The flow speed of aether near earth's surface, at that particular time and at that particular location, will be equivalent to one half of the difference in the two speeds calculated. Since, by carrying / dragging both light waves downwards, automatically, the flow speed of aether will be added to the speed of the light waves travelling towards earth (when the light source was at the top), while it will be deducted from the speed of the light waves travelling away from earth (when the light source was at the bottom). In other words,

Such an experimental setup allows for accurate measurement of the flow speed of aether that is approaching earth and is giving rise to its force of gravity.

Note that, the accelerated flow speed of aether that is giving rise to earth's gravity is expected to be equivalent to the escape velocity from earth's gravity at that particular location (altitude with respect to earth's center of mass), **where the experiment is conducted.**

As shown in the figure, this experiment can also be carried out using two light sources and two light sensors positioned at the opposite ends of two tubes which are mounted side by side on one frame. In this case, while the tubes are horizontal, the two sine waves shown on the multichannel analyzer screen must be adjusted so that they are superimposed on each other.

Then, as the tube's angle is adjusted from horizontal to vertical, the resulting shifts in the positions of the two sine waves on the multichannel analyzer screen can be readily compared. Also, if needed, the very same effects can be confirmed to take place in reverse by the two light waves travelling in the opposite directions, by simply turning the frame by 180 degrees.

Note that, the two tubes experience the same amount of length contraction, due to being in line with the accelerated flow of aether towards earth.

Very important note, since gravitational forces are due to movements of the medium they do not lead to any red shifts or blue shifts in the spectrum/frequency of the light waves received. Instead, they cause the propagation speed of such waves to vary accordingly, depending on the relative motion of the medium of aether to the direction of propagation of those waves.

However, if the light source and the light sensor were to move relative to one another, their relative motion would lead to red shift or blue shift in the frequencies of light waves received, depending on whether they be moving away from each other or they be moving towards each other, respectively.

Of course, by taking into account the positions of the moon, the sun and the center of the Milky Way galaxy while performing such experiments at proper timing, as well as using proper angles (the tilt angle of the tubes), the flow speeds of aether that are giving rise to their respective forces of gravity, at any given location on earth and at any given time, can also be detected and measured accurately.

Effects of Magnetic Field
on the Speed of Light
and its Direction of Propagation

Magnetic field is basically one type of aether flow which completes a round trip. The following experiments demonstrate how light waves are carried by the flow of aether that manifests as magnetic field.

- **First experiment:**
  (Effect of magnetic field on the speed of light)

This experiment is designed to demonstrate the effect of magnetic field on the speed of light or any other electromagnetic wave. The design details of the proposed apparatus to be used for this experiment are shown below.

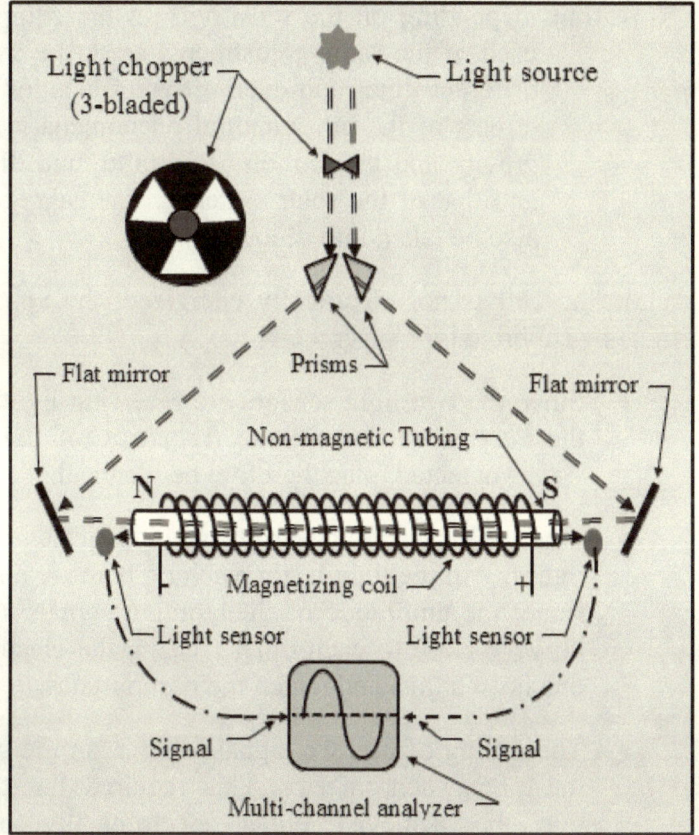

It is of great importance to mention that:

1- In order to prevent the two light waves from affecting each other's wave forms, as they propagate through the magnetic field in the opposite direction of each other, a simple chopper mechanism is added right after the light source. This way, only one light beam will be allowed to cross the magnetic field at a time.

   This potential effect is explored in another experiment titled "Crossflow Experiments for Two Monochromatic Light Waves or Two Monotonic Sound Waves".

2- This experimental setup may be assembled at any time of day or night. However, the final adjustments and fine-tuning needed for the light beam must be done when the test is to be performed.

Note that, depending on the sensitivity of the equipment used, if the same adjustments are to be used at different times and even different locations, the effects of the other natural phenomena such as gravity and the motion of earth around the sun and that of the solar system in the galaxy must also be taken into account.

3- While the coil is not electrically energized, the apparatus must be calibrated for several effects:

- Since the two light sensors are receiving light from the very same source, the frequency of the light waves detected must therefore be identical.

- Since the two light sensors are identical, and the intensity of the light beam reaching both are also the same, the amplitude of their output signals should be very close to each other. Using the electronics one must adjust and match their amplitudes.

- The timing of the two signals shown on the screen must be synchronized. This required fine-tuning may be achieved either electronically or by adjusting the lengths of the two wires delivering the signals from the light sensors to the multichannel analyzer, by using sliding contacts.

4- This experiment must be repeated with different magnetic field strengths generated by the electromagnet using different DC voltage settings on the power supply.

The first part of this experiment should be performed without energizing the electromagnet. In this case, the two sine waves observed on the monitor of the multichannel analyzer, representing the signals received from the two light sensors, will be superimposed on each other.

Next, using the minimal voltage setting available, the electromagnet must be energized. The induced magnetic field is in fact an induced flow in the local aether medium, from one end of the electromagnet to its other end. This induced aether flow will cause a difference in the speed at which the two light beams will propagate in opposite directions.

The presence of the magnetic field will speed up the light beam travelling in one direction, while it will slow down the other beam travelling in the opposite direction. Therefore, the two sine waves corresponding to the signals received from the two light sensors, shown simultaneously on the screen, will become out of phase. They will be shifted **by almost the same amount** but in opposite directions.

Note that, the two signals will shift in opposite directions by slightly different amounts. Because, the one associated with the light beam travelling upstream of aether flow will lag a bit more in its timing as compared to the amount of timing gained by the signal associated with the light beam travelling downstream.

Such differences between the amounts of time gained and lost by the two light beams travelling in opposite directions (upstream and downstream) become more noticeable for faster flow speeds of aether, when stronger magnetic fields are generated.

An experiment with sound waves in a medium such as air which is flowing would clearly demonstrate such an effect. Even two airplanes flying at equal and constant speed relative to their local air medium, one heading downwind and the other heading upwind, will gain and lose different amounts of time, respectively, as they travel the same ground distance.

The amount of the induced difference between the propagation speed of light with and against the induced flow of aether and the apparent shift in the position of the two sine waves on the screen will directly depend on the strength of the magnetic field generated. This relationship can be readily confirmed as the magnetic coil is energized by different output voltage settings on the power supply.

Note that, the shifts in the sine waves will only be in the timing of their signals and not in their frequencies. No red shift or blue shift will be detected in these experiments. Since, the emitters (which in this case are the two flat mirrors) and the receivers (the two sensors) are stationary relative to each other.

Since the multichannel analyzer can also provide the frequency of the light signals received, one can create a chart demonstrating the amount of shift in the sine waves, relative to each other, as a function of the strength of the magnetic field generated by different voltages supplied to the coil.

The amount of shift generated in the two sine waves, as compared to the width of a full sine wave will allow the calculation of the amount of variations induced in the propagation speed of the two light beams travelling in the opposite directions through the magnetic field of certain strength. Since, the other required pieces of information, namely the length of the magnetized region (along the tube), and the width of the individual wavelength of the light beam used, are readily known.

Note that, the amount of shift recorded will provide the information needed to determine the strength of the magnetic field necessary to slow down the propagation of the light waves by any desired amount. In the very extreme case, even if it may be for only a moment,

"By generating a magnetic field that is sufficiently strong, one can even try to literally stop the propagation of light waves, in the desired direction."

- **Second experiment:**
(Effect of magnetic field on the direction of light's propagation)

This experiment is designed to demonstrate the effect of the magnetic field on the direction travelled by light or other electromagnetic waves. The basic idea in this experiment is to demonstrate how the locally induced motion of aether (manifesting as the magnetic field) can drag the light waves downstream and hence cause a shift in the projection of the light spot observed on a calibrated screen. The design details of the proposed apparatus to be used for this experiment are shown below.

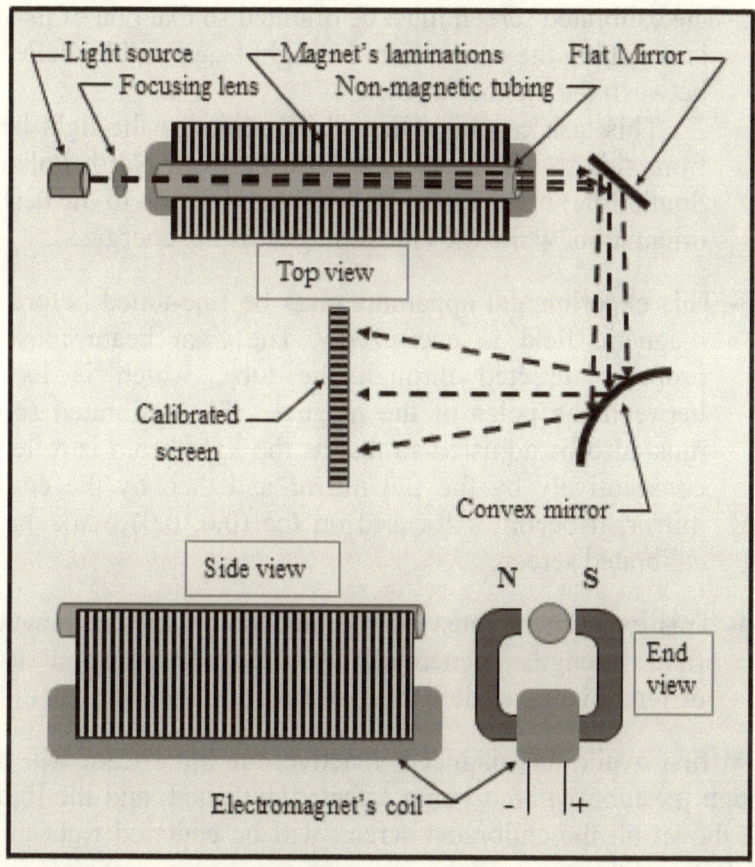

It is important to mention that:

1- This experimental setup may be assembled at any time of day or night. However, the final adjustments and fine-tuning needed for the light beam must be done when the experiment is to be performed.

   Note that, depending on the sensitivity of the equipment used, if the same adjustments are to be used at different times and even different locations, the effects of the other natural phenomena such as gravity and the motion of earth around the sun and that of the solar system in the galaxy must also be taken into account.

2- The calibrated screen must be oriented so that one of its axes is matching the orientation that light beam will be deflected between the magnetic poles.

This task can be achieved by adjusting the light beam from side to side (back and forth between North Pole and South Pole) and rotating the calibrated screen to the desired orientation, while the electromagnet is not energized.

3- This experimental apparatus must be fine-tuned before the magnetic field is energized. The light beam must be properly directed through the tube, which is located between the poles of the magnet. The calibrated screen must also be adjusted so that as the light beam is reflected consecutively by the flat mirror and then by the convex mirror, it becomes focused on the (0.0, 0.0) mark on the calibrated screen.

4- This experiment must be repeated with different magnetic field strengths generated by the electromagnet using different voltages delivered by the power supply.

At first, while the magnet is inactive, the light beam will pass through the tube without being affected/deflected, and the lighted spot shown on the calibrated screen will be centered right at the (0.0, 0.0) position.

Next, using the minimal voltage setting available, the electromagnet must be energized. The induced magnetic field will cause the light beam to deviate from its straight path. The amount of this deviation will directly depend on the strength of the magnetic field generated. This effect can be easily demonstrated by energizing the electromagnet with different voltages.

The deflected light beam will hit the convex mirror at different spots. Due to the nature of the convex shape of the mirror, it will exaggerate the angle at which the light beam will be reflected towards the calibrated screen. Therefore, as the electromagnet is energized by different voltages, the light beam will show its presence at different spots on the calibrated screen,

This experiment can also be repeated with the poles of the magnet reversed. This can be achieved by simply reversing the wires supplying the electrical power to the magnet's coil. The change in the direction of the magnetic field will automatically reverse the light show observed on the calibrated screen, earlier.

- Such an experiment can also be performed using the light beam generated by a Laser, or focused electromagnetic waves of any frequency. However, in case the waves used are not in (or near) the visible light range of the electromagnetic waves spectrum, suitable deflection as well as detection devices must be employed.

During this type of experiment, where the effects associated with the cross flow of aether is examined, it can be demonstrated that magnetic fields drag anything and everything, including neutral particles such as neutrons, that happen to be in their path, just as the force of gravity does.

Once the length of time the light waves spend in the magnetized region of the tube and the length of time neutral particles (of any kind) spend in the magnetized region of the tube are calculated and compared, it can be confirmed that the amounts of deviations introduced in their flight path are consistent with one another. Meaning, the more time any wave or particle spends in the magnetic field, while passing through, the more it will get deflected from its original path.

The magnitudes of such deflections can be represented by a common acceleration factor which is along the direction of the induced aether flow, manifesting as the magnetic field present. Such acceleration can be readily likened to the acceleration associated with the local force of gravity of earth which is represented by the gravitational constant (g).

Effects of Electric Field
on the Speed of Light
and its Direction of Propagation

Electric field is a locally induced one-way flow in the aether medium, a flow that is from the negatively charged particles towards the positively charged particles. The following experiments demonstrate how light waves are carried by the flow of aether that manifests as electric field.

> • First experiment:
> (Effect of electric field on the speed of light)

This experiment is designed to demonstrate the effect of electric field on the speed of light or other electromagnetic waves. The design details of the proposed experimental setup are shown below.

While the electric current is disconnected from the two plates, the apparatus must be calibrated for several effects:

- The light beam must be directed towards one of the plates at a predetermined angle, as shown in the figure, so that after it is reflected a known number of times, between the two plates, it is received by the light sensor.

- The position of the signal (sine wave) shown on the screen of the multichannel analyzer must be centered on the screen. This required task of fine-tuning may be achieved either electronically or by adjusting the length of the wire delivering the signal from the light sensor to the multichannel analyzer, using a sliding contact.

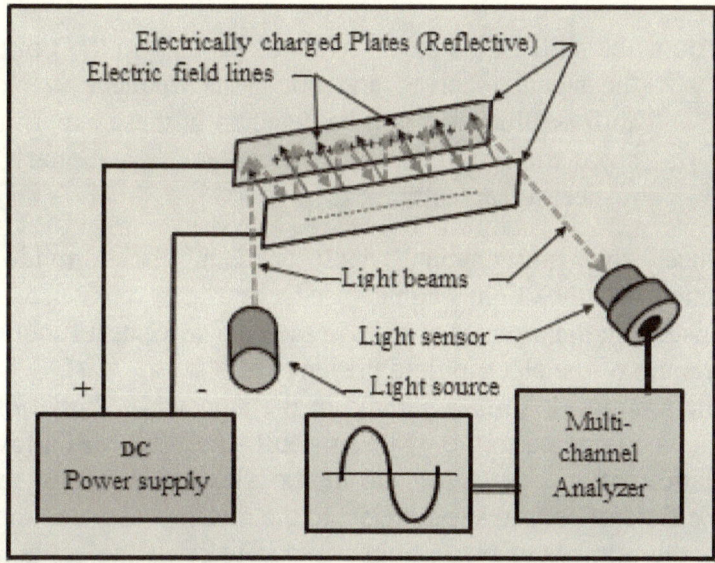

This experiment should begin without electrically activating the plates. In this case, the sine wave representing the signal received from the light sensor will be positioned right at the center of the multichannel analyzer's screen.

Next, the two plates must be charged with the minimal voltage deliverable by the power supply. The induced electric field will cause a difference in the speed at which the light beam will propagate in one direction as compared to the other direction, as it is repeatedly reflected back and forth between the two plates. Since, the presence of the electric field will speed up the light beam as it travels in one direction, while it will slow it down as it travels in the opposite direction.

However, the amount of signal shift generated in one direction, due to every upstream passage, will be greater than the amount of signal shift generated in the opposite direction, due to every downstream passage. This is simply due to the amount of time it takes for the light waves to travel across the gap, while going upstream as compared to the amount of time it takes for the light waves to travel across the same gap, while going downstream.

Therefore, the sine wave signal shown on the screen of the multichannel analyzer will shift towards one side. The amount of this shift will directly depend on the strength of the electric field present.

Note that, the shift in the sine wave will be only in the timing of the signal received and not in its frequency. No red shift or blue shift will be detected in these experiments, since the emitter and the receiver are stationary with respect to each other.

Since, the multichannel analyzer can also provide the frequency of the light signal used, one can create a chart demonstrating the amount of shift in the sine wave as a function of the strength of the electric field present.

The amount of shift generated in the sign wave shown on the screen, as compared to the width of a full sine wave will allow the calculation of the speed of the aether flow manifesting as the electric field of certain strength, since the distance between the two plates, the number of times light is made to cross the gap and the width of the individual wavelength of the light wave used are readily known.

Note that, the amount of shift recorded, during such experiments, will provide the needed information to determine the strength of the electric field necessary to slow the propagation of the light waves by any desired amount.

By charging the two plates with a sufficiently strong power supply (possibly from a very large bank of capacitors), one can try to literally stop the propagation of the light waves in a desired direction. The light waves have to fight their way upstream against the flow of aether which is manifesting as the induced electric field.

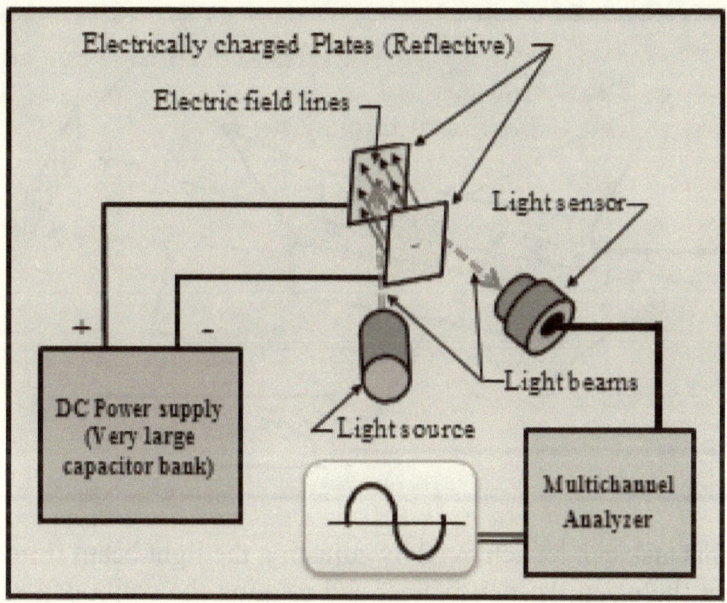

In such an experiment, as shown in the above figure, only one pass of light waves through the induced aether flow will suffice to demonstrate the desired effect, since light will be literally stopped for a moment as it tries to propagate upstream.

> - Second experiment:
>   (Effect of electric field on the direction of light's propagation)

This experiment is designed to demonstrate the effect of electric field on the direction travelled by light or other electromagnetic waves. The design details of the proposed experimental setup to be used are shown below.

First, before energizing the plates, the calibrated screen must be positioned so that it is in-line with the light beam passing between the two plates. Also, the orientation of its calibrated marks must be so that they will indicate the amount of deflection experienced by the light beam, during each of the following steps of the experiment.

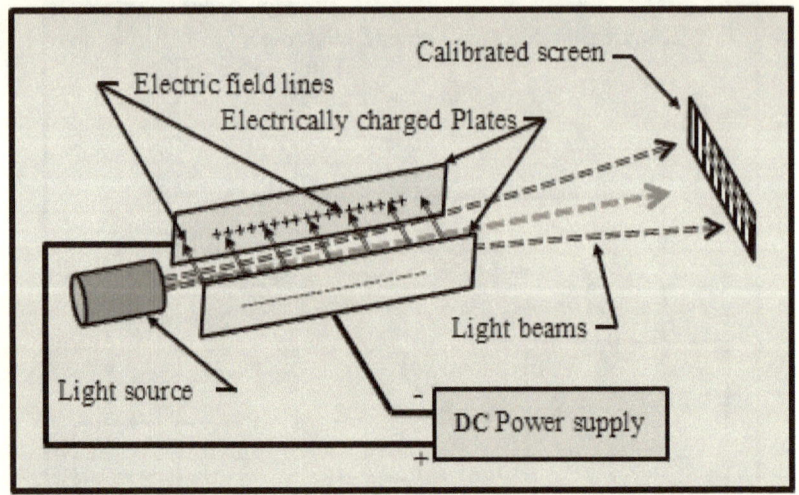

This task can be achieved by adjusting the light beam from side to side (between the two plates) and rotating the calibrated screen to the desired orientation, while the electric current is disconnected from the two plates. The apparatus must be fine-tuned so that the light beam passing between the two plates is focused on the (0.0) mark of the calibrated screen.

Next, the two plates must be charged with the minimal DC voltage setting available. The induced electric field will cause the light beam to deviate from its straight path. The amount of this deviation will directly depend on the strength of the electric field generated. This effect can be easily demonstrated by energizing the two plates with different voltages. Since, the deflected light beam will hit the calibrated screen at different spots, as the two plates are charged with different voltages.

This experiment can also be repeated after reversing the electric charge on the two plates. This can be achieved by simply reversing the wires supplying the electrical power to them. The change in the direction of the electric field will automatically reverse the light show observed on the calibrated screen, in the previous part of this experiment.

Crossflow Experiments
for Two Monochromatic Light Waves
or Two Monotonic Sound Waves

As shown below, consider two beams of light (or any other kind of electromagnetic waves) of the very exact same frequency directed so that they cross each other's path and are picked up by separate sensors and fed into a multichannel analyzer. The monitor on the multichannel analyzer will allow the simultaneous observation of the two sine waves.

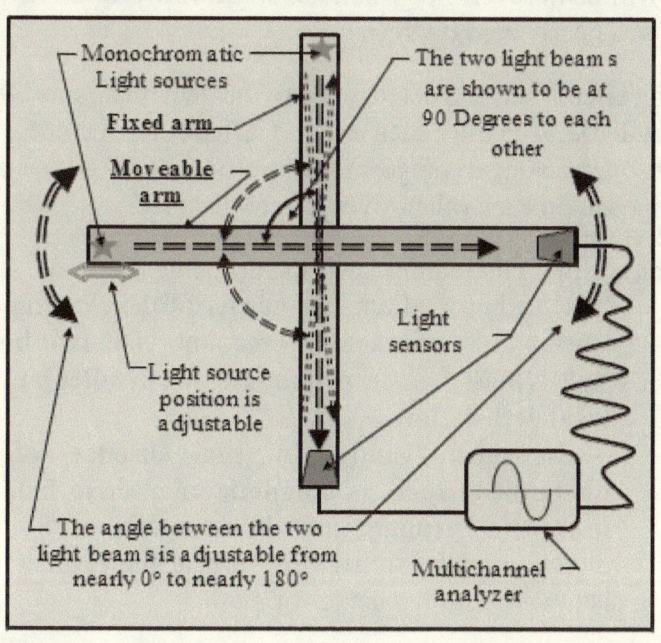

By adjusting the timing (shifting) of the sine wave form of one wave along its propagation route (equivalent of one full wavelength, by simply adjusting the position of one of the two light sources), one can expect to observe a variation in the timing of the wave cycle of the second light beam.

By repeating this experiment for different angles between the two beams, it should become clear that:

- The maximum effects will occur when the angle between the propagation lines of the two beams of light is nearest to zero degrees, as possible.

- As the angle between their propagation lines is gradually increased towards 90 degrees, the observed effect will approach its minimum (zero).

- Further increase in the angle between the propagation lines of the two light beams, from 90 degrees towards 180 degrees, will gradually direct them to propagate in the opposite direction of each other. In this case, the observed effect will approach its maximum again, but this time it will be in reverse of when the angle between the two beams was nearly zero degrees.

Such effects are expected due to the fact that, the two light waves will be affecting each other's effects on the local aether medium over a longer range as their propagation routes become superimposed on each other, over longer distances.

Note that, when conducting such experiments, the motion of the local aether medium (the carrier of the electromagnetic waves) must be taken into account. **The two beams of light (both) must propagate perpendicular to the local aether flow.**

In other words, **in the absence of other phenomena such as magnetic or electric fields, they must be propagating horizontally.** Since, the direction of the aether flow at/near the surface of earth is towards earth's center of mass.

The very same type of experiments can be readily conducted with monotonic sound waves in a medium such as air. Again, the motion of the air medium must be taken into account. Also, in the case of sound waves in mediums such as air, there must be no echoing effect of the sound waves. To properly conduct such experiments with monotonic sound waves in a medium such as air, they must be performed only inside sound proofed environments.

Michelson and Morley Experiment
(Revised)

As it was mention in the beginning of this chapter, the nineteenth century physicists such as Mr. Maxwell had accepted the existence of a stationary medium, namely aether, through which light and other electromagnetic waves were assumed to propagate. Since they knew that earth is orbiting the sun at about 107,000 kilometers per hour, various attempts were made to detect this drift velocity. The assumed motion of earth in the medium of aether is shown in the following figure.

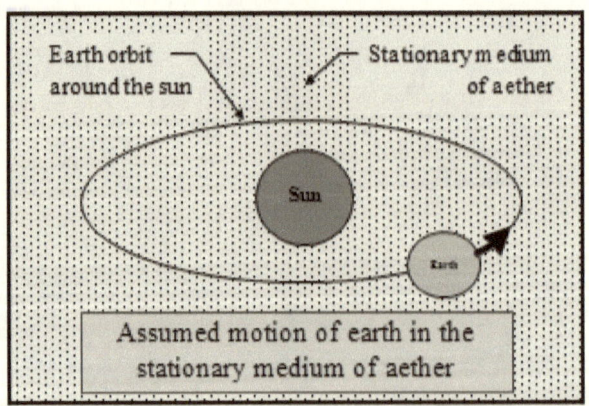

The most promising of all the experiments were designed and performed by Mr. Michelson and Mr. Morley during 1887. They had designed their apparatus with sufficient accuracy to detect such a minute velocity variation caused by earth's motion around the sun. However, these two gentlemen had also assumed that aether was stationary in space and earth was moving through it.

Their experiment was based on the comparison of wave patterns created by the same monochromatic light beam passing through two different paths, one in the direction of earth's motion around the sun and the other at 90 degrees to the said direction. By reflecting back and superimposing the two light branches, Mr. Michelson and Mr. Morley were expecting to see the formation of some sort of interference pattern. The following figure shows the experimental setup used by Mr. Michelson and Mr. Morley.

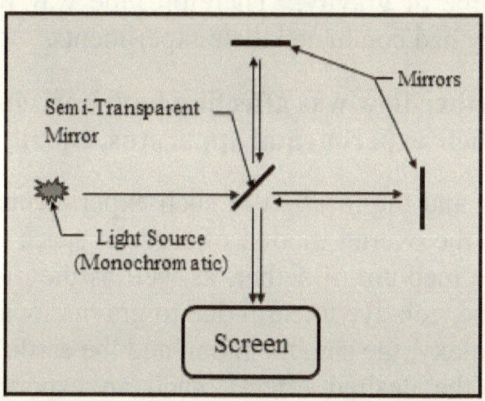

They tried their experiment on the surface of earth. They also conducted their experiment up at high altitudes, with the help of a balloon. However, no interference pattern was formed in any of their attempts. Some scientists suggested that the planet earth can be carrying a thin layer of this aether medium with it, as it goes around the sun. This would explain why no effect was detected.

The aim of this section is to demonstrate that Michelson and Morley experiments would have confirmed the existence of aether if they were performed differently and other affecting phenomena were also taken into account.

According to the theories presented here in this book, aether is responsible for a variety of effects and forces in nature. The force of gravity, the magnetic field and the electric field, among other phenomena, are literally due to a variety of motions induced in the medium of aether. Therefore, as one tries to detect aether's motion due to one phenomenon he/she must also take other affecting factors into account.

The apparatus used by Mr. Michelson and Mr. Morley was set up on a level/horizontal plane (floating on a bath of Mercury).

Their experiments were performed without paying any attention to the other phenomena influencing the local flow of aether and hence directly affecting the task performed by their apparatus. In those conditions, no matter which two orthogonal horizontal directions were picked, there would not have been any observable effects. Because, as it is explained in details in the chapter titled "What is Gravity?", the direction of the flow of aether at/near earth's surface is in fact almost vertically downwards, the very same flow that induces the force of gravity. Therefore, the way Mr. Michelson and Mr. Morley had conducted their experiments,

The local aether flow was affecting both horizontal arms of their experimental apparatus, equally.

The timing and the location of such experiments are also very crucial, due to the overall motion of earth in space (in the galaxy) which is in the medium of aether, as well as the motions that are induced in aether relative to earth due to gravitational forces of the center of the galaxy, the sun, the moon and the earth.

To obtain the desired effects, such an experiment must be conducted **at about midnight**, when:

- The center of the galaxy, the sun, the moon, and earth (in that order) are in line, and hence **their gravitational effects encourage motions in aether that are complementary to each other.**

- The motion of the surface of earth (due to its spin) at that particular location, earth's orbital motion around the sun, and motion of the solar system around the center of the Milky Way galaxy are nearly in the same direction, leading to **motions performed in aether** that complement each other.

These two overall motions of earth relative to its local aether (one due to gravity and the other due to the physical motion of earth in space) will be at almost 90 degrees to each other. The one due to gravity would be perpendicular to the surface, while the one due to various motions of earth in space would be nearly tangent to the surface of earth, and nearly in the east-west direction.

Therefore, one arm of the Michelson & Morley experimental setup should have been nearly in the vertical direction, in the overall (resultant) direction of the two motions of aether relative to earth. The other arm should have been horizontal (level to the surface) and yet perpendicular to both of those two types of relative motions. In other words, the other arm should have been nearly in the North/South direction, or nearly perpendicular to the planes of the solar system and the Milky Way galaxy.

The following figure shows the proper timing to perform such experiments.

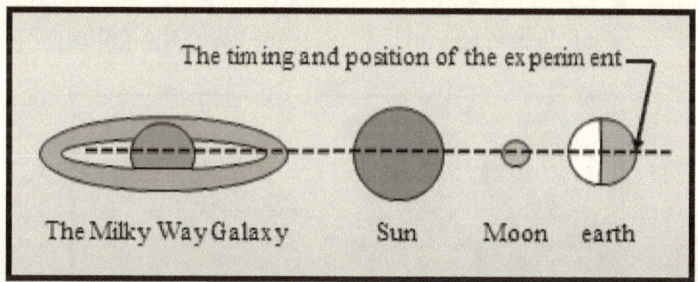

Under these conditions, as well as the proper timing being around midnight, the apparatus which would have one arm raised up vertically (but tilted towards the direction of earth's rotation around its own axis) and one arm horizontal (nearly perpendicular to the rotational planes of the solar system and the galaxy), would clearly demonstrate the desired effect.

The output signals from both arms should be connected to a multichannel analyzer capable of showing both sine waves simultaneously.

The accurate direction (the vertical tilt angle) pointing towards the resultant of the two motions of earth relative to aether can be determined by adjusting the tilt angle until the maximum effect is observed.

The signal that is traveling back and forth in the direction that corresponds to the highest speed relative to aether **will not experience a frequency change** but rather **a change in the length of the time of travel,** leading to a delay in the time of arrival of the signal to the multichannel analyzer.

Therefore, the signal shown on the multichannel analyzer's monitor will indicate a shift in the location (timing) of the sine wave on the screen, but no changes in its frequency will be detected.

The general setup indicating the directions of the apparatus' arms, as it is proposed here, is shown below.

Note that, the North-South direction indicated in the figure should be chosen so that it also takes into account:

- The tilt angle of earth's axis relative to its orbital plane around the sun, and

- The tilt angle of earth's orbital plane around the sun relative to sun's orbital plane in the galaxy.

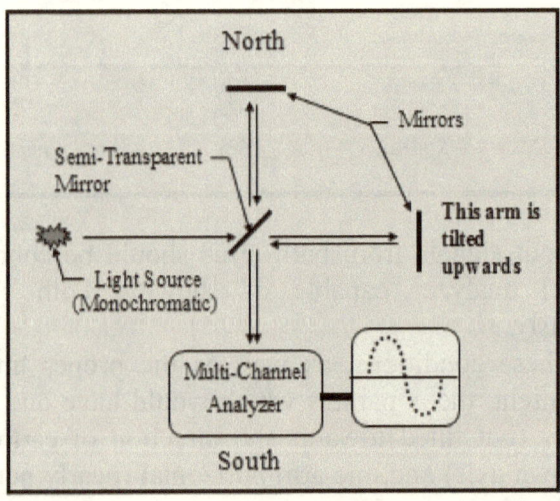

As shown below, if such experiments are performed using a multichannel analyzer, the apparatus used will no longer need to have two orthogonal arms, sophisticated split mirrors and precise timing of the two beams of light and so on. **It would need to have only one arm.**

The apparatus shown below includes two independent yet parallel beams of light travelling in the opposite directions, each with its own independent light source as well as light sensor.

By setting the single arm of the apparatus on a double axis pivot-mount (just like an astronomical telescope), one can point the device towards any desired direction and instantly observe the amount of shift in the location of the sine wave shown on the multichannel analyzer's screen.

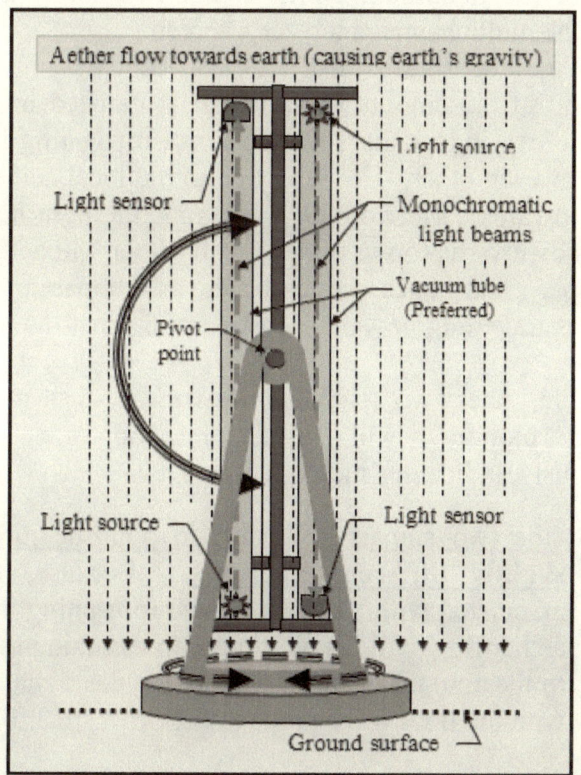

In case needed, one can install the two light sources and the two light detectors inside one tube, rather than two separate tubes. However, each light beam must have its own source and its own light sensor located at the opposite ends of the common tube.

To use such a setup the signals received from the two light beams (shown on the multichannel analyzer's monitor) must be electronically adjusted so that they are superimposed on each other, while the apparatus is in the horizontal position and oriented nearly along the north-south direction. Next, the apparatus should be tilted upright, so that one light beam travels downwards, while the other travels upwards, in their respective contained, vacuumed tubes.

By rotating the stand horizontally and aiming the arm at a variety of vertical angles/directions, the direction corresponding to earth's overall motion relative to its local aether medium (at that particular location and at that particular time) can be readily identified. Since, that particular orientation of the arm will demonstrate the maximum shifts in the two sine wave signals shown on the multichannel analyzer's screen.

This kind of setup also allows for the instantaneous confirmation of the amount of shift that is induced in the arrival times of the two independent light beams propagating in opposite directions of each other. Because, due to the local aether medium travelling towards earth from all directions, the light beam that is travelling downwards (downstream) with aether will actually travel faster in space and hence arrive earlier, as compared to the light beam that is travelling upwards (upstream) against the local aether flow.

Therefore, the two signals shown on the multichannel analyzer's monitor will shift in opposite directions, simultaneously, indicating such effects.

Note that, the two signals will shift in opposite directions by **slightly different amounts**. Because, the signal associated with the light wave propagating upstream of aether flow will lag a bit more in its timing as compared to the amount of timing gained by the signal associated with the light wave propagating downstream.

By rotating the arm of the apparatus by 180 degrees, on its horizontal axis, one can readily observe the two signals exchanging their positions on the multichannel analyzer's screen.

The overall aether's speed relative to earth (at that particular location and at that particular time) can be calculated directly from the maximum shifts generated in the sine wave signals shown on the monitor. The other pieces of information, such as the exact frequencies and hence the wavelengths of the light beams used and the distance between the light sources and their respective light sensors which are needed for such calculations are readily known.

Conclusions

According to the theory presented in this chapter, along with a readily doable experiment which conclusively supports its validity, light and other electromagnetic waves are proposed to be only and only phase vibrations in the medium of aether, just as sound waves are phase vibrations in a medium such as air. In other words,

Light waves and other phase vibrations are waves and do not consist of any kind of particles.

Such properties as color, reflection, refraction, diffraction, dispersion, interference and polarization of light have already been well documented, explained and understood as being due to wave nature of light.

Here in this chapter, other behaviors of light that were the bases for the acceptance of its particle theory, particularly the photoelectric effect and the amplification of light beams by lasers are readily explained using the wave theory of light.

In the absence of any matter particles, the apparent speed of light is equal to its actual speed in aether. In such cases, the light waves travel in a straight line. However, in any other scenario where there are matter particles present, particularly when they are nearby, due to local cross flows of aether induced towards individual matter particles, the path followed by the light waves becomes zigzagged.

Denser matter mediums give rise to more frequent drags / deflections that are mandatorily enforced onto the light waves, as they pass through. Therefore, light waves require more time to cross the medium, since they follow a path that has become ever more lengthened by the more rigorously zigzagged patterns that are formed due to cross flows in the local aether medium.

Light waves make their presence known to an observer as they reach his/her eyes. Light waves will go unnoticed if they just pass by an observer. In other words,

Light waves themselves are not visible, when they are not directed at the observer's eyes.

The very phenomenon of spark is explained to be due to phase vibrations induced in the local aether medium. Also, through a variety of experiments, it is shown how gravity, magnetic field as well as electric field affect the speed of light and the direction it propagates.

It should be emphasized that,

"To obtain accurate results from theories that were based on the existence of aether, one needs to use the instantaneous speed of the object relative to its local aether medium and not in relation to another object or even space."

Also,

"The speed of light used in such theories must be the instantaneous speed of light in the local aether medium, a medium which can have its own motion. Therefore, depending on the motion of the local aether being in the same direction or in the opposite direction, with respect to the direction of the light beam, the effects observed will be weakened or strengthened, respectively."

Note that, the speed of phase vibrations in any medium apparently acts as an upper limit or barrier for objects that intend to move faster in that medium. However, such speed limits are breakable, given the proper requirements are met. For example, the speed of sound in air was viewed as an upper speed barrier for objects. Yet, varieties of objects such as supersonic airplanes and rockets have gone and can go through this speed barrier.

Due to the ongoing decrease in the density of aether and the internal pressure of its medium,

"The speed of light and other electromagnetic waves, as well as other phase vibrations in that medium is gradually increasing."

Currently, the speed of phase vibrations is much higher in the accompanying universe, as compared to what it is in this universe, since the pressure and the density of aether in that universe are much lower than those of aether in this universe.

However, as aether is leaking from this universe into the accompanying universe, gradually the pressure and the density of aether in this universe will become equal to those of aether in the accompanying universe, and will stay equal, as they all gradually decrease towards zero. Hence,

The speed of phase vibrations in this universe will gradually become equal to the speed of phase vibrations in the accompanying universe, and will stay the same afterwards, as it will gradually increase towards infinity in both universes.

What
is
Gravity?

Introduction

Gravity is one of the main forces in nature, a force that has clearly demonstrated its abilities in shaping the contents of this universe, as well as organizing their motions relative to each other. It is also the most common force that man has had to deal with in his daily life.

Particularly over the last 400 years, he has developed relatively solid and reliable theories regarding the laws that govern its various effects.

Mr. Galileo proposed that all objects, regardless of the differences in their densities, in the absence of air resistance, accelerate towards earth at exactly the same rate. Mr. Keppler formulated laws that govern the motions of the planets around the sun. Mr. Newton, with the help of an apple, provided the basic laws of gravity. Using such laws, man has gained remarkable amount of knowledge and insight about the overall structure of this universe.

Through his general theory of relativity, Mr. Einstein explained the force of gravity in terms of curvature in the spatial dimensions.

Some other proposed theories on gravity are based on the exchange of certain particles, namely gravitons, between subatomic particles. According to such theories, all particles in this universe exert gravitational forces on each other by exchanging gravitons.

In 1933, Mr. Zwicky published his findings regarding a peculiar gravitational anomaly concerning the speed of stars in the outer regions of spiral galaxies. They seemed to defy Mr. Newton's universal law of gravity. To explain this anomaly, he proposed the existence of much more matter spread out inside galaxies, matter that is not readily visible and yet gravitationally affects the orbital speeds of stars in the outer regions of galaxies. Later on, the term dark matter was introduced to refer to such types of matter.

Also, as this universe was confirmed to be accelerating in its expansion process (in 1998) another term, namely dark energy, was introduced to take responsibility for enforcing such an expansion process on a universal scale.

It has been decades since astrophysicists and physicists have been using such terms as dark matter and dark energy to refer to certain phenomena which have managed to stay unknown, to this day. Yet, scientists are quite persuaded of the existence of dark matter and dark energy, since they can observe their effects on the visible contents of this universe.

Dark matter and dark energy are found to be everywhere. The amount of dark matter in the universe is calculated to be more than five times the amount of regular matter from which all stars and planets are made.

Furthermore, the equivalent amount of dark energy that is present in this universe is estimated to be nearly three times the amount of dark matter present. In other words,

There is no such thing as a vacuum in this universe, since dark matter and dark energy are literally everywhere.

Among effects associated with the force of gravity, one can mention the following:

- Black holes exert such strong gravitational forces on their surroundings that even light or any other type of electromagnetic waves cannot escape their grip, once they come within a certain distance.

- The force of gravity is gradually becoming weaker and its proven effects include the widening of earth's orbit around the sun.

One of the basic mysteries about the force of gravity is the speed at which it acts. All gravitational computations regarding the navigation of the inter-planetary probes are done using the instantaneous location of the planets and not where they look to be, from the probe's position, at any given point in time. Also, according to direct observations and collected data, at any given instant planet earth is pulled almost towards the true location of the sun and not where it looks to be. In other words, even though it takes about 8.3 minutes for sun's light to reach earth, apparently it takes almost no time at all for its force of gravity to travel this very same distance.

This is in direct contradiction with the underlying principle of special theory of relativity. According to Mr. Einstein's special theory of relativity, nothing can travel faster than the speed of light in a vacuum, while the force of gravity has been doing so all along and will keep on doing so from now on, as well. Of course, by introducing the force of gravity as being due to the curvature of the spatial dimensions, in his general theory of relativity, Mr. Einstein had attempted to provide an explanation for the relatively instantaneous effect of gravity in this universe. However, there are no curvatures in the spatial dimensions. Therefore, again, one needs to ask,

What is gravity?

Any newly proposed theory of gravity must be able to explain all of the known facts, as well as unexplained mysteries concerning gravity in this universe. It must also be able to provide certain verifiable predictions.

A New Theory of Gravity

The new theory of gravity presented here is based on two assumptions:

1- The whole universe is occupied by fluid aether, which is a compressible, non-viscous and quite a dynamic medium.

2- Each subatomic particle in this universe is a tiny 3-D drain hole, through which aether that is under very high pressure in this universe escapes and enters an accompanying universe where its pressure is much lower.

As a first step, for simplicity, one can consider a subatomic particle that is freely floating in space and it is relatively stationary. The incoming flow of aether which is accelerating towards this subatomic particle will be from all spatial directions. The inbound flow of aether will extend literally to the end of space, and it will be directed towards the current position of the subatomic particle.

As it is shown in the following figure, any other particle in existence in this universe will be automatically affected by the flow of aether which is accelerating towards the first subatomic particle and will experience a dragging force directly towards that particle's position.

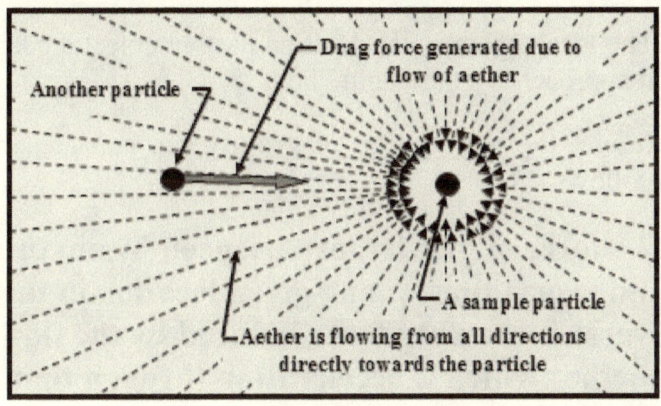

This drag force is the force of gravity that is experienced by anything and everything in this universe.

It must be emphasized early on that,

- **It is the acceleration of aether relative to / through any given location in this universe that gives rise to the drag force which is experienced as the force of gravity by any object that may be positioned at that particular location.**

- **Steady motion of aether relative to an object, regardless of its rate of speed, does not generate any drag force, whatsoever.**

In other words,

> **It is the instantaneous change in the motion of an object relative its local aether medium that gives rise to the drag force that is exerted on that object, the same drag force that is known as the force of gravity.**

That is why, the very same phenomenon of **weight**, is induced when:

- An object is in a vehicle that is accelerating in the aether medium,

- The local aether is accelerating past an object's position, as it approaches a gravitating body.

Therefore, **the new theory of gravity** can be stated in a nut shell as follows:

"The force of gravity, experienced by anything and everything, at any given location in this universe is the drag force induced by the flow of aether which is accelerating through that particular location."

Notes,

- **For the force of gravity to be manifested, matter and/or anti-matter particles must exist.**

  Various types of energy, including phase vibrations in the medium of aether, such as light and other electromagnetic waves are not made of any kind of matter particles. Hence, regardless of their amounts / intensities / magnitudes, they **cannot generate any gravitational force.**

  Also, **the fluid aether** which occupies the whole universe is not made of any kind of particles and hence **does not generate any gravitational force.**

- **"Matter and anti-matter particles are like bubbles that are formed as a result of induced 'cavitation' in the very fabric of fluid aether medium, as the local phase vibrations resonate into sufficiently high amplitudes."**

  The formation of matter and anti-matter particles (bubbles) in the aether medium by resonating phase vibrations is analogous to the formation of bubbles in a fluid medium such as water due to cavitation induced by ultrasound waves which are phase vibrations in that medium.

  Bubbles (particles) formed by cavitation in the medium of fluid aether are openings into the accompanying universe. They act as drain holes (openings / tubes) through which fluid aether flows from this universe into the accompanying universe.

The following picture shows the entrance of a water-shoot inside the reservoir of a dam. Such water-shoots prevent the water that is behind the dam to rise beyond a certain safe level. They perform their important task by simply allowing the excess water to fall in them and get transferred downstream, bypassing the dam structure altogether.

The flow of aether towards and through matter and anti-matter particles, as well as black holes, is quite the same as the flow of water towards and through such openings. However, the flow of aether towards any given particle is from the whole volume of space surrounding that particle, a volume of space which is three dimensional, while the flow of water towards the inlet of such a water-shoot is from a planar surface which is only two dimensional.

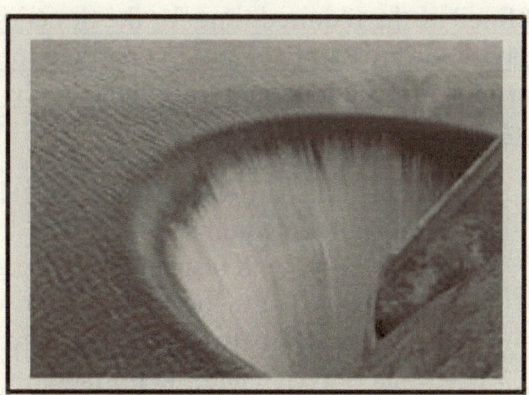

Such water intakes clearly demonstrate how the flow of water changes direction, as it literally disappears from the two dimensional planar surface in which it existed before. The very same type of a process takes place with the fluid aether. Since, as it reaches a given matter or anti-matter particle (or a black hole) it literally disappears from the familiar three dimensional volume of space that forms this universe and enters the physical dimensions corresponding to the accompanying universe.

Note that, in the two-dimensional surface of water, the inlet part of such a water-shoot is observed as a void.

The very same is the case for the matter and anti-matter particles in the aether medium, since they too are bubbles, voids in the three-dimensional space which is this universe.

For details on how matter and anti-matter particles were/are formed in this universe, please refer to chapters titled "Formation and Development of the Universe" or "What is Aether?".

- In the case of a particle that is moving in space, the flow of aether towards that particle automatically adjusts its direction and closely follows that particle's current position, since the driving force for aether's flow towards particles is its internal pressure. The amount of **'lag time'**, defined as the required time for the direction of the force of gravity towards a given moving particle (or an aggregate of particles) to catch up and truly indicate the true position of that particle(s), depends on:

  1- How fast the particle(s) is moving with respect to the local aether medium,
  2- Its trajectory, with respect to the location of interest,
  3- The distance to the location of interest, as well as
  4- The difference between the pressure of aether that is in this universe and the pressure of aether that is in the accompanying universe.

- The drag force induced due to the flow of aether is directly proportional to the rate at which aether is accelerating and it is not like the drag force induced by the flow of a medium such as air on a solid object, since particles are not solid obstacles in the path of aether's flow.

  It is very important to emphasize that, it is the relative motion/acceleration of aether approaching from the opposing directions that generates the drag force, since all particles have the tendency to have uniform motions with respect to the fluid aether that is in their respective immediate vicinities.

- The proposed curvature of spatial dimensions near a gravitating celestial body, as described by the general theory of relativity, corresponds to the gradient that exists in the speed of aether as it accelerates while approaching that object. Therefore,

The very same formulations derived in the General Theory of Relativity which describe the force of gravity in terms of the curvature of spatial dimensions can be readily modified and applied to represent the drag force induced by the flow of aether as it accelerates towards such celestial bodies.

In other words,

Each and every phenomenon that has been or can be explained by the General Theory of Relativity can be readily inherited by this new theory.

Comment:

One could state that the very expansion process of the aether medium must also give rise to a force of gravity of some sort which is actually acting opposite of the normal force of gravity. Since, instead of encouraging particles (as well as their aggregates) to come closer together it is pulling them apart from each other, as it is expanding the domain of space in this universe.

The term gravity, as it is defined in this book, refers to **the drag force at any given location, in this universe, that is induced by the flow of aether accelerating through that particular location**, a motion that is towards matter and/or anti-matter particles. The motion of aether that is due to its expansion process, at any given location in this universe, does not contribute to any flow of aether at that location (in space), let alone being an accelerated flow, because space itself is literally expanding with the aether medium that is occupying it. Hence, no flow of aether is generated by the expansion of its medium.

The spreading action / motion of aether that is due to its medium's expansion can be readily likened to the spreading of the molecules in a piece of rubber as it is being stretched. Since, as shown below, even though two dots drawn on the surface of such a rubber piece will recede from each other, if the rubber piece is stretched, there will be no flow of rubber molecules through the position of either one of these two dots.

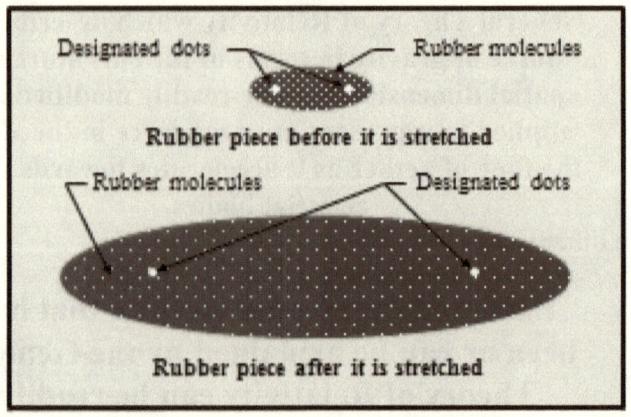

Explanations and Predictions Provided by the New Theory of Gravity

The newly proposed theory of gravity provides the following consistent explanations for the known gravity related phenomena. The new theory also makes certain predictions, a few of which are also included and described in detail.

1- Mr. Galileo's experiment

All objects are made of the very same three elementary particles, namely protons, neutrons and electrons. The atoms in all of the elements are only different combinations of these three particles. Different objects in nature (table, chair, water, Iron, wood, meat and so on including celestial bodies) are also made possible by different combinations of various elements.

According to the newly proposed theory (presented in this chapter), each and every particle experiences the force of gravity (such as that associated with earth) independent of all other particles.

Therefore, regardless of them being as individual particles or as parts of an atom, a molecule, or an object of any size or type, when they are allowed to freely fall towards earth, they all will accelerate towards earth at the very same rate, given there be no air resistance (no mid-way collisions). This is why Mr. Galileo's proposal that, in the absence of air resistance, even a feather and an Iron ball will fall at the very same rate, is correct and accurate.

Also should be noted that,

> "The constancy of the gravitational acceleration (g) at any given elevation is also due to the constancy of the drag force induced by the flow of aether which is accelerating towards earth's center of mass at that particular elevation (measured from the center of mass of earth), regardless of the longitude and latitude of the location."

2- Mr. Newton's law of gravity

The strength of the force of gravity between any two objects is directly proportional to their masses and inversely proportional to the square of their distance. This relationship was derived by Mr. Newton and it is known as the inverse square law of gravity,

$$F = Gm_1m_2 \, / \, x^{2.000}$$

Where, **F** is the gravitational force that two objects of masses m_1 and m_2 separated by a distance of **x** exert on each other. And, **G** is the universal gravitational constant.

The effect associated with the masses of the two objects and the effect associated with their separation distance should be considered separately, as follows.

The mass effect:

As shown below, each of the particles in one object directly and independently experiences the force of gravity induced by each individual particle that exists in the other object. Therefore, the overall gravitational force that is experienced between any two objects is in direct relation with the number of their particles multiplied together. Hence, the term (**m₁** x **m₂)** appears in the above equation.

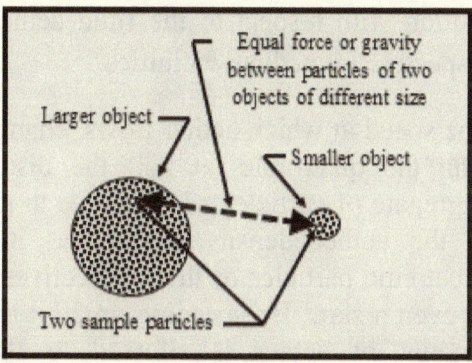

The distance effect:

Aether flows towards subatomic particles from their surrounding volume of space. Since the surface area of a sphere covering all around any particle is given by the formula, $(4\pi r^2)$, the total surface area of the sphere changes by the square of the distance (radius, $r = \mathbf{x}$) to the particle.

Therefore, in order for the flow of aether to stay consistent, the speed at which aether moves changes accordingly, as it approaches the particle. For example, at half the distance the spherical surface area surrounding a given particle reduces to a quarter which means at that point the fluid aether must be moving (accelerating) four times faster. Consequently, as the speed/acceleration of aether quadruples to compensate for the reduction in its flow cross sectional area, the drag force generated by the motion of aether also quadruples. Hence, the square term (x^2) in the equation provided by Mr. Newton.

Notes,

- The drag force induced due to the flow of aether is directly proportional to the rate at which aether is accelerating and it is not like the drag force induced by the flow of a medium such as air on a solid object, since particles are not solid obstacles in the path of aether's flow.

  It is very important to emphasize that, it is the relative motion/acceleration of aether approaching from the opposing directions that generates the drag force, since all particles have the tendency to have a uniform motion with respect to the fluid aether that is in their respective immediate vicinities.

- The speed at which aether flows, changes by a bit more than the quadruple, at half the distance to a given aggregate of particles. This is due to minute reductions in the aether density/pressure as it approaches the subatomic particles or their collectives such as a planet or even a star. By speeding up a bit more, a consistent amount of aether is allowed to flow towards the particles, at all distances.

3- Equivalency of the gravitational forces exchanged between two objects such as earth and sun, which are different in size

The inequality of two objects in size does not affect the equality of the gravitational forces they exert on one another. For example, the following figure shows an object which is made of 1,000 particles and another object which is made of 150,000 particles. In this case, each of the 1,000 particles in the first object experiences the force of gravity induced by each and every one of the 150,000 particles in the second object. And, correspondingly, each of the 150,000 particles in the second object experiences the force of gravity induced by each and every one of the 1,000 particles in the first object.

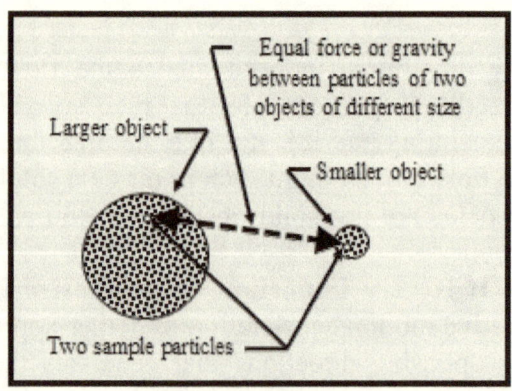

As shown, in the figure, each of the particles in one object directly and independently experiences the force of gravity induced by each individual particle that exists in the other object. And, the forces exchanged between any two particles are equal in both directions. Therefore, the totals of the gravitational forces that are experienced by all of the particles within any two objects, relative to each other, are automatically equalized.

Note that, the force of gravity experienced between any two particles that are relatively stationary in their common background aether medium is equal in both directions, even if the two particles happen to be of different sizes.

As it is shown in the following figure, the flow intensities of the fluid aether towards the two particles are in direct relation with their respective sizes (cross sectional areas which are in direct relation to their respective masses) and automatically cause the two forces to become equal.

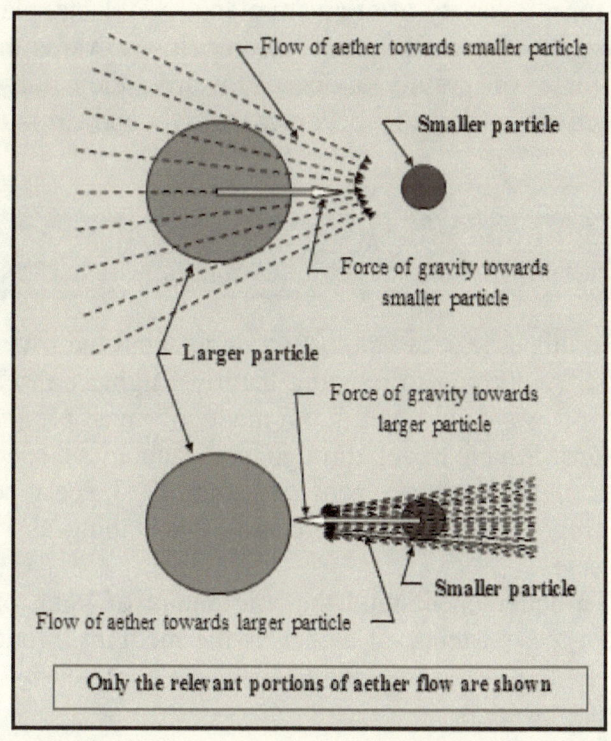

Flow of aether towards smaller particle

Smaller particle

Force of gravity towards smaller particle

Larger particle

Force of gravity towards larger particle

Smaller particle

Flow of aether towards larger particle

Only the relevant portions of aether flow are shown

Even if a single particle is confronted by a black hole at a fair distance the very same effect still holds, since the ratio between their masses will be exactly that of their cross sectional areas.

4- Accumulative effect of the forces of gravity due to several celestial bodies in one region of space

The speed of aether rushing towards any given region of space directly depends on the overall number as well as sizes of available openings or drain holes in that region. Therefore, the greater is the number of subatomic particles present in one region of space the faster fluid aether is automatically encouraged to flow towards that region. This is analogous to the flow of a fluid through a strainer, since more open holes automatically allow more fluid to flow towards and through the strainer, as a whole.

Therefore, as more subatomic particles are grouped together and form a variety of atoms, molecules, objects, planets, stars and galaxies, the overall induced aether flow towards them and hence the force of gravity induced towards their larger and larger collectives, as a whole, becomes stronger and stronger.

5- Bending of light's trajectory, as it passes by a star or a galaxy

In this case, one has to first understand the true nature of light itself. According to existing theories, light is an electromagnetic type of wave and yet it is made of massless particles called photons, which travel through the vacuum of space. However, based on the evidences provided since 1933, space between stars is not empty. Therefore, there is no such thing as vacuum in this universe.

To better understand the true nature of light, one has to first accept the existence of aether as the medium through which light travels, the very same aether that was accepted by the nineteenth century physicists.

Light and all other electromagnetic waves are phase vibrations in the medium of aether which is everywhere. Therefore, laws governing the propagation of sound waves in a medium such as air also apply to the propagation of light waves in aether. The similarities between the bending of sound waves and light waves are shown in the following figure.

Just as sound waves get deflected due to being dragged while passing through a cross wind, so do electromagnetic waves as they pass by a star. Since, every star induces a relatively high speed aether flow towards its center of mass. And, as electromagnetic waves pass by a star they go through this induced cross flow of aether which happens to be along their paths.

Light waves dragged towards stars and galaxies, from all around, give rise to what is known as the **gravitational lens effect**. This lens effect becomes particularly obvious as distant galaxies are observed next to galaxies that are relatively closer to the observer.

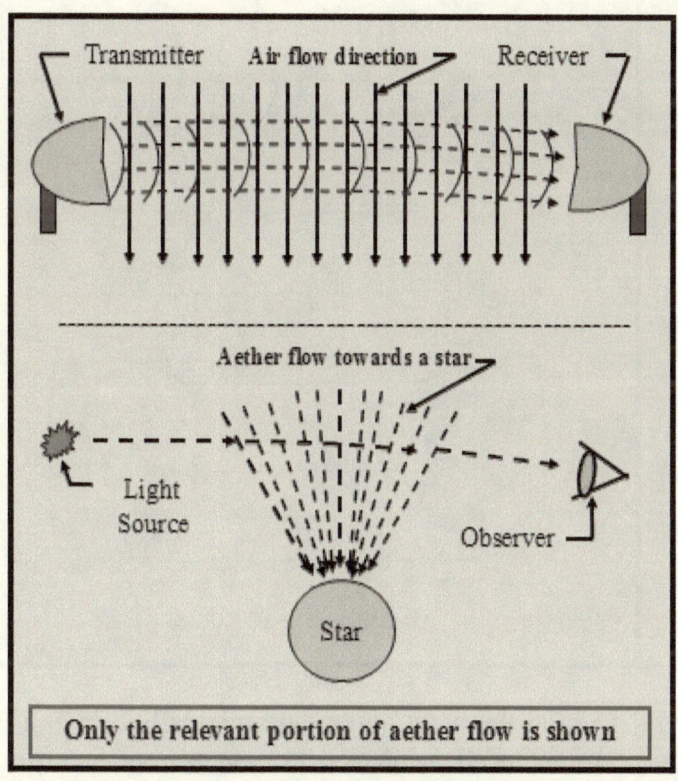

6- Extreme force of gravity near black holes

The following two examples can help the reader to visualize the strength of the force of gravity that exists in the immediate vicinity of a black hole.

- **One can consider the propagation of sound waves in air.** Using a supersonic wind tunnel, a variety of experiments can be performed to demonstrate the desired effects.

   To perform such experiments, as shown below, one needs to install a speaker (which is attached to a stereo system) near the outlet and a microphone (which is connected to a headphone set) near the inlet of the supersonic portion of the wind tunnel.

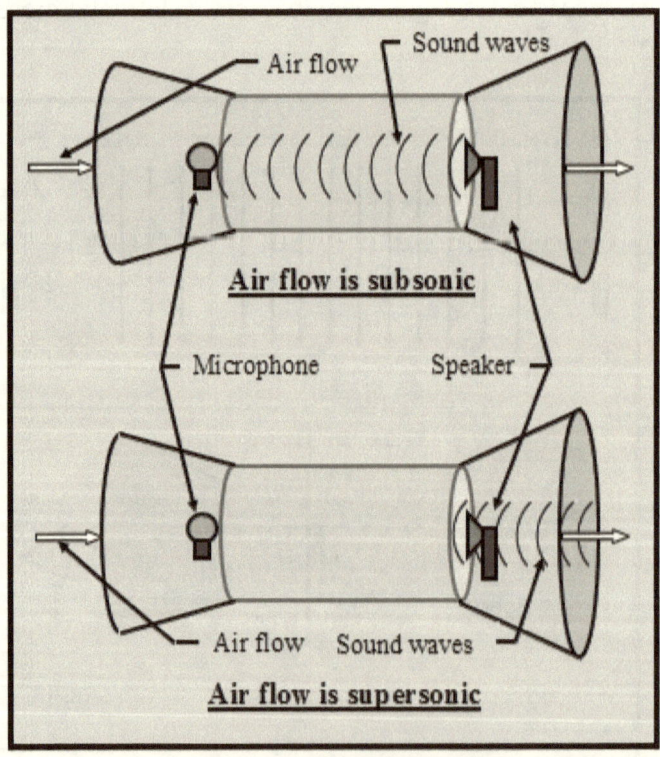

By starting the wind tunnel and gradually bringing it up to its supersonic speed, while listening to his/her favorite music, he/she will notice that the sound heard through the headphone set gradually becomes weaker and weaker, as if it is broadcast from farther and farther away. So long as the speed of the wind generated by the wind tunnel is less than the speed of sound waves in that air medium, sound waves generated by the speaker can propagate against the flow of air and reach the microphone.

However, once the sonic barrier is crossed, he/she will not hear the music at all, because the sound waves can no longer compete against the speed at which air is flowing in the opposite direction. In fact, they will get dragged downstream, even farther away from the microphone.

- **One can consider a giant drain hole at the bottom of an ocean.** In this case, as shown below, if the diameter of the hole is big enough and the ocean is sufficiently deep, due to the enormous pressure that will be present, the flow speed of water towards the drain hole at a certain distance, which can be readily calculated, will reach and exceed the speed of sound waves in water.

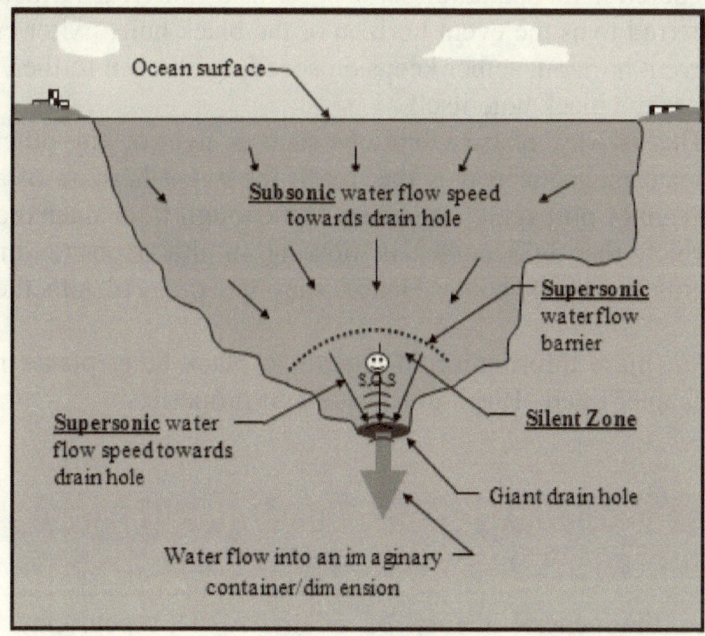

If someone who is being pulled into such a drain hole and is already passed that certain distance, tries to broadcast a sound wave type of an S.O.S. message towards outside, his cry for help will never be able to propagate outwards and it will be dragged into the drain hole. This is due to the fact that, the sound waves will not be propagating fast enough to counter the speed at which water is flowing towards the drain hole. Such a region in any ocean can be referred to as **"Esmailzadeh Silent Zone"**, since no sound waves will ever be heard coming out of it.

These examples clearly demonstrate exactly what is taking place in the immediate vicinity of a black hole. A black hole is made up of an enormous number of matter particles (drain holes) which have literally joined together and have formed a <u>single unified spherical gateway</u>. Such gateways allow unrestricted flow of fluid aether from this universe into the accompanying universe. At a certain distance from such giant gateways the induced inward flow of aether from the surrounding volume of space reaches the speed of phase vibrations in that medium. That distance, when considered in all spatial directions, forms a spherical surface which is referred to as the event horizon of the black hole. After crossing the event horizon, aether keeps on speeding up even further, until it reaches the black hole itself.

This is why, phase vibrations such as light or any other kinds of electromagnetic waves that reach the event horizon of a black hole cannot propagate up-stream fast enough to counter the speed at which the fluid aether is flowing in the opposite direction, towards the black hole. Hence, they get dragged into the black hole.

For more information in regards to black holes please refer to the chapter titled "Black holes and their properties".

7- Topographical presentation of the gravitational force near stars, planets and black holes

In the general theory of relativity the topography of the gravitational force near massive stars, galaxies and particularly near black holes are interpreted as the curvature of the spatial dimensions. In fact, the curvatures / slopes calculated and shown represent the gradients that exist in the flow speeds of the local aether that is headed towards the star, the galaxy or even the black hole.

Stars, galaxies and black holes (on their own scales) are dense matter concentrations in relatively small or limited regions of space. Denser matter concentrations mean more available drain holes, in a given volume of space, through which aether can escape from this universe. This effect automatically translates into faster flow rates of aether towards those particular regions of space.

The strength of the gravitational force of any celestial body, particularly that of a black hole, as shown in the following figure, can be represented by a funnel type of a shape because the strength of the gravitational force is directly proportional to the rate at which the flow of aether is accelerating at any given location (distance from such celestial bodies).

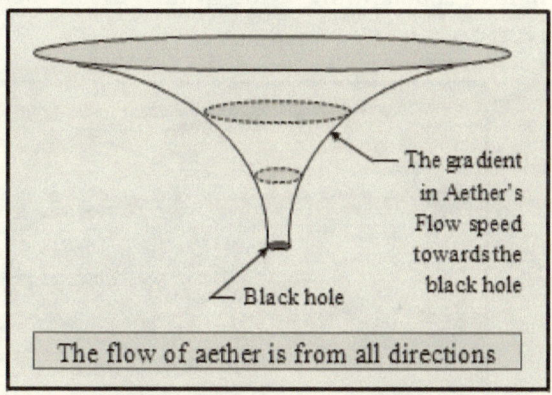

In other words, the progressively steeper slopes of the smooth surface of the funnel shape shown in the figure actually represent the gradual increases in the speed of aether that is approaching the black hole. Since, the closer aether gets to the black hole, the faster it accelerates.

The very same is also true about the gravitational influences of other celestial bodies such as those of planets, stars and galaxies. However, in their cases, the overall slope of the central portion does not become so pronounced.

Note that, the experiment that was performed in 1919, during a total solar eclipse, had apparently confirmed the prediction made by the general theory of relativity. However, such experiments did not and do not indicate any curvature in the fabric of space (or space-time), but rather simply indicate how light waves were and are dragged by the flow of aether which is accelerating towards gravitating bodies such as stars. The following figure shows such an effect as it is induced by aether flow, in a space that consists of straight dimensions.

The prediction made by the general theory of relativity agrees with the observed amount that light waves bend as they pass near a star, simply because the very same equations derived in general theory of relativity apply to the proposed theory of gravity presented in this book, a theory that introduces the force of gravity as the drag force induced by the flow of aether which is accelerating through any given location in this universe.

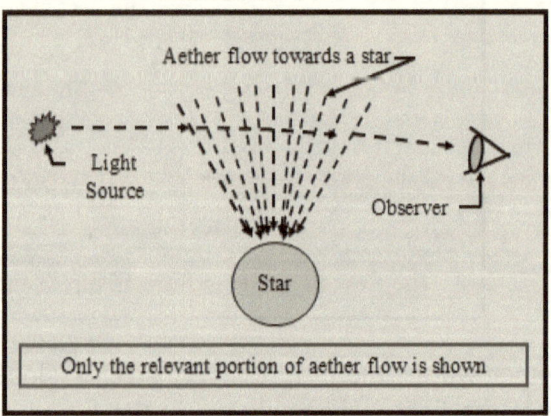

Only the relevant portion of aether flow is shown

8- Do light waves, other electromagnetic waves and other phase vibrations in the aether medium, generate a force of gravity of their own?

For the force of gravity to be manifested, matter and/or anti-matter particles must be present, since it is the accelerated flow of aether towards matter and/or anti-matter particles (bubbles in the medium of aether) that exerts the drag force on anything and everything that happens to be in its path, the very same drag force that manifests as the force of gravity experienced by other particles and/or their aggregates.

Various types of energy, including phase vibrations in the medium of aether, such as light and other electromagnetic waves are not made of any kind of matter particles. Hence, regardless of their amounts / intensities / magnitudes, they cannot generate any gravitational force, whatsoever.

Note that, it was due to this very simple and yet very important fact that when all of the energy content of this universe was concentrated in a very small volume of space, during the early stages of this universe's life (when matter and anti-matter particles were not formed yet), no force of gravity existed. And, also that was why the aether medium could continue with its rapid expansion until matter and anti-matter particles started to form in that medium.

9- Does fluid aether generate a force of gravity of its own?

For the force of gravity to be manifested, matter and/or anti-matter particles must exist. Since, it is the accelerated flow of aether towards such bubbles that exerts the drag force on anything and everything that happens to be in its path, the very same drag force that manifests as the force of gravity experienced by other particles and/or their aggregates.

The fluid aether which occupies the whole of this universe, as well as the accompanying universe, is not made of any kind of particles and hence cannot and does not generate any gravitational force, whatsoever.

Note that, it was due to this very simple and yet very important fact that when all of the aether content of this universe was concentrated in a very small volume of space, during the early stages of this universe's life (when matter and anti-matter particles were not formed yet), no force of gravity existed. And, also that was why the aether medium could continue with its rapid expansion until matter and anti-matter particles started to form in that medium.

10- Existence of dark matter in the entire universe

As it is described in detail in the chapter titled "What is Aether?", dark matter particles are comprised of a variety of matter and anti-matter particles that are not readily visible. They can be in such forms as neutron stars, black holes, but mainly as individual particles, particularly a variety of particles which are either quite short lived due to their continuous production and destruction by resonances/spikes forming in phase vibrations propagating in the aether medium, or are quite small in size, among many others.

Dark matter exists in vast quantities throughout this universe, but its existence can only be detected through its side effects such as its force of gravity. Since, each and every dark matter particle generates an accelerated aether flow towards itself. The induced aether flow, in turn, gives rise to a drag force which is the gravitational force exerted on any and all other particles/objects that happen to be in its path.

According to the above mentioned sources/types of dark matter within galaxies, the dark matter content (in fact, the percentage of the overall mass) of the older galaxies must be greater than that of the younger ones. To confirm this overall proposal, one can perform the following experiment.

> • Experiment:
> (Relation between age of galaxies and the percentage of their overall mass that is in the form of dark matter)

To demonstrate the relation between the age of galaxies and the percentage of their overall mass that is in the form of dark matter within them one must study the orbital velocities of stars in many galaxies which are of a variety of shapes and forms. Particularly, the galaxies selected must include younger ones which are still spherical / elliptical in their overall geometries, as well as those that are much older, well developed, demonstrating well-defined flat geometries such as spiral galaxies.

The data needed is the magnitude of the redshifts associated with many stars that are following their orbital paths around their respective galaxy's center of mass, at different distances.

Note that, such sets of data can be extracted from the existing data which have been collected for stars of many galaxies, already.

The following figure shows the relative motions of several typical stars in two different types of galaxies, namely a spherical / elliptical galaxy and a spiral / flat galaxy.

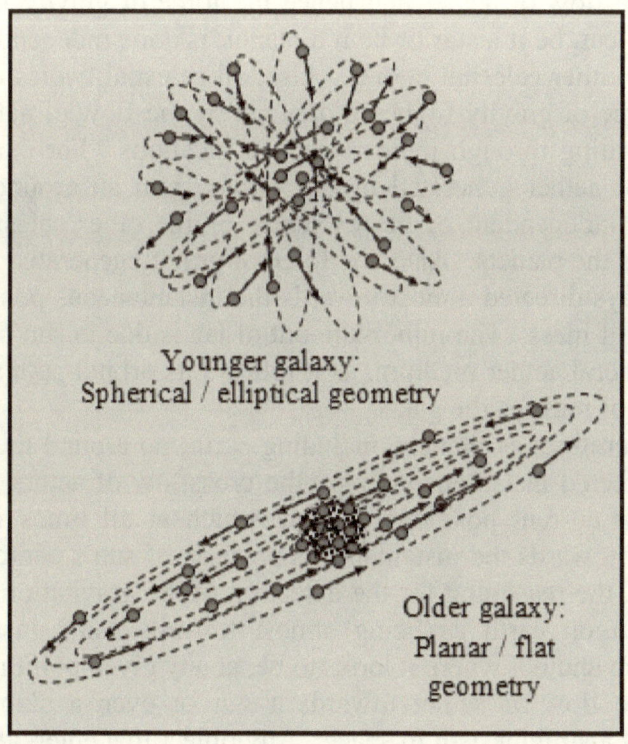

Younger galaxy:
Spherical / elliptical geometry

Older galaxy:
Planar / flat
geometry

The extra gravitational forces associated with the dark matter buildup within star systems, even though are relatively quite weak, they cause faster than expected deceleration of probes that are trying to explore the outer reaches of star systems such as our solar system.

In fact, this is exactly what has been experienced by the two probes, namely Pioneer 10 and Pioneer 11, which were sent towards the outer regions of our solar system. The extra (unexpected) decelerations of these two probes were confirmed since the early 1980s, and are well documented.

11- Nearly instantaneous effect of the gravitational force of the sun on earth

The flow of aether and hence the force of gravity induced by any object, be it a star or be it a planet, is done independent of any and all other celestial bodies. Also, all celestial bodies experience the force of gravity by being dragged by the flow of aether that is accelerating through their respective locations. For example, the flow of aether generated by the sun is from all around and it is totally independent of the existence of the other celestial bodies such as the planets. Also, the force of gravity generated by the sun is always directed almost towards the instantaneous position of its center of mass. The minor amount of lag is due to sun's motion in the general aether medium, as it follows its orbital path around the center of mass of the galaxy.

Therefore, as planets, including earth, go around the sun, they are affected instantaneously by the crossflow of aether that exists at their current positions, a flow which at all times is directed almost towards the instantaneous position of sun's center of mass. Hence, the reasoning for the detection of the gravitational force of the sun on earth as being almost towards sun's instantaneous location and not where it looks to be, at any given point in time.

The flow of aether towards a star or even a planet follows **almost** a straight path in space. Any object that enters this flow, it is affected instantly, and not after a few seconds or minutes. It instantly experiences the drag force that is induced by the local flow of aether. The exerted force is always **almost** directly towards the current location of the planet or the star. **Of course,** the faster a given star (or a planet) moves in any direction that is nearly perpendicular to the line of sight from a given object, the more lag-time will be detected before the direction of the gravitational force exerted by that star on the object advances and points directly at the instantaneous location of the star.

The motion of earth around the sun and how earth experiences sun's gravity is shown in the following figure.

In fact, the orbital motion of earth around the sun is screw shaped, since sun and its associated planets are moving at a fairly high rate of speed around the center of mass of the galaxy, and earth's orbital plane around the sun and sun's orbital plane around the center of mass of the galaxy form an angle which is about 60 degrees.

Therefore, earth is actually constantly dragged not towards the current position of sun's center of mass, but rather towards a point in sun's path around the center of mass of the galaxy, where sun used to be a short while ago.

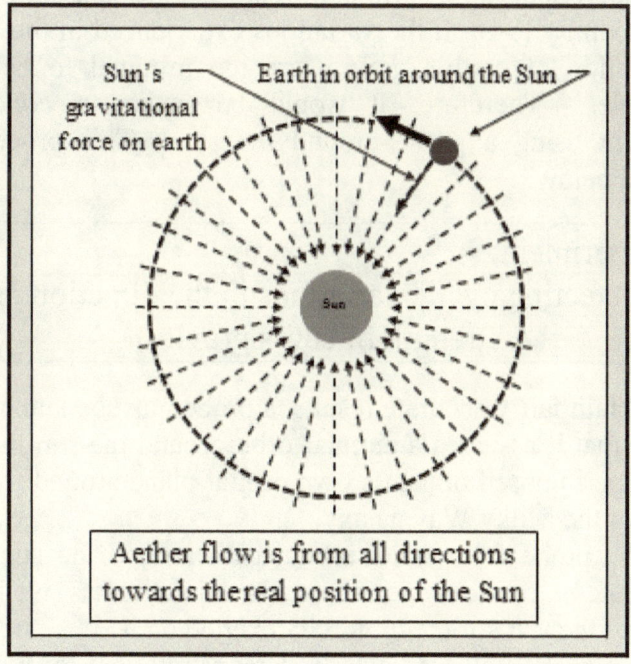

Aether flow is from all directions towards the real position of the Sun

Note that, the flow of aether and the resulting force of gravity due to the existence of planets and stars is always present, regardless of the other physical objects being present or not.

In fact, **sun and earth as well as all of the other celestial bodies in this universe don't know and don't need to know that any other celestial body of any size even exists.** At any given instant, any celestial body independently generates an influx of aether towards itself, a flow that is gradually accelerating towards its position and induces a drag force on objects that happen to be in its path.

12- How fast variations in the strength of the force of gravity are experienced at a given distance from a star or a planet?

Even though the moon is orbiting the earth in a plane which is nearly superimposed on earth's orbit around the sun, but due to its close proximity to earth the variations experienced in the strength and direction of earth's gravity are too minimal to be readily measurable. Therefore, it would be better to conduct an experiment using a probe around the sun. This procedure is described below.

> - Experiment:
>   (Detecting cyclic variations in the direction and strength of sun's gravity)

To obtain fairly accurate results, a probe must be launched into space so that it assumes a circular orbit around the sun in a plane that is superimposed on sun's own orbital plane around the center of mass of the Milky Way galaxy.

As the probe follows its orbital path around the sun, it will demonstrate a cyclic variation in the direction as well as the strength of the sun's force of gravity exerted on it.

Such a probe will also clearly demonstrate that, **only when it crosses sun's orbital path around the center of mass of the galaxy, it actually experiences the force of sun's gravity in the**

exact same direction that also points towards sun's instantaneous location.

However, at all other parts of its orbital path the probe will indicate that it is pulled towards points where sun used to be at some short time in the past. And, as the probe gets farther away from sun's orbital path it will be dragged towards where sun used to be further back in time.

Also, **as the probe crosses sun's orbital path in front of the sun, where sun has not passed through yet, it will experience a weaker force of gravity as compared to when it crosses sun's orbital path behind the sun, where sun has already passed through.**

The difference in the strength of the gravitational forces of the sun as it is experienced by such a probe at those two very special spots along its orbital path, as compared to what they are expected to be based on the distance to the sun's center of mass, can be used to calculate the speed at which sun's gravitational force varies at those particular distances from the sun. The orbital path of such a probe is shown in the following figure.

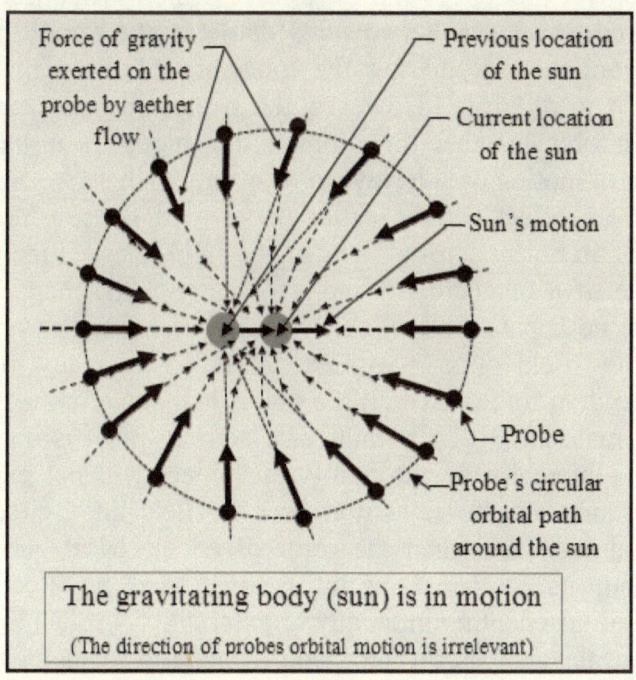

The gravitating body (sun) is in motion

(The direction of probes orbital motion is irrelevant)

Note that, once sufficiently sensitive equipment is developed, such an experiment can be performed using moon as a probe which is already orbiting earth in a plane that is nearly superimposed on earth's orbital plane around the sun.

13- Why fluid aether does not exert any drag force on planets, causing them to slow down?

Planets and other objects that are in motion relative to their local aether medium, regardless of their instantaneous speeds relative to their local aether, do not experience any drag force along their respective direction of uniform motion in the local fluid aether. They only experience the drag force which is due to the accelerated flow of aether through their respective immediate vicinities. Since, **aether exerts a drag force on objects, as well as waves, only and only in the direction that it is in an accelerated motion with respect to them.**

Therefore, as a planet follows its orbital path around its nearby star, it only experiences the drag force which is induced by the accelerated flow of aether towards that star, the same drag force that is continuously altering the trajectory of the planet and is keeping it in its orbit. However, the planet does not experience any drag force in other directions, either along its instantaneous direction of motion or sideways to its orbital path.

Such an effect can be better understood by considering the motion of an object across the magnetic field lines. If the object is not made of a magnetically inclined material the magnetic field will have no apparent effect on it whatsoever. Such an object can cross the field lines without experiencing any type of resistance/drag force. Even if the object is made of a magnetically inclined material <u>BUT</u> the induced electricity in it is not used in any way, internally or externally, the object will not experience any resistance/drag force as it crosses the field lines, either. This case is almost like when the rotor of an electrical generator is turned but its electrical circuit is open and no electricity is transferred / used either internally or externally. In such a case, the rotation of the generator's rotor would not demand any power from its driving power source.

However, if the object is made of a magnetically inclined material <u>AND</u> the electricity induced inside it is either transferred to some outside application or it is somehow consumed internally (due to induced eddy currents), the magnetic field will affect that object's motion. Since, as the induced electricity is used by either an internal loss or by an external application, some energy is extracted from the magnetic field, hence the object will experience some resistance to its motion as it moves across the field lines.

The very same can be stated in regards to sails mounted on sailboats. Since, only when sails are open and are extracting energy from the motion of air, sailboats and their motions become affected by the local air flow.

14- Formation of the force of gravity, temporary and rapid expansion of space and its slowing down

In the beginning, the medium of the fluid aether was experiencing its highest pressure/density. As somehow phase vibrations were introduced in that medium, the aether medium started to expand. Even though aether was experiencing its highest pressures ever, and was experiencing its fastest expansion rate ever, it could not exert any such force as gravity, because **there was no aether flow**. In other words,

"When this universe was born, the very lack of the force of gravity allowed the unrestricted expansion of the aether medium which defines space."

This growth burst which is referred to as the inflationary period, in a relatively short period of time (based on then-current time scale) spread the fluid aether and hence expanded space in this universe from a very small region to a volume that based on present-time scales was millions of light years across.

During its rapid expansion process, aether was not confronted by any kind of resistive force, until its density was reduced to a certain level and allowed the formation of matter and anti-matter particles.

The formation of matter and anti-matter particles in aether and the timing of their appearance all across the whole volume of the aether medium that was hosting phase vibrations can be likened to the formation of icicles from individual groups of molecules of water, as the temperature of the medium of water is lowered. Because, in this case also, as a certain temperature is reached (depending on the pressure), the entire volume/body of water starts to crystallize, simultaneously.

This is basically what happened within the medium of aether. Because, as its density and pressure were decreased due to its expansion process, the needed environment was provided for stable eddies to form and eventually develop into matter and anti-matter particles.

Note that, in the beginning, phase vibrations were distributed uniformly in the aether medium, except for its outer regions which were (and still are) void of any kind of phase vibrations. Therefore, as the pressure and the density of the fluid aether medium were decreasing, due to its expansion, the same circumstance was simultaneously provided throughout its volume that was hosting phase vibrations. Consequently, phase vibrations present in the fluid aether resulted in the formation of matter and anti-matter particles everywhere, at about the same time. In other words,

"The whole universe simultaneously bloomed with matter and anti-matter particles across its volume that was hosting phase vibrations."

Matter and anti-matter particles acted as drain holes which allowed aether to escape from this universe and accumulate in a different environment, where it gave meaning to space in the accompanying universe. That was when and how **aether started to flow** as it was literally pushed towards each and every particle present and hence started to exert a drag force on everything that happened to be in its path, and gave meaning to the force of gravity. In other words,

"The very formation of matter and anti-matter particles marked the birth of the force of Gravity."

The very formation of matter and anti-matter particles and their sudden abundance, to the point of saturating the whole central region of the aether medium, caused two major side effects:

1- Due to their abundance, matter and anti-matter particles allowed unrestricted flow of aether from this universe into the accompanying universe. Hence, in a relatively short period of time, aether pressure in this universe dropped drastically, the same pressure that was the driving force for aether's expansion.

2- The pressure driven flow of aether towards matter and anti-matter particles exerted a mutual drag force on the neighboring particles and eventually on all existing particles in this universe. Due to their abundance, matter and anti-matter particles generated and imposed such strong gravitational forces on each other that, together they acted like an effective braking system that drastically slowed down the receding motions of all matter and anti-matter particles from each other in the whole universe.

In other words,

"By their very formation, matter and anti-matter particles drastically reduced the aether pressure in this universe which consequently slowed down the expansion of the aether medium and the expansion of space, hence effectively ended the rapid expansion era, or the inflationary period."

The inflationary period of this universe can be likened to the very operation of **airbags**, installed in vehicles. Because, as shown in the following figure, when an airbag is deployed the internal gas pressure causes the volume of the airbag to increase quite rapidly. The internal gas pressure causes the cover fabric to stretch, but only to a certain point when holes appear in the cover fabric and allow the gas to leak through causing the internal pressure to drop, in a short time.

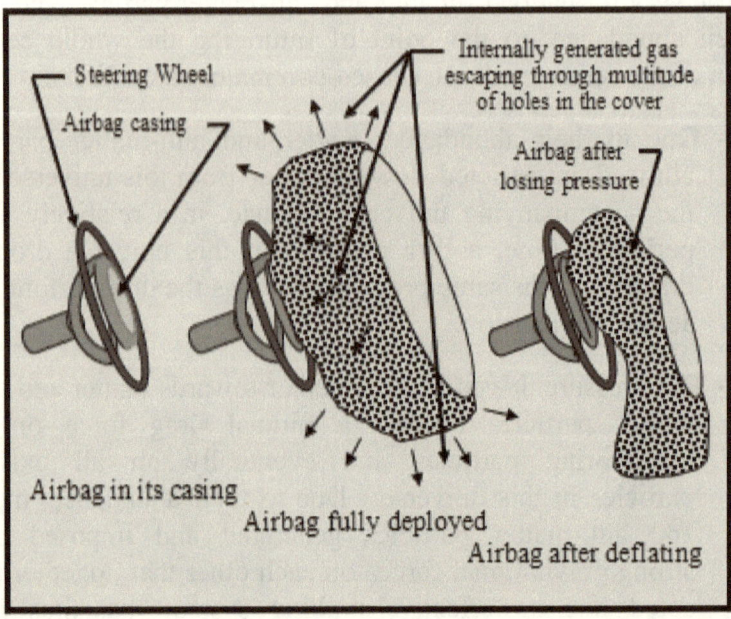

One could say, airbags experience a miniaturized inflationary process of their own, as compared to what the whole universe experienced on a grand scale, during its infancy.

15- How do neutrons enable protons to bond together in the nuclei of atoms?

Protons have positive electrical charge and so when they happen to get close to each other, they repel one another. In the case of complex nuclei, where two or more protons are necessarily coexisting within relatively speaking short distances of each other, there is a need for an outside help to encourage protons to get along. Neutrons are the particles that basically do just that in any and all complex nuclei.

The way protons resist being in close proximity of each other and how the very presence of neutrons persuades them to get along can readily be demonstrated by two strong magnets and a chunk of magnetic but not magnetized material. This is shown in the following figure.

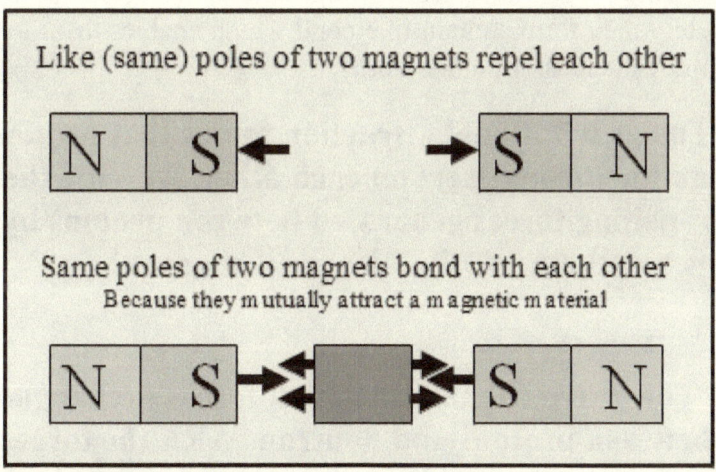

Shown in the figure are two different circumstances. In one case, the like poles of the two magnets are made to directly face each other. In such a scenario, the like poles of the two magnets definitely repel one another. While in the other case, a chunk of magnetic material is placed in the space between the two magnets. In such a scenario, the two like poles of the two magnets literally get distracted by the presence of the magnetic material and both attract that same magnetic material towards themselves. In so doing, they actually bond with each other through the magnetic material. In other words,

The attraction forces that like poles of two magnets exert on a magnetic material that is placed in between them overrides their mutually repelling force towards each other.

The very same effect takes place as two protons are made to get close to each other. If there are no neutrons in between them, the two protons definitely repel one another from far away distances. Yet, when there are one or more neutrons present in the space between the two protons, those two protons literally ignore each other's electrical charges and simply pay attention to the one or more neutrons that happen to be present between them. One could say, **protons concentrate their efforts in exchanging gravitational forces with those neutrons.** In so doing, each and every proton gets attracted towards the existing neutrons and ultimately they form a chain like bond which enables them to form a complex nucleus. In other words,

> **"The gravitational attraction forces that protons and neutrons exert on each other override the repelling forces generated between protons by their positive electrical charges."**

Therefore,

> **"The gravitational attraction forces exchanged between protons and neutrons ARE the forces responsible for holding the nucleons together, in complex atomic nuclei.**

Hence, according to aether-based theory of gravity,

> **"There is no such a force as the weak nuclear force in this universe."**

The following figure shows three different scenarios. In the first one, there are only two protons that resist getting too close to each other. In the second one, the two protons notice that there is a single neutron between them and try to exchange gravitational forces with that neutron, and in so doing, they form a semi bond between themselves, as well. In the third case, the two protons notice that there are two neutrons in between them, and as each proton tries to attract the two neutrons and also by being attracted by them both, the four particles form a much better and more stable bond together, as a whole.

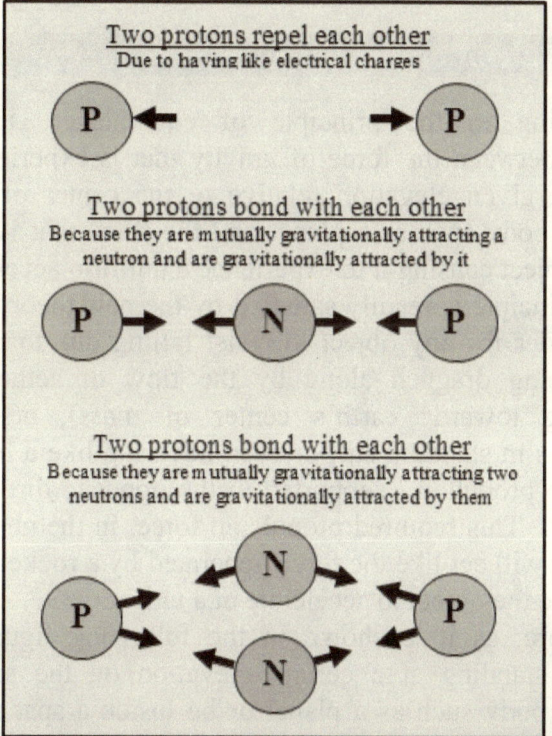

Two protons repel each other
Due to having like electrical charges

Two protons bond with each other
Because they are mutually gravitationally attracting a
neutron and are gravitationally attracted by it

Two protons bond with each other
Because they are mutually gravitationally attracting two
neutrons and are gravitationally attracted by them

Note that, <u>it is due to the lack of sufficient number of neutrons or excess number of neutrons that certain nuclei (isotopes) are unstable.</u>

The lack of sufficient number of neutrons in a given complex nucleus, necessary to properly bond the existing protons together, eventually causes that nucleus to lose its overall structural integrity and split in a variety of fashions, as the particle constituents of such a nucleus get maneuvered around by the local phase vibrations, as well as their harmonics.

Correspondingly, extra neutrons in a given nucleus also motivate instability in the overall structure of that nucleus. Since, the excess neutrons tend to cause protons to move in an abnormal / unstable fashion, as they try to uphold their bonds with too many neutrons, concurrently. Hence, they are literally encouraged to get out/off of their stable tracks, eventually leading to the overall instability of the nucleus, as a whole.

16- Principle of equivalence

According to the principle of equivalence, there is no difference between the force of gravity that is experienced by an object at a given elevation relative to the center of mass of a gravitating body such as a planet and the force that is exerted on that very object causing it to experience a uniform acceleration.

This principle is readily satisfied by the new theory of gravity, since in order for any object to resist falling due to the force of gravity (being dragged along by the flow of aether which is accelerating towards earth's center of mass), at any given elevation, as in standing still or even hovering like a helicopter, it needs to be propelled / supported in the opposite direction of the aether flow. This required propulsion force, in the absence of the aether flow will act like the force generated by a rocket engine that would cause the object to accelerate at a uniform rate.

Therefore, as it is shown in the following figure, either a person be standing at a certain elevation on the surface of a gravitating body such as a planet or be inside a spaceship that is accelerating at a constant rate, he/she will experience the very same effect, which is being pulled towards a specific direction by a constant force.

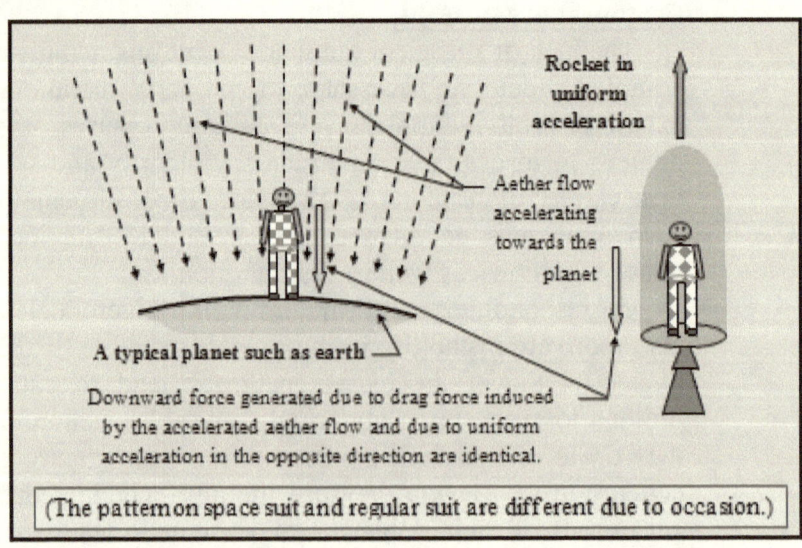

Rocket in uniform acceleration

Aether flow accelerating towards the planet

A typical planet such as earth

Downward force generated due to drag force induced by the accelerated aether flow and due to uniform acceleration in the opposite direction are identical.

(The pattern on space suit and regular suit are different due to occasion.)

It must be emphasized early on that,

- **It is the acceleration of aether relative to / through any given location in this universe that gives rise to the drag force which is experienced as the force of gravity by any object that may be positioned at that particular location.**

- **Steady motion of aether relative to an object, regardless of its rate of speed, does not generate any drag force, whatsoever.**

In other words,

It is the instantaneous change in the motion of an object relative its local aether medium that gives rise to the drag force that is exerted on that object, the same drag force that is known as the force of gravity.

That is why, the very same phenomenon of **weight**, is induced when:

- An object is in a vehicle that is accelerating in the aether medium,

- The local aether is accelerating past an object's position, as it approaches a gravitating body.

It must be particularly emphasized that,

According to the information presented in the chapter titled "What is Time?" along with the results of the experiments proposed to verify their validities,

"Even though force of gravity and uniform rate of acceleration give rise to certain identical effects, such as the weight of objects, they have nothing in common as far as inducing 'Time Dilation' is concerned."
Therefore,

"The paradoxes / twins stories about accelerating objects / living beings that experience time at a slower

pace, as compared to those left relatively stationary on
their home planet, are fundamentally flawed. Since,
such paradox stories totally ignore the actual speeds
at which such objects / living beings are moving
relative to their respective local aether mediums."

It is the instantaneous speed of an object relative to its
local aether that gives rise to 'Time Dilation' effect
and not its instantaneous accelerations.

17- Can acceleration, gravity and high speed lead to experiencing 'Time' at a faster pace?

The two experiments described below explore the possibility of
combining the effects of acceleration, gravity and high speed on
how "Time" is experienced, as indicated by three identical atomic
clocks.

- First experiment:
      (The balloon version)

To perform this experiment, three identical atomic clocks ("A",
"B" and "C") should be synchronized, while all three are
positioned side by side, on the ground level. The three clocks
should be equipped with transmitters so that they can
independently transmit their respective "Time" experiences to a
computer located at a receiving station, fixed on the ground level,
nearby. The computer should be programmed to graphically show
the differences in the rates at which "Time" is experienced by the
three clocks, simultaneously, while taking into account the effects
due to its instantaneous distance to each one of the clocks. In other
words, it should simultaneously show three curves,

- One corresponding to the difference in the rates at which
  "Time" is experienced by clocks "A" and "B"

- One corresponding to the difference in the rates at which
  "Time" is experienced by clocks "A" and "C"

- One corresponding to the difference in the rates at which "Time" is experienced by clocks "B" and "C"

Clocks "A" and "B" should be lifted up into a fairly high altitude, by separate balloons. Clock "C" should be kept near the receiving station on the ground level to act as a reference. Clock "A" should be lifted up while it is housed in a rocket. Therefore, due to the weight of the rocket, the balloon carrying clock "A" must be considerably larger in size, as compared to the balloon carrying clock "B".

Once the two balloons reach the desired altitude, the balloon carrying clock "B" should be kept hovering at that altitude, while clock "A" (along with its rocket) should be released from its balloon and with the help of its rocket engine should be propelled directly towards earth, at a high rate of acceleration.

The three clocks should be continuously transmitting the rate at which they experience the passage of "Time", to the computer, from the instant they are synchronized up until clock "A" either hits the ground level or melts / explodes due to generation of excessive heat as it travels through the lower layers of the atmospheric.

The following figure shows the overall concept of this experiment in a simple fashion.

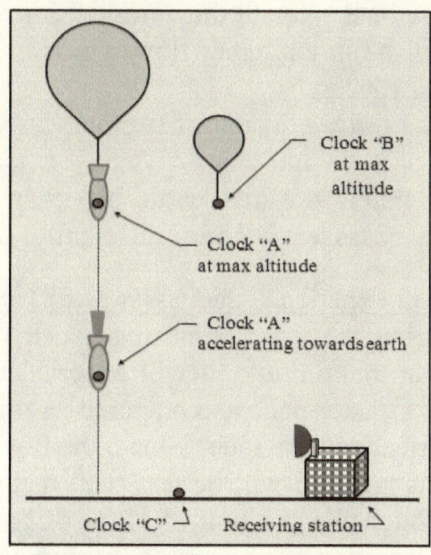

Clock "B"
at max
altitude

Clock "A"
at max altitude

Clock "A"
accelerating towards earth

Clock "C"   Receiving station

Note that, the acceleration of the rocket carrying clock "A" back towards earth can be of any rate, yet it is preferable that it be as high and over the longest period of time as possible.

According to aether-based theories, presented in this book, the following results are expected to be shown on the computer screen, during such an experiment:

Clock "C" will experience the passage of 'Time' at a constant pace, during the whole period of this particular experiment, due to its speed relative to its local aether being constant. Since, clock "C" will be in the path of an aether flow which is going through its position at a constant rate of speed equivalent to the vector sum of earth's rotational speed at its fixed latitude and altitude and the flow speed of aether which is moving directly towards earth's center of mass and is giving rise to earth's force of gravity, at that particular altitude measured from earth's center of mass.

Clocks "A" and "B", at first, will mutually experience the passage of 'Time' at gradually faster paces, as compared to clock "C", as their altitude from the center of mass of earth is gradually increased, that is until they reach the desired altitude. Since, as clocks "A" and "B" get farther away from earth's center of mass they will be in an aether flow which is moving towards earth at slower speeds.

However, from the instant that clocks "A" and "B" reach their highest altitude and clock "B" hovers there, while clock "A" is accelerated towards earth by its rocket, they will experience the passage of "Time" differently, since,

Clock "B" will experience the passage of "Time" at a constant rate. During the whole time that clock "B" can be kept hovering at that altitude, it will experience the passage of 'Time' at a faster pace as compared to the rate clock "C" will experience its passage. Since, the flow speed of aether through its position will be less than that of aether that is going through the position of clock "C", the same aether flow that is giving rise to earth's force of gravity.

For this part, the prediction made by the General Theory of Relativity is in agreement with that of the aether-based theories, but due to different reasoning. Since, according to the General Theory of Relativity clock "B" will experience the passage of 'Time' at a faster pace because of being exposed to a weaker gravitational force, as compared to clock "C", due to being farther away from the center of mass of earth.

Clock "A" will experience the passage of "Time" at gradually faster paces, as compared to clock "B" (and clock "C"). Since, its acceleration towards earth due to the rocket engine will be over and above its acceleration which is due to being dragged by the accelerated flow of aether towards earth. Therefore, the speed of the rocket (clock "A") relative to its local aether will gradually decrease. In short,

The rocket, along with clock "A", will be gradually catching up with the local aether flow which is accelerating towards earth's center of mass.

Consequently, as the speed of clock "A" relative to its local aether gradually decreases, it will experience the passage of 'Time' at faster and faster paces. Since, it is the speed of that clock relative to its local aether that dictates how fast it will experience the passage of 'Time'. In other words,

As clock "A" is accelerated towards earth, even though,

- **Its speed relative to earth and clock "B" will be continuously increasing, and**
- **It will be in an accelerated state of motion, going away from clock "B", and**

- **It will be experiencing a stronger gravitational force dragging it towards earth, than what is experienced by clock "B",**

yet, it will gradually experience the passage of 'Time' at faster and faster paces, as compared to clock "B", let alone compared to clock "C".

The following figure is a very simple and yet exaggerated presentation of the chart that will be shown on the monitor of the computer receiving and calculating the differences between the rates at which 'Time' is experienced by the three atomic clocks used. The timing of the chart starts from the instant clock "A" starts its accelerated (powered) fall directly towards earth.

As shown in the figure, the amount of 'Time' difference between the two clocks "B" and "C" will increase at a constant rate, since those two clocks will experience the passage of 'Time' at slightly different but fixed rates.

However, as clock "A" starts its powered acceleration towards earth it will gradually experience the passage of 'Time' at faster and faster rates, as compared to clock "B" which is already experiencing 'Time' at a slightly faster pace as compared to clock "C".

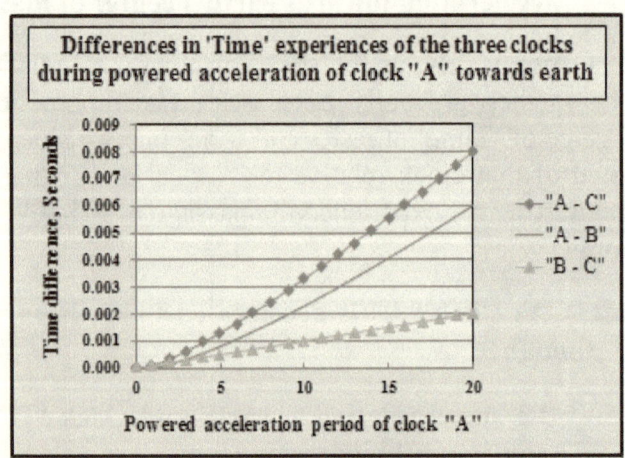

Such an outcome is clearly in contradiction with what is expected / predicted by the Special

Theory of Relativity, the Equivalence Principle and the General Theory of Relativity. Since, they all predict that in such a scenario, clock "A" must be experiencing the passage of 'Time' at gradually slower paces, as compared to the rate clock "B" or clock "C" experiencing its passage.

If the rocket used to accelerate clock "A" be powerful enough so that it can catch up with the local aether that is flowing towards earth, clock "A" will experience the passage of 'Time' the fastest. In fact, at that instant, clock 'A" will experience the passage of 'Time' as if it were positioned at an infinite distance from earth.

Depending on the precision / accuracy of the atomic clocks used, as well as those of the timings of the signals received / processed by the computer, such an experiment can be performed on a much smaller scale, since only the difference in the rates at which the two clocks "A" and "B" experience the passage of 'Time' is of prime interest.

Even a small rocket launched from a helicopter would be quite adequate to perform such an experiment, accelerating clock "A" towards earth, while clock "B" is kept onboard the helicopter, so long as the rocket used has sufficient propellant to accelerate directly towards earth's center of mass at a fairly high rate, until it hits the ground surface.

- **Second experiment:**
  (The orbital version)

To perform this experiment, an atomic clock ("A") should be sent into a very elongated elliptical orbit around earth, while another atomic clock ("B") should be kept on the surface of earth as a reference. The altitude of the aphelion of the orbital path chosen may need to be roughly twice that of a geosynchronous orbit, due to limited length of time that this experiment will take to perform.

As the rocket approaches the aphelion of its elliptical orbit, it must reverse its thrust and slow its orbital speed down, change direction and **accelerate directly towards earth**. The powered acceleration of the rocket can be of any rate, yet it is preferable that it be at such a rate that, as it runs out of propellant, its speed towards earth becomes equivalent to that of scape velocity from earth's gravity (at its then current altitude), before it reaches earth's atmosphere. The following figure shows such a scenario.

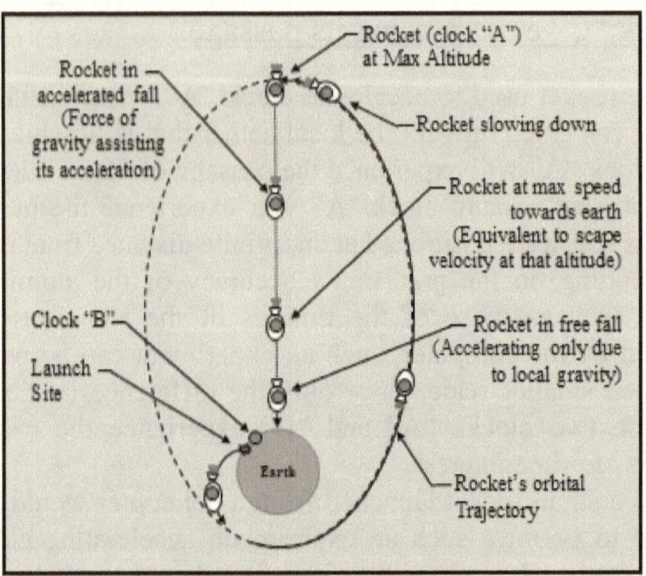

As clock "A" reaches the aphelion of its orbit and starts its powered accelerated descent directly towards earth, the two clocks must transmit their current 'Times' to a computer which should be positioned next to clock "B". The computer needs to calculate the true difference between the rates at which the two clocks experience the passage of 'Time', by taking into account the instantaneous distance of clock "A" as it accelerates towards earth (as well as earth's rotational speed at the altitude and latitude of the location where the computer and clock "B" are positioned), and show the result as a graph on its screen.

According to aether-based theories, presented in this book, the following results are expected to be graphically shown on the computer screen, during such an experiment:

Clock "B" will experience the passage of 'Time' at a constant pace, due to its speed relative to its local aether being constant.

Since, clock "B" is in the path of an aether flow which is going through its position at a speed equivalent to the vector sum of earth's rotational speed at its fixed latitude and altitude and the flow speed of aether which is directly towards earth's center of mass and is giving rise to earth's force of gravity at that particular altitude measured from earth's center of mass.

Clock "A" will experience the passage of 'Time' at varying rates. Since, its speed relative to its local aether will be continuously changing, from the instant it starts its powered accelerated descent directly towards earth until it runs out of propellant, shortly before it hits earth's atmosphere and burns up. However, its experiencing 'Time' can be divided into three major parts.

1- While being relatively at rest with respect to earth, at the aphelion of the orbit chosen, clock "A" will experience the passage of time at a faster pace, as compared to clock "B". Since, the speed of clock "A" relative to its local aether will be less than that of clock "B" relative to its local aether. Because, the speed of aether rushing by them will be equivalent to scape velocity from earth's gravity at their respective altitudes measured from earth's center of mass.

For this part, the prediction made by the General Theory of Relativity is in agreement with that of the aether-based theories, but due to different reasoning. Since, according to the General Theory of Relativity clock "A" will experience the passage of 'Time' at a faster pace because of being exposed to a weaker gravitational force, as compared to clock "B", due to being farther away from the center of mass of earth.

2- Once the rocket starts to accelerate towards earth, due to being pulled by earth's gravity and also being pushed by its own engines, clock "A" will experience the passage of time at gradually faster paces as compared to clock "B". Since, the acceleration of the rocket due to its engines will cause it to speed up over and above its acceleration which is due to being dragged by the accelerated flow of aether towards earth. Therefore, the speed of the rocket (clock "A") relative to its local aether will gradually decrease. In short,

The rocket, along with clock "A", will gradually catch up with the local aether which is accelerating towards earth's center of mass.

Consequently, as the speed of clock "A" relative to its local aether gradually decreases, it will experience the passage of 'Time' at faster and faster paces. Since, it is the speed of that clock relative to its local aether that dictates how fast it will experience the passage of 'Time'. In other words,

As clock "A" is accelerated towards earth and its speed is increased relative to earth, and while is in an accelerated state of motion, it will experience the passage of 'Time' at faster and faster paces, as compared to clock "B", which is stationary on the surface of earth.

The following figure is a very simple presentation of the chart that will be shown on the monitor.

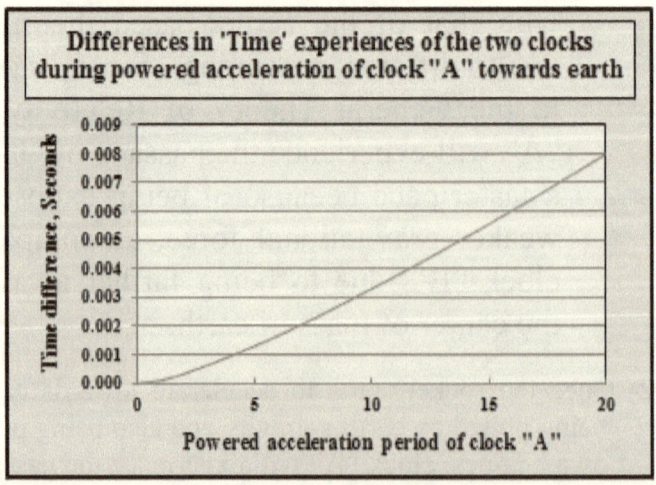

Such an outcome is clearly in contradiction with what is expected / predicted by the Special Theory of Relativity, the Equivalence Principle and the General Theory of Relativity. Since, they all predict that in such a scenario, clock "A" will experience the passage of 'Time' at gradually slower paces, to the point that it will be even slower than the rate clock "B" experiences its passage. Because, speed of clock "A" in space will be much faster than that of clock "B". Clock "A" will also be in a state of acceleration, over and above that of the local force of gravity.

3- Once the speed of the rocket towards earth becomes exactly equal to that of aether flowing towards earth, clock "A" will experience the passage of 'Time' the fastest. In fact, at that instant, clock 'A' will experience the passage of 'Time' as if it were positioned at an infinite distance from earth.

Since rocket engines will run out of propellant, as its speed becomes equivalent to that of its local aether flow towards earth, clock "A" will continue with its accelerated descent towards earth but in a state of free fall, as it will be dragged along by the accelerated flow of aether towards earth's center of mass.

During this portion of its journey towards earth, clock "A" will keep on experiencing the passage of 'Time' the fastest, as if it were hovering at an infinite distance from earth's center of mass. Since, its speed relative to its local aether will be equal to zero. In other words, clock "A" will keep on experiencing the passage of 'Time' faster than clock "B" until it hits earth's atmosphere and burns up.

Such an outcome is clearly in contradiction with what is expected / predicted by the Special Theory of Relativity. Since, it predicts that in such a scenario clock "A" should experience 'Time' at gradually slower paces, as compared to clock "B" which is stationary on the surface of earth, that is until it hits the atmosphere and burns up. Because, speed of clock "A" in space will be much faster than that of clock "B".

Mr. Einstein is quoted as having said,

"No amount of experimentation can ever prove me right; a single experiment can prove me wrong."

From the above quotation it is clear that, even though performing such experiments were not quite feasible during the first half of the twentieth century, Mr. Einstein was already fully aware of such potential scenarios.

18- Why does gravity induce 'Time Dilation'?

According to the new theory of 'gravity' presented in this chapter,

"The force of gravity, experienced by anything and everything, at any given location in this universe is the drag force induced by the flow of aether that is accelerating through that particular location."

In other words, the very existence of the force of gravity at any given location in this universe implies that there is a flow of aether which is accelerating relative to anything and everything that may exist at that particular location. The speed at which the local aether is flowing towards a gravitating body depends on the overall mass of that gravitating body as well as the distance to it, since as shown below, the flow of aether towards any and all gravitating bodies is from all spatial directions.

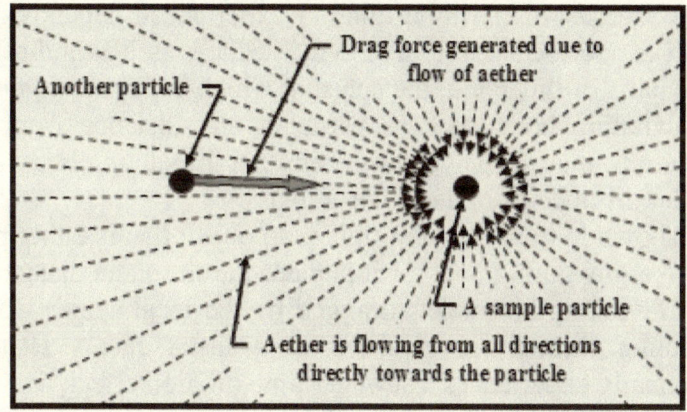

Also, according to the theory of 'Time' presented in the chapter titled "What is Time?",

"The rate at which 'Time' is experienced by an object depends on that object's speed relative to its local aether medium, as compared to the propagation speed of phase vibrations in that medium."

As the speed of an object, relative to its local aether, approaches that of phase vibrations in that medium, the object experiences the passage of time at progressively slower paces. In case the speed of the object reaches the speed of phase vibrations in its local aether medium, the passage of time will no longer be experienced by that particular object.

Therefore, the very presence of the force of gravity at any given location in this universe automatically indicates that anything and everything that may exist at that particular location experiences 'Time' at a slower pace as compared to objects that are located far away from any matter or anti-matter particles, let alone away from any of their aggregates. In other words,

"The very presence of the force of gravity at any given location in this universe automatically implies that anything and everything that may exist at that particular location experiences 'Time dilation'."

The severity of the time dilation experienced depends on the strength of the local force of gravity. Since, as it is indicated in various parts of this book, the speed at which aether flows towards any gravitating body such as earth, at any distance from it, is equivalent to the escape velocity from that gravitating body at that particular distance from its center of mass.

As shown in the following figure, an object that is either placed closer to a gravitating body or it is placed at the same distance to a larger gravitating body, gets dragged by the local aether which is accelerating through its position at a faster rate. Hence, it automatically experiences a stronger time dilation effect, as well.

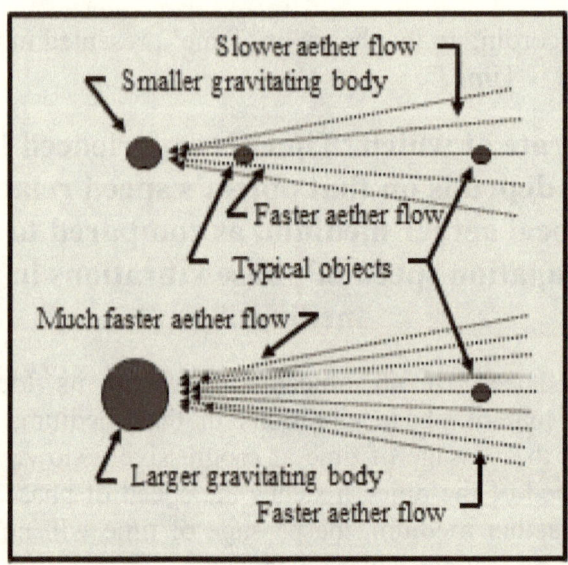

In the extreme case where an object may be placed right at the event horizon of a black hole, that object will not experience the passage of time at all. Since, its local aether will be flowing through its position at exactly the same speed as that of phase vibrations in its local aether medium.

It must be particularly emphasized that,

According to the information presented in the chapter titled "What is Time?" along with the results of the experiments proposed to verify their validities,

"Even though force of gravity and uniform rate of acceleration give rise to certain identical effects, such as the weight of objects, they have nothing in common as far as inducing 'Time Dilation' is concerned."

Therefore,

"The paradoxes / twins stories about accelerating objects / living beings that experience time at a slower pace, as compared to those left relatively stationary on their home planet,

are fundamentally flawed.

Since, such paradox stories totally ignore the actual speeds at which such objects / living beings are moving relative to their respective local aether mediums."

It is the instantaneous speed of an object relative to its local aether that gives rise to 'Time Dilation' effect and not its instantaneous accelerations.

19- How does gravity affect the flow of 'Time'?

A variety of experiments can be designed and conducted that will demonstrate how the acceleration of the local aether relative to objects (or living beings), manifesting as the local force of gravity, affects the rate at which they experience the passage of time. Details of four experiments are provided in the chapter titled "What is Time?", section "How does gravity affect the flow of 'Time'?".

- First experiment: (Orbital free fall vs. direct free fall)

- <u>Second experiment:</u> (How does an object experience the passage of 'Time' as it <u>directly</u> approaches a black hole, in a state of free fall?)

- <u>Third experiment:</u> (Esmailzadeh paradox)

- <u>Fourth experiment:</u> (Esmailzadeh Orbital Altitude Range)

These experiments are proposed to confirm the validity of the aether-based theory of 'gravity' which states,

"The force of gravity, experienced by anything and everything, at any given location in this universe is the drag force induced by the flow of aether that is accelerating through that particular location."

And, also confirm the validity of the aether-based theory of 'Time' which states,

"The rate at which 'time' is experienced by an object (or a living being) depends on its speed relative to its local aether, as compared to the propagation speed of phase vibrations in that aether medium."

Such experiments also clearly confirm the fact that, in order to obtain true/accurate (**<u>absolute</u>**) results,

The speed of objects in the Lorentz transformation equations, used to calculate the time dilation (and other relativistic effects), must be relative to their local aether, not relative to space, and definitely not relative to another object such as a planet, a star, a galaxy or even the birthplace of this universe.

20-Precession of the perihelion of Mercury's orbit

According to the new theory of gravity presented in this chapter, the precession of Mercury's orbital perihelion is **almost**

exactly equivalent to what has already been predicted by the general theory of relativity. Since, the variations in the strength of the force of gravity, as a function of distance from a gravitating body, which are interpreted by the general theory of relativity as the curvature of the spatial dimensions, are the gradients that exist in the speed of the fluid aether approaching that gravitating object. The following figure shows a simplified presentation of the precession of Mercury's perihelion as it orbits the sun.

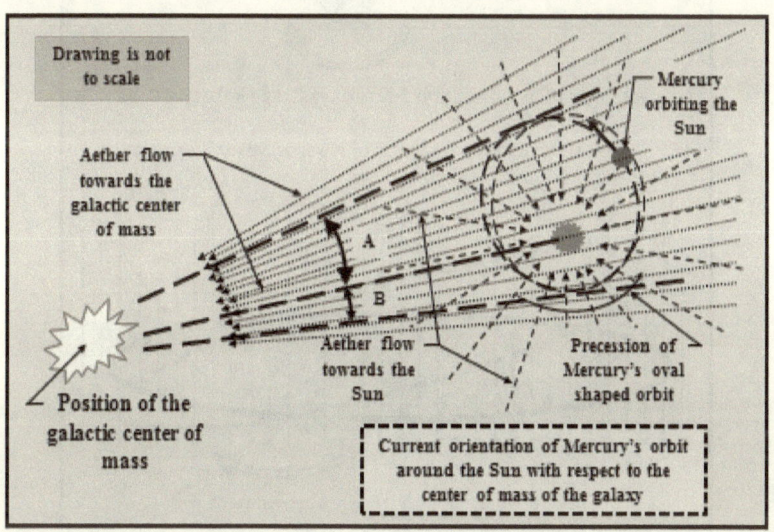

Note that, the orbital plane of Mercury around the sun is not the same as the plane in which sun is following its trajectory around the center of mass of the Milky Way galaxy.

In the above paragraph, it is stated that, the prediction of the newly proposed theory is **almost exactly** equivalent to that of the general theory of relativity, because based on the new theory, the current orientation of Mercury's elliptical orbit around the sun with respect to the location of the center of mass of the Milky Way galaxy (as shown in the figure) necessitates the addition of a new correction.

This new correction is due to the sun NOT being at the center of Mercury's orbital path, as viewed from the center of mass of the galaxy. Sun was at the center of Mercury's orbital path, as viewed from the center of mass of the galaxy in the past, when Mercury's orbit was at about 90 degrees to its current orientation, and it will be again in the future, when the orientation of Mercury's orbital path advances forward by about 90 degrees.

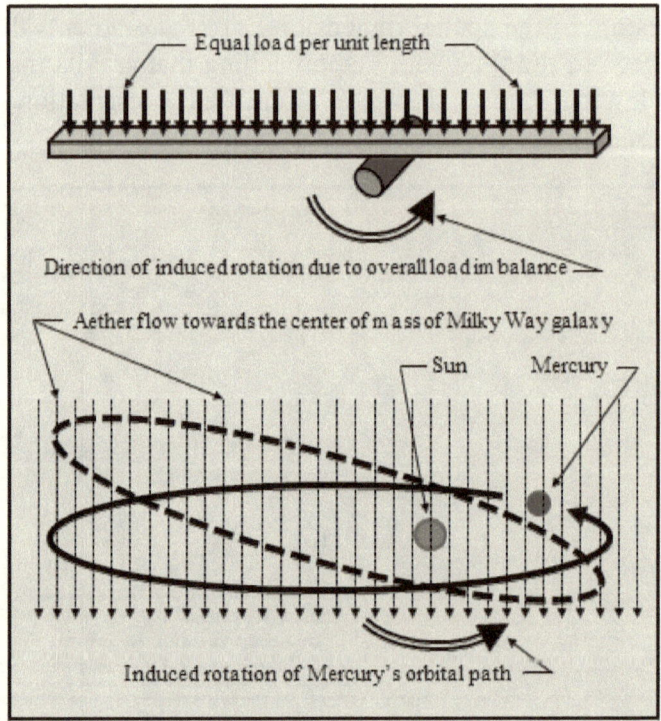

The accelerated flow of aether towards the center of mass of the galaxy exerts a consistent drag force on Mercury. However, due to Mercury spending larger fraction of its orbital period travelling farther away from the sun on one side of the sun as compared to its other side, the drag force exerted becomes unequal. Hence, currently, the orientation of Mercury's orbit is encouraged to rotate by the net force that is exerted, just like the imbalance on a teeter-totter bar that is loaded equally all along its full length but its pivot point is not right in the middle, as shown in the above figure.

Once the effects due to such imbalance in the drag force induced by the aether flow that is accelerating towards the center of mass of the galaxy is taken into account, the most precise results will be obtained not just for precession of Mercury's perihelion but also for those of the other planets, as well.

The angular momentum transferred to the rotation of the orbital path of Mercury by such an imbalance in the drag force exerted on that planet, is accumulative. Therefore, currently, the magnitude of its contribution apparently increases with each pass, as Mercury completes full orbits around the sun. Hence,

"To obtain the most accurate results for the precession of Mercury's perihelion, the effects contributed during the previous rotations of Mercury around the sun must also be added and included in what is contributed during its current pass around the sun."

Since,

"Currently, every time Mercury goes around the sun, the overall contribution of the net imbalance in the drag force, induced by the accelerated flow of aether that is towards the center of mass of the galaxy, on the precession of Mercury's perihelion is gradually increasing. This is due to the current orientation of the orbital path of Mercury around the sun with respect to the direction pointing towards the center of mass of the Milky Way galaxy."

And,

"This effect continues to contribute positively to the precession of Mercury's perihelion until the longitudinal axis of Mercury's orbital path becomes parallel to the line along which the gravitational force of the center of mass of the galaxy is acting on Mercury. After which point, the effect becomes negative and gradually decreases the magnitude of the rotational speed of Mercury's orbital path, during each pass it makes around the sun."

The very same accumulative effect can be readily observed in the case of the teeter-totter (see-saw) bar. Since, the induced rotational speed (angular speed, hence angular momentum) increases with time until the length of the bar becomes parallel to the line along which the gravitational force of earth is acting on it. After which point, the effect is reversed and causes the rotational speed of the bar around its axis to decrease with time.

21- What are gravitational waves?

The flow of aether towards any given particle or aggregate of particles is pressure driven. The driving pressure is actually the difference between the pressure of aether that is in this universe and the pressure of aether that is in the accompanying universe.

Therefore, as shown in the figure, if a particle (or an aggregate of particles) does not move or is moving at a very slow rate of speed with respect to its local aether medium, the flow of aether from all directions will always be directly towards its instantaneous position.

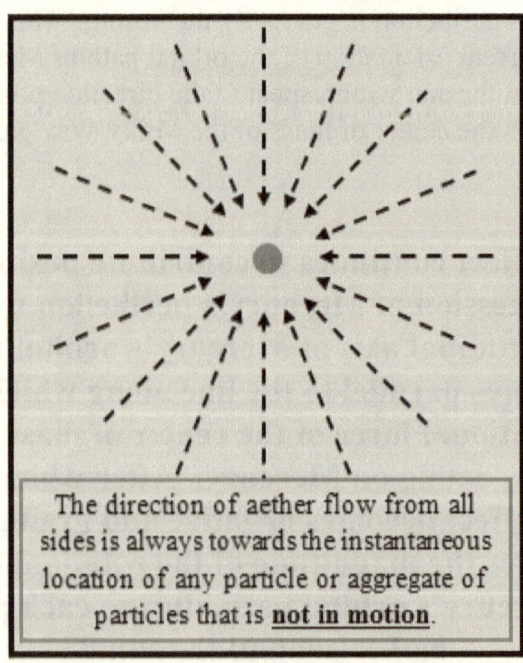

The direction of aether flow from all sides is always towards the instantaneous location of any particle or aggregate of particles that is **not in motion**.

However, as shown in the following figure, if the particle (or an aggregate of particles) be moving fast with respect to its local aether medium, the direction of the flow of aether that is approaching it will be automatically affected, particularly the portions that are approaching it from the sides (away from its line of motion). The direction of the aether flow will be automatically adjusted so that it is towards the location of that particle (or the aggregate of particles), as close as possible.

The amount of deviation between the direction of aether flow induced at any given location in the surrounding space and the direction pointing towards the actual instantaneous location of the particle (or aggregate) depends on:

- The relative speed and direction of motion of that particle (or aggregate of particles) with respect to the line of sight from that particular location, as well as

- The distance in between that particle and the location of interest.

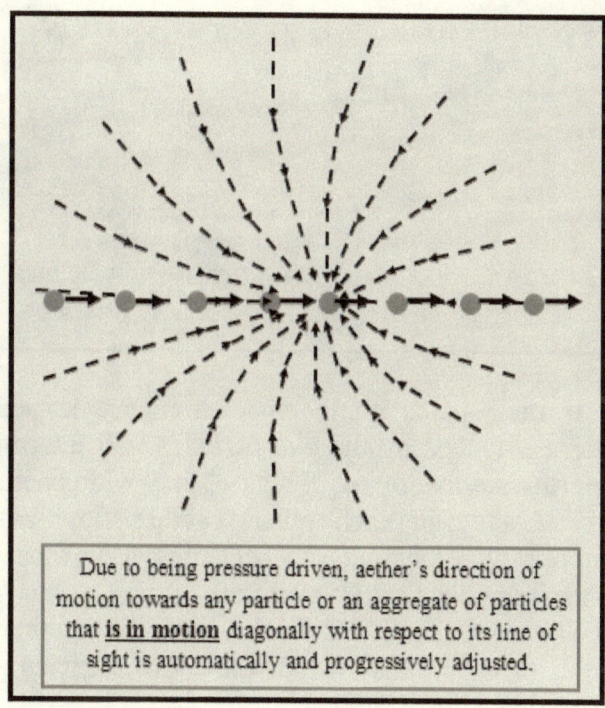

Due to being pressure driven, aether's direction of motion towards any particle or an aggregate of particles that **is in motion** diagonally with respect to its line of sight is automatically and progressively adjusted.

The faster a particle moves in a direction that is closer to the perpendicular to the line of sight from the location of interest and/or the farther away the particle is, will lead to a more pronounced lag time or delay time for aether course corrections to take effect.

If two particles (or two aggregates of particles) move past each other at a fairly close distance, during their close encounter, they can follow one of three main scenarios described below:

- **The two particles (or aggregates of particles) can move in parallel and yet in the opposite direction of each other**, just like two vehicles travelling in the opposite direction of each other, along a two way road. As shown below, such particles will approach each other and then recede from one another.

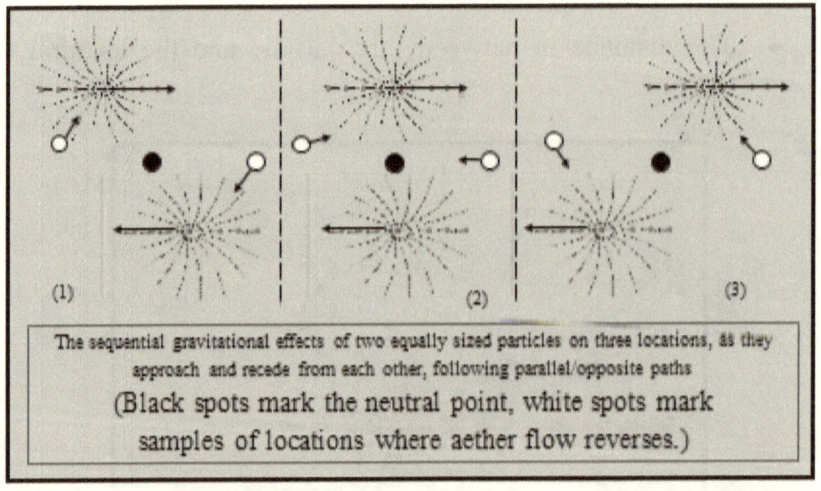

The sequential gravitational effects of two equally sized particles on three locations, as they approach and recede from each other, following parallel/opposite paths

(Black spots mark the neutral point, white spots mark samples of locations where aether flow reverses.)

If their paths come close enough, the aether flow patterns induced by the two particles will intermingle and **generate a wave form**. Because, they will generate a peak and a valley in each other's aether flow patterns by temporarily resonating and canceling each other's effects on the immediate aether medium.

The imposed effects on the immediate aether medium and its flow patterns towards these two particles will result in two distinct side effects:

1- At a certain distance between the two particles, the flow of aether will actually stagnate as the two particles approach and recede from each other. The exact location of where aether will stagnate with respect to the two particles (or aggregates of particles) will be inversely proportional to those particles respective masses. That location corresponds to a neutral point or point of zero gravity, as far as the accelerated flow of aether towards those two particles (or the two aggregates of particles) is concerned.

Note that, this stagnation location can be either stationary or moving in space. It will be stationary only if the two particles are of the same mass and both are moving along parallel straight lines at exactly the same rate of speed relative to their local aether medium, but in opposite directions.

Otherwise, the stagnation location can be moving in a variety of directions. If both particles are moving along parallel straight lines, but either are moving at different rates of speeds relative to their local aether medium or are not of the same mass, then the stagnation location will be moving in two dimensions, on the very same plane in which the two particles are moving.

However, if the two particles are also following curvilinear paths, then the stagnation location will be moving in three dimensions.

2- At all other points in between the two particles' pathways aether flow will experience a flow reversal, as the two particles pass each other and interfere in the aether flow induced by the other, due to being closer to those locations in space.

This type of effect is opposite of the effect observed as two ships or boats go past each other, in opposite directions, at close distance. In such a scenario, water which is the local medium is pushed away from the boats rather than being pulled towards them. However, the effect is similar since water that is at first pushed away in one direction due to the motion of one boat, is subsequently pushed in the opposite direction as one boat goes away and the other boat comes closer.

- **The two particles (or aggregates of particles) can be following paths that are perpendicular to each other.** In this case, they will approach each other and then recede from one another. If their paths come close enough or cross each other the induced aether flow patterns will again intermingle with each other and **generate a wave form**. Because, they will generate a peak and a valley in each other's aether flow patterns by resonating and canceling each other's effects on the immediate aether medium.

  Note that, in this case also, both of the effects as described above will be manifested, except that aether stagnation location will be moving in space, in accordance with the motion of the two particles.

- **The two particles (or aggregates of particles) can be spinning around each other.** In this case, the effects described above will be experienced, but in a rotating and continuous fashion. The neutral point will follow a circular path of its own around the mutual center of mass of the two particles.

  The flow of aether that is approaching from outside of the two particles circular paths will be repeatedly experiencing higher and lower rates of speeds, due to one particle (at a time) coming closer as compared to both particles being away from that particular location.

  Such variations in the aether flow speeds at any given point just outside the two particles circular paths will result in higher and lower pressures and densities in the immediate aether medium. The following figure shows this effect.

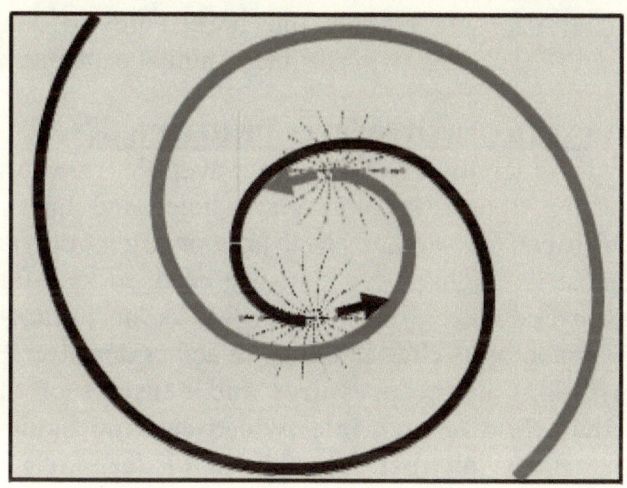

The variations induced in the local aether medium's pressure and density will have a rotational speed of their own around the two particles, in the very same direction as the particles themselves are moving. The speed at which the variations induced in the aether medium will progress around the particles' collective will directly depend on the distance to the perimeter of the circular paths of the particles.

These strong nearby cyclic variations in the pressure and density of the aether medium will propagate outwards, directly away from the mutual center of mass of the two particles. Because, the flow of aether in the surrounding volume of space is pressure driven, the aether medium responds by compensating for any variations induced. This transfer of response to the outer layers automatically generates a wave pattern that can be referred to as **gravity waves**.

Note that, aether that is farther away will experience such variations in its pressure and density at lower amplitude due to both distance as well as intermixing of aether flow towards the two particles. These variations will gradually smooth out with distance, since at great distances aether flow towards the two particles will act as one unified flow of aether towards the twin particles' mutual center of mass.

- ## How can gravity waves be detected?

In order to detect gravity waves that are formed around massive objects such as black holes and their companions which can be another black hole or a neutron star or even a regular live/light-giving star, one needs to look for temporary, yet cyclic lens effects as in waves or mirage-like effects generated by cyclic variations in aether density. **The induced variations in the pressures and densities of the layers of aether close to such intertwined celestial bodies will cause the objects on their far-side to be seen in a lively wavy fashion.**

Note that, the wavy images of the objects on the far side of two massive celestial objects with intermingling gravity forces can be likened to the wavy images seen of objects that are located on the far side of the surface of a paved road or the surfaces of the hoods or even the metal roofs of cars in a hot summer day. The observed images become wavy due to constantly changing air density above such surfaces. Variations in the air density make the local air medium act like a constantly varying lens.

Gravity waves temporarily forming near massive celestial objects that happen to pass by each other can be observed or detected, as well. However, such observations can only occur by chance, due to rare occurrence of such instances when the telescope will be pointed towards the exact direction, at the exact right time, and it be recording what it is focused on.

22- Why is the expansion of this universe accelerating rather than decelerating? And, what is the source of energy driving such a mechanism?

Based on the calculations performed by two independent groups of scientists in 1998, using data collected through direct observation of various supernovas in distant galaxies, the universe was found to be accelerating rather than deceleration in its expansion process. The overall history of the expansion of the universe is shown below, in a simple fashion.

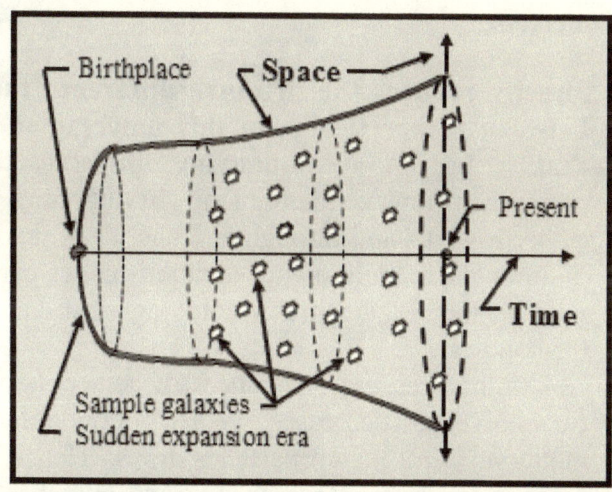

Even though, various theories have been put forward suggesting that some unknown form of energy, namely dark energy, is responsible for the current accelerated expansion process of this universe, there have not been any explanations on what can potentially be the source of this dark energy.

However, once the effects of the gradual decreases in both the density and the pressure of the aether medium in this universe on the overall expansion process of aether medium and also on the strength of the force of gravity are taken into account, on a universal scale, the reason for such accelerated expansion process can be readily understood and such findings can be well expected.

Notes:

- **The internal pressure of the medium of aether is the driving force for its expansion,** and **the expansion rate of the aether medium is directly proportional to its internal pressure.** This characteristic of the aether medium is analogous to that of a volume of pressurized gas which is allowed to expand uniformly in all spatial directions, into its surrounding environment.

- **The gravitational force that celestial bodies exert on one another is roughly proportional to the square of any variations in the pressure of aether that is in this universe**, because any decrease in the pressure of aether in this universe:

  1- **Directly reduces the pressure difference that exists between aether that is in this universe and aether that is in the accompanying universe,** the same pressure difference that is the driving force pushing aether towards and through any and all of the available drain holes, namely matter and anti-matter particles, as well as their aggregates (including neutron stars and black holes).

     As the pressure difference (of aether) between the two universes decreases, aether's flow rate towards matter and anti-matter particles drops. Hence, the drag force exerted by aether on any and all other particles that happen to be in its path gets reduced accordingly, as well.

  2- **Directly reduces the density of aether in this universe,** because aether is a compressible fluid. This reduction in aether density in turn leads to even weaker drag force (weaker force of gravity) induced by the accelerated flow of aether, anywhere and everywhere in this universe.

  Therefore, if the pressure of aether in this universe is reduced to one half, the force of gravity will be reduced to **about** one quarter. Since, as aether is transferred from this universe into the accompanying universe the overall pressure of aether that is in the accompanying universe is gradually rising.

- Other than due to aether's expansion in this universe, **aether's density and pressure are also decreasing further because of aether leaking from this universe**, through matter and anti-matter particles, as well as their aggregates (including neutron stars and black holes). Aether's leakage from this universe directly leads to a more pronounced effect due to expediting weakening both the expansion force and the force of gravity.

- **Due to the ongoing expansion of this universe the separation distances between the galaxies are increasing.** As a result, galaxies are gradually exerting weaker and weaker gravitational forces on each other, that is even if the pressure and the density of aether were to stay constant. It should be noted that, the receding motion of galaxies is due to two types of motions, namely:

1- The motion of galaxies IN the aether medium.
   The **general** direction of this motion of galaxies is away from the birthplace of the universe.
   This motion of the galaxies can be likened to the movements of individual ants running away from a central point which is drawn on the surface of a balloon that is inflating.

2- The motion of galaxies due to being carried by the expanding aether medium.
   The direction of this particular motion of galaxies is **exactly** away from the birthplace of the universe.
   This motion of the galaxies can be likened to the motion of raisins in a raisin cake/dough relative to the center of that cake/dough, as it is baking.

- The aether medium in the accompanying universe is expanding at a much slower pace as compared the aether medium in this universe. Since, its pressure is much lower than the aether pressure in this universe. Therefore, the pressure of aether in the accompanying universe **is rising as it is receiving aether from this universe**.
   In fact, due to aether's pressure and volume being much greater in this universe as compared to those of aether in the accompanying universe, aether's pressure in the accompanying universe is rising at a much faster rate as

compared to the rate at which aether's pressure is dropping in this universe.

Therefore, <u>as the overall volume of aether is doubled in this universe, even though the pressure of aether which is the driving force for the expansion of this universe, as a whole, is reduced to roughly one half, the gravitational force which is responsible for slowing down the existing expansion rate of this universe, is reduced to less than one quarter.</u>

The following figure shows the variations in the relative magnitudes of the force of gravity and the force of expansion in this universe (both normalized to their current values), as the overall volume of this universe is gradually doubled in size.

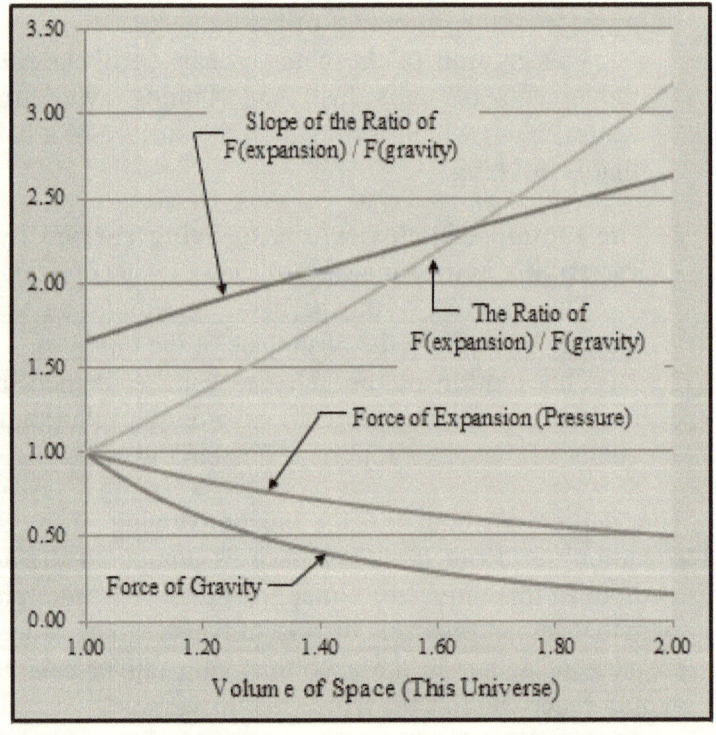

As it is obvious in the figure, the ratio of the force of expansion to the force of gravity has a positive slope. Also, the magnitude of this slope is gradually increasing as well. This is indicated by the curve marked as "Slope of the Ratio of F(expansion) / F(gravity)". In other words,

"Over time, the ratio between the force that is responsible for the expansion of this universe and the force that is trying to slow the rate of its expansion down is gradually increasing. Hence, <u>the expansion of the universe</u> (the matter medium) <u>is accelerating instead of decelerating</u>."

Note that, the universe (the matter medium) is accelerating in its expansion rate, due to:

- The overall weakening of the gravitational force, and

- Being dragged at faster and faster paces by the local aether which is experiencing an overall expansion of its own.

However,

"Due to the gradual decrease in aether's internal pressure which is the driving force for that medium's expansion, the aether medium in this universe, as a whole, is continuing to slow down in its overall expansion rate."

The overall expansion rate of the aether medium in this universe is still much faster than the speed of light in that medium. The current rate of expansion of the matter medium is only a fraction of the speed at which the fluid aether medium is expanding. However, as time goes on, the aether medium is dragging the matter contents of this universe at faster and faster paces. This is while the overall expansion rate of the aether medium is gradually slowing down, due to gradual decrease in aether's internal pressure.

Eventually, the expansion rate of the matter medium will become equal to that of the aether medium. Afterwards, both will gradually slow down towards a complete halt, as the overall aether pressure gradually approaches zero.

The following figure shows the accelerating expansion process of the matter medium inside the decelerating expansion of the fluid aether medium, as a whole, which defines 'space' in this universe.

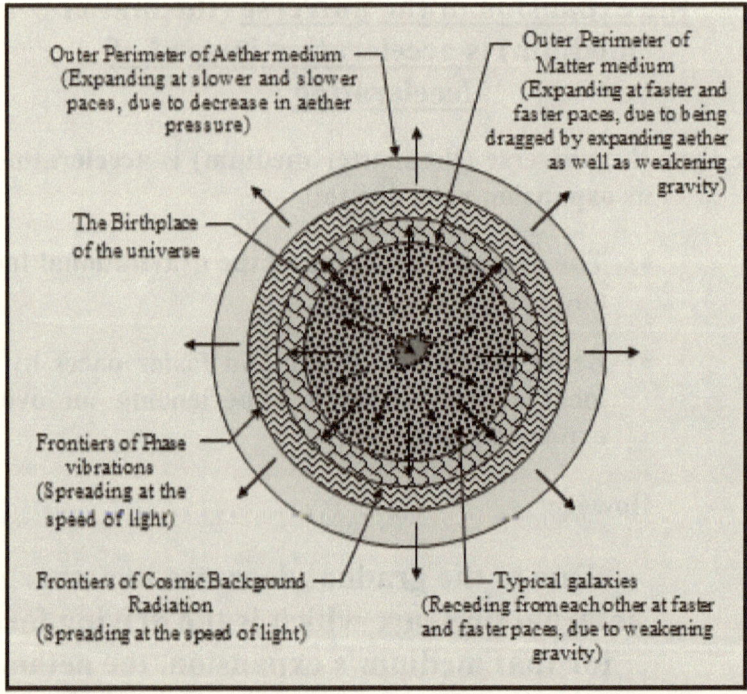

The following figure shows a general overview (NOT to scale) of **the relative expansion rates** of the fluid aether medium and the matter medium, as well as the expansion rates of the cosmic microwave background radiation due to the initial phase vibrations that were introduced in the aether medium and the cosmic microwave background radiation due to phase vibrations which were generated later on, as most of the matter and anti-matter particles annihilated each other.

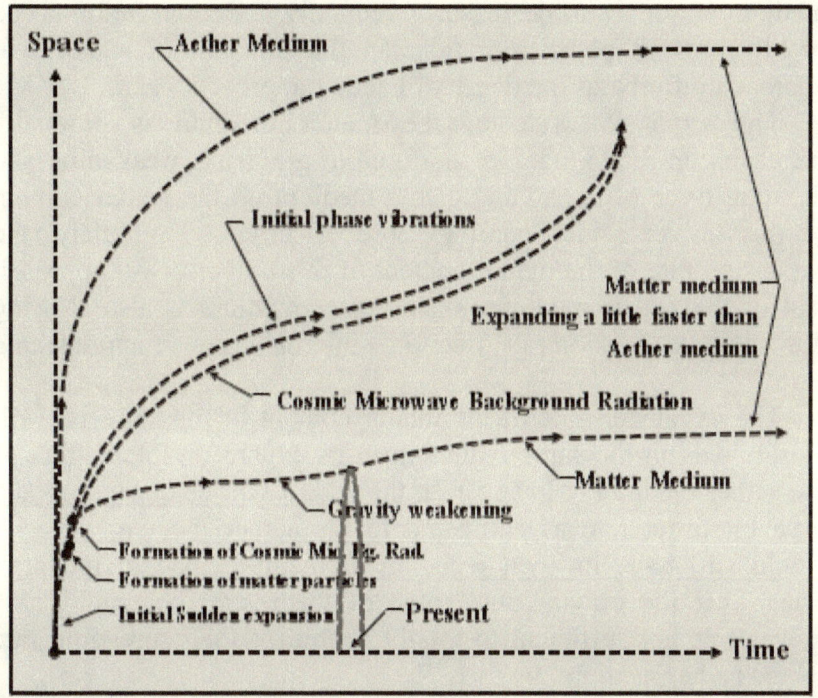

This figure shows how the initial rapid expansion of the aether medium (space) ended, as matter (and anti-matter) particles gave rise to the force of gravity with their very formation. By the time most of the matter and anti-matter particles had annihilated each other, the overall spreading rate of the leftover matter particles and the expansion rate of the whole aether medium were drastically reduced.

As it is shown, the expansion rate of the portion of the cosmic microwave background radiation (CMBR) which was due to the initial phase vibrations was changed according to the reduction in the expansion rate of the aether medium, as a whole. The portion of the cosmic microwave background radiation (CMBR) which was due to the matter and anti-matter annihilations experienced the same fate as the initial phase vibrations, since both were phase vibrations and both had the same speed of propagation in that medium.

Over time, both types of CMBR have gradually slowed down, due to the reduction in the expansion rate of the aether medium. But, they will gradually speed up, as the density of the fluid aether medium continues to decrease. Eventually, the propagation speed of all types of phase vibrations in the medium of aether will approach infinity, as the density of aether approaches zero.

The expansion rate of the matter medium is gradually increasing in magnitude, as the force of gravity is weakening and the drag force of expanding aether medium on the matter content of this universe is becoming more effective. Eventually, the expansion rate of the matter content of this universe will approach that of the aether medium, since aether medium is also slowing down in its expansion rate, due to its internal pressure approaching zero.

The expansion rate of the matter content of this universe will closely exhibit the same reduction in its expansion rate as that of the aether medium. However, <u>at the end, it will be expanding at a little bit faster rate as compared to the aether medium, since it would still have its own receding momentum IN that medium.</u> That is, if the gravitational force of the matter content of this universe is not sufficient to totally neutralize their spreading rate IN the aether medium, by then.

23- Gradual weakening of the force of gravity and its consequences

The force of gravity at any given location in this universe is directly dependent on aether's density and pressure at that particular location. Therefore, as the overall density and pressure of the aether medium are decreasing with time, due to aether's expansion and leakage, the flow of aether is gradually becoming less effective in inducing the force of gravity. Consequently,

- **The Universal Gravitational Constant is gradually decreasing**

    At the present time, the magnitude of the universal gravitational constant is **(G=6.67384 × 10^{-11} m^3 kg^{-1} s^{-2})**. The rate at which this value is decreasing at the present time can be roughly estimated using the current value of the Hubble's constant. Hubble's constant is a direct indication of the rate at which the physical contents of this universe (<u>not the overall</u>

aether medium) are spreading by receding from each other. The spreading of the physical contents of this universe is due to the expansion of the aether medium, as a whole, as well as the motion of galaxies IN that medium.

Therefore, the speed at which matter contents of this universe are dragged at any given location is proportional to (but not equal to) the speed at which aether itself is moving due to its ongoing expansion process.

The relative motion of matter contents of this universe with respect to aether in the overall picture of this universe is literally like the relative motion of objects that are allowed to free fall towards a gravitating celestial body. Such objects will be dragged by the flow of aether, yet they will not be moving at the same rate of speed as the local aether itself. However, if those same objects were allowed to start their free fall at infinite distance from the gravitating body, during their entire free fall journey, they will be moving at just about the same rate of speed as that of their local aether.

As the matter contents of this universe were formed, they fell behind from the phase vibrations that had formed them. Even more so, they fell behind with respect to aether that was expanding. Yet, they all were moving IN that medium. The general direction of their motion / momentum was away from the birthplace of this universe, since the phase vibrations that caused the generation of the matter and anti-matter particles were propagating away from where this universe had started its existence.

Gradually, all of the matter contents of this universe are being dragged by their local aether at faster and faster paces, due to the weakening of the force of gravity which is directly opposing their receding motions from each other.

According to the most recent measurements, Hubble's constant is **about** 70 km/sec/Mega parsec. Since one Mega parsec is equal to the distance traversed by light in free space in 3.26 years, the current value of Hubble's constant can be rewritten as,

$$(HC = 7.157 \times 10^{-11} \text{ km/year//km}).$$

Therefore,

The current annual rate of <u>increase</u> in the volume of aether that is in this universe <u>is greater than</u>

$$((R_2)^3 - (R_1)^3) / (R_1)^3 = 2.1477 \times 10^{-10}$$

Where, R_1 and R_2 are the radii of a given amount of aether, as of last year and as of this year, respectively.

The annual variations in the density of aether that is in this universe are numerically the same but in reverse of the variations in its aether's volume. In other words,

The current annual rate of <u>decrease</u> in density and pressure of aether in this universe <u>is greater than</u>

$$(2.1477 \times 10^{-10})$$

The universal gravitational constant is affected by both aether's density, as well as its pressure. Hence, the rate at which the value of the universal gravitational constant is decreasing is at least twice the above variation rate. Therefore,

$$(dG/G > 4.295 \times 10^{-10} \text{ per year})$$

Note that, this rate of variation in aether density (calculated using Hubble's constant) is **only** based on the receding motion of galaxies from each other. It is not based on the expansion of the aether medium, as a whole. It does not take into account the rate at which aether is gradually leaking through matter and anti-matter particles (including neutron stars or black holes), either.

- **Gradual widening of the planetary orbits and expansion of the galaxies**

  With the gradual weakening of the force of gravity, stars are gradually exerting weaker gravitational forces on their respective planets. As a result, planets' orbits are gradually becoming wider. This side effect of decrease in aether's density has already been detected in regards to earth's orbit.

According to the collected data, earth's orbit is widening by about 7 meters (about 23 feet) per century.

Of course, a small portion of this observed amount of expansion in the orbital path of earth is justifiably because:

1- Sun's gravity is becoming weaker due to losing mass, as it is continuously transforming some matter into energy and also as it is literally throwing huge amounts of matter into space as solar flares / solar storms.

2- Earth is being pushed away, as it is absorbing and/or deflecting particles that are thrown at it by the sun, in the form of solar storms.

Also, due to the very same effect, the orbital paths of all stars around the center of mass of their respective galaxies are widening. As a result, not just the solar systems but also all of the galaxies are gradually expanding in diameter.

Note that, the widening of the orbital path of planets around their respective stars will lead to eventual and definite **GLOBAL COOLING** on all planets. This global cooling, which will take place simultaneously on all planets existing in this universe, can be referred to as,

"Scharback Universal Planetary Cooling"

- **All black holes will eventually lose their designations as being black holes**

    All black holes will eventually cease to be black holes due to a natural phenomenon. That natural phenomenon is the overall pressure difference that exists between aether that is in this universe and aether that is in the accompanying universe.

    As time goes on, aether's pressure in this universe is decreasing due to the expansion of the aether medium that forms this universe and also due to aether's leakage through matter and anti-matter particles, as well as their aggregates such as planets, stars, neutron stars and black holes. In the meantime, even though the accompanying universe is also expanding in size, aether's pressure in it is continuously raising due to receiving aether from this universe. Therefore, the difference between aether pressures in these two universes is gradually decreasing.

At a certain point in the expansion process of the universe, the difference between the aether pressure in this universe and the aether pressure in the accompanying universe will approach a critical value. That value of the pressure difference corresponds to the minimum required pressure difference that can promote the flow of aether towards the largest possible black holes to reach the speed of light (the speed of phase vibrations in aether). This critical aether pressure difference can be referred to as:

"Esmailzadeh Critical Aether Pressure Difference"

Three factors affect how fast/soon this critical aether pressure difference will be experienced. These factors are:

1- The rate at which the overall pressure of aether is dropping in this universe, due to aether's expansion and leakage,

2- The rate at which the overall pressure of aether is rising in the accompanying universe (while that universe is expanding), due to receiving aether from this universe, and

3- The rate at which the speed of phase vibrations in aether (the speed of light and other electromagnetic waves) is increasing in this universe, due to decreasing aether density.

As this critical aether pressure difference is approached, all of the black holes, starting with the smallest ones, will lose their status as a black hole since the local aether pressure difference will no longer be sufficient to cause the flow speed of aether to reach the speed of light by the time it reaches the surface of those black holes.

Once this critical value of the aether pressure difference has been reached, all of the black holes, regardless of their sizes, will cease to act as black holes, since the speed of the incoming aether reaching their surfaces will be just under the speed of light. This in turn means that, none of the black holes will possess an event horizon, and light and other electromagnetic waves will be able to propagate against the incoming flow of aether and escape. In other words,

"Eventually, all black holes will become visible."

The following figure shows a typical black hole's fate, as the pressure difference between aether that is in this universe and aether that is in the accompanying universe is decreased passed its critical value.

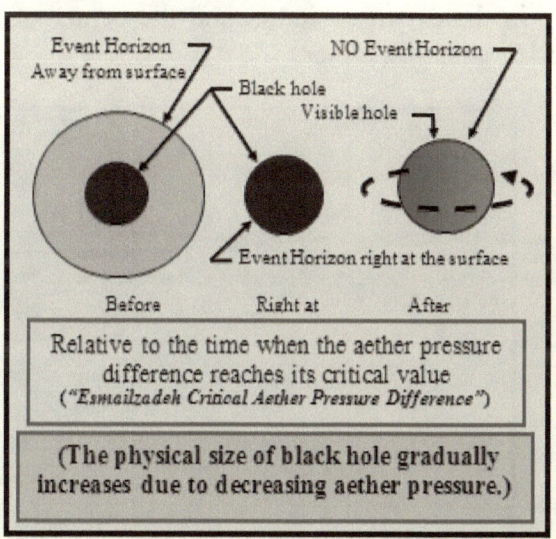

By **'visible hole'** it is meant that electromagnetic type of radiations of various frequencies can and will escape to the surrounding space. The colors chosen demonstrate the red shift and the blue shift generated due to the rotation of the 'then-visible' black hole.

Note that, as it is indicated in the figure, the physical sizes of black holes are gradually increasing due to decreasing aether pressure in this universe. That is, even if they stop devouring their peaceful neighbors, they will still grow in size.

The very same effect will also be experienced by any and all of the particles in this universe, since they are bubbles in the medium of aether. Hence, as the internal pressure of aether is gradually decreasing the physical sizes of particles and those of black holes will be increasing.

- **All of the star systems and galaxies will lose their structural integrities**

  As shown below, as the pressure and the density of aether in this universe are reduced further and the force of gravity becomes even weaker, stars will no longer be able to hold on to their respective planets. Stars also will no longer be motivated to stay in their own orbits around the center of mass of their respective galaxies, either.

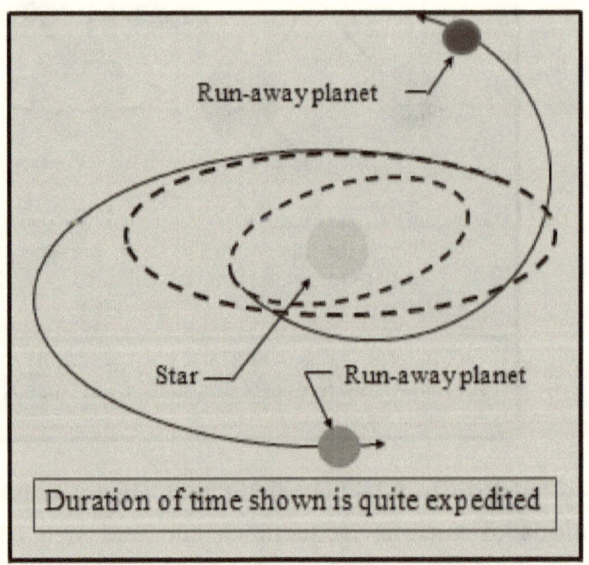

Duration of time shown is quite expedited

Therefore, star systems and galaxies will lose their structural integrities and all planets and stars will float relatively freely in space. In other words,

"At some point in the future, all of the celestial bodies will become independent wanderers in this universe."

- **All of the stars will turn off and cool down**

  As the force of gravity becomes even weaker, all of the stars will expand, as their outer layers will no longer weigh as much. Consequently, the outer layers will not exert sufficient pressure on the inner layers to promote the ongoing nuclear fusion reactions. At that point, stars will literally turn off and start to cool down. They would basically become cold gas giants. Therefore,

  "As all of the stars, one by one, turn off, the lights in the night sky will gradually become less in numbers and eventually all of them will disappear, forever."

  Once all of the stars are turned off, the whole universe will start its eternal dark era. In other words,

  "The universe will become pitch dark."

- **Stars and planets will gradually expand and disperse in space**

  As the pressure and the density of aether in this universe are reduced even further and the force of gravity becomes weaker, all of the celestial bodies will lose their own structural integrities. At that point, all of the planets and stars in the universe will expand, fall apart and disperse in space.

- **The force of gravity will gradually fade away, altogether**

  The pressure of aether that is in this universe will keep on dropping until it equalizes with the rising pressure of aether that is in the accompanying universe. At that point in time, there will be no motivation (pressure difference) for aether to flow towards matter particles or black holes, and enter the accompanying universe. Therefore, as the flow of aether towards matter particles, and even black hole, comes to a complete halt, it automatically implies that the force of gravity will gradually approach zero. In other words,

  "The force of gravity will gradually fade away."

Therefore,

"Just as the very formation of matter and anti-matter particles had given birth to the force of gravity, their continued existence will eventually cause it to fade away."

At that point in time, if the universe has any momentum left in its expansion, it will expand forever, since due to the lack of gravitational force, the expanding universe will literally lack its braking instincts. This era in the existence of this universe can be referred to as,

"Esmailzadeh Zero Gravity Era"

As the zero gravity era is approached, reached and beyond into the future, the occurrences of the following will be unavoidable, since the density of aether will gradually decrease in this universe:

- The mass of any and all particles in this universe will gradually decrease.

- The physical size of any and all particles in this universe will gradually increase.

- The speed of phase vibrations will gradually increase.

- The rate at which time is experienced will gradually increase.

24- Variable dependence of the force of gravity on distance

The strength of the force of gravity is inversely proportional to the square of the distance to any gravitating object which may be a particle, a planet, a star or any other type of celestial body. This relationship was given by Mr. Newton and it is known as the inverse square law,

$$F = Gm_1m_2 / x^{2.000}$$

Where, F is the gravitational force that two objects of masses m_1 and m_2 separated by a distance of x exert on each other. And, G is the universal gravitational constant.

In the future, more accurate measurements will indicate that the force of gravity associated with aggregates of matter particles such as stars and planets (in relatively short ranges, such as within a given galaxy) changes slightly faster than it is predicted by the inverse square law. This is due to the gradient that exists in the density profile of aether as it approaches an aggregate of matter particles, since its density decreases by a minute amount. Hence, its speed increases to compensate, in order to allow for the same amount of aether to flow through, at all distances from the matter particle aggregate.

Therefore, since the same amount of aether is passing through but at slightly higher speeds than expected (as compared to if aether density were to stay constant), the induced drag force in close proximity of the gravitating body becomes slightly greater than it is predicted by the inverse square law of Mr. Newton. In other words,

On relatively smaller scales (within galaxies), <u>the dependence of the force of gravity on distance</u> is slightly stronger than it is dictated by Newton's inverse square law. For example,

$$F \approx Gm_1m_2 / x^{2.001}$$

Also, the separations of galaxies (particularly between galaxies that are more distant from each other) are changing <u>due to their receding motions</u> with respect to each other, receding motions that are <u>in the aether medium</u>. This receding motion of galaxies is over and above their receding motion that is due to their being carried by the expanding aether medium (space), as a whole. The motions of galaxies <u>in the aether medium</u> expectedly lead to a stronger force of gravity than expected over long distances, particularly between distant galaxies.

This is due to the fact that, longer distances are observed after the passage of longer periods of time which include older times, when galaxies were closer to each other than the overall expansion rate of the aether medium would predict, and also <u>when aether was denser</u>. Hence, back then galaxies were justifiably exerting stronger gravitational forces on one another. In other words,

On intergalactic scales, <u>the dependence of the force of gravity on distance</u> is slightly weaker than it is dictated by Newton's inverse square law. For example,

$$F \approx Gm_1 m_2 / x^{1.999}$$

25- Effect of gravity on the speed of light

The force of gravity is the drag force that the accelerated flow of aether (towards matter and anti-matter particles) exerts on other particles that happen to be in its path. The very same flow of aether also carries any and all kinds of phase vibrations such as light and other electromagnetic waves that may be in its path.

The effect of gravity on the direction of propagation of light coming towards earth from distant galaxies has already been demonstrated in a variety of experiments. However, the effect of gravity on the speed at which light propagates was never attempted.

Effect of gravity on the speed of light is presented in detail in the chapter titled "What is Light?", along with an **Experiment**: (Effect of gravity on the speed of light) particularly designed to explore the effect of gravity (earth's gravity) on the propagation speed of light waves.

26- In most cases, the forces of gravity exchanged between two celestial bodies are not along the line connecting their two centers of masses

The direction of the gravitational force of a celestial body experienced by objects located in its surroundings depends on the motion of that gravitating body with respect to its local aether medium. If the gravitating body is relatively stationary with respect to its local aether medium, its generated force of gravity which is the drag force due to the accelerated flow of aether from all around will be directly aimed at its instantaneous position.

However, if the gravitating body is moving with respect to its local aether medium the direction of its gravitational force experienced by objects located in its surroundings will be lagging behind, relative to the line of sight from the position of those objects towards the instantaneous position of the gravitating body. Therefore, not all objects including other celestial bodies will be pulled towards its instantaneous position.

The following figure shows these two distinct scenarios. The directions of the gravitational force of a relatively stationary celestial body are shown on the left. The directions of the gravitational force of a moving celestial body are shown on the right, in an exaggerated fashion.

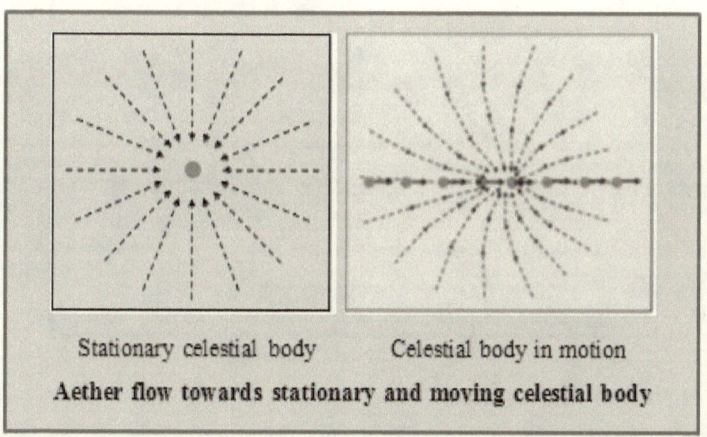

Stationary celestial body   Celestial body in motion

Aether flow towards stationary and moving celestial body

Therefore, as two celestial bodies fly passed each other, the direction in which each one experiences the gravitational pull of the other will directly depend on the motion of the other celestial body with respect to their common local aether medium.

An example of such a scenario is the case of earth and sun. Since, even though sun is moving through the aether medium, as it is following its orbit in the Milky Way galaxy, and it is carrying its associated planets along, earth is moving in addition to that motion, as it orbits the sun. The following figure shows how this extra motion of earth affects the directions that these two celestial bodies experience each other's force of gravity.

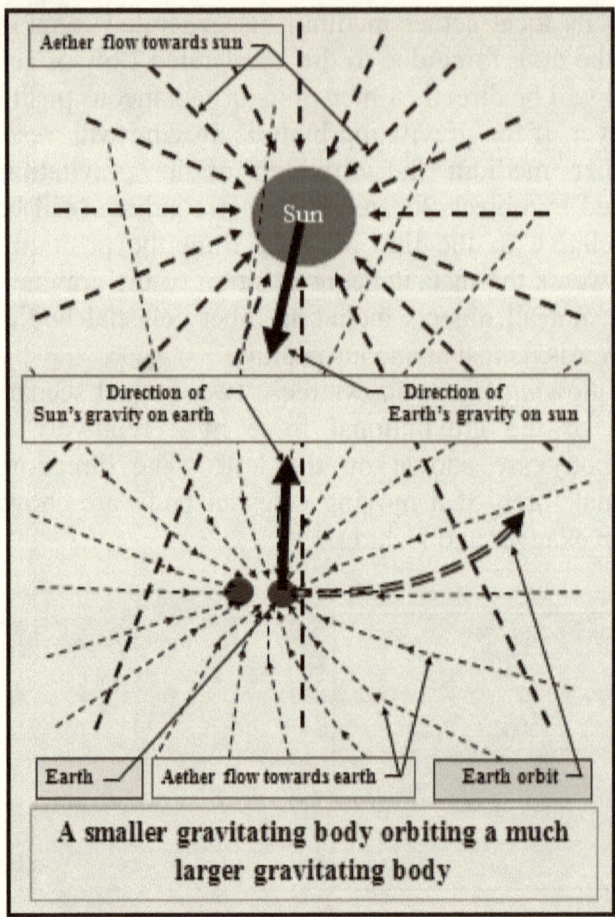

A smaller gravitating body orbiting a much larger gravitating body

As it is shown in the figure,

Earth and sun are not pulled towards each other along the same line.

Earth is continuously dragged towards where sun used to be only a very short while ago. Yet, sun is dragged towards where earth used to be a longer time in the past. This is due to the fact that, the flow of aether towards the sun, and hence its response time at earth's distance from the sun, is much faster than the flow of aether towards earth, and its response time at sun's distance from earth.

Therefore, the direction of aether flow towards the sun is quite readily adjusted, as earth experiences it. However, the direction of the aether flow towards earth, as it is experienced by sun is showing a more pronounced lag time. Hence,

"The directions of the gravitational forces of earth and sun on each other are not along the same line."

In cases where both of the celestial bodies are moving with respect to their common local aether medium, the directions of the forces of gravity experienced by both celestial bodies will be lagging with respect to the line connecting their instantaneous positions. Such a scenario, where two stars orbiting the position of their common center of mass, is shown below.

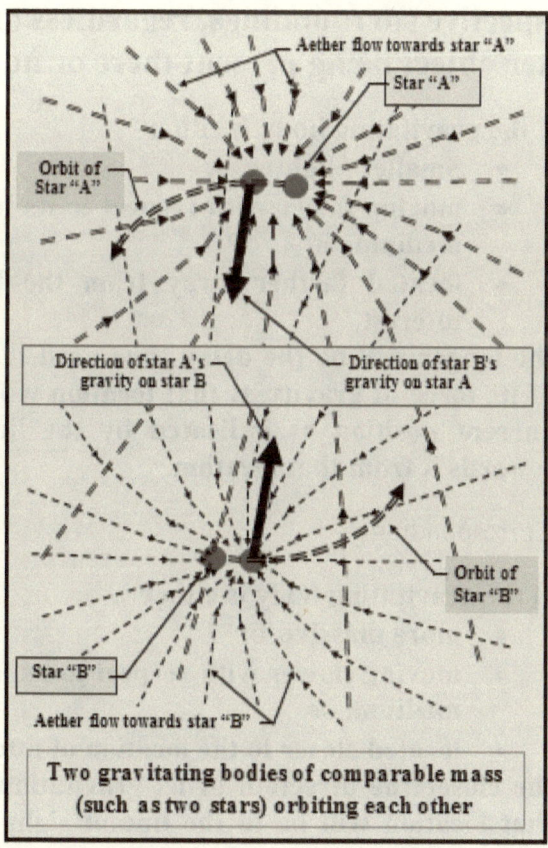

Two gravitating bodies of comparable mass (such as two stars) orbiting each other

As it was stated earlier in this chapter, earth and sun (as well as all of the other celestial bodies in this universe) do not exert any force on each other. They only allow the passage of the fluid aether from this universe into the accompanying universe. In so doing, they induce an accelerated flow in the aether medium towards themselves.

The force of gravity due to the existence of each and every celestial body is simply the drag force that their corresponding accelerated aether flows exert on any and all other celestial bodies, objects and even waves that happen to be in their paths. In fact,

"Regardless of their sizes, celestial bodies are not even aware of each other's existences. Each one independently induces an inward flow of aether and hence generates a force of gravity in its respective surroundings, regardless of any other object being present there or not."

Note that, if the gravitating body is either
- **Smaller / lighter**, or
- **moving faster with respect to its local aether medium,** or
- **located farther away from the location of interest,**

the longer will be the delay time until the direction of its force of gravity at that location will match its current position as indicated by the line of sight towards it from that location.

Correspondingly,

If the gravitating body is either
- **more massive,** or
- **moving slower with respect to its local aether medium**, or
- **located closer to the location of interest,**

the closer the direction of its gravitational force at that location will be to the line of sight towards it from that location.

27- Magnitude of the force of gravity, experienced between two objects depends not only on their motions relative to each other, but also on their motions relative to their local aether medium

The following figure shows a gravitating body approaching a given location. In such a scenario, the exerted force of gravity on any object that happens to be at that location, at any given instant, is slightly weaker than it is expected to be based on their instantaneous distance, <u>and using the inverse square law</u>.

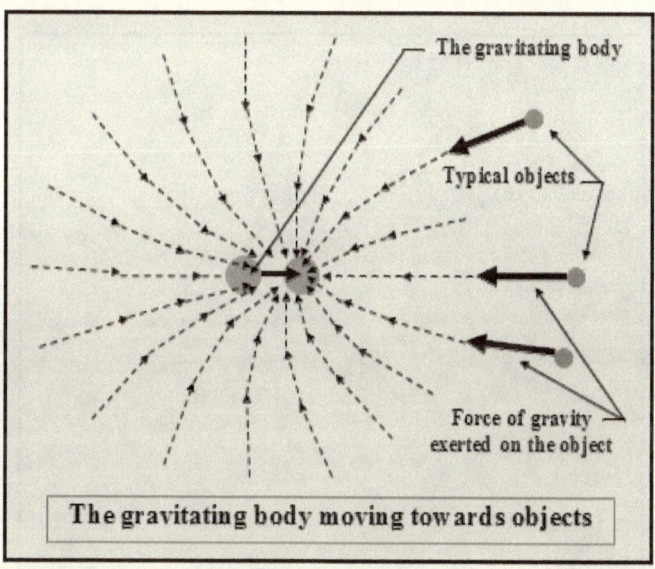

The gravitating body

Typical objects

Force of gravity exerted on the object

The gravitating body moving towards objects

In this case, the instantaneous aether flow induced by the presence of the gravitating body, giving rise to the force of gravity at the location of interest, is dictated by the distance to the gravitating body at some time in the past, **when it was farther away**. This is due to the fact that,

"Aether flow towards any and all objects in this universe (including the largest black holes) has a finite speed."

Therefore, the force of gravity due to an approaching gravitating body, at any given instant, is weaker than what it would be if it (gravitating body) was relatively stationary at its position, at that instant.

Correspondingly, as a gravitating body moves away from a given location, its exerted force of gravity on any object that happens to be at that location, at any given instant, is stronger than it is expected to be <u>according to the inverse square law,</u> based on their instantaneous distance. In this case, the instantaneous aether flow speed induced by the presence of the gravitating body giving rise to the force of gravity at the location of interest is dictated by the distance to the gravitating body at some time in the past, **when it was closer by**. This scenario is shown in the following figure.

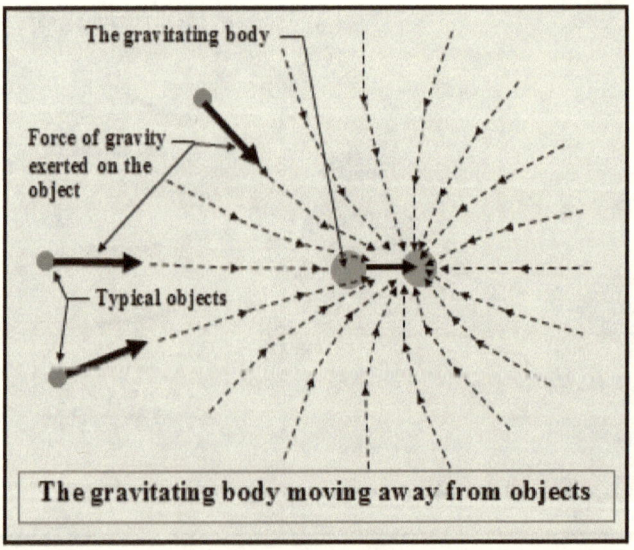

The gravitating body moving away from objects

However, **if the gravitating body is relatively stationary** with respect to its local aether medium, the strength of the force of gravity it exerts on any object located in its surrounding will be exactly equal to what is predicted by the inverse square law, based on its instantaneous distance to that object.

In such a scenario, the motion of the object towards or away from the gravitating body does not affect the strength of the force of gravity it experiences at any given distance from it. This is due to the fact that, the gravitating body is not moving in the local aether medium, hence the accelerated flow of aether which is giving rise to the induced drag force (force of gravity) at any given distance from the gravitating body is not varying with time.

In fact, if the object moves in any direction, including towards or away from the gravitating body, it will continuously experience the drag force induced by the accelerated flow of aether that is already at its new locations. The following two figures show an object moving towards and away from a stationary gravitating body, respectively.

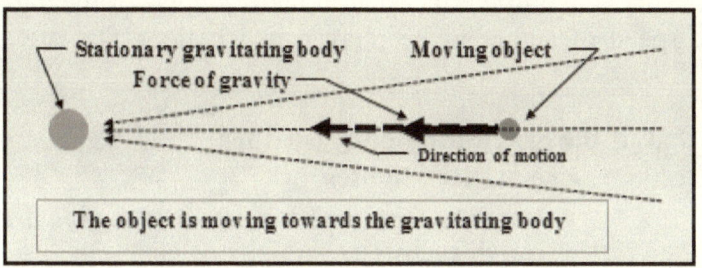

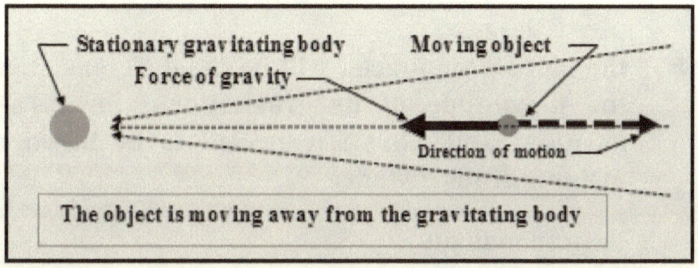

Note that, in cases where two celestial bodies move at different rates of speeds with respect to their common local aether medium and move either towards or away from each other, particularly when they are of different sizes (masses), they will attract each other with unequal gravitational forces.

This is due to their motions with respect to the local aether medium which affects their respective gravitational force delay times in two different ways, as it is explained earlier.

It must be emphasized that,

"The only case in which the gravitational forces exchanged between two gravitating bodies will be exactly the same in

**magnitude is when both objects are in a
complete rest with respect to their
common local aether medium."**

The motion dependence of the magnitude of the force of gravity is dependent on the current distance to the gravitating body. This is due to the fact that, it takes longer for the variations in the speed of aether to be relayed to locations that are farther away.

Note that, if the gravitating body is either
- **smaller / lighter**, or
- **moving faster with respect to its local aether medium,** or
- **located farther away from the location of interest,**

the more pronounced will be the difference between the magnitude of its gravitational force (at that location) and what is expected to be based on its instantaneous distance.

Correspondingly,

If the gravitating body is either
- **more massive,** or
- **moving slower with respect to its local aether medium**, or
- **located closer to the location of interest,**

the less will be the difference between the strength of its gravitational force (at that location) and what is expected to be based on its instantaneous distance.

28- Gravity in the accompanying universe

According to the new theory of gravity proposed in this chapter, the force of gravity at any given location in this universe is the drag force induced by the flow of aether which is accelerating through that particular location. Once aether reaches its destination which may be either matter or anti-matter particles (or their aggregates such as planets, stars, neutron stars or even black holes) it gets transferred into the accompanying universe.

Therefore, as viewed from this universe, matter (as well as anti-matter) particles and black holes are like drain holes at the bottom of a swimming pool through which aether is leaking out. While, as viewed from the accompanying universe, matter (as well as anti-matter) particles and black holes are like water spring jets in a swimming pool through which aether is flowing into and is joining the aether that is already there.

In this universe, the accelerated flow of aether towards matter and anti-matter particles gives meaning to the force of gravity. However, in the accompanying universe there is no such phenomenon as accelerated aether flow towards matter and anti-matter particles. Therefore,

"There is no force of gravity in the accompanying universe."

Note that, even though there is no force of gravity in the accompanying universe, aggregates of matter and anti-matter particles hold on to their structures due to the force of gravity that exists in this universe.

In fact, as aether in the accompanying universe is flowing away from matter and anti-matter particles, in a decelerated fashion, it gives rise to a phenomenon in that environment which can be truly referred to as **'anti-gravity'**.

Such anti-gravitating force becomes stronger near larger aggregates of matter and anti-matter particles that correspond to planets, stars and particularly black holes in this universe.

The following figure clearly shows that in this universe the flow of aether is towards matter and anti-matter particles, an accelerating flow which gives rise to the force known as the force of gravity in this universe. Correspondingly, the flow of aether is shown to be away from such particles, as viewed from the accompanying universe, a decelerating flow which gives rise to what can be rightfully called the force of anti-gravity.

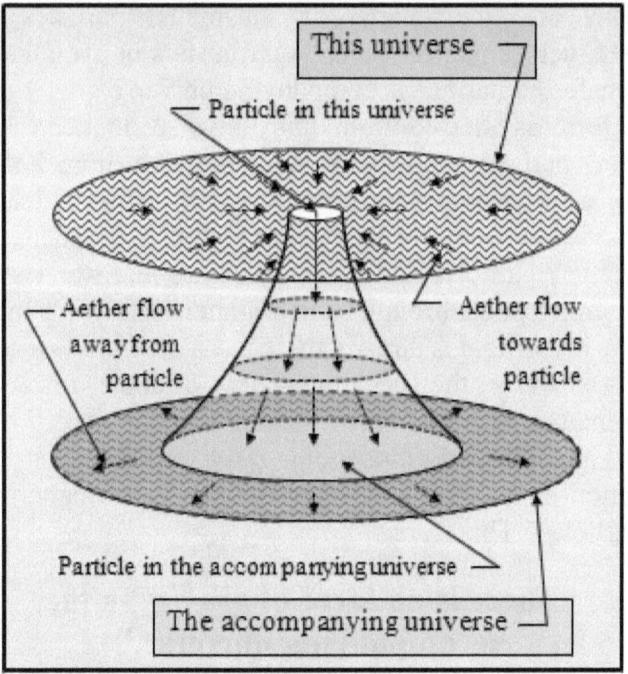

Note that, the difference in the physical size of the particle in the two universes is due to the difference that exists between the pressures of aether in those universes.

29- How can the current pressure difference between aether that is in this universe and aether that is in the accompanying universe be calculated or even estimated?

The current pressure difference between aether that is in this universe and aether that is the accompanying universe can be calculated or estimated using two different methods.

- First method:
   (Nozzle effect on gas flow)

It is known that the speed of air flowing out of a compressor's tank, through an opening such as a valve or a nozzle, is dependent on the pressure difference between air that is inside the tank and air that is outside, as well as the size of the opening. The required air pressure difference that would promote the speed of air escaping the tank through a given size of an opening to reach the speed of sound can be determined experimentally, using a setup such as the one shown below.

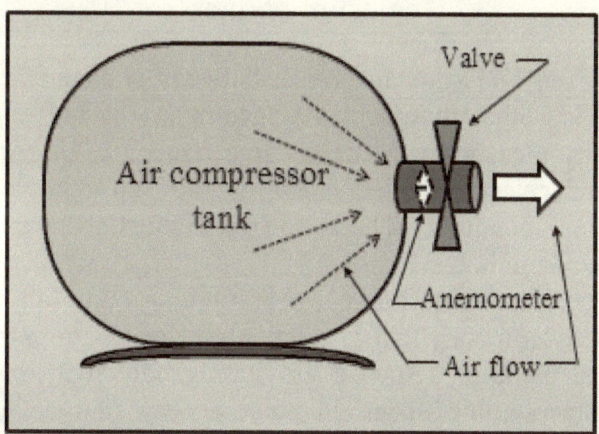

One can use the very same methodology to calculate the pressure difference that currently exists between aether that is in this universe and aether that is in the accompanying universe. The needed pieces of information are:

- Aether has zero viscosity.

- The very existence of black holes implies that the pressure difference is sufficiently high that it can promote the speed of the aether flow to reach that of light (the speed of phase vibrations) in fluid aether, as it reaches any black hole's event horizon.

- The minimum size star that can form a stable black hole is known. This implies that the physical size of the event horizon corresponding to the smallest stable black hole that can possibly form at the present time is known.

Using the information already known regarding the minimum possible size of a stable black hole (in fact the size of its event horizon) and knowing that the flow has to reach the speed of light as it reaches the event horizon, one can readily calculate the minimum pressure difference that must exist between aether that is in this universe and aether that is in the accompanying universe.

- Second Method:
  (Delay time / lag time in gravitational response of celestial bodies)

The current pressure difference between aether that is in this universe and aether that is in the accompanying universe can be estimated by measuring the delay time (lag time) in gravitational response of celestial bodies such as the sun relative to their instantaneous locations. The following figures show the concept behind such a method.

The figure on the left shows a gravitating body (an individual particle or even a star) that is relatively at rest with respect to its local aether medium. Such a gravitating body will draw aether from its surrounding space in such a way that aether's flow direction at all locations is exactly towards its current / instantaneous position.

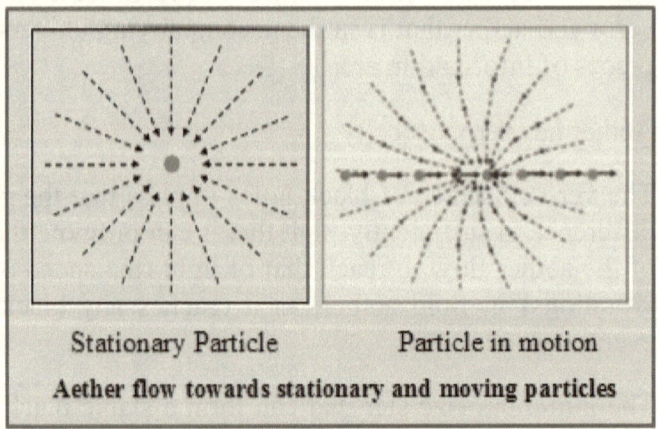

Stationary Particle   Particle in motion

Aether flow towards stationary and moving particles

Correspondingly, the figure on the right shows a gravitating body (an individual particle or even a star) that is in motion at a relatively high rate of speed with respect to its local aether medium. In this case, the flow direction of aether at any given location that is not directly along the gravitating body's line of motion is automatically and progressively adjusted. Since, the driving force for aether flow towards any opening is the difference between its internal pressure in this universe and its internal pressure in the accompanying universe.

The difference between the instantaneous aether flow direction at any particular location and the line of sight that directly points towards the instantaneous position of the gravitating body can be used to define a **gravitational lag time.**

Note that, the gravitational lag time is dependent on the motion of the gravitating body with respect to its local aether medium, since it literally generates a river-like flow towards itself in that medium. The faster it moves in that medium the more pronounced will be the curvature in the direction of aether flow induced at different distances. This is the very effect shown in the right side of the figure.

The gravitational lag time experienced / measured at any particular location away from a moving gravitating body depends on:

- The pressure difference between aether that is in this universe and aether that is in the accompanying universe.

- The size of the gravitating body,

- How fast the gravitating body is moving with respect to its local aether medium,

- How close the gravitating body's direction of motion is to the perpendicular to the line of sight from the location of interest,

- How far the gravitating body is from the location of interest.

Using a compressible gas environment, one can simulate and study such lag times and derive the empirical equations that will take into account the effects of such factors. The following figure shows a simple experimental setup that can be used to perform such a task.

The compressible fluid medium used can be either air or any other gaseous medium. **The size of the tank should be much, much larger than shown.**

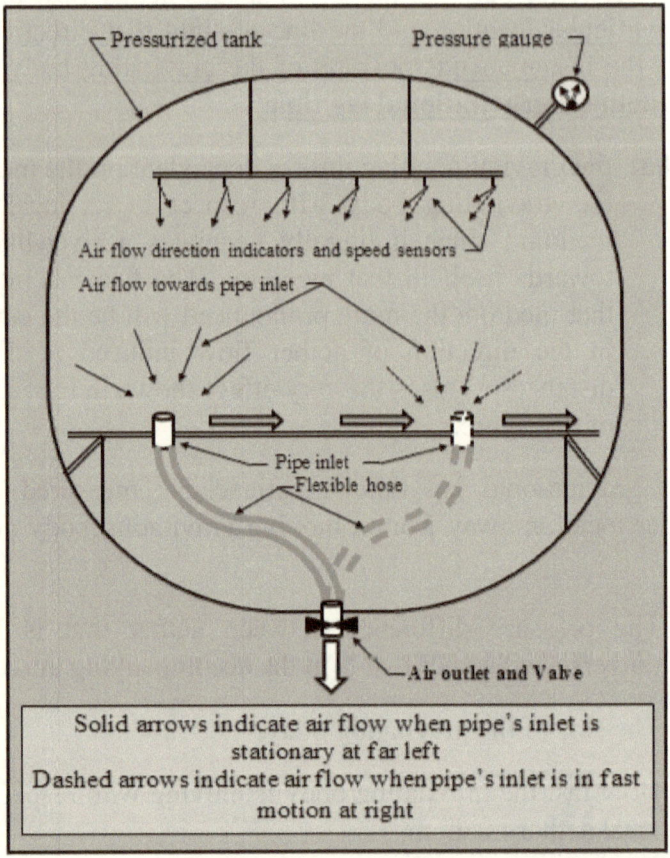

Solid arrows indicate air flow when pipe's inlet is stationary at far left
Dashed arrows indicate air flow when pipe's inlet is in fast motion at right

As shown in the figure, this setup includes a pressurized air tank. The compressed air inside the tank is allowed to go through the open-end of a flexible hose. The open end (the inlet) of the hose is attached to a slide that can move along a track. The speed at which the slide can move is adjustable. Inside the tank, there are several strips of light weight material which can freely move and hence indicate the direction in which the local medium is flowing.

At first, the slide and the open end of the hose are positioned at the far left end of the track. The solid arrows show aether flow directions, as indicated by the individual strips, as well as the arrows which are near slide's position.

Then, as slide is made to carry the open end of the hose towards the right end of the track at different speeds, the flow of aether is automatically adjusted trying to point towards the instantaneous position of the open end of the hose. Such flow directions are indicated by the dashed arrows.

This experiment must be repeated with various gas pressure levels, as well as different rates of speeds for the motion of the slide along the track. The effects of variations introduced in:
- The air pressure inside the tank,
- The speed of the slide on the track,
- The size of the opening (valve setting used at the outlet end of the hose), as well as,
- The distance to the strips,

on the delay time that it takes for the strips indicate the instantaneous position of the open end of the hose on the track, can be tabulated. Then, the results can be formulated into an empirical equation.

Once the needed empirical equation is derived, it can be applied to the lag time of the gravitational force of the moon, as measured from earth, to determine the current pressure difference that exists between aether that is in this universe and aether that is in the accompanying universe.

If need be, such an empirical equation can be applied to the lag time of the gravitational force of the sun on a probe that is sent into an orbit around it, in the same plane as sun's orbital plane around the center of mass of the galaxy.

Regardless of which celestial body is considered as the gravitating body for this method, to obtain more accurate results, the overall motion of that celestial body in its local aether medium must be used / accounted for.

30- Probes' near-earth flyby anomaly explained (Unexpected speed variations)

Over the last few decades, as different probes were directed to execute near earth flybys, to gain speed while changing direction towards their desired destinations, an anomaly was noticed. Starting with 'Galileo' spacecraft (flybys made during 1990 and

1992), and later on 'NEAR' (1998), 'Cassini' (1999), 'Messenger' (2005) and 'Rosetta' (2005, 2007 and 2009) some unexplained variations were detected in their respective speeds, as they had left earth behind.

The variations detected in the speeds of such probes were quite small in their magnitudes, the maximum being that of 'NEAR' at 13.48 mm/s. However, such variations were simply not expected, since all possible affecting forces were already accounted for. Also, the amounts of variations in the speeds of various probes were different. It should also be mentioned that, each flyby was executed at a different angle with respect to earth's equator and at a different altitude from earth's center of mass.

In the chapter titled "What is Magnetic Field?" (in the section with the same title as above), it is clearly shown in details that, once the effects of aether flow associated with earth's magnetic field are taken into account, not only such a variation in the speed of any probe performing a flyby routine around earth, or any other celestial body which possesses a magnetic field, becomes explicable but also their occurrences become quite expected and their magnitudes can even be predicted, in advance. Since, any probe performing a sling-shot maneuver around planet earth not only passes through the aether flow that is giving rise to earth's gravity, but also through a secondary aether flow which is associated with earth's magnetic field. Hence, the probe gets dragged by this secondary aether flow, also.

Such a secondary aether flow is mainly from the Magnetic North Pole towards the Magnetic South Pole. However, there is also a minor aether flow associated with earth's magnetic field which is towards magnetic / geographic east, due to its rotation with earth, as a whole. Therefore,

"Any probe that executes a near earth flyby routine, regardless of its trajectory, experiences variations in its speed as well as its direction of motion. However, in most cases, depending on its specific trajectory, it may experience more of a change in its speed or more severe alteration in its direction of motion."

31- Was the ancient science of astrology based on the fact that each celestial body has a unique gravitational signature?

Astrology is an ancient science with origins that can be traced back to the first civilizations located in the Mesopotamia, present-day Southern Iraq. And, it has gradually spread globally, over thousands of years.

Astrology uses the locations of various planets, as well as twelve specific constellations in deep space to formulate its predictions. Therefore, it encompasses astronomy.

At the present time, astronomy has become restricted to looking at different planets, stars and even galaxies, as well as studying their motions relative to each other. However, in the distant past, astronomy was a much more detailed science. It was pursued to understand the effects that various celestial bodies have on the physical as well as mental capabilities and shortcomings of living beings on planet earth. In other words,

"The real science of astrology encompasses not only astronomy but also biology/physiology as well as psychology/spiritology."

To use astrology certain specific information including the position of the sun, the moon, and the five nearby planets, namely Mercury, Venus, Mars, Jupiter and Saturn, all relative to earth, must be known. The exact time, date and place of birth on earth, also specifies the constellation that is rising over the eastern horizon, as seen from the birthplace.

The twelve constellations used in the science of astrology are in fact twelve **relatively fixed reference points** in space around earth which are used to specify the exact direction pointing towards the center of the Milky Way galaxy from the location of birth on earth's surface, at the time of birth. Constellations also specify the directions pointing towards the positions of various planets, as well as the directions pointing towards the positions of the sun and the moon.

Note that, the very selection of stars that, by forming recognizable constellations, can serve as long term reference markers in space around earth is a clear indication of in-depth astronomical knowledge of the initiators of the science of astrology. Since, the chosen stars had to be clearly visible from earth, and had to be far enough in space so that neither the movement of earth around the sun, nor the movement of the solar system as a whole in the galaxy would affect their formations as seen from earth. And, they had to be recognizable by earthlings over thousands of years into the future.

As shown below, the twelve constellations used in astrology are almost evenly spread like a belt around the plane of the solar system. They are spaced just like the twelve numbers on the face of an old-fashioned clock.

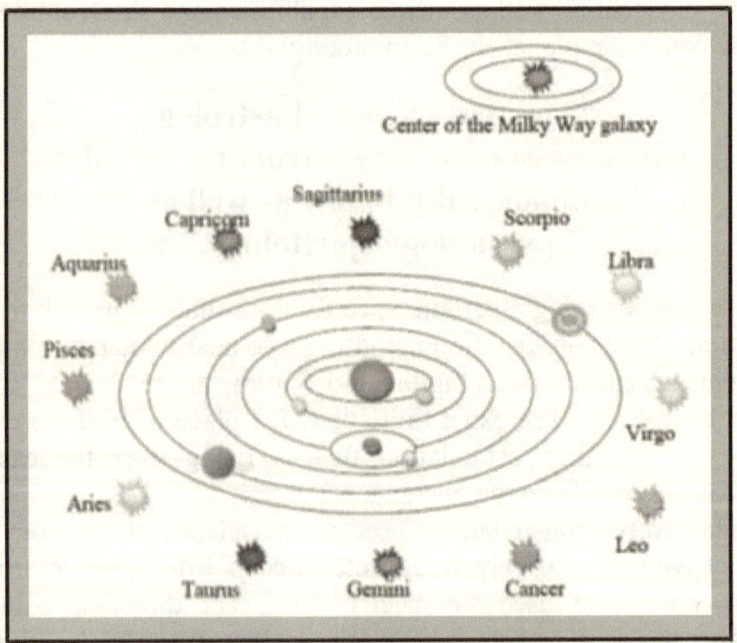

The dependence of an individual's physical and mental abilities, as well as limitations, on the positions of the center of the galaxy, the five nearby planets, the sun and the moon, with respect to earth, at the time and place of his/her birth, are all indicative of the effects of one universal physical phenomenon. **That physical phenomenon is the force of gravity.**

The gravitational effects of the nearby planets are quite noticeable in the motion of earth, as it follows its orbital path around the sun. The gravitational effects of the more distant planets are relatively weaker. The gravitational force of the moon (due to its being very close to earth) and that of the sun (due to its being very massive) are quite strong and readily affect the liquids that are on and under the surface of earth. The occurrence of the tides is a clear indication of such effects imposed on planet earth by the gravitational forces of these two celestial bodies.

The gravitational forces of the moon and the sun have direct effects on the living beings on earth, as well, effects such as the period in women. Many more of such regular physiological effects can be readily observed in the lives of various animals, as they go through their regulated pregnancy timings / periods, year after year.

The effects of gravitational forces associate with different celestial bodies on various fluids that form inside the body of any living being can be compared to such effects as tides in the oceans. And, the exact time of birth in fact marks the commencement of such gravitational influences as the newborn's body starts its independent and unshielded existence, by separating from the mother's body. During the first moments of its independent existence, all of the liquids and various solutions in the body are suddenly directly exposed to a variety of gravitational forces from different directions, particularly those of the Moon and the Sun, due to their being closer and more massive, respectively.

As the body starts its independent existence, it becomes exposed to these extra gravitational forces that were not as effective before, while it was inside the womb. Just like positioning a long magnet in parallel and close to a multi-walled (multi layered, insulated from each other) Iron pipe through which a fluid containing iron shavings is flowing. The fluid and the iron shavings floating in it are protected from the outside magnetic field, by the pipe itself. However, as the end of the pipe is reached and the fluid exits the pipe, the outside magnetic field will directly affect the iron shavings that are floating in the fluid. Such a scenario is shown in the following figure.

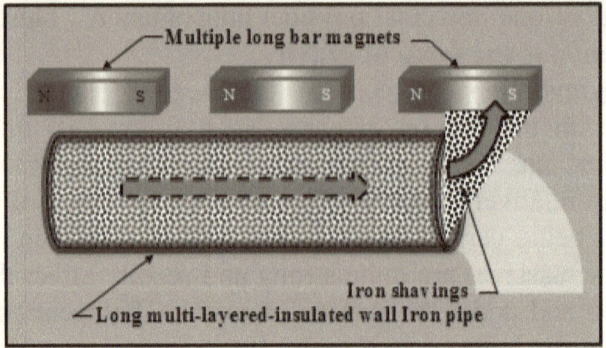

Presented in sufficient details at the end of this section are two experiments which explore the differences between the gravitational effects of various celestial bodies on a variety of elements and bodily fluids, over a long period of time.

The awareness of the members of the ancient civilizations of such effects, and their comprehensive knowledge of them, clearly demonstrate their in depth understanding of physiological and spiritual characteristics of living beings, particularly those of humans.

Gravitational forces affect the spiritual abilities and limitations of living beings in a completely different way as compared to how they affect their physical abilities and limitations. In order to understand such effects on spirits that reside in individual living beings, one needs to be familiar with the spiritual world and particularly the internal structure of complex spirits. The author has presented such information in great details, in a separate book titled "The Evolution of Spirits", (2012). A summary of which is also presented in a different book titled "Purpose of Life in This Universe", (2012). A brief description is provided here.

All of the spirits have started their existences at the very same instant, when an original unified spirit split into infinite number of tiny pieces. These tiny spirits have been and still are evolving through various stages, as they reside in progressively more complex physical bodies and simulate more capable life-forms. They start with plant type of life simulations. After many plant type life-form simulations, they progress into simulating animal life-forms. In between life-simulations, they unite with one or more other tiny spirits which used to be their immediate neighbors in the original unified spirit, just as the pieces of a jigsaw puzzle can fit together. In so doing, they form ever more complex spirits.

As a given complex spirit starts a new physical life simulation, different parts of it take over the development and operation of various body parts. In the case of plant type of life-form simulations, each complex spirit stays intact and occupies a single leaf (not the whole plant, or tree), and takes over its internal operations. However, in the case of animal life-forms, each complex spirit divides itself into two major parts upon taking over a physical body which is forming inside a womb or an egg. Each major part is composed of several simpler spirits.

The constituents of one part (which is referred to as the "Soul") take over the internal functions of the physical body, as they try to control the operations of individual body parts in a coordinated fashion. The constituents of the second part (which is referred to as the "Spirit") take over the external functions of the living being, including communication and analyzing the information gathered through various senses.

The proper coordination of the tasks performed by different body parts (just like different departments in a complex organization / factory) ensures the proper health of that physical body. The abilities and limitations of any given complex spirit are dictated by the previous life experiences of its constituents / spiritual members. However, depending on which spiritual constituent is selected as the leader / organizer in a complex spirit, the complex spirit (the living being) will demonstrate different sets of abilities and limitations. It will even have different personalities, since the personality of the leading spiritual constituent will be the dominant one.

In regards to how the abilities and limitations of complex spirits residing in living beings are affected by gravitational forces, one needs to be reminded that spirits do not have any mass associated with them. Therefore, the force of gravity cannot affect them directly. However, their residing in a physical body (which includes the whole volume of the physical body) and particularly their connection points with the physical body which is controlled by the part referred to as the Soul, has characteristics that are similar to material characteristics and therefore are influenced by the force of gravity.

The spiritual components of any complex spirit always try to preserve their configurations. However, as they enter a physical body, due to the conditions and situations they encounter in the body and also how the body can be guided to develop into its final form, they have to improvise in their configuration so that they can adapt to the physical body in which they are residing. These changes can even include choosing a different leader, which automatically implies that a different personality and a different set of characteristics will be exhibited by the individual, later on in life.

Therefore, the very same complex spirit can and will demonstrate possessing different abilities and limitations as it resides in different bodies or even the same body but under different circumstances. Such potential variations in the outcome of the union between a given complex spirit and a physical body can be likened to different scenarios that can arise as an individual is allowed to drive a bicycle, a motorcycle, a car, a pickup truck, or a semi-truck, on roads that might be in quite different conditions.

Spirits that simulate animal life-forms are very complex spirits that are formed from the unification of many simpler spirits. During their previous lives, these simpler spirits had identified each other as adjacent neighbors in the original unified spirit and had managed to unite, step by step (a few at a time) after they had finished their concurrent physical lives. By joining together, step by step they have formed more complex and more capable spirits. Any newly formed more complex spirit has to adapt itself to the next animal body in which it is going to reside, so that it may benefit from all of its potential abilities.

Even though, it is going to be limited to the physical body, the complex spirit must try to benefit from all of the potential abilities that it has access to due to each and every one of its spiritual components. In short, the internal structure of any complex spirit depends on the previous life experiences of individual spiritual components involved in the collective, as well as the internal structure and the capabilities of the physical body in which the complex spirit resides.

As the spiritual components join together they choose one as the leader. Then, the newly formed complex spirit will wander in the spiritual world for a while, as the spiritual components coordinate their positions in the complex spirit. This is exactly like several individuals who get together to plan for a joint trip to some faraway destination. As the abilities of each and every spiritual component are identified, a specific trip is picked, a trip which is in fact experiencing another physical life in this physical world.

From the time it enters a new physical body, the complex spirit has to try to benefit from any and all of that body's abilities. This is exactly like the scenario in which a group of friends decide to go for a distant trip and depending on the type of vehicle they are provided with, which may be a car, a truck, a plane, a train, a ship or even a bicycle, they have to choose the right member to act as the leader or the driver of the vehicle, so that they can get to their designated destination.

After accepting the vehicle, which is the physical body, the spiritual components must cooperate and try to perform their assigned tasks in a coordinated fashion. The physical bodies provided for complex spirits are directly affected by the gravitational forces associated with various planets, as well as those of the moon, the sun and the center of the galaxy.

The aether medium in this universe is also hosting a variety of phase vibrations. Therefore,

**"As celestial bodies move in the aether medium, they
are in direct contact with the phase vibrations that
are present everywhere."**

Consequently, as each celestial body follows its respective orbital path in the aether medium, it generates a unique echo / reflection of the phase vibrations with which it comes in contact. The reflections of such phase vibrations are in turn broadcast in their surrounding environments.

The very same effect takes place due to the existence of boats and ships on the surface of a lake. Since, as shown below, the floating body of a boat or a ship, even if it is anchored down in one spot, causes a unique variation in the overall texture of the waves that already exist on the surface of water in its surrounding.

The variations induced depend on the size of the vessel, its overall weight and the shape of its haul, among other factors. The variations induced by the same boat or ship also changes when it starts to move. The exact details of the variations induced, in this case, will depend on its direction of motion and its rate of speed.

Celestial bodies are floating in the medium of fluid aether, just like submarines that float in the medium of water. Therefore,

Just as submarines reflect the phase vibrations that are present in the medium of water, planets also reflect the phase vibrations that exist in the medium of aether.

Note that, the overall shape / size of a given celestial body, along with the texture of its surface dictates how that particular celestial body reflects the cosmic background phase vibrations with which it comes into direct contact. Even the very motion of that celestial body in the aether medium, as well as its surface temperature (which dictates how intensely the constituent particles that are at or near its surface are vibrating) affect the overall frequencies of the echoed / modified phase vibrations.

The individual atoms / molecules present, as well as their vibrations, also directly affect the frequencies of phase vibrations that are reflected back into the surrounding medium. This is why any given object reflects different ranges of visible light waves, as its temperature is raised.

The consequences of atoms / molecules present can be readily observed in the way light waves from a common source such as a regular household light bulb are reflected differently by walls / doors / furniture / carpets / clothing materials and so on. Since, each object has a unique surface texture as well as atomic / molecular constituents that are different. In a sense, one can state that,

"Each celestial body has its own gravitational color / frequency, signature."

The reflected phase vibrations by various planets are in turn broadcast in their respective surrounding aether medium. In time, these newly formed phase vibrations interact and blend with other phase vibrations in the aether medium, as they propagate away from their respective sources.

Such interactions between various phase vibrations lead to the formation of harmonics / resonances at certain distances from each celestial body. They also cancel each other's wave forms at certain distances, as well. **The overall effect is very much like the formation of interference patterns, comprised of resonances and cancelations in the wave forms of phase vibrations as they propagate in the medium of aether.**

Note that, the separations between consecutive resonances generated in the aether medium by different celestial bodies are different. Since, no two celestial bodies have the very same surface texture or atomic / molecular structure, let alone being of the very same overall dimension / diameter. In other words,

"Each and every celestial body has its own unique way of affecting the texture of the phase vibrations that exist in the medium of fluid aether."

Modified phase vibrations (in fact, newly formed vibrations) induced by different celestial bodies spread in all directions.

Eventually, they all reach earth, which was/is of prime interest, in the science of astrology.

Earth is continuously moving through all of these modified phase vibrations, including their resonance regions and their cancelation regions. However, due to the fine size of such vibrations, many of their resonance regions and cancelation regions are constantly sweeping across the entire volume of earth, concurrently.

Therefore, at any given instant, different locations on the surface of earth are exposed to different wave forms that are generated due to the intermixing of phase vibrations which are modified by different nearby celestial bodies. That is why, **it is of prime importance to know when (date and time) and where (longitude and latitude on earth) the newborn was/is born.**

The following figure shows the directions pointing towards nearest five planets from earth, as well as those pointing towards the sun, the moon and the center of the Milky Way galaxy, at a specific time and day in the year, all of which were accounted for in the original science of astrology.

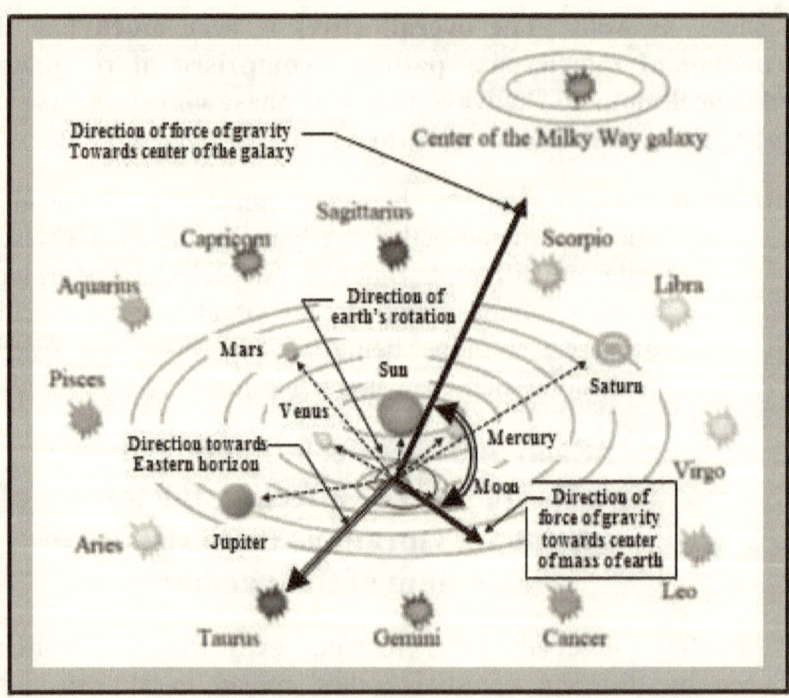

Every newborn's body can be treated as an imaginary focal point of interest where the consequences of the interactions between the varieties of modified phase vibrations are to be considered. Therefore, by knowing in which constellations the individual celestial bodies of interest are located, as viewed from an individual's birthplace on earth, at the time of his/her birth, one can calculate the angles between the directions pointing towards them. In other words, one can calculate the angles between the propagation trajectories of the incoming phase vibrations that are modified (generated / echoed) by different nearby celestial bodies.

As different modified phase vibrations interact at the birthplace of a newborn, they give rise to certain general wave formations which can include various harmonics / resonances, at specific intervals. Such wave formations, in turn, in their own specific ways, affect the individual local matter particles that make up different elements in various body parts and bodily fluids, in the newborn's body.

Such wave formations can either encourage or prevent the formation of certain atomic or molecular bonds in various body parts. They can also affect the composition of certain compounds / solutions, and either encourage or prevent the formation of certain other solutions, that are supposed to function as bodily fluids.

All of these effects have consequences on the future development of the newborn's body, and hence affect his/her abilities in performing various tasks, both physically as well as mentally. Since, the elemental constituents of the brain get affected by such waves just as those in other parts of the body. Hence, as different parts of the brain, as well as other body parts, are affected differently in different individuals, who are born at different places, at different times and on different dates, those individuals, demonstrate possessing various levels of mental capabilities and/or shortcomings in performing certain tasks, as they grow older. It short,

It is the force of gravity of various local celestial bodies and the reflected phase vibrations by them that are affecting the physical and the mental capabilities of newborns in mysterious ways.

Reasons for the science of astrology losing its accuracy, over time

Based on the above information, it is readily understandable why the predictions provided by the science of astrology, regarding physical and mental abilities of an individual, used to be more precise in the distant past as compared to what they are these days. The following are four of the reasons why such predictions have become less accurate, over time:

1- **Correct direction pointing towards the position of the center of the galaxy**

    The direction which points towards the center of the galaxy is constantly changing, due to the motion of the solar system in the galaxy. Currently, the position of the center of the galaxy is no longer in the same direction as it used to be when the science of astrology was first established.

2- **Constancy of the location where the expected mothers lived**

    Living in one region for the whole duration of the pregnancy encourages regulated cyclic effects of various gravitational forces exerted particularly by the sun and the moon on the solutions within the mother's womb and the body of the fetus.

    In the distant past, expected mothers used to spend the entire period of their pregnancies in one locality. However, in recent years, they travel to anywhere and everywhere, using cars, trains, airplanes, ships and so on. Therefore, by travelling to distant places, they expose their fetuses to irregular combinations of gravitational forces, while in the older days, due to staying in one place, such gravitational forces had well defined cycles.

3- **Spread of the human population all over the surface of earth**

    Astrology was originally developed by civilizations that were settled in the Mesopotamia, present-day Southern Iraq, which is located at 30 degrees latitude north. The initiators of astrology had to have taken into account the effects due to 3-dimensional angles that were formed between the directions of the gravitational forces associated with various celestial bodies with respect to the direction of earth's gravity at that latitude.

Naturally, astrology was specifically customized for individuals who were born at about 30 degrees latitude north. Therefore, as it is, astrology cannot be applied to individuals who are born at other latitudes. The following figure shows how the angles between the gravitational force of earth, which is towards its center of mass, and those of the other celestial bodies change as the latitude of the birthplace is varied, even though the longitude and the time and date of birth can be identical to each other.

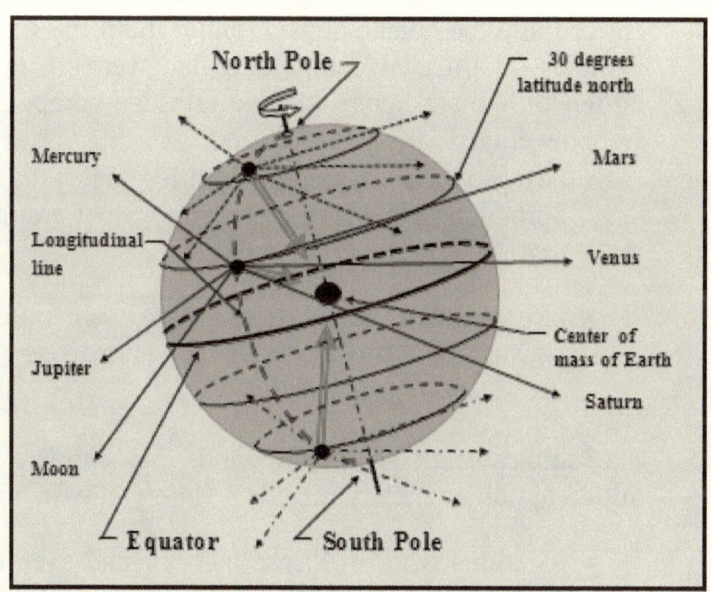

Now-a-days, astrologers take the longitude of the birthplace into account, as they try to identify the rising constellation on the eastern horizon. However, they do not pay any attention to the latitude of the birthplace, as they write their charts for the angles of where the planets fall in the zodiac signs. According to them, a baby that is born at the equator and another one that is born near the North Pole or near the South Pole will have the same chart as the one who is born at 30 degrees latitude north, as long as they share the same date and time of birth.

Such astrologers unknowingly introduce a great uncertainty in their own calculations. Such uncertainties consequently lead to inaccurate predictions of their clients' physical as well as spiritual strong and weak points.

4- Dietary based on natural foods

In the past, all of the chemical compounds in human diet used to be natural. Therefore, the human body (as well as those of other animals) was based on such natural building blocks, hence it had well adapted to such raw materials.

However, particularly over the last hundred years or so, in different ways, a variety of artificial compounds have entered into the human dietary, either indirectly or directly. Samples of artificial compounds that were/are **indirectly added** to human dietary include artificial compounds that were developed / used / introduced:

- To produce genetically modified seeds, to boost the production yields of different types of grains and so on,

- To produce animal feed which were modified to hasten the growth rates of such animals as chickens, cows and pigs and so on.

Samples of artificial compounds that were/are **directly added** to human dietary include a variety of extras such as:

- Various types of drinks, snacks, and even chewing gums,

- Various medications, as well as vitamins,

- Various air-borne compounds that are breathed in, due to a variety of chemical / industrial activities, nearby.

Even though new food products seem to be more beneficial in certain ways, in short term, but they have unknown long term effects on the overall structure of living beings' bodies. This issue can be likened to using concrete blocks to build buildings, as compared to stones which were used in the construction of the older buildings. Buildings made of concrete blocks can be put together much faster and easier, but they definitely lack longevity as compare to buildings made of stones.

Human bodies (and those of the other living beings) were/are developed in different regions on this planet, based on what were/are naturally available to them, as food in their localities. Once the composition of the food is altered, as it has been over the past hundred years or so, the internal structure, composition and functionality of each and every body part are affected, directly.

Such apparent effects are quite known in regards to certain foods / diets that are recommended for athletes in various fields, such as weight lifting and particularly body building. Even though such variations in the food intake can be beneficial in certain ways, they can and do have adverse effects in other ways.

In other words, the consumption of certain foods that are either made of artificial compounds or have been artificially prepared, directly affects the predictions made based on astrology. Since, that ancient science was based on the consumption of natural foods.

These are four of the major reasons why astrology has lost its accuracy over the millennia. The new-age astrologers apply their skills in astrology to predict their clients' future attributes, but without paying proper attention to the original rules that were incorporated in the very development / formulation of the science of astrology.

For the predictions made through astrology to become as precise and accurate as they used to be in the distant past:

- The proper location (direction) of the center of the Milky Way galaxy must be taken into account.

- The expected mothers should reside in one locality for the whole duration of their pregnancies.

- The effects due to 3-dimensional angles between various gravitational forces must be taken into account, or if in doubt, the expected mothers should live in regions which are at or near 30 degrees latitude north.

- The expected mothers should stick to natural diets. Also, their newborns should try to stick to natural diet, as well.

Astrology can be as accurate in its predictions, as it was in the distant past, particularly if its underlying rules are properly obeyed.

By knowing the details of the rules incorporated in the development of astrology, one can even run hypothetical scenarios covering thousands of years into the past, as well as into the future. The very important aim will be finding the best possible combination of time/date and place required for an individual to be born with super unique physical and/or spiritual capabilities. In other words,

"Using the science of astrology, one can specify the exact time, date and place of birth for a child who will have certain desired physical and/or spiritual capabilities."

The following two experiments are designed to detect any particular influence that the gravitational forces and echoed phase vibrations associated with different planets, the sun, the moon and the center of mass of the Milky Way galaxy may have on a variety of elements, and on different bodily fluids, respectively.

The results of such experiments will confirm why the science of astrology was based on so strict rules regarding the positions of various celestial bodies, and that of the center of the Milky Way galaxy, with respect to the birthplace of a newborn.

The results of such experiments will also demonstrate why different celestial bodies affect different attributes in an individual's body, as well as his/her spirit. Since, each organ in a living body is composed of certain types of elements and contains / interacts with certain types of bodily fluids. Therefore, forces of gravity as well as phase vibrations associated with different planets, the sun, the moon and the center of the galaxy, affect how each and every organ in a living body develops and functions. The very same applies to the spiritual part of a living being.

- **First experiment:**
  (Gravitational effects of various celestial bodies on different elements)

This experiment is designed to determine the overall effects of forces of gravity of different celestial bodies on different elements.

As shown below, the experimental setup can consist of many plumbulbs hanging from a common support structure. In such a setup, each of the plumbulbs is made of a different element. They are all housed inside a very large glass container so that the local air movements do not affect the positions indicated by the plumbulbs' tips. The floor of the glass container is calibrated so that movements of the plumbulbs' tips, in any direction, can be readily observed and recorded.

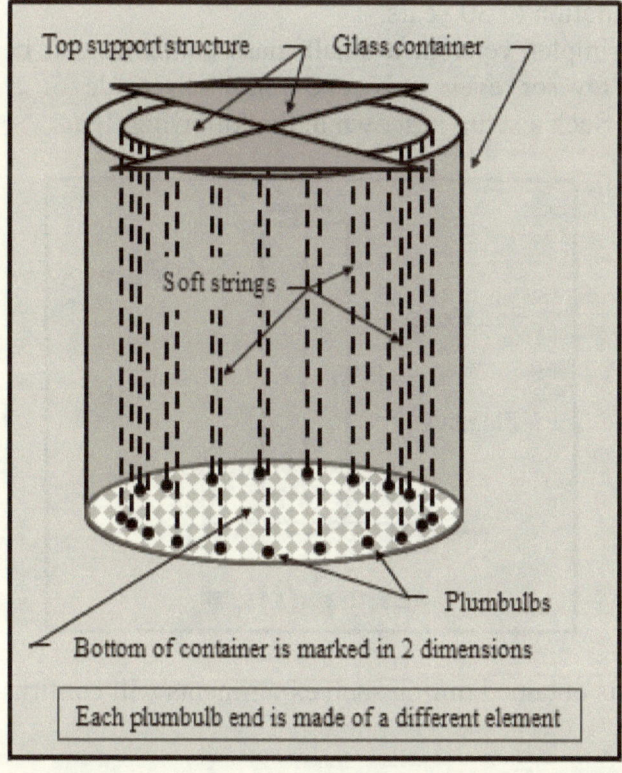

Top support structure — Glass container —

Soft strings —

Plumbulbs —

Bottom of container is marked in 2 dimensions

Each plumbulb end is made of a different element

This particular setup can be used to observe potential effects of the force of gravity of different planets on various elements, concurrently. Such an experiment needs to be conducted over a period of several years, since in order to obtain comprehensive results, each and every one of the nearby five planets used in astrology must complete one full orbital pass around the sun. The longest orbital period belongs to Saturn which is the farthest of the five planets from the sun. Therefore, the required length of time to properly conduct such an experiment would be about 30 earth years, since Saturn's year is equivalent to about 29.5 earth years.

However, after a few years, the relative effects of the forces of gravity associated with different celestial bodies can be noticed. Once the observed effects are found to be consistent with the movements of the five planets, the moon, the sun and the center of the galaxy, the experimental results can be extrapolated to cover the full range of the 30 years. If need be, and even if it is to confirm the extrapolated parts of the data, the experiment can and should be allowed to run its course, and the results be recorded for the full duration of 30 years.

In a simpler version, a small glass container can be used to house a few or even only one plumbulb made of a specific element. Such a setup is shown in the following figure.

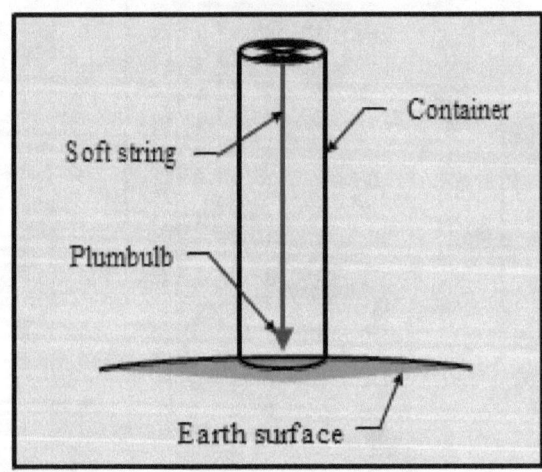

The results obtained during such experiments will confirm that,

"The force of gravity associated with any given celestial body (or a collection of them, as in a galaxy) has its own unique frequency.

And,

That is how they can affect certain elements more strongly as compared to the way they affect other elements."

Such preferences can be readily likened to the way magnets interact with different elements. They interact strongly with ferromagnetic materials, partly interact with paramagnetic materials and they nearly totally ignore the existence of all the other materials.

It should also be mentioned that,

"All eggs belonging to various living beings have calcium based shells."

Other than being a fairly strong material suitable for protecting the contents of eggs which are turning into life-forms, such a commonality in the composition of the exterior / protective layer of all eggs is due to **calcium** having certain resistive capabilities towards gravitational forces of various celestial bodies. Such resistive (isolating) capabilities can be expected to manifest during such experiments, as well.

- Second experiment:
(Gravitational effects of various celestial bodies on different bodily fluids)

To conduct such an experiment, as shown in the following figure one needs to use multiple, identical, small size containers in which samples of various known bodily fluids can be placed. Such containers should be placed in a common quarantined environment.

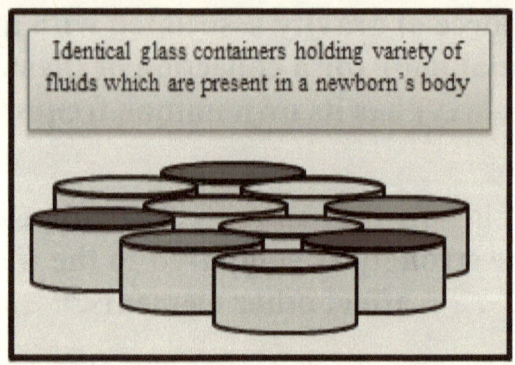

Identical glass containers holding variety of fluids which are present in a newborn's body

Each container must host only one type of bodily fluid. The composition of the overall cover of the quarantined environment should closely resemble the composition of egg shells or the mother's body, as the case may be for the type of life-form being investigated.

Note that, this experiment has to be conducted over a time period which would match the incubation period or the pregnancy period of the life-form of interest.

Conclusions

According to the new theory of gravity introduced in this chapter, all of the matter and anti-matter particles are literally bubbles in the medium of fluid aether, in this universe. They act as drain holes through which the fluid aether is escaping from this universe and is entering and accumulating in the accompanying universe. The driving force for the flow of aether towards and through the particles is the pressure difference, between the pressure of aether in this universe and the pressure of aether in the accompanying universe. Since, aether that is in this universe is under a much higher pressure as compared to aether that is in the accompanying universe.

The accelerated flow of aether towards each and every particle induces a drag force on all other particles and objects that happen to be in its path. This drag force is the force of gravity which everyone is familiar with. In other words,

"Matter and anti-matter particles, with their very formation, gave birth to the force of gravity in this universe."

In summary,

"The force of gravity, experienced at any given location in this universe, is due to the drag force induced by the flow of aether which is accelerating through that particular location."

Note that, for the force of gravity to be manifested, matter and/or anti-matter particles must exist. Therefore,

- **Various types of energy, including phase vibrations in the medium of aether,** such as light and other electromagnetic waves are not made of any kind of matter particles. Hence, regardless of their amounts / intensities, they **cannot generate any gravitational force.**

- Also, **the fluid aether** which occupies the whole universe is not made of any kind of particles and hence **does not generate any gravitational force.**

The newly proposed theory provides consistent explanations for various known phenomena related to the force gravity. It also offers many predictions which can be readily verified using current technology. Several experiments are proposed and described in detail, as well, experiments that will confirm the validity of this new theory.

According to this new theory, as the aether medium in this universe is gradually expanding, and also as aether is escaping from this universe by going through matter and anti-matter particles, the overall pressure and consequently the overall density of aether that is in this universe are gradually decreasing. Such gradual reductions in aether's pressure and density are automatically causing the fluid aether to become less effective in inducing the drag force which is the force of gravity. Since such an effect is taking place on a universal scale, one can state that,

"The force of gravity is gradually becoming weaker, on a universal scale."

In other words,

"The universal gravitational constant is gradually decreasing in magnitude."

As mentioned in the chapter titled "What is Aether?", **as the density of aether is decreasing in this universe the 'mass' of**

each and every particle is also gradually decreasing. One side effect of such a decrease in the strength of the force of gravity on a universal scale is expected to be the gradual weakening of the gravitational forces that stars exert on their respective planets. This effect has already been verified in the case of the planet earth. Since, it has been proven that earth's orbit around the sun is gradually widening by an amount that is over and above what can be contributed to the gradual reduction in sun's mass, due to nuclear fusion reactions and solar flares / solar storms.

Note that, the gradual decrease in the universal gravitational constant, as well as the decrease in the mass of particles, support the theory presented in this chapter. Since, if it were not for the flow of aether, which is gradually losing pressure and is thinning out, the gravitational force of individual matter particles and their aggregates, such as stars and galaxies, would not have been weakening over time.

The topographical presentation of the gravitational force near massive stars and particularly near black holes are interpreted by the general theory of relativity as curvatures in the spatial dimensions, while they are not. The calculated curvatures actually represent the gradients that exist in the flow speeds of the local aether that is headed towards the star or the black hole. Since, the closer aether gets to any particle (or aggregate of particles) the faster it accelerates towards that particle (or aggregate of particles).

It must be emphasized at this point that,

"The individual particles such as electrons, protons and neutrons, as well as their corresponding anti-particles, are simply bubbles in the medium of aether, bubbles which are not composed of any smaller constituents."

In other words,

"There are no such things as quarks in this universe."

Therefore,

"There is no such a force as the strong nuclear force in this universe."

Also, according to the aether-based theory of gravity presented here,

"The gravitational attraction forces that protons and neutrons exert on each other override the repelling forces generated between protons by their positive electrical charges."

Therefore,

"The gravitational attraction forces exchanged between protons and neutrons ARE the forces responsible for holding the nucleons together, in complex atomic nuclei.

Hence, according to aether-based theory of gravity,

"There is no such a force as the weak nuclear force in this universe."

The pressure of the fluid aether in the accompanying universe is gradually rising. This is while the pressure of aether is gradually dropping in this universe, due to both expansion of the overall aether medium and loss of aether to the accompanying universe. Therefore, as the pressure of aether that is in this universe and the pressure of aether that is in the accompanying universe are gradually becoming equal in magnitude, the force of gravity is gradually fading away. Since, the pressure difference between aether that is in this universe and aether that is in the accompanying universe is the driving force that encourages aether to accelerate towards and go through various particles.

Therefore, as this pressure difference approaches and equals to zero, the force of gravity gradually fades away. In other words,

"Just as the very formation of matter and anti-matter particles had given birth to the force of gravity, their continued existence will cause it to gradually weaken and eventually fade away."

Black holes
and
Their Properties

Introduction

The possibility of the existence of black holes was first derived from the general theory of relativity, nearly one hundred years ago. Mathematically, it was shown that, there could be such places in this universe where the gravitational force is so strong that even light waves are dragged in and are absorbed, once they get close enough.

Due to follow up research and theoretical advancements, it is generally accepted that black holes can form as massive stars reach the end of their useful life-cycle. Since, at that stage, they collapse due to their own force of gravity. During such a process which is known as supernova, the outer layers are thrown into the surrounding space, while the inner layers are compressed into a much, much smaller volume. If the compressed inner portion is gravitationally strong enough it will form a black hole, otherwise it can become a neutron star.

Various theories have speculated on black holes' characteristics as well as their overall behaviors. However, there are many issues that still need to be addressed.

In order to understand what black holes really are, one needs to start with a solid theory of gravity. The following is a brief description of a new theory of gravity that is presented in detail in chapter titled, "What is Gravity?".

A New Theory of Gravity

The new theory of gravity presented here is based on two assumptions:

1- **The whole universe is occupied by fluid aether,** which is a compressible, non-viscous and quite a dynamic medium.

2- **Each subatomic particle in this universe is a tiny 3-D drain hole,** through which aether that is under very high pressure in this universe escapes and enters an accompanying universe where its pressure is much lower.

As a first step, for simplicity, one can consider a subatomic particle that is freely floating in space and it is relatively stationary. The incoming flow of aether which is accelerating towards this subatomic particle will be from all spatial directions. The inbound flow of aether will extend literally to the end of space, and it will be directed towards the current position of the subatomic particle.

As it is shown in the following figure, any other particle in existence in this universe will be automatically affected by the flow of aether which is accelerating towards the first subatomic particle and will experience a dragging force directly towards that particle's position.

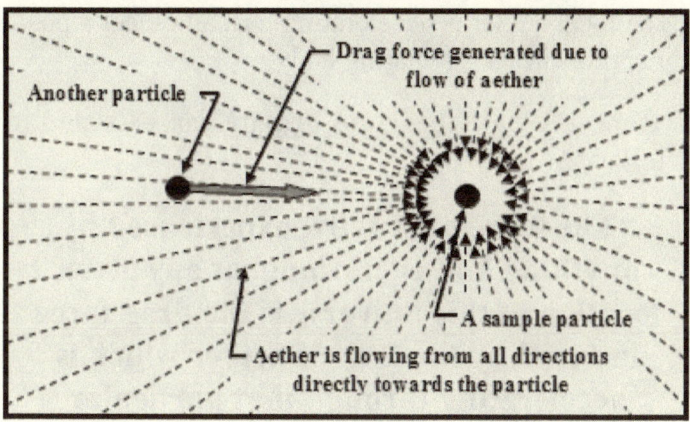

Another particle

Drag force generated due to flow of aether

A sample particle

Aether is flowing from all directions directly towards the particle

This drag force is the force of gravity that is experienced by anything and everything in this universe.

It must be emphasized early on that,

- **It is the acceleration of aether relative to / through any given location in this universe that gives rise to the drag force which is experienced as the force of gravity by any object that may be positioned at that particular location.**
- **Steady motion of aether relative to an object, regardless of its rate of speed, does not generate any drag force, whatsoever.**

In other words,

It is the instantaneous change in the motion of an object relative its local aether medium that gives rise to the drag force that is exerted on that object, the same drag force that is known as the force of gravity.

That is why, the very same phenomenon of **weight**, is induced when:

- An object is in a vehicle that is accelerating in the aether medium,

- The local aether is accelerating past an object's position, as it approaches a gravitating body.

Therefore, **the new theory of gravity** can be stated in a nut shell as follows:

"The force of gravity, experienced by anything and everything, at any given location in this universe is the drag force induced by the flow of aether which is accelerating through that particular location."

Black Holes
Based on the New Theory of Gravity

The newly proposed theory of gravity, which is based on the existence of aether in this universe, provides the following explanations in regards to black holes.

1- What is a black hole?

One can first consider the propagation of sound waves in air. Using a supersonic wind tunnel a variety of experiments can be performed to demonstrate the desired effects.

To perform such experiments, as shown below, one needs to install a speaker (which is attached to a stereo system) near the outlet and a microphone (which is connected to a headphone set) near the inlet of the supersonic portion of the wind tunnel.

By starting the wind tunnel and gradually bringing it up to its supersonic speed, while listening to his/her favorite music, he/she will notice that the sound heard through the headphone set gradually becomes weaker and weaker, as if it is broadcast from farther and farther away. So long as the speed of the wind generated by the wind tunnel is less than the speed of sound waves in that air medium, the sound waves generated by the speaker will propagate against the flow of air and reach the microphone.

However, once the sonic barrier is crossed, he/she will not hear the music at all, because the sound waves can no longer compete against the speed at which air is flowing in the opposite direction.

In fact, they will get dragged downstream, even farther away from the microphone.

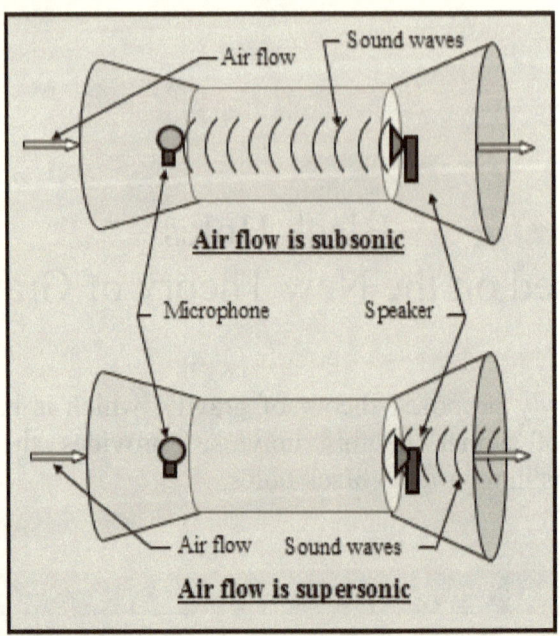

To visualize the force of gravity near a black hole, one can also consider a giant drain hole at the bottom of an ocean, as shown below.

In this case, if the diameter of the drain hole is big enough and the ocean is sufficiently deep, due to the enormous amount of pressure that will be there, at a certain distance which can be readily calculated, the flow speed of water towards the drain hole will reach and even exceed the propagation speed of sound waves in water.

If someone who is being pulled into such a drain hole, and is already passed that certain distance, tries to broadcast a sound wave type of an S.O.S. message towards outside, his cry for help will not be able to propagate outwards and it will be dragged into the drain hole. This is due to the fact that, the sound waves will not be moving fast enough to counter the speed at which water is flowing towards the drain hole. This kind of a place at the bottom of any ocean can be referred to as, **"Esmailzadeh Silent**

Zone", since no sound waves will ever be heard coming out of them.

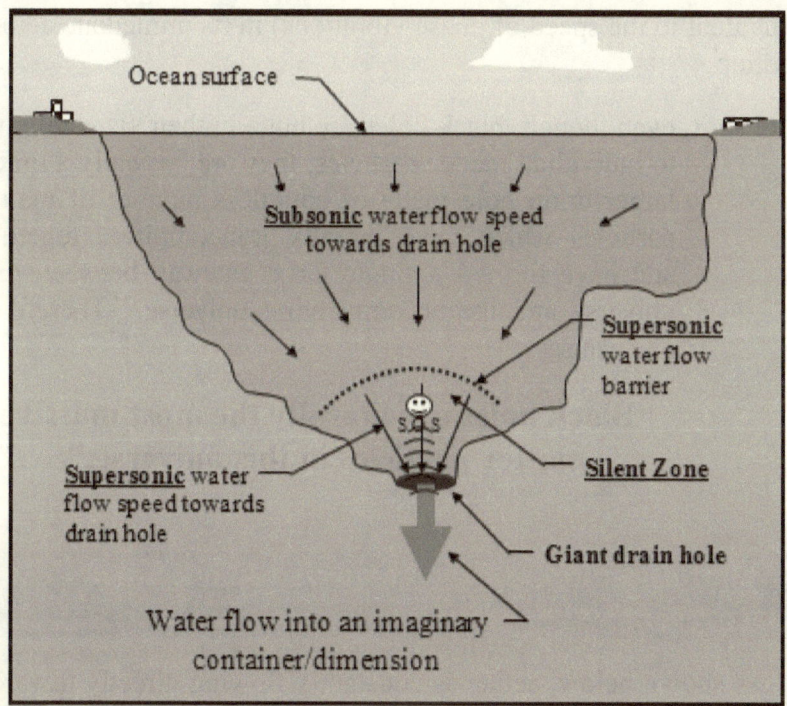

These examples clearly demonstrate exactly what is taking place in the immediate vicinity of a black hole. A black hole is made up of an enormous amount of matter particles (drain holes) gathered together in a very small volume of space. In fact, they are joined together and have formed a single large unified spherical gateway allowing unrestricted flow of aether from this universe into the accompanying universe.

The induced inward flow of aether towards such a gateway reaches the speed of phase vibrations as it crosses the event horizon (explained in the following section). And, it keeps on speeding up even further, as it approaches the black hole itself. This is why, phase vibrations such as light or any other kinds of electromagnetic waves that reach the event horizon of a black hole, cannot travel up-stream fast enough to counter the speed at which aether is flowing in the opposite direction towards the black hole. Hence, they are dragged in and are absorbed by the black hole.

Therefore, the minimum requirement for a celestial body to be called a black hole is that it must be capable of speeding the flow of aether towards itself to at least the speed of light (which is equivalent to the speed of phase vibrations) in its immediate aether medium.

Note that, even though, black holes are huge in their sizes relative to individual matter particles, they are basically a much larger drain hole made of countless number of matter particles which have literally joined/unified together and have formed a single large gateway between this universe and the accompanying universe. Therefore, one can say,

"Black holes are literally the most massive matter particles in this universe."

2- What does the event horizon of a black hole represent?

As shown below, aether is constantly flowing directly towards black holes from all spatial directions.

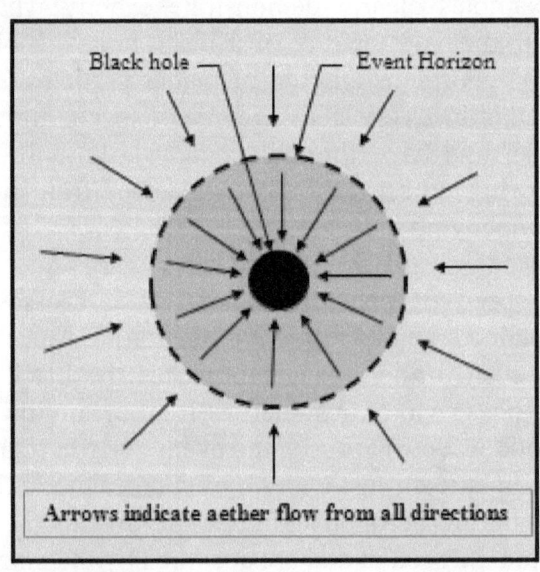

As aether flow speeds up in its approach towards a given black hole it eventually reaches the speed of phase vibrations which is the speed at which light and other electromagnetic waves travel through aether. The distance (radius) from the center of a black hole at which the influx of aether reaches the speed of phase vibrations in that medium, is referred to as the **Event Horizon**. The above drawing shows a black hole with its corresponding event horizon.

The event horizon can also be referred to as **the point (radius) of no-return** for light or any other type of electromagnetic waves, since once they reach the event horizon of a black hole, they can no longer go back.

Note that, for all practical purposes, one may propose that, the event horizon of a black hole can be considered as to be the black hole, since no particles or waves of any kind can escape the interior region of an event horizon. Hence, the event horizon of any given black hole basically represents that black hole's outer limit.

3- Black holes, viewed from the accompanying universe

Black holes devour all of the phase vibrations (of all types and all frequencies) that reach their event horizons, and transfer them into the accompanying universe. Therefore, they not only appear and act as fountains of fluid aether but also as strong sources of a variety of phase vibrations including visible light, in the accompanying universe. In other words, as shown in the following figure,

"Contrary to what they pretend / appear to be in this universe, black holes act as brilliantly shining spherical light sources in the accompanying universe."

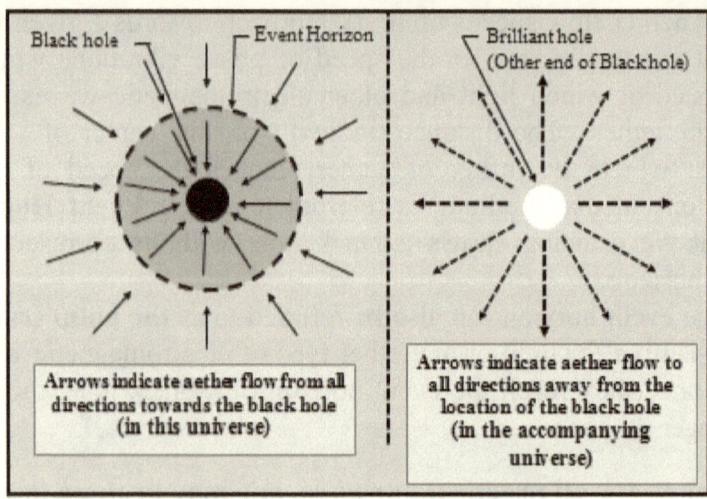

The following figure shows how galaxies may appear to a hypothetical observer in the accompanying universe.

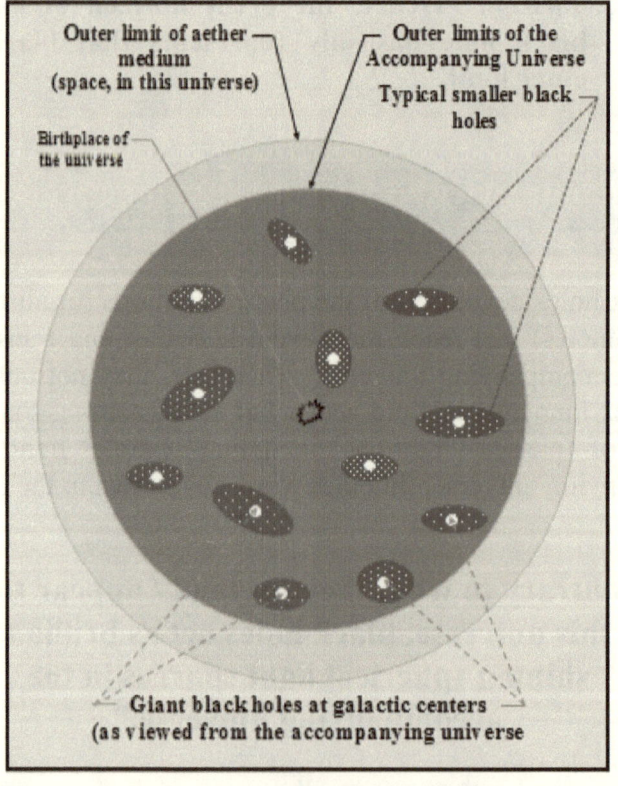

The bright spots shown are the many black holes that exist in each galaxy. The sizes of the giant black holes, in the center of galaxies, are exaggerated to clearly indicate their existence.

Currently, **the speed at which the fluid aether is flowing into the accompanying universe** varies, since it depends on the passageway it takes. It is slowest as it flows through individual matter and anti-matter particles (tiny drain holes), and it is the fastest (**but still less than** the speed of phase vibrations in the aether medium that is occupying the accompanying universe) through back holes (the giant gateways).

The maximum entry speed of fluid aether into the accompanying universe is stated as being **less than** the speed of phase vibrations in that medium, since due to the density of aether being much, much lower in that universe, as compared to what it is in this universe, the speed of phase vibrations in that medium is currently much, much faster than that of phase vibrations in this universe. In other words,

> **"Even though any black hole possesses such a special domain referred to as its event horizon in this universe, it does not possess such a special zone in the accompanying universe. Since, as the fluid aether flows out of the black hole its speed is less than that of phase vibrations in that medium."**

4- How are black holes formed?

Black holes can form in at least two different ways, both of which are due to the implosion of the inner layers of massive stars, as they go nova. During nova explosions, the outer layers of the stars are thrown into the surrounding space and are eventually absorbed and reused for a variety of applications by other stars and/or planets. While, the inner layers of the stars are squeezed into a relatively solid chunk of matter.

1- If the imploded portion of the star is massive enough, its matter particle constituents (the tiny drain holes) will literally unite and form a single large gateway that will allow aether to flow through so freely that the incoming speed of aether will be in excess of the speed of light (speed of the phase vibrations in the local fluid aether medium). In other words, the newly born entity will possess an event horizon and can be called a black hole.

The following figure shows such a unification of matter particles (tiny drain holes) into one giant matter particle (gateway).

As it is shown, <u>when matter particles inside a star coalesce to form a black hole, their surface areas are added together and not their volumes.</u> Because, each and every matter particle is literally a spherical shell that is empty inside, through which aether flows into the accompanying universe.

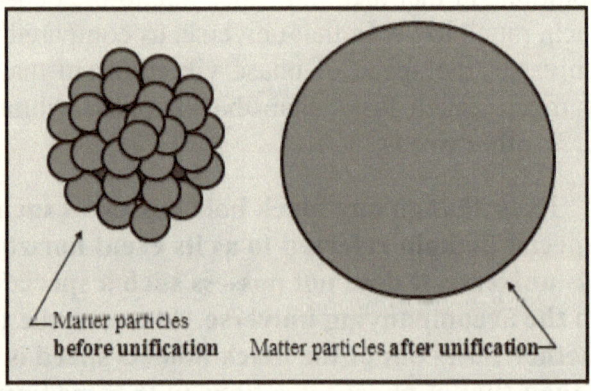

Matter particles before unification   Matter particles after unification

Note that, as a black hole is formed in one step, from a sufficiently massive star that has gone nova, the event horizon will already be separated from the surface of the black hole. However, the initial diameter of the event horizon will directly depend on the mass of the imploded portion of the star. And, over time this diameter can only grow larger as the black hole literally feeds on whatever is dragged into it by the influx of aether.

2- If the imploded portion of the star is not sufficiently massive, it will become a neutron star. During such processes, the electrons and protons unite and form more neutrons, while the existing neutrons keep their identities. In other words, all of the matter particles get so packed that they actually resemble a single giant nucleus, made of trillions of trillions of trillions of trillions of nucleons. One can say, <u>a neutron star is basically composed of matter particles that are packed close to each other, but still act as individual drain holes for aether to flow through.</u>

Neutron stars possess a very strong gravitational field, since equivalent of two to three times the mass of the sun is squeezed within a very small ball that is only a few Kilometers in radius. The flow speed of aether towards neutron stars is less than the speed of phase vibrations in the medium of aether. Therefore, phase vibrations in the medium of aether, such as light and other electromagnetic waves, can and do escape their gravitational grip.

As neutron stars attract and devour other celestial bodies that happen to get too close, they grow in size and hence their gravitational forces become even stronger until they can no longer stand their own gravitational forces and their matter particles become squeezed against each other. At that stage,

"All of the particles that used to form the neutron star, starting with the ones located right at its very center, literally merge together and basically form a single, much more massive particle."

As shown in the previous figure, such an oversized unified giant matter particle acts as a giant gateway for aether to freely escape from this universe and enter the accompanying universe. The flow of aether towards these newly manufactured giant gateways reaches the speed of phase vibrations in the aether medium either as it reaches their surfaces or before it reaches their surfaces. At this point, the neutron star has officially turned into a **black hole**, since not even light can propagate against the incoming stream of aether and escape.

Note that, as a neutron star becomes a black hole, at first, its event horizon is right at its surface level. However, over time, as it continues to gain more mass, its event horizon separates from its surface and spreads wider and wider into its surrounding space.

The joining of the matter particles together as they reach a black hole can be better understood if they were visualized as bubbles in a fluid medium such as water. As bubbles reach and touch each other, they simply unite and form a larger bubble. In other words,

Black holes are basically drain holes which are much larger than individual matter particles.

Therefore, one can say,

"Black holes are literally the most massive matter particles in this universe."

5- How does aether pressure affect the size of a given black hole?

Since black holes are giant bubbles in the aether medium in this universe, their overall sizes are dictated by the local aether pressure. Therefore, as the internal pressure of the fluid aether gradually decreases, due to its ongoing expansion process, the overall sizes of all black holes are gradually increasing.

Since black holes connect this universe to the accompanying universe, they are automatically exposed to two different aether pressures from their opposite ends, due to the pressure of aether that is in the accompanying universe being much, much lower than that of aether that is in this universe. Therefore, one can reason that,

"Any given black hole, when considered as a whole, is shaped like a funnel, with its narrower end facing this universe and its wider end facing the accompanying universe."

However, as the pressure of aether in this universe is gradually decreasing and the pressure of aether in the accompanying universe is increasing, eventually the two pressures will equalize and the overall shape/geometry of black holes will approach that of a large diameter tube with uniform diameter along its full length.

Later on, as the overall aether pressure continues to decrease in both universes, <u>the diameter of the black holes will gradually become larger and larger. That is, until they dissolve due to aether pressure and density reaching certain critical low magnitudes.</u>

6- Formation of giant black holes at galactic centers

Over billions of years that it takes for a typical spherically (or elliptically) shaped galaxy evolve into a more or less flat geometry, its central region continuously receives (or should say, it is literally bombarded with) matter particles due to three distinct processes:

- All along their lives, each and every star in any given galaxy constantly throws matter particles (as in particle storms) in all spatial directions, including towards that galaxy's central region, where they are absorbed by the local celestial bodies.

- As large stars reach the end of their life-cycles they experience a nova moment. During such a process, their inner layers are imploded into a relatively solid ball made of matter particles, while their outer layers are thrown into the surrounding space. A portion of the outer layers of each and every star are thrown towards the center of the galaxy and are absorbed by the local stars in that region.

- Many of the stars in any galaxy collide with each other, as they cross each other's orbital paths around their common galactic center of mass. During such collisions, some of the matter particles that used to form such stars gain some momentum and are thrown outwards due to their excess velocities, while some other particles lose some of their orbital momentums and are subsequently drawn towards the galactic center, where they are absorbed.

As shown in the following figure, the galactic centers are continuously becoming more populated with new stars, as well as hosting more massive stars, since matter particles that are thrown towards the galactic center get trapped in that region and either encourage the growth of existing celestial bodies or contribute towards the formation of new stars, there. While, the outer galactic regions are gradually becoming thinner in their matter contents and are becoming more spread out.

The higher concentrations of stars in the galactic centers automatically lead to more frequent collisions between stars, as well as more frequent supernova moments contributing towards higher concentrations of matter particles in the core regions of galactic centers. Such high concentrations of matter particles at the core of galaxies which have evolved into a more or less flat geometry encourage the fast growth of black holes already formed there, due to large stars which have already experienced their respective supernova moments.

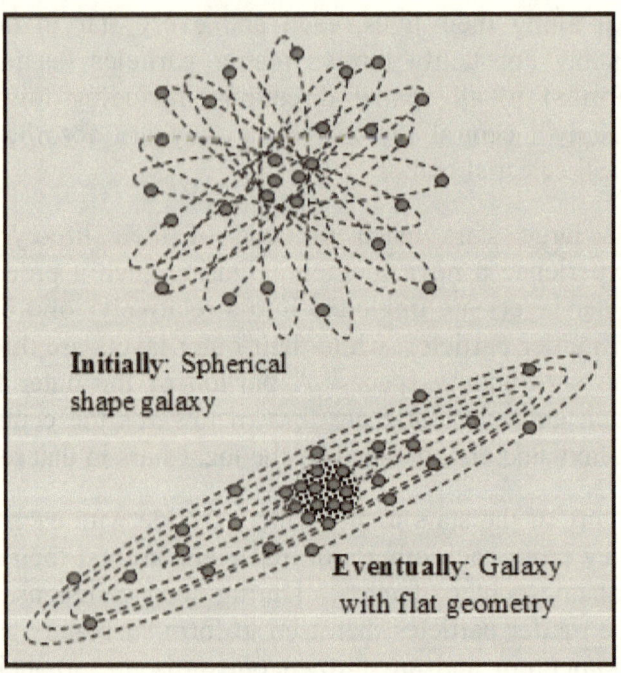

Initially: Spherical shape galaxy

Eventually: Galaxy with flat geometry

Over time, such centralized black holes merge together and form gigantic black holes which simply keep on becoming more massive, as more matter particles are literally thrown at them by stars in their surroundings. In other words,

"All of the galaxies which have more or less evolved into their flat geometries must be hosting super massive black holes at their respective centers."

7- Relation between the mass of a black hole and the size (radius) of its event horizon

The physical size (radius) of a black hole's event horizon is a direct indication of how massive that black hole is. However, as a given black hole devours more matter particles (stars, planets and so on) and becomes more massive the volume of its event horizon does not increase linearly with its mass.

As it was explained earlier in this chapter, as a black hole becomes more massive its surface area (**not its volume**) increases by the same exact ratio. Therefore, its radius increases proportional to the square root of the increase in its mass, since surface area of a sphere is defined by $(4\pi r^2)$.

Since, aether's flow speed, hence its flow rate per unit surface area, through any opening between this universe and the accompanying universe is dictated by the local pressure difference between aether that is in this universe and aether that is in the accompanying universe, one can make the following statement regarding the maximum aether flow speed into black holes.

"As the size of a black hole increases, the amount of aether flowing through it increases in correspondence with its surface area."

Therefore, as the mass of a given black hole increases by 100 folds, and its surface area (**not its volume**) increases by 100 folds, the amount of aether flowing through its overall surface area increases by 100 folds.

Since black hole's radius will become 10 times larger than it used to be, the flow of aether will reach its maximum speed at a distance 10 times farther away from the center of the black hole, as it used to.

Such a relationship holds for the flow speed of aether at all distances from the center of the black hole. In other words,

"As a given black hole becomes 100 times more massive, the flow speed of aether will reach the speed of phase vibrations at a distance that is 10 times farther from the center of the black hole, as compared to before."

In other words,

"As a black hole becomes 100 times more massive, the radius of its event horizon becomes 10 times larger."

Here is the proper place to make another statement regarding how the size of the event horizon of a given black hole changes as that black hole's mass increases.

"The radius of the event horizon of a given black hole increases in direct proportion to the square root of its mass increase."

Also, as the mass of a given black hole increases 100 folds and its radius grows by a factor of 10 (for example from 5 Km to 50 Km), the radius of its event horizon which may be around 20 Km will grow to be 200 Km.

Therefore, the distance between the surface of the black hole and its event horizon will grow from its previous value of 15 Km to 150 Km. In other words,

"As a black hole becomes more massive, the distance between its surface and its event horizon grows proportional to the square root of its mass increase."

8- Does a black hole's spin affect the incoming flow of aether and hence the trajectories followed by in-falling objects?

All of the black holes start with some initial spin, due to the rotational momentum of the interior layers of their parent stars. As a given massive star goes nova, the mass of its inner layers is squeezed into a very small volume, increasing its rotational speed and causing the newly formed neutron star (which eventually becomes a black hole) or the black hole to spin at quite a high rate, just as a figure skater spins faster on ice by bringing his/her arms closer to his/her body's rotational axis.

In general, all of the black holes exhibit complex (3-D) spinning actions which are due to the arrival of the planets and stars from different directions. As shown in the next figure, each planet or star contributes some rotational momentum in a certain direction, depending on its approach trajectory.

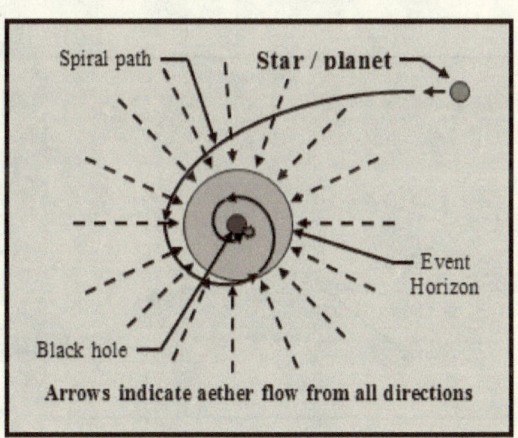

If most of the stars and planets approach a given black hole in a common plane and direction, they will form a disk around that black hole which can resemble the rings of the planet Saturn in the solar system. As the matter contents of such stars and planets are devoured by the black hole, they contribute to the rotation of the black hole causing it to spin faster in that particular direction. However, if enough stars and planets arrive at angles and directions that cancel the black hole's spin, the black hole will gradually slow down and eventually stop rotating altogether.

The best way to answer the question raised is by conducting an experiment.

- Experiment:
  (Effect of the spin of a black hole on the in-falling objects)

Consider a swimming pool full of water. A pipe which is covered by another pipe as a sleeve-like cover, as shown in the following figure, can be placed inside the water so that its open end can be oriented towards different directions, as in up, down and sideways. The other end of the inner pipe, through a swivel joint is either connected to the pool's drainage system or to the intake of a water pump.

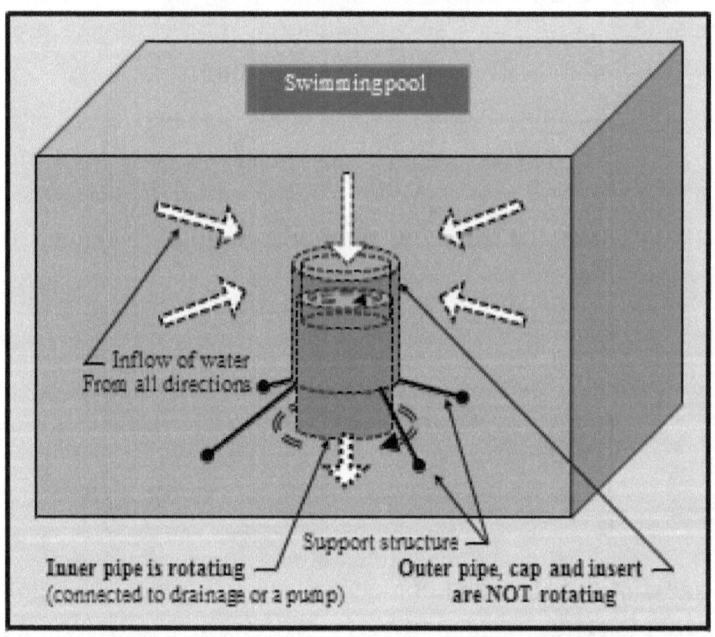

Note that, the shorter outer pipe **(with a short lip)** which is used as the cover for the inner one, must be secured so that it does not rotate with the inner pipe during the experiment and does not cause the water in its vicinity to rotate.

For the first part of the experiment, the inner pipe should be kept stationary, as the valve is opened (or the water pump is turned on) so that water can flow through the inner pipe, towards outside (either through the drainage piping or the water pump). In this case, water will approach the inner pipe's open end (inlet) from all directions, in straight lines. Of course, if the open end of the pipe is facing up or down the inflow of water will start to form a rotational motion due to the Coriolis Effect which is caused by earth's rotation around its own axis.

Therefore, at this point the direction of the pipe should be selected so that there would not be any rotational motion in the water that is flowing towards the pipe that is detectable. If all directions tested still show some rotational motion of water towards the inlet of the pipe (which it should, regardless of the direction chosen) the open end of the pipe should be pointed towards UP, due to simplicity and direct visual access. Then, the rate at which water is rotating should be recorded.

For the second part of the experiment, while the valve is still open and water is flowing through the inner pipe, the drive mechanism should be turned on and the inner pipe should be brought to a fairly high rate of rotational speed, in the same direction as the water is rotating due to the Coriolis Effect.

In this case, <u>if the inflowing water, from the pool into the pipe, starts to demonstrate any change in its rotational speed around the inlet of the spinning inner pipe, it would mean that the aether flow which is dragging everything into a spinning black hole will also be forming a spinning disk.</u>

To confirm this effect, all is needed to be done is to reverse the direction of rotation of the inner pipe, by running its drive mechanism in the opposite direction. At a certain speed of pipe's rotation, the turning motion of water should come to a halt.

However, <u>if the incoming water, from the pool into the inner pipe, does not show any effects that can be related to the spinning pipe, it would indicate that the aether flow which is dragging everything into a spinning black hole will not be forming a spinning disk, either.</u>

Note that, even though, during the second part of this experiment, the inner pipe will be spinning at a high rate of rotational speed the outer pipe and the cap which overlaps the tip of the inner pipe do not spin. Their stationary mode eliminates any potential rotational motion which can be transferred to the surrounding water by the spinning motion of the inner pipe, due to skin friction.

9- Do black holes have electrical charge or any kind of external field?

Black holes form as matter and/or anti-matter particles merge together and form unified, giant gateways, connecting this universe to the accompanying universe. **As particles merge together they lose their identities.** Therefore, black holes are electrically neutral. However, any given black hole may seem to possess one type of electrical charge or another, due to the charged particles that are in the process of falling into them, but have not crossed their event horizons, yet. Therefore,

"Black holes can only be of neutral charge. However, they may seem to be positively or negatively charged, due to the in-falling charged particles that are still around them."

Also, none of the various types of phase vibrations, or properties associated with phase vibrations or associated with aether flow can exit the domain of any black hole's event horizon. In other words,

"Black holes cannot have any kind of external field that is due to what is inside their respective event horizons."

However, various electric and magnetic fields generated by the presence and motion of the charged particles that are being dragged inwards by the aether flow, towards the event horizon of any black hole, will be present in the surrounding volume of space.

10- What is the physical size of a black hole?

As particles unite and form a black hole and become more massive, the size of their aggregate is dictated by their total surface areas added together and not by their total volumes. This is so, because particles are basically bubbles in the aether medium, just like ping pong balls that can be floating in a medium such as water. There is nothing inside bubbles that manifest as matter and anti-matter particles. It is only their surface area that counts.

Therefore, as a black hole grows by devouring more matter particles, the increase in its overall surface area is equivalent to the sum of its original surface area and the surface areas of the added particles.

For example, as a black hole grows to be 100 times as massive as it used to be, its surface area will be 100 times its previous surface area. This means, its radius will be increased to 10 times its previous radius, since surface area of a sphere is defined by (4π r^2). However, due to this 10 fold increase in its radius, its volume will increase by a factor of 1000, since volume of a sphere is defined by ($(4/3)\,\pi\,r^3$).

Therefore, the density of the black hole will decrease by a factor of 10, the same ratio as its radius is increased, or equivalent to the square root of the increase in its overall mass. In other words,

"As any given black hole becomes more massive it actually becomes less dense, since its density decreases proportional to the square root of the overall increase in its mass."

In short,

"Black holes are not single points in space, they can be thousands even millions of kilometers across."

11- Can light stay in a stable orbit around a black hole?

Light waves have a chance of staying in a stable orbit around a given black hole while they are still outside of that black hole's event horizon. However, to do so, light waves must follow a circular path around the black hole, at a distance so that their speed relative to the local aether which is flowing into the black hole is exactly equal to the speed of phase vibrations in that immediate aether medium. The drawing below shows such an orbital path around a black hole.

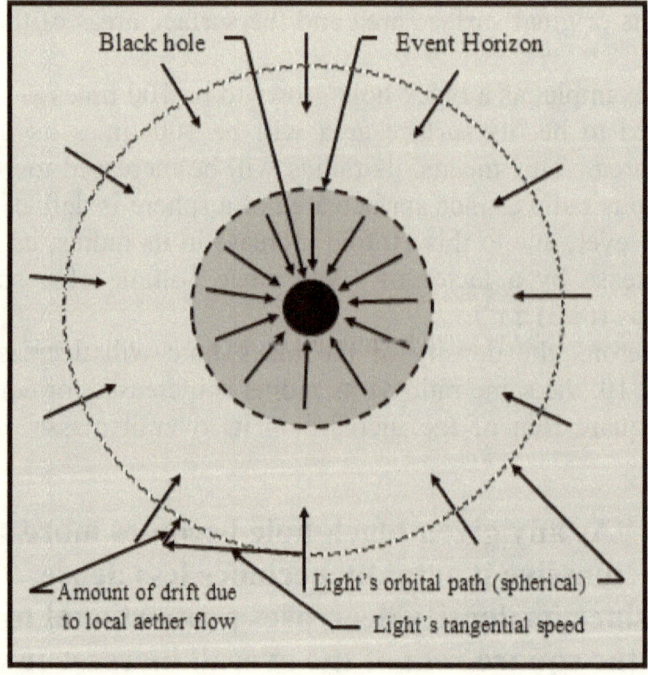

Note that, since light waves can and do arrive from all directions, if their incoming trajectories are right, they do get trapped in such a stable orbit, and will never reach the event horizon. Together they form an orbital path that is spherical in shape, a very thin sphere that its surface is literally made of light and other types of phase vibrations in the medium of aether.

12- How does an object experience the passage of 'Time' as it approaches a black hole and gets captured by it?

According to the information provided in the chapter titled "What is Time?", the passage of 'time' is experienced by any object that is moving with respect to its local aether at a speed that is less than the speed of phase vibrations in that medium. However, as its speed approaches that of phase vibrations in its local aether medium, it experiences time at slower and slower paces. And, if its speed reaches the speed of phase vibrations in that medium, the object will stop experiencing the passage of time, altogether.

Since black holes are large gateways for aether to pass through, aether is naturally flowing directly towards black holes from all spatial directions. Therefore, the passage of time experienced by an object that is captured by a black hole will depend on the manner in which it approaches the black hole. There are two distinct methods of approach that can be followed by an object:

- **Tangential Approach**

    If an object (a spaceship) passes close enough to a black hole, as it is shown below, it will get captured by that black hole's gravitational force. In this case, it will start a spiral motion around the black hole and will eventually get devoured by it.

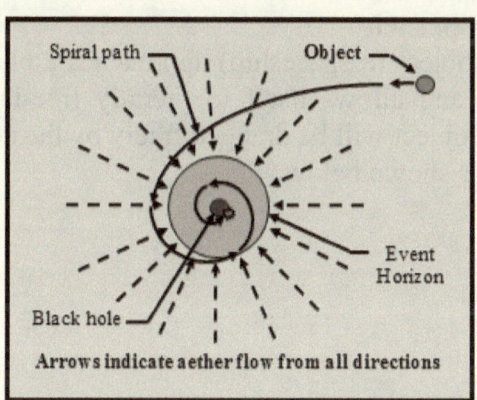

Arrows indicate aether flow from all directions

During its spiral motion, the object will continuously gain speed relative to its local aether medium. So long as that object's speed relative to its local aether which is flowing directly towards the black hole remains less than the speed of phase vibrations in that medium, the object will experience the passage of time. But, the pace at which the passage of time is experienced will be increasingly slower, since that object will be continuously gaining speed relative to its local aether.

At some point, it will be moving at the speed of light which corresponds to the speed of phase vibrations in the local aether medium. At that point, the object will stop experiencing the passage of time. In fact, **the object will reach this speed before it even reaches the event horizon of the black hole, since it will be moving almost perpendicular to the aether flow which is headed directly towards the black hole.** And, it is the overall resultant speed of an object relative to its local aether that counts, as far as experiencing time is concerned. In other words,

"As an object follows a tangential trajectory, in its approach towards a black hole, it will experience all of the special effects associated with relativistic speeds, because it will be moving relative to the local aether and not with it."

- **Direct Approach**

If an object (a spaceship) approaches a black hole directly head on, and allows itself to literally free-fall into it, at all times the object will be dragged freely by the aether flow. This scenario is shown below.

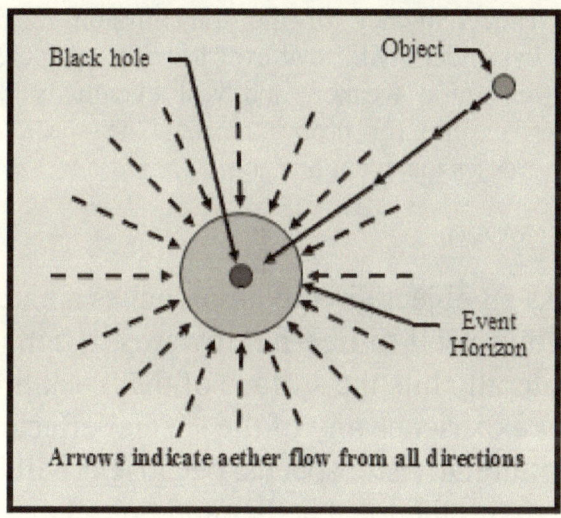

Arrows indicate aether flow from all directions

In this case, the speed of the object relative to its local aether that is dragging it directly into the black hole will remain to be equivalent to the escape velocity from that black hole at the location/distance where that object starts its free fall (minus its initial speed towards the black hole). And, for the full duration of its free-fall, the speed of the object relative to its local aether will remain constant. This means that, the speed of the object will eventually become equivalent to the speed of phase vibrations in space but without experiencing any of the special effects associated with relativistic speeds, whatsoever.

Notes, in such a scenario,

1- The occupants of the spaceship (or the scientific probe) that is headed directly into a black hole will keep on experiencing time at the very same pace as the individuals who are stationed far away, anxiously waiting to receive exciting news (telemetry).

2- The spaceship occupants will also be in their normal sizes, as well, since they will not shrink in the direction their ship is moving. They will not experience any of the special effects associated with traveling at relativistic speeds.

3- The frequency of the transmission received by the bystanders will be at ever lower frequencies, as well as becoming weaker, and will eventually cease, as the spaceship (or probe) reaches the speed of light and crosses the event horizon.

In other words,

"As an object directly approaches a black hole and allows itself to free-fall towards it, until that object literally hits the surface of the black hole, it will not experience any of the special effects associated with relativistic speeds, because it will be moving with the local aether and not relative to it."

13- Does time exist inside the event horizons of black holes?

As it is explained in a separate chapter of this book titled "What is Time?", time is a property / side effect of aether and the existence of phase vibrations in that medium. Since, the rate at which objects experience the passage of time is dependent on their instantaneous speeds relative to their local aether, as compared to the propagation speed of phase vibrations in that medium. The faster they move relative to their local aether the slower they experience the passage of time. If their speed relative to their local aether reaches that of phase vibrations in that medium, they will no longer experience the passage of time.

Therefore, in regards to whether objects that happen to be inside the event horizon of black holes experience 'time' or not, one has to state that, it all depends on the motion of such objects relative to their respective local aether which is already flowing towards those black holes at higher speeds than the speed of light (speed of phase vibrations) in those mediums.

There are two distinct ways that objects can approach a given black hole. These two approach methods result in totally different outcomes, as far as experiencing time is concerned:

- **In the tangential approach method**, the speed of the object relative to its local aether medium reaches the speed of phase vibrations in aether, before it even crosses the event horizon. Therefore, objects that follow such an approach path experience the passage of time at slower and slower paces as they gain more speed relative to aether. And, as their speeds relative to their local aether reach the speed of phase vibrations in that medium, they no longer experience the passage of time.

  They will experience this timeless state/effect starting right before crossing the event horizon and will continue to do so until they reach the surface of the black hole.

- **In the direct approach method**, which is in fact a free-fall scenario, the speed of the object relative to its local aether medium is minimal, even as it crosses the event horizon of the black hole. Therefore, objects (probes) that follow such an approach path will not experience any of the effects associated with moving at relativistic speeds. These objects will continuously experience the passage of time at the very same rate as other objects (or beings) that may be positioned far away (where the probe's free-fall starts) and patiently waiting and hoping to receive telemetry data from such probes.

  Of course, the duration of such a period will be on the order of milliseconds (or few seconds, in the case of giant black holes), due to their high speeds and the shortness of the distance that they have left to travel between the event horizon and the surface of the black hole.

  Note that, even though any probe that follows such a path may stay intact, as it crosses the event horizon of a black hole, its sent messages will propagate towards outside in the local aether medium at the speed of light relative to the probe (or the local aether). Because, such a probe will be relatively at rest with respect to its immediate aether medium that is dragging it towards the black hole.

  However, the messages will still be dragged towards the black hole, since the local aether is flowing towards the black hole faster than the speed at which these messages are propagating in the opposite direction.

A variety of experiments can be designed and conducted that can explore such effects. The following two experiments which are designed to explore such effects are described in detail in the chapter titled "What is Time?".

- First experiment: (Orbital free fall vs. direct free fall)

- Second experiment: (How does an object experience the passage of 'Time' as it directly approaches a black hole, in a state of free fall?)

The results of such experiments will clearly confirm the following answer to such an important question/issue in regards to how the passage of time is experienced by objects that directly approach black holes, in a state of free fall and cross their event horizons:

"As an object allows itself to free fall directly towards a black hole, it will continuously experience the passage of time at a CONSTANT FIXED RATE, as if it were still hovering at the same distance to the black hole from where it starts its free fall. Such an experience will last for the entire duration of that object's free fall, not just until it crosses the event horizon, but rather until it literally hits the surface of the black hole."

14- What becomes of matter particles or objects as they reach the surface of a black hole?

Matter and anti-matter particles are in fact tiny bubbles in the fluid aether medium. There is nothing in the interiors of any of the matter particles. It is the surface area associated with each matter particle that is directly proportional to its mass. Since, the size of a particle dictates the severity of the wake-like wave that it forms in the aether medium as it moves in that medium, the same effect that is interpreted as being the mass of that particle.

As matter and anti-matter particles reach the surface of a black hole, which is only a much, much larger bubble in the aether medium, they simply merge with it and contribute towards the expansion of its overall surface area. Therefore,

> ## "As matter (as well as anti-matter) particles reach the surface of a black hole, their surface areas are added to that of the black hole and so form a larger bubble."

The direct consequence of such an addition of surface areas instead of volumes of matter particles, as they join a black hole, is shown in the following figure.

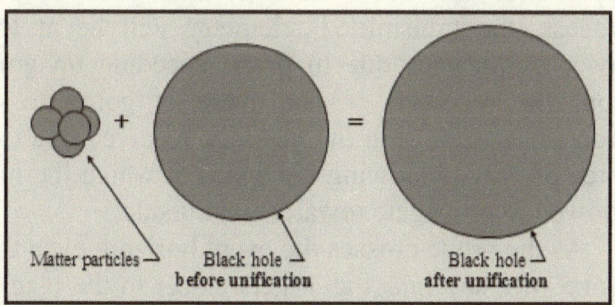

15- From how close a distance to the event horizon of a black hole a probe can transmit information to its creators who are waiting far away?

A probe sent towards a black hole will experience different scenarios, depending on the method of approach used.

- **In the case of tangential approach,** the probe will take longer time to reach the event horizon. But, its speed relative to the local aether will reach the speed of phase vibrations in aether long before that.

In this case, the probe will be able to transmit electromagnetic type of signals towards the outside. But, gradually, there is going to be more pronounced delays in the message received due to stronger force of gravity slowing the speed at which the message is allowed to propagate towards the outside.

However, in this scenario, **if the probe can manage to reach the speed of phase vibrations relative to its local aether, before reaching the black hole's event horizon, it will be orbiting the black hole at the same distance as light waves forming a sphere around that particular black hole. In such a case, the probe will be able to stay in that orbit and transmit telemetry to its designers / creators / fans, for a long time.**

- **In the case of direct approach**, as the probe directly approaches the event horizon and gains speed in the process, the transmitted telemetry will be at lower and lower frequencies due to probe speeding up going away from the receiver. Also, there is going to be more pronounced delays in the message received due to stronger force of gravity slowing the speed at which the message is allowed to propagate towards the outside.

  As the probe crosses the event horizon, even though the probe will stay intact as it gets closer to the black hole, as compared to the tangential method of approach, but no electromagnetic type of signals can be sent outward, since they will get dragged in, towards the black hole, as well.

16- Is the information that crosses a black hole's event horizon preserved or is it lost forever?

Information can be carried across the event horizon and into a black hole in two different ways, one by phase vibrations such as electromagnetic type of waves and the other by matter and/or anti-matter particles.

- **In the case of the electromagnetic type of waves**

Electromagnetic waves are simply phase vibrations in the medium of aether, just as sound waves are in a medium such as air. Therefore, as the carrier medium, namely aether, flows passed the event horizon of a black hole the information will be carried in as well, and it will still be fully intact.

However, as aether goes through the black hole and enters the accompanying universe which is like an ocean full of aether, phase vibrations will get dispersed in all directions. The very same effect can be observed in the case of a river stream that reaches an ocean. Because, the waves that are present in the water that is running downstream in the river get dispersed in the ocean water, as the river merges with it.

Note that, all of the phase vibrations that go through black holes and reach the other ocean of aether, namely the accompanying universe, will be there forever, but their strength / amplitudes / intensities will be weakened so much that their detection will gradually become more difficult.

As shown below, **this scenario can actually be physically tested using simple and readily available equipment.**

> - Experiment:
> (Are electromagnetic waves reaching a black hole preserved, or are they lost forever?)

In order to perform such an experiment, one needs to use an air compressor, a speaker connected to a stereo system and a sound receiving system with frequency selection capability. As shown in the following figure, the speaker should be installed inside the compressor's tank so that it is facing an opening which is equipped with a valve. And, the microphone which is connected to the receiving system should be positioned far away from the tank, but inside the very same large room wherein the compressor's tank is located.

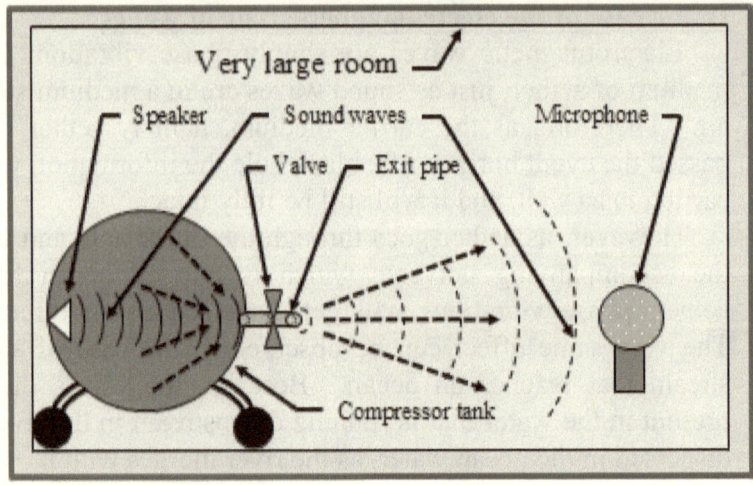

Note that, the interior of the compressor's tank should be acoustically non-reflective, to reduce the amount of echo/noise as much as possible.

After the compressor's tank is filled to the highest possible pressure, the compressor should be turned off and the whole compressor should be left alone for a few minutes. This amount of time will allow the compressed air inside the tank to settle down and stop carrying any kind of noise that is related to the operation of the compressor, as well as the physical internal commotion that is due to the initial momentum of air as it was pumped into the tank.

Next, while the microphone and the related sound system is prepared to receive and record the sound waves, the valve should be opened to allow the compressed air inside the tank to flow through and get mixed with the air that is in the room. Then, the stereo should be turned on to broadcast a monotonic sound of a certain frequency.

The microphone will pick up the vibrations corresponding to the monotonic sound waves that are generated by the speaker that is inside the air tank. Of course, the receiving system that the microphone is attached to, must have the capability to filter out the static sound waves generated due to the flow of air through the opening and also any other sound waves that might be generated due to any echoing effects.

The frequency of the sound waves received will be affected by the differences between the pressure (and the density) of the air that is inside the tank and the pressure (and the density) of the air that is in the room. However, as the air pressure inside the tank drops and equalizes with the air pressure in the room, the frequency received will gradually approach and exactly match the frequency of the monotonic sound that is broadcast by the speaker.

Note that, the very same type of experiment can be carried out for two different cases, namely **Subsonic flow of air through the opening** which will correspond to the flow of aether leaving this universe through individual particles, including their aggregates as large as planets, stars and even neutron stars, and **Supersonic flow of air through the opening** (at the valve) which will correspond to the flow of aether leaving this universe through black holes.

Such an experiment can also include gradual frequency changes, every few seconds or so. This way, the older frequencies and not just the current frequency can be detected and recorded even minutes later, while they continue to echo, spread and weaken in their new environment.

Once knowing the empirical relationship between the density / pressure difference between air that is inside and outside of the compressor's tank and the frequency that is received, by programming the receiving system to adjust automatically, one can even listen to and clearly recognize Paganini's second violin concerto or a speech by an alien being saying "What the heck you think you are up to?". In other words,

"Any information that is carried inside a black hole by electromagnetic type of waves is preserved, indefinitely. However, as time progresses, the waves

**become more and more spread out and
therefore more difficult to detect."**

- **<u>In the case of matter and/or anti-matter particles</u>**

    In this case, as particles, which are basically bubbles in the fluid aether medium, are dragged across the event horizon of a given black hole by the flow of aether, they stay intact only until they reach the surface of that black hole. Once they arrive at the surface of the black hole they lose their identities. Because at that point they become part of the black hole's perimeter wall by literally becoming pasted unto its wall and causing the opening to widen.

    However, while particles are being dragged towards a given black hole, and across its event horizon, **their information (which can be in the form of various types of vibrations) is transferred to the fluid aether which is dragging them.** The fluid aether in turn carry's such vibrations as it crosses the surface of the black hole and flows into the accompanying universe.

    Therefore, **even though particles themselves literally lose their identities at the instant they reach the surface of black holes, their vibrations which are indicative of their previous existences and carry information about them individually, are saved due to being transferred to the medium of aether in their immediate vicinity, and are then carried into the accompanying universe.**

In short,

The information that is carried across the event horizon of a black hole, either by electromagnetic type of waves or by any kind of matter or anti-matter particles, in one way or another is transferred to the aether medium that is dragging them and is consequently carried into the accompanying universe.

Once the fluid aether arrives in the accompanying universe, it disperses in the ocean of aether that is already there, just as the water in a river reaches and merges with the water that is already in a sea or an ocean. Therefore,

"As the waves spread in the accompanying universe, they become weaker and more difficult to recognize, yet they will be preserved there for eternity."

17- Can a black hole be kept in one place or be moved to another location?

Imagine a magnet (any shape or type) freely floating in the outer space. To manipulate its location, motion or even its orientation (directional effects), one needs to use something that will encourage it to respond/react in the desired manner. In order to accomplish such a task, one needs to use either another magnet or a magnetically inclined piece of material, at a proper distance. In so doing, he/she can make the freely floating magnet to perform a variety of maneuvers in a controlled fashion. The very same type of setup / experiment can be performed using a magnet that is resting on a flat piece of wood which is floating on the surface of water. The rate of speed, direction of motion as well as the magnetic orientation of such a magnet can be readily manipulated by using another magnet or a piece of magnetically inclined material.

The very same types of actions can be thought of as one considers black holes. Black holes love to devour matter and anti-matter particles. Therefore, to cause a given black holes to change its position and/or direction of motion, it literally needs to be fed objects of different sizes from different directions and with different initial momentums. Small black holes can be readily controlled in this fashion. For example:

- By sending a moon or a planet, with proper speed and at a predetermined angle, towards a small black hole, that black hole can be pulled or pushed towards the desired direction, as it is encouraged to literally feed on that moon or planet. The figure below shows such a feeding maneuver.

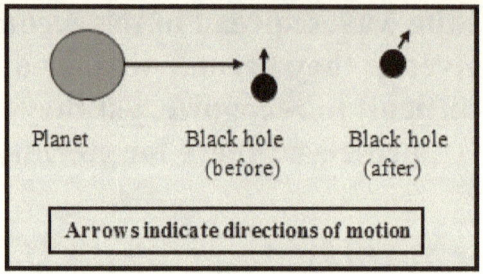

- A planet can also be made to follow a slingshot trajectory around a black hole. As the planet makes its U-turn around the black hole, the black hole actually receives twice the planet's initial momentum in the planet's original direction of motion, since the black hole has to stop the planet's motion, along its original direction, and provide it with a reverse momentum. The following drawing shows a simple presentation of such an undertaking.

  In this case, the black hole will attract the planet but it will not be close enough to pull the planet in. That is, if the minimum distance is properly chosen and the speed is kept above escape velocity at the nearest intended distance to the black hole. Therefore, the black hole will be literally directed towards the desired direction.

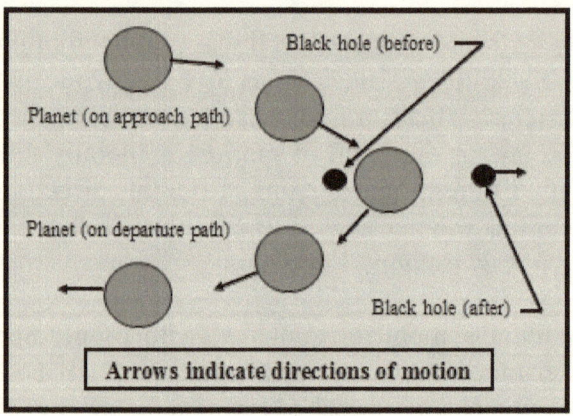

In the case of large black holes of any size, it is wiser to leave them alone. In most cases, would be much easier to simply get out of their path. That is, if the exact time of their passage are known sufficiently in advance. However,

"In most cases of close encounter with black holes of any size, it is actually wiser to relax and use the remaining time to seriously contemplate about the meaning of life and the reasoning for its existence in this universe."

18- Do known laws of physics apply inside the event horizons of black holes?

The validity of the currently known laws of physics depends on 'time' being experienced. Therefore,

"From the instant any object's speed relative to aether in its immediate vicinity reaches the speed of phase vibrations in that medium, the laws of physics lose their validity and can no longer be applied to describe that object's status, since 'time' will become meaningless to that object."

In other words, the currently known laws of physics can be applied inside the event horizons of black holes only to describe the status of objects that directly approach and cross their event horizons in a state of voluntary free-fall. Because, in this case, the speed of such objects relative to their local aether will be less than that of phase vibrations in that medium, even after they cross the event horizon.

However, currently known laws of physics cannot be applied to describe the status of objects that approach black holes following a tangential trajectory, because long before they even reach the event horizon of those black holes their speed relative to their local aether reaches that of phase vibrations in that medium. Since, aether itself is already moving towards the black hole at a very high rate of speed, at nearly 90 degrees to the trajectory of such objects.

19- How can the strength of the gravitational force of a black hole be presented graphically?

The topographical presentation of the strength of the gravitational forces near a given black hole, as shown in the following figure, can be represented by a funnel type of a shape, with gradually increasing slopes towards its center.

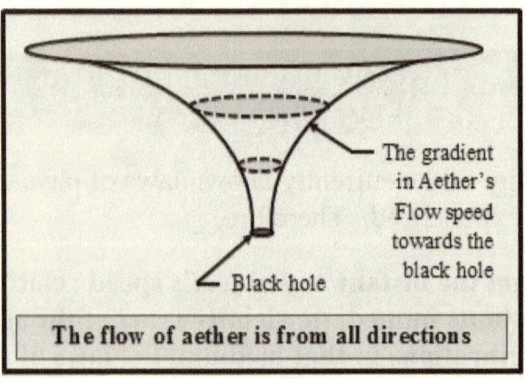

The gradient in Aether's Flow speed towards the black hole

Black hole

The flow of aether is from all directions

The slopes of this funnel shape surface are interpreted by the General Theory of Relativity as being the curvature of spatial dimensions. While, in fact they represent the gradients that exist in the flow speed of the local fluid aether that is accelerating directly towards that black hole. Because, the strength of the gravitational force at any given location is directly affected by the rate at which the local aether is accelerating through that location.

The very same is also true about the gravitational forces of other celestial bodies such as stars, galaxies and even planets. However, in their cases, the slope of the central portion does not become so pronounced.

The experiment that was performed in 1919, during a total solar eclipse, had apparently confirmed the prediction made by the General Theory of Relativity. However, such experiments did not and do not indicate any curvature in the fabric of space (or space-time), but rather simply indicate how light waves were dragged by the accelerated flow of aether towards gravitating bodies which was a star.

The following figure shows such an effect as it is induced by aether flow, in a space that consists of straight dimensions.

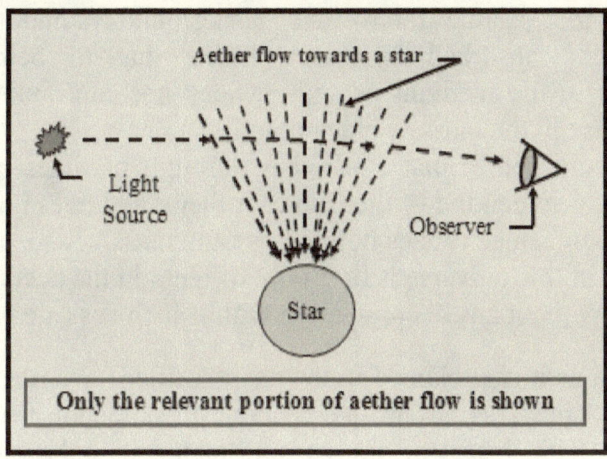

Aether flow towards a star

Light
Source

Observer

Star

Only the relevant portion of aether flow is shown

Note that, the prediction made by the General Theory of Relativity agrees with the observed amount that light waves bend as they pass near a star, simply because the very same equations derived in the General Theory of Relativity apply to the newly proposed theory of gravity presented in this book, a theory that introduces the force of gravity, at any given location in this universe, as the drag force induced by the flow of aether which is accelerating through that particular location.

20- Do black holes gain mass just by sitting around and not devouring any matter particles?

The flow of aether into black holes is a continuous process, regardless of any matter particles or objects being present nearby to attract those black holes' attention, or not. Black holes and matter particles that exist in this universe, are continuously receiving aether from their respective surroundings.

Just as none of the matter particles such as protons or neutrons are gaining any mass due to the continuous flow of aether into them, none of the black holes get more massive just by devouring fluid aether alone, either. Black holes gain mass only as more matter (as well as anti-matter) particles are literally dragged into them by the flow of aether.

Even the cosmic microwave background radiation that is continuously absorbed by black holes, due to being phase vibrations in the medium of aether, does not contribute towards any increase in the mass of black holes.

The fluid aether that is flowing through particles and black holes is like some kind of diet food for them, and does not add any mass to them, since it does not possess any mass. Also, as it flows into them in this universe it flows out of them in the accompanying universe, without any accumulation within such passage ways.

Note that, as it is explained in the chapter titled "What is aether?", 'mass' is not an independent phenomenon from aether. The 'mass' of any particle manifests only and only as that particle tends to change its rate of speed relative to its local aether medium. Since, 'mass' is a side effect of the formation of some sort of 'wake-like wave' in the aether medium by individual particles (or their aggregates).

21- Is there any direct way of observing and collecting data from two black holes approaching and merging together?

As two black holes approach and unite with each other, they will not reveal any details to any observers / instruments that are in this universe. Since, their separate event horizons (before they join) and their enhanced event horizon (after they join) will prohibit the leakage of any information into this universe, by any type of waves or any kind of particles.

However, **black holes can be directly observed and readily tracked in the accompanying universe**, since they act as giant fountains of fluid aether in that universe. As fluid aether flows into the accompanying universe, it carries with it a variety of phase vibrations, which include electromagnetic waves such as visible light waves, of all frequencies.

In other words, from the view point of potential observers (in fact, their spirits) wandering in the accompanying universe, black holes would appear as brilliantly shinning markers in the expanse of space.

Therefore, such observers can clearly observe the two black holes as they approach each other. They can directly witness and record the whole merging process of black holes in as much detail as they please.

Note that, spiritually observing the merging process of two black holes would be quite an awe inspiring experience.

22- Why there were no black holes present right after this universe was born, when the whole energy content of this universe was concentrated in a very small volume of space?

As it is defined in this book, the force of gravity at any given location in this universe is the drag force induced by the flow of the fluid aether which is accelerating through that particular location. Aether flow is accelerating towards matter or anti-matter particles (or their aggregates) which act as drain holes leading into the accompanying universe. In other words,

"For the force of gravity to manifest, matter and/or anti-matter particles must exist."

It was due to the very lack of the existence of matter and anti-matter particles that, **there were no black holes of any size present and none could have formed**, when this universe had just started its existence. Since,

- Various types of energy, including phase vibrations in the medium of aether, such as light and other electromagnetic waves are not made of any kind of particles. Hence, regardless of their amounts / concentrations / intensities, they could not and cannot generate any gravitational force.

- Also, the fluid aether which occupies the whole universe is not made of any kind of particles and hence it could not and cannot generate any gravitational force.

In other words,

"When this universe was born, even though all of its energy content and all of its aether content were concentrated in a very small volume of space, force of gravity and black holes simply did not exist, since matter and anti-matter particles were not born, yet."

23- What is the eventual fate of all black holes?

As the universe expands and the pressure difference between aether that is in this universe and aether that is in the accompanying universe approaches a **critical aether pressure difference**, all of the black holes, starting with the smaller ones, will cease to act as black holes. Once this critical aether pressure difference has been reached, the induced influx of aether reaching the surface of even the largest black holes will be flowing at speeds that are barely equal to the speed of the phase vibrations in the local aether medium. At that instant, the event horizon of the largest black holes will be right at their respective surfaces. This critical aether pressure difference can be referred to as,

"Esmailzadeh Critical Aether Pressure Difference"

The following figure shows a typical black hole's fate, as the pressure difference between aether in this universe and aether in the accompanying universe is decreased passed its critical value.

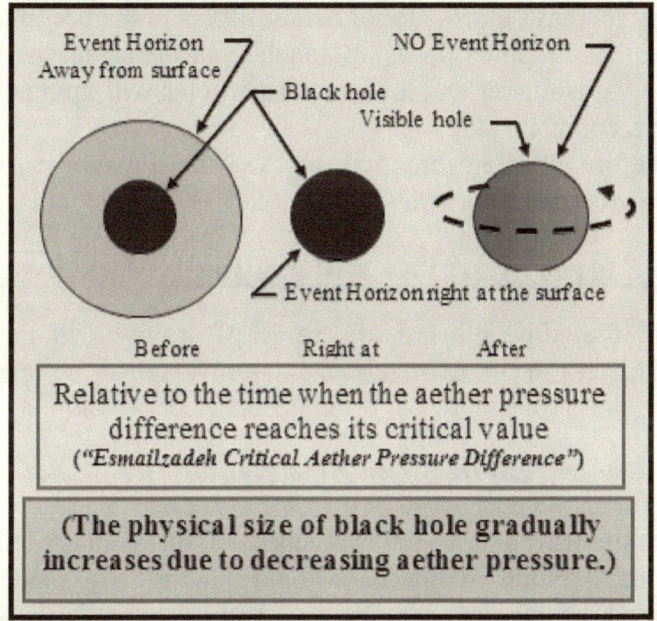

Event Horizon Away from surface

NO Event Horizon

Black hole

Visible hole

Event Horizon right at the surface

Before   Right at   After

Relative to the time when the aether pressure
difference reaches its critical value
("*Esmailzadeh Critical Aether Pressure Difference*")

(The physical size of black hole gradually
increases due to decreasing aether pressure.)

As shown in the figure, as the aether pressure difference between this universe and the accompanying universe continues to decrease, and the speed of aether flowing into black holes becomes less than that of phase vibrations in this universe, electromagnetic waves such as light will be able to escape the black holes. Therefore, all black holes will gradually lose their status as being black holes. Hence,

"Eventually, all of the black holes will become visible."

As it is indicated in the figure, the physical size of black holes gradually increases. Since, due to gradual decrease in aether pressure in this universe, even if black holes do not devour any of their peaceful neighbors, they will still grow in size.

By 'visible hole', in the figure, it is meant that electromagnetic type of radiations of various frequencies can and will escape to the surrounding space. The colors chosen demonstrate the red shift and the blue shift generated due to the rotation of the 'now-visible' black hole.

As the internal pressure of the aether medium in this universe decreases even further, eventually, it will become equal to gradually increasing pressure of aether that is in the accompanying universe. As this pressure is approached and reached, the flow of aether towards matter and anti-matter particles will gradually slow down to a complete stop.

Therefore, no drag force will be exerted on matter particles or on phase vibrations. In other words,

"The force of gravity will gradually, fade away."

From that time onward, the force of gravity will no longer exist. This era in the existence of this universe can be referred to as,

"Esmailzadeh Zero Gravity Era"

Once the zero gravity era is reached, all of the black holes will remain as bubbles in space, without having any tendency to capture and devour stars and planets. And,

"As aether density and pressure gradually decrease in this universe, eventually they will reach certain low magnitudes at which point, all of the black holes will gradually dissolve in the background fluid aether medium and vanish."

Conclusions

Black holes are most commonly formed as individual massive stars experience an unforgettable nova moment of their own. Black holes are basically matter particles (and/or anti-matter particles) that have united and have formed a single but giant gateway between this universe and the accompanying universe.

Once formed, black holes can only grow in size, as they feed on anything and everything that is delivered to them by the inflow of the fluid aether towards them, over time. In other words, black holes are not just a single point in space, but rather have a definite physical size which in the case of giant black holes at the center of galaxies can be millions of kilometers across.

Note that, one can even **propose** that,

"The event horizon of any given black hole represents that black hole's true outer <u>physical boundary</u>."

And, since apparently anything and everything that crosses the event horizon of any black hole disappears from the spatial dimensions of this universe and enters those of the accompanying universe, one may even suggest that,

"The event horizon of any black hole represents the actual surface of the giant bubble that is made of unified

**matter particles, forming that
particular black hole."**

In other words,

**"The event horizon of a given black hole
IS that black hole."**

However, <u>such a proposal cannot be correct</u>,
since its most evident consequence will be that,

**The maximum speed at which the fluid
aether can flow in this universe is limited
to the propagation speed of phase
vibrations in that medium.**

Even, air does not abide by such a limitation on
how fast it can flow. Since, air can be readily
motivated to flow at supersonic speeds in wind
tunnels. The flow of air can also be made to reach
supersonic speeds as it is allowed to escape from
highly pressurized tanks, through nozzles / valves.

As a black hole becomes more massive, the increase in its
surface area (and not its volume) is proportional to the increase in
its mass. Therefore, as a black hole gains more mass, the increase
in its radius is proportional to the square root of the increase in its
mass. In other words,

**"As the mass of a black hole increases, its
density decreases."**

The radius of the event horizon of a given black hole also
stretches proportional to the square root of any increase in its mass.
In other words,

**"As the mass of a black hole increases, its
own radius, the radius of its event horizon
and the distance between its surface and**

its event horizon increase proportional to
the square root of its mass increase."

Since black holes connect this universe to the accompanying universe, they are automatically exposed to two totally different aether pressures from their opposite ends, due to the pressure of aether that is in the accompanying universe being much, much lower than that of aether that is in this universe.

Therefore, one can reason that the overall physical shape of black holes, when considered as a whole, is not a simple narrow tube with a uniform diameter along its full length. But rather, it is shaped like a funnel, with its narrower end facing this universe and its wider end facing the accompanying universe.

However, as the pressure of aether in this universe is gradually decreasing and the pressure of aether in the accompanying universe is increasing, the overall shape/geometry of black holes is approaching that of a tube with a uniform diameter along its full length. Afterwards, the overall aether pressure will continue to simultaneously decrease in both universes and the diameter of the black holes will gradually become larger and larger.

Black holes can only be of neutral charge. However, any given black hole may seem to possess one type of electrical charge or another, due to the charged particles that are in the process of falling into them, but have not crossed their event horizons, yet, let alone joining the black hole. In other words,

"Black holes can only be of neutral charge.
However, they may seem to be positively or
negatively charged, due to charged particles
that are currently around them."

As the pressure difference between aether that is in this universe and aether that is in the accompanying universe is gradually reduced below a certain critical magnitude, namely the Esmailzadeh critical aether pressure difference, all of the black holes will lose their status as being black holes. Because, after that point in time, various types of phase vibrations such as electromagnetic waves (light) will be able to propagate against the incoming flow of aether which is towards black holes and hence escape into the surrounding space. In other words,

"Eventually all black holes will become visible."

As the pressure of aether in this universe continues to drop due to the expansion of the aether medium, as well as its leakage into the accompanying universe, eventually its pressure will become equal to that of aether in the accompanying universe. At that point in time, due to the lack of any kind of aether flow towards any matter (or anti-matter) particles or their aggregates (including black holes) the force of gravity will no longer exist.

Once the zero gravity era is reached, all of the black holes will remain as bubbles in space, without having any tendency to capture and devour stars and planets. In other words,

"As aether density and pressure gradually decrease in this universe, eventually they will reach certain low magnitudes at which point, all of the black holes will gradually dissolve in the background fluid aether medium and vanish."

What
is
Magnetic Field?

Introduction

In 1865, Mr. Maxwell proposed his theory on electromagnetism which was based on the existence of a medium through which light and other electromagnetic waves were assumed to propagate. At that time, this medium was referred to as aether. However, what aether was made of and what kinds of properties it had were not known. Aether was accepted as a needed medium through which electromagnetic waves propagate. To electromagnetic waves the stationary aether medium was thought to be just as a medium such as air is to sound waves.

However, contrary to what was accepted back then, aether is not a stationary medium. The varieties of motions that exist in that medium can be likened to the motions of air molecules in the atmosphere or the motions of water molecules in the oceans. The varieties of motions in the aether medium give rise to and manifest as such phenomena as gravity, magnetic field and electric field, among many others.

Electric field and magnetic field are very closely related to each other. The most important common point between these two fields is that, both owe their existences to the existence of charged particles in this universe. Since, the very existence of a charged particle at any location induces an electric field in its surrounding environment. Also, if a charged particle happens to pass by a given location, it induces a magnetic field at that particular location. In other words,

"The historical existence of electric field and magnetic field goes back to when the first charged particles were formed, in this universe."

Magnetic field and electric field are two of many kinds of flow patterns that can exist in the aether medium. They are stated to be aether flows of some sort, because they exhibit preferred directions, as shown below. In the case of the magnetic field, shown on the left, the aether flow inside a given magnet is from its South Pole towards its North Pole and after aether exits the magnet at its North Pole it continues through the surrounding environment until it returns to the South Pole where it enters the magnet, hence completing a loop. And, in the case of the electric field, shown on the right, the aether flow is away from the negatively charged particles (as if aether is pushed away due to being under higher pressure near such particles) and flows towards the nearest positively charged particles (as if there is a pressure gradient encouraging aether to do so).

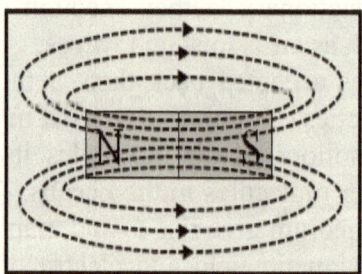 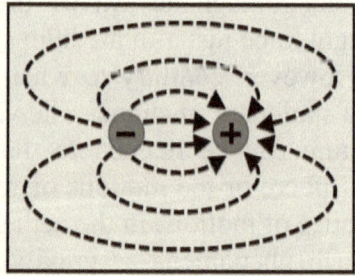

Such aether flow patterns can be readily likened to the flow pattern an operating household fan generates in a **medium** such as air, shown below.

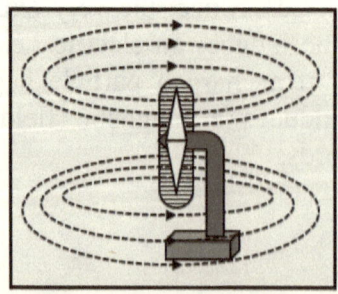

In order to properly understand what magnetic field is, one needs to understand what electric field is. In turn, a good understanding of electric field requires a better understanding of what charged particles are.

Charged particles act as if they are the cutoff faces of a hose through which aether is flowing either out of or into, depending on the charged particle being a negatively charged particle or a positively charged particle, respectively. To better visualize this hose analogy one can consider a hose which is used to connect the outlet of an air pump directly to its inlet, as shown below.

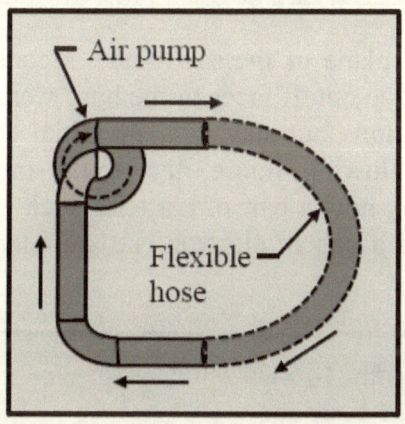

If the hose is cut off by a saw, at a certain point along its length, while the air pump is operating, as shown in the following figure, even though the two cutoff faces of the hose will be of exact same size and shape, they will act opposite of each other in one very important and distinctive way. Since, as soon as the hose is cut off, the cutoff face of the section of the hose that is connected to the inlet of the air pump will be sucking air from its surrounding, while the cutoff face of the section of the hose that is connected to the outlet of the air pump will be blowing air into its surrounding. The two cutoff faces of the hose will basically act like a positively charged particle and a negatively charged particle, respectively.

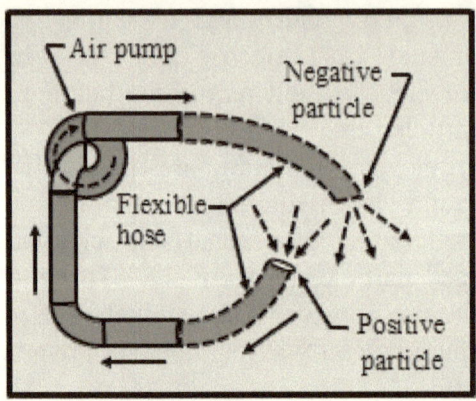

In fact, if the plane of the saw blade is considered to be this universe and the two cutoff faces of the hose were allowed to move in different directions but stay in their initial common plane, as shown in the following figure, the above mentioned scenario describes precisely how a pair of particles such as a pair of proton and anti-proton or a pair of electron and anti-electron (positron) is produced in nature.

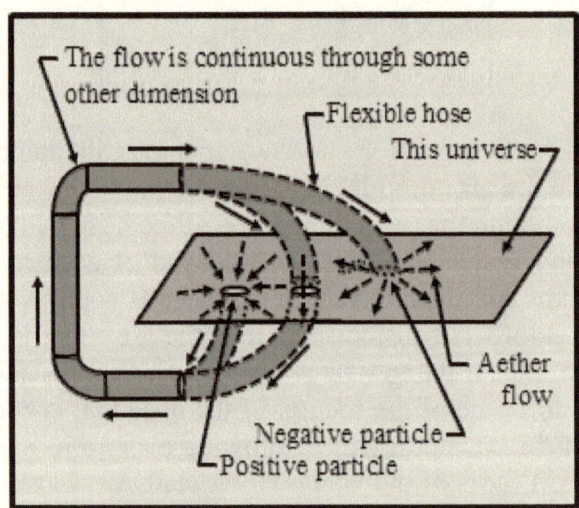

Note that, the two-dimensional cutoff faces of the hose in the above example resemble the three-dimensional bubbles / voids which are the charged particles in this universe which is shown as a two-dimensional plane in the figure.

Also, the **"other dimensions"**, indicated in the figure, do not refer to the accompanying universe but to yet another universe which will be explored in detail in a separate book.

Once formed, charged particles continuously induce a steady flow of aether which manifests as their electric fields, in their surroundings. And, as they come close to each other, depending on the two particles having the same type of charge or the opposite types, they repel or attract each other, respectively, since they mandatorily interfere / interact with each other's induced aether flow patterns.

The induced aether flow by any given charged particle is at a steady pace. Therefore, the rate at which aether is flowing towards a positively charged particle (or away from a negatively charged particle) is inversely proportional to the square of the distance between the location of interest and the charged particle.

If a charged particle happens to be stationary at a given location, its induced aether flow at any other location in its surrounding will be at a fixed rate. However, if the charged particle happens to be passing by a given location its induced aether flow rate will gradually increase at that particular location as it approaches that location and then will gradually decrease as it goes away from that location. Also, if a particle approaches a relatively stationary charged particle and then recedes from its location, that particle will experience an electric field which becomes gradually stronger during its approach and then becomes gradually weaker as it goes away from that charged particle's location.

It is the variation in the electric field at any given location in this universe that gives rise to what is known as the magnetic field at that particular location. This is why, as a charged particle moves in this universe, regardless of its speed or the presence of any other charged particles in its vicinity, it induces a magnetic field around itself. Since, as the charged particle approaches any given location in space and as it goes away from it, the strength of its electric field at that particular location changes. The variable electric field, in turn, induces a magnetic field.

This is the reason why as charged particles, namely electrons, move along a wire, forming an electrical current, they induce a concentric magnetic field around that wire. Since, <u>it is the variation in the acceleration of aether towards or away from a given charged particle that manifests as its magnetic field.</u> The intensity of the induced magnetic field is proportional to the number density of electrons (charged particles) passing through the wire.

Again, any charged particle in motion generates a concentric magnetic field around its path. The direction of the magnetic field induced depends on the charge of the particle being positive or negative. In other words,

"By their very motion in the aether medium, charged particles encourage the local aether to be stirred in one direction or another, depending on their charges being positive or negative."

Note that, as it was stated earlier, it is proposed in this chapter that, magnetic field is a form of motion induced in the medium of aether. Therefore, its strength is dependent on the density of the local aether. Hence, as aether's density is gradually decreasing in this universe, the strength of the magnetic field induced by the motion of any kind of charged particle is correspondingly weakening with time.

The most effective way of demonstrating the fact that magnetic field is a type of aether flow is to experimentally show its various effects. The following experiments are specifically designed to explore different effects of a certain type of aether flow which manifests as magnetic field.

Effects of Magnetic Field
on
the Passage of 'Time'

A variety of experiments can be designed and conducted that will demonstrate how the flow of the local aether relative to objects (or living beings), manifesting as the local magnetic field, affects the rate at which they experience the passage of time. Details of four experiments are provided in the chapter titled "What is Time?".

- First experiment: (Magnetically induced 'Time Dilation', on a planetary scale)

- Second experiment: (Time anomaly experienced by clocks onboard satellites in polar orbits)

- Third experiment: (Magnetically induced 'Time Dilation' on Laboratory scale)

- Fourth experiment: (Magnetic field powered time dilation machine)

These experiments are proposed to confirm the validity of the aether-based theories of 'Magnetic field' and 'Time'.

Effects of Magnetic Field
on the Speed of Light
and its Direction of Propagation

Magnetic field is basically one type of aether flow which completes a round trip. The effects of magnetic field on the propagation speed of light and its direction of propagation are described in great detail in the chapter titled "What is Light?".

The following two experiments are also described in detail in that chapter, experiments which are specifically designed to demonstrate the effect of magnetic field on the speed of light or any other electromagnetic wave and the direction of their propagation, respectively.

- First experiment: (Effect of magnetic field on the speed of light)

- Second experiment: (Effect of magnetic field on the direction of light's propagation)

These experiments are proposed to confirm the validity of the aether-based theories of 'Magnetic field' and 'Light'.

Effects of Magnetic Field on the Strength of the Force of Gravity

> - First experiment:
> (Effect of magnetic field on Mr. Galileo's experiment)

Suppose a very strong coil type electromagnet is positioned vertically so that the field lines inside the coil are in parallel with the flow of aether towards earth, the same flow of aether which is giving rise to earth's gravity. Such a setup is shown in the following figure.

Two plugs or caps are used to close to two ends of the glass tube. The top cap should be equipped with three specific features:

- **A release mechanism** so that different small objects can be released to pursue a free fall journey downward towards earth.

- **A tube connected to a vacuum pump** to allow the air inside the glass tube to be vacuumed out to nearly eliminate air resistance.

- **An air intake valve** to allow the tube to be refilled with air so that the lower cap can be removed to retrieve the sample after it has fallen. This feature will also allow the top cap to be removed and reloaded with a new sample object.

The interior of the coil must be wide open and readily accessible from both ends so that various small non-magnetic objects can pass through it, as they will be literally dropped in from the top. This can be accomplished by winding the coil around a glass tube which is capped at both ends.

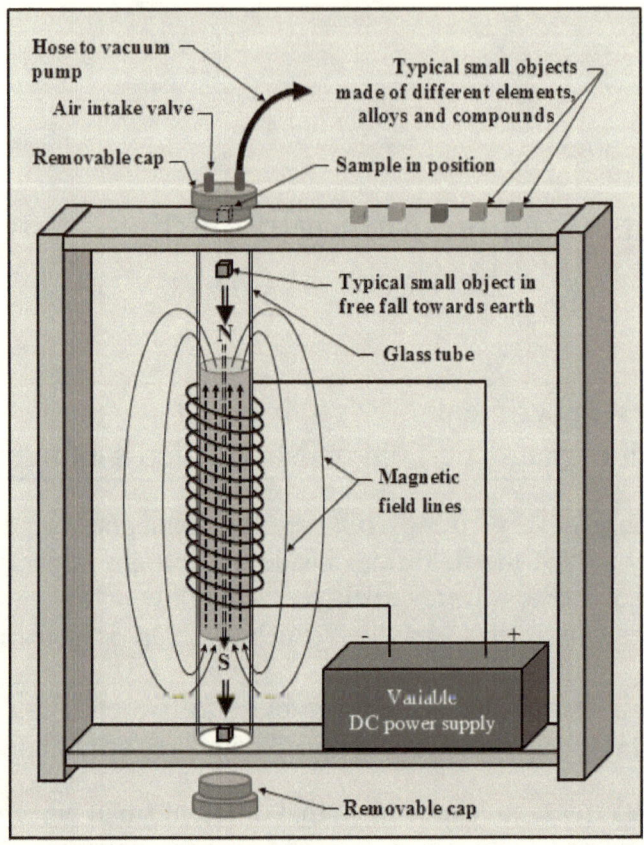

At first, before connecting the power supply to the coil, one by one a few of the small non-magnetic, not-electrically charged objects should be placed inside the top cap. Then, after the air inside the tube is vacuumed out, they should be released to experience their respective free falls towards earth. The amount of elapsed time as they pass through the length of the electromagnet coil must be measured and recorded. It will be shown that, in the absence of air resistance (and, without the presence of a magnetic field), all objects regardless of their densities and/or compositions will experience free fall towards earth at an identical rate of acceleration.

During this part of the experiment, the small objects used will experience the very same force that was experienced by the objects dropped by Mr. Galileo Galilee, namely the drag force associated with the accelerated flow of aether towards earth which is giving rise to earth's gravity.

Note that, the measured elapsed time must be between the instant each sample object enters and the instant it exists the core of the electromagnet's winding. Since, it is the flow of aether associated with the magnetic field within the electromagnet's winding that will be of interest.

In the next step, the electrical power of a known high voltage level should be selected. Also, the duration of the time for the voltage to automatically / gradually increase from zero to its preselected maximum limit should be set to be equal to the elapsed time objects had taken to pass through the electromagnet's coil, in the previous part of this experiment.

The following step of the experiment should be repeated for two totally different scenarios:

1- With the North Pole of the induced magnetic field at the top end and its South Pole at the bottom end (as shown in the figure)

In this case, once the electrical power is connected to the electromagnet, and is (gradually) increased from zero volts to its maximum voltage level, the induced accelerated aether flow manifesting as the induced magnetic field inside the glass tube will be upwards. Therefore, it will partially counteract the downward accelerated flow of aether which is giving rise to earth's force of gravity. Hence, regardless of their compositions, as sample objects are one by one dropped from the top cap, all of them will accelerate towards earth at a **slower pace**, as compared to when the electromagnet was not energized. The amount of elapsed time as each object passes through the length of the electromagnet's coil should be measured and recorded.

This procedure can be repeated as the electromagnet is energized with different maximum voltage settings. The higher maximum voltage settings on the power supply will encourage the local aether inside the glass tube to accelerate upwards at higher rates, as every time the voltage level is raised from zero to its higher maximum voltage levels, within the fixed time period, measured during the first part of this experiment.

Note that, if a sufficiently high voltage setting is selected as maximum on the power supply, and it is applied to the coil of the electromagnet, the induced upwardly accelerating aether flow will cancel out the downwardly accelerating aether flow which is responsible for inducing the drag force called gravity. Hence, **the dropped sample objects momentarily will not experience any acceleration towards earth. And, if the sample objects were positioned at the North Pole of the electromagnet, while hanging from a string attached to some sort of elastic acting as a scale, their weight would temporarily drop to zero, as the elastic would temporarily contract, indicating that they are experiencing weightlessness / levitation.**

On the surface of earth, the rate at which aether is accelerating towards earth is about 9.8 meters per second squared. Therefore,

"Varying the electrical voltage delivered to the coil of the electromagnet at a rate that corresponds to an acceleration rate of the local aether of about 9.8 meter per second squared in the upward direction, will encourage any sample object that may be hanging from a string to momentarily experience a state of weightlessness, as it will float inside the tube within the length of the electromagnet's coil."

2- **With the South Pole of the induced magnetic field at the top end and its North Pole at the bottom end** (opposite of what is shown in the figure)

In this case, once the electrical power is connected to the electromagnet and is (gradually) increased to its maximum level, the induced accelerated aether flow manifesting as the induced magnetic field inside the glass tube will be downwards. Therefore, it will complement the downward accelerated flow of aether which is giving rise to earth's force of gravity. Hence, regardless of their compositions, as sample objects are one by one dropped from the top cap, all of them will accelerate towards earth at a **faster pace**, as compared to when the electromagnet was not energized. The amount of elapsed time as each object passes through the length of the electromagnet's coil should be measured and recorded.

This procedure can be repeated as the electromagnet is energized using different maximum voltage settings. The higher maximum voltage settings on the power supply will encourage the local aether inside the glass tube to accelerate downwards at higher rates, as every time the voltage level is raised from zero to its higher maximum voltage levels, within the fixed time period, measured during the first part of this experiment.

If the sample objects were positioned at the North Pole of the electromagnet, while hanging from a string attached to some sort of elastic acting as a scale, their weight would temporarily increase every time power is connected to the coil of the electromagnet, as the elastic would temporarily stretch, indicating that they are experiencing having excess weight.

The results obtained during either parts of such an experiment will indicate that, wherever the acceleration of the induced flow in the local aether (manifesting as the magnetic field) is comparable to the acceleration of the local downward aether flow (manifesting as earth's force of gravity), it can directly affect the strength of the local force of gravity.

Note that, if the induced accelerated aether flow (manifesting as the magnetic field) and the accelerated aether flow associated with the force of gravity cross each other's paths, the local force of gravity will act at an angle other than directly perpendicular to the horizon (towards the center of mass of earth). The severity of such a deviation or the magnitude of such a tilt angle will depend on:

- The relative magnitude of the two induced accelerations in the local aether, giving rise to the magnetic field and the force of gravity, and

- The relative angle between these two induced aether flows.

It should be emphasized that, if the electrical power of a constant voltage be connected to the coil of the electromagnet, during the whole period each of the samples are released by the top cap and when they are stopped by the bottom cap, the induced flow of aether manifesting as the **constant magnetic field would not have any effect, whatsoever, on the rate at which those sample objects will accelerate towards earth.** Since,

"Aether flow generates a drag force only when it is accelerating."

- ## Second experiment:
  ### (Magnetic Hill Effect)

The previous experiment titled "Effect of magnetic field on Mr. Galileo's experiments" was designed to demonstrate how the presence of a variable magnetic field, which in itself is a type of accelerated flow induced in the local aether medium, affects the local gravity.

Application of such an effect can readily explain why at certain locations around the globe (referred to as **'magnetic hills'** or **'gravity hills'**) it seems like, the force of gravity does not act in the vertical direction towards earth (perpendicular to the horizon) and hence causing objects (including cars, when put in neutral) to roll towards higher rather than lower elevations. Even water will run uphill, and will show a surface level that does not match the distant horizon. In other words, it can be proven that,

"The observed phenomenon is NOT an optical illusion."

Such effects are due to the fact that, certain places on the surface of earth either are more effectively connected (magnetically) to the magnetic material that forms earth's interior magnet, or are the very pointed tops of very massive chunks of magnetically inclined material that are buried deep underneath very large areas. The following figure shows one such location in the Northern magnetic hemisphere where the localized Pole is of North Pole type. Hence, its outflow of aether is directed towards the Southern magnetic Pole.

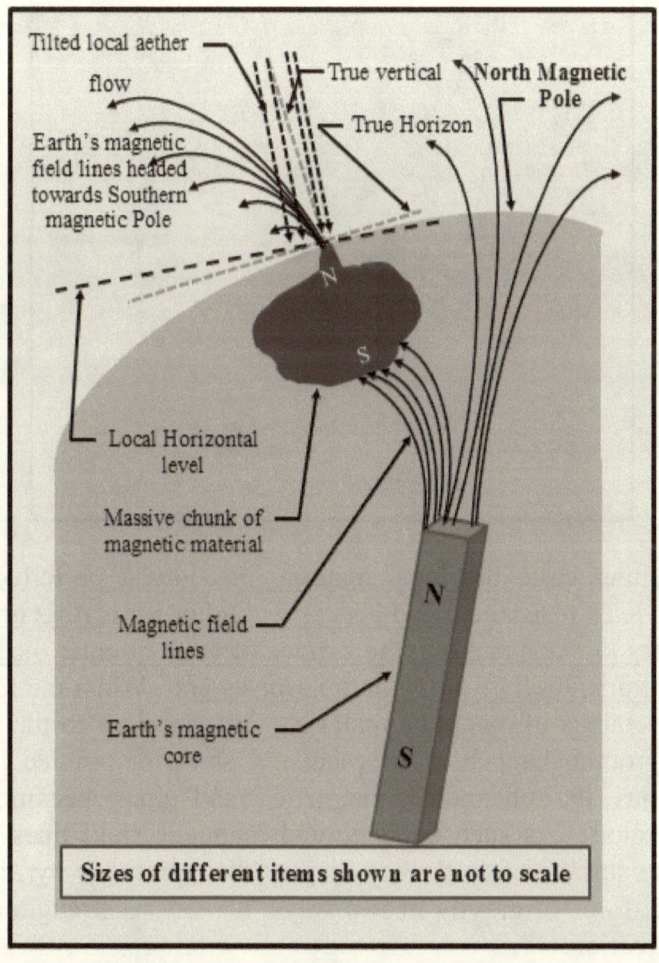

The following figure shows one such location in the Southern magnetic hemisphere where the localized Pole is of South Pole type. Hence, its inflow of aether is coming from the North magnetic Pole.

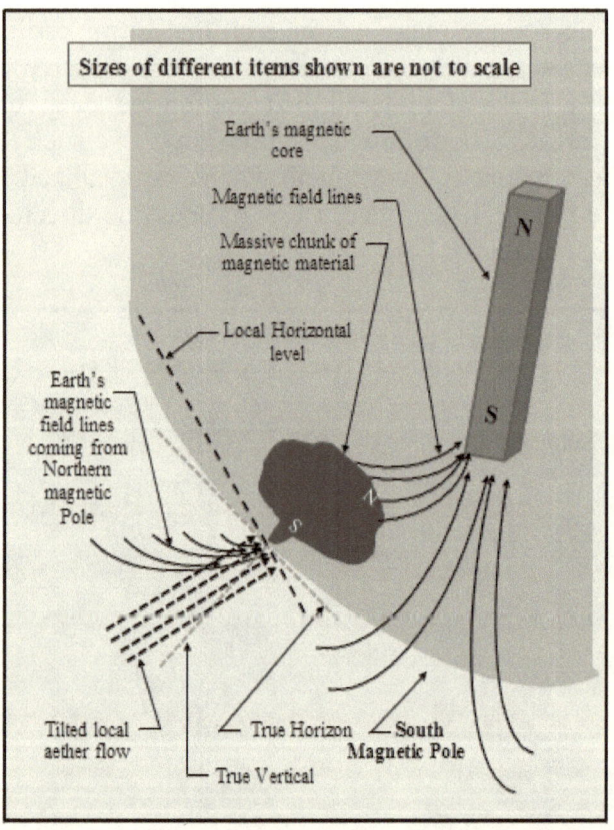

In either case, <u>the local material provides a preferred route</u> (path of least resistance) <u>and acts as a conduit for earth's magnetic field lines to reach and exit the surface of earth or enter the surface and get transferred towards the interior layers.</u> When the top ends of such chunks of magnetic materials that are close to the surface of the ground, happen to be relatively sharp or pointed in their geometries the channeled magnetic field lines become quite concentrated. As such concentrated magnetic field lines exit or enter the surface of earth they display their presence by affecting the local force of gravity in two ways, namely its strength as well as its direction.

Note that, in both cases shown in the figures the local horizontal level tilts upwards on the Southern side. This is due to the fact that, in the first one the downward aether flow (which is giving rise to earth's gravity) in the Northern hemisphere is **pushed** towards the Southern magnetic Pole. Also, in the second one the downward aether flow (which is giving rise to earth's gravity) in the Southern hemisphere is **pulled** towards the Southern magnetic Pole. Therefore, in both cases,

"The top of any object standing upright at such locations will be tilted towards the magnetic North Pole (NOT to be mistaken with the Geographic North Pole)."

This is why all of the known magnetic hill sites located in Canada, United States, France, Korea, China, Iran and so on, demonstrate their effects on various objects towards the general southerly direction. Since, at such locations, the local ground surface is between the true distant horizon and the locally modified horizon. Therefore, to any object such a surface is experienced as being downhill, towards the southern magnetic pole.

The severity of the influence induced on the strength and the direction of the local force of gravity depends on:

- Such locations being magnetic poles of the North type or being magnetic poles of the South type,

- Such locations being in the Northern magnetic hemisphere or in the Southern magnetic hemisphere,

- The distance between such locations to the nearest earth's magnetic pole,

- The quality of the connection between the local magnetic material and the inner magnetic core of earth, and

- The size of the magnetic region close to the surface as compared to the size of the chunk of magnetic material which is underneath, deeper underground. Since, the smaller area closer to the surface automatically means that the local magnetic material is more finely pointed (narrowed) close to the surface of earth. The wider spread upper layers will generate a weaker effect due to the spreading of the magnetic field lines at the location where they exit / enter the surface.

At such locations, any tilt in the local ground surface with respect to the local (magnetically affected) horizontal level would naturally be felt as either UPHILL or DOWNHILL, depending on the direction aimed at. Therefore, any object (including liquids such as water and other none magnetic objects/fluids) will have the tendency to roll/flow in the localized downhill direction, even though it might be uphill with respect to the overall curvature of earth (centered at earth's center of mass) at that particular place. This is why objects such as cars (when put in neutral and breaks are released) have the tendency to roll towards higher elevations at such locations. Hence, titles of 'magnetic hill' or 'gravity hill' are well suited to literally describe the phenomenon experienced.

The closer such magnetically intensified locations are to the magnetic North Pole or to the magnetic South Pole of earth, the stronger will be their effects on the strength of the local force of gravity. While, their effect on the deviation (tilt) of the direction along which the force of gravity is sensed by the local objects will be weaker. This is due to the fact that, the closer such locations are to the magnetic North Pole or the magnetic South Pole of earth the closer will be the direction of their associated aether flow to the true local vertical line (locally perpendicular to the curvature of a sphere centered at the center of mass of earth).

However, as the latitudes of such locations decrease and approach the magnetic equator of earth, the direction of the induced aether flow (magnetic field lines) in the immediate vicinity of such places will deviate greater and greater from the vertical. Therefore, the noticeable effects will include stronger variations in the angle at which objects are pulled towards earth's surface (the local horizon will be tilted more). But, their effect on the strength of the local force of gravity will be weaker. This is the very effect which is manifested at places referred to as 'magnetic hills' or 'gravity hills' that are close to the magnetic equator.

One can perform a very simple and readily doable test that will verify the variation of an object's weight as a result of being placed at such a location (known as a "magnetic hill", or a "gravity hill"). Such a test can be performed using any non-magnetic weighing scale, even a non-magnetic elastic material (a bungee cord, without the metal hooks at its two ends), a few known non-magnetic weights and a non-magnetic measuring tape.

By comparing the weight of each object, as they are individually weighed at such popular locations and also at locations with normal geomagnetic characteristics, one can readily confirm whether such locations are actually what they are publicized to be or are not.

Note that, as shown below, **the non-magnetic weighing scale such as a simple bungee cord and several non-magnetic weights on it, as a whole, should be freely hanging down, so that the weights are measured (by measuring the variations in the length of the elastic material) while the elastic material is truly perpendicular to the local horizon.**

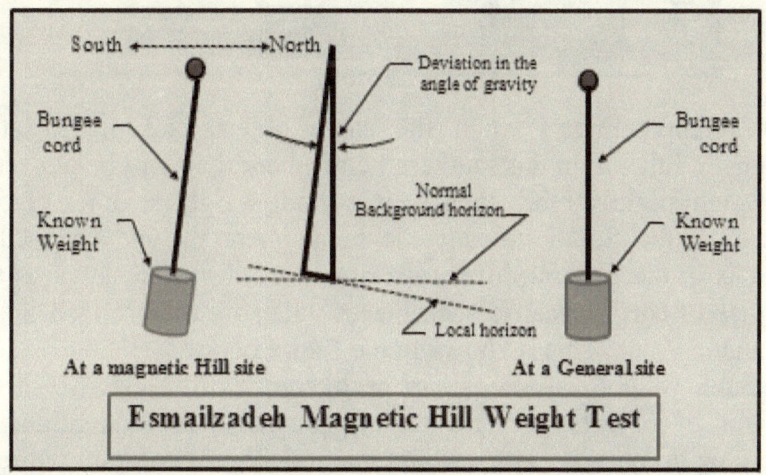

Esmailzadeh Magnetic Hill Weight Test

It can also be proven that, the amount of variation in an object's weight at such locations, as compared to its weight at other normal locations, will be in direct correspondence with the angle at which the local force of gravity is acting on that object. Such weighing tests can be referred to as,

"Esmailzadeh Magnetic Hill Weight Test"

At a magnetic hill site in the northern hemisphere, objects are expected to weigh slightly less than their normal weights at other locations. This is due to the fact that, in the Northern magnetic hemisphere the induced aether flow is in the general upward or outward direction, away from the surface of earth, because such places will pose as localized North magnetic Poles. This is shown in the figure below, in an exaggerated fashion.

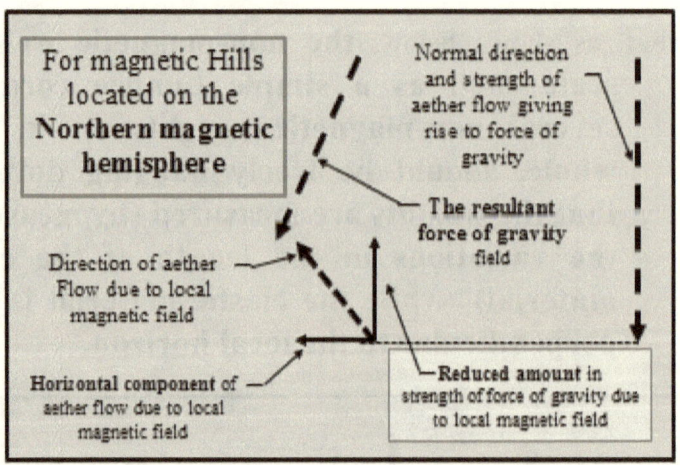

Correspondingly, when the same objects are placed at a magnetic hill site in the southern hemisphere they will be expected to weigh slightly more than their normal weight at other places. Since, in the Southern magnetic hemisphere the induced aether flow is in the general downward or inward direction, towards the surface of earth, because such places will pose as localized South magnetic Poles. This is shown in the figure below.

Such weight variations can be likened to the increase in the weight of objects onboard an airplane that is banked and is pursuing a curved path, while maintaining a constant altitude during its flight. In such a case, not only the force of gravity is acting on the objects but also the centrifugal force which is due to airplane changing its flight direction.

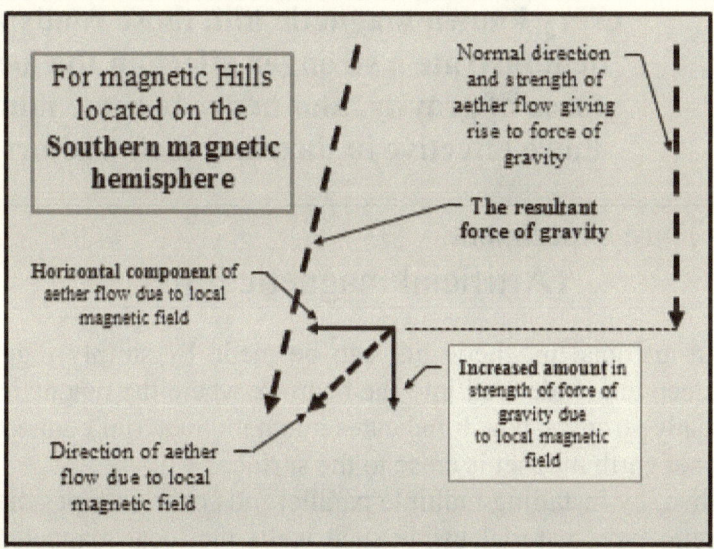

However, the centrifugal force in the case of the airplane scenario does not have a vertical component. Whereas, as shown in the figures, magnetic hills depending on being on the northern hemisphere or the southern hemisphere do possess a vertical component which is upward or downward, respectively.

The amount of such variations in the weight of objects, as they are placed at the magnetic hills, is dependent on the strength of the locally concentrated magnetic field. To calculate the expected percentage of variation in the weight of objects placed at a specific magnetic hill the strength of the local magnetic field and its angle with respect to the line of sight towards the center of mass of earth must be measured / known.

Note that, since nearly all of the known magnetic hills are located along existing roadways, which are not necessarily directed towards the true opposite magnetic pole of earth, it has only been by chance that their effects and hence their very existences have been noticed by the local people. However,

"If a passageway (roadway) pointing directly towards the opposite magnetic pole of earth is constructed from the focal point of each and every known magnetic hill, those roads will demonstrate a stronger effect on the local force of gravity, and hence become much more effective in dazzling their visitors."

- Third experiment:
  ## (Artificial magnetic hills)

An artificial magnetic hill can be made by simply digging a very deep and wide well into the bedrock where the magnetic field is already stronger which indicates magnetic material connected to the inner earth magnet is close to the surface.

Then, by installing multiple parallel rods or even pipes made of magnetic type material inside such wells the local magnetic field lines can be guided (mandatorily channeled) through from the interior of earth towards the surface (in the Northern magnetic hemisphere) or from the surface towards the interior layers of earth (in the Southern magnetic hemisphere). The internal structure of such an artificial magnetic hill is shown in the following figure.

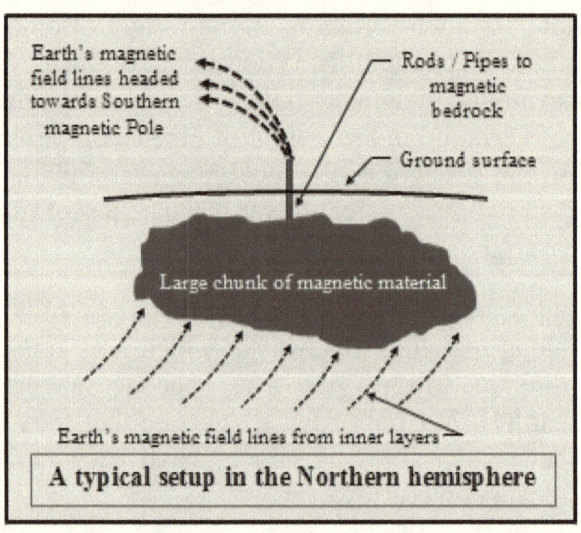

A typical setup in the Northern hemisphere

The strength of such an artificially made magnetic hill will depend on the following:

- Such locations being magnetic Poles of the North type or being magnetic Poles of the South type,

- Such locations being in the Northern magnetic hemisphere or in the Southern magnetic hemisphere,

- The distance between such locations to the nearest earth's magnetic Pole,

- The quality of the connection between the local magnetic material and the inner magnetic core of earth, and

- The size of the magnetic region close to the surface as compared to the overall size of the chunk of magnetic material which is underneath, deeper underground. The smaller area closer to the surface automatically means that the local magnetic material is more finely pointed (narrowed) close to the surface of earth. The wider spread upper layers will generate a weaker effect due to the spreading of the magnetic field lines at the location where they exit/ enter the surface.

The attachment at the top (ground surface) will depend on the specific application in mind. For example,

- The top plate above ground level can be used as a designated spot where any **weighing scale would defy its own reading** of a given weight at other places. Such an application better be located closer to the magnetic poles, since the induced aether flow just above such plates will be closely lined up with the vertical. Hence, it will be more effective in counter acting (or assisting) the inflow of aether responsible for earth's local gravity.

- Can be used to **create a localized tilt in the horizontal level**. Such a place will make standing upright in the vertical direction a challenge. Such an application would be more effective if it is installed at lower latitudes, closer to the magnetic equator, since at such places the aether exiting or entering the surface of earth will be nearly in the horizontal direction. Therefore, the induced tilt in the direction in which the force of gravity acts upon objects will be more pronounced.

- Artificially made (or even naturally occurring) magnetic hills (localized magnetic poles) can have a variety of useful applications such as generating power, as well. The following figure shows such a setup.

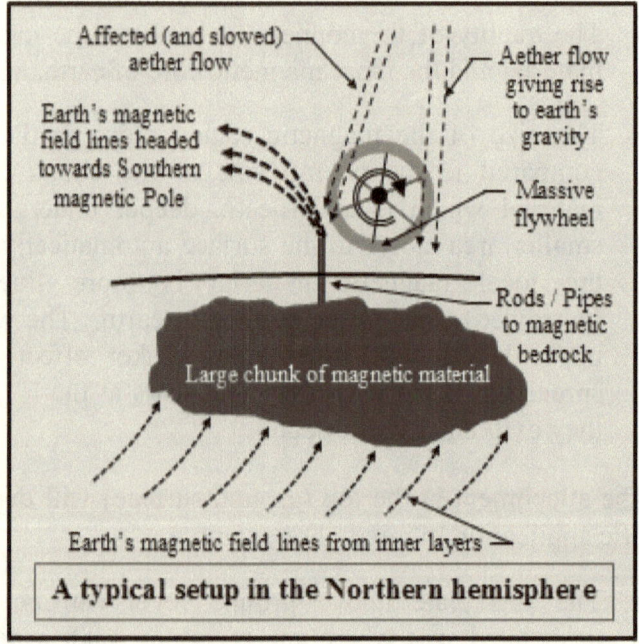

Affected (and slowed) aether flow

Aether flow giving rise to earth's gravity

Earth's magnetic field lines headed towards Southern magnetic Pole

Massive flywheel

Rods / Pipes to magnetic bedrock

Large chunk of magnetic material

Earth's magnetic field lines from inner layers

A typical setup in the Northern hemisphere

- A power generating station can be made to benefit from the weight difference experienced in two adjacent regions. Using a normally balanced but very large diameter and quite heavy wheel will allow such extraction of energy without adding any pollution to the local environment.

In fact, such a power take-off unit will work just like a water wheel, simply due to the difference that would be induced in the weight of the two sides of the wheel. The same type of setup can be made for applications in the Southern magnetic hemisphere where the aether flow is downward towards the surface rather than upward and away from it.

- By attaching a wide and long plate of the same type of material to the top of the rods/pipes the magnetic field lines will be allowed to spread over a pre-designated area, which is forming an uphill, can serve as a **magnetic hill ramp for cars and objects to roll uphill**.

  One can also make a **continuous circular path for water to freely run** on its own, as shown below. The flowing water can even be made to experience short falls for creating some pleasant scenery along its path, as well.

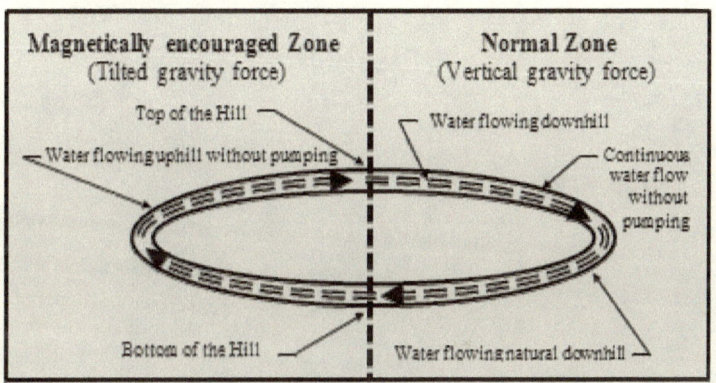

In such a setup, water will gain elevation as it passes through the region where the localized magnetic field has tilted the horizontal level. Therefore, as far as water is concerned it is continuously running downhill, totally on its own, and yet completing a loop. Such a setup, as shown below, will be a pleasant addition or feature in an amusement park, as it will definitely dazzle the visitors.

Also, while allowing a larger volume of water to flow in such a manner, a water wheel can be used to generate some shaft power to either generate electricity or be applied locally to perform certain other exciting task.

- A train, as long as the whole length of a circular track, can be set up so that it can take the visitors for a ride, without using any external power delivered to it. Only a braking system is required so that people can get on and get off safely.

Note that, if it is left unattended, such a train will keep on gaining speed until it derails itself, since it will continuously be rolling downhill. The only resistances it will experience will be that of air and its wheel bearings.

Magnetic Field Propulsion System

As it is shown in the following figure, the main components in this type of propulsion system are one (or more) coil type electromagnet, one variable pulse generator, and one electrical supply of DC type.

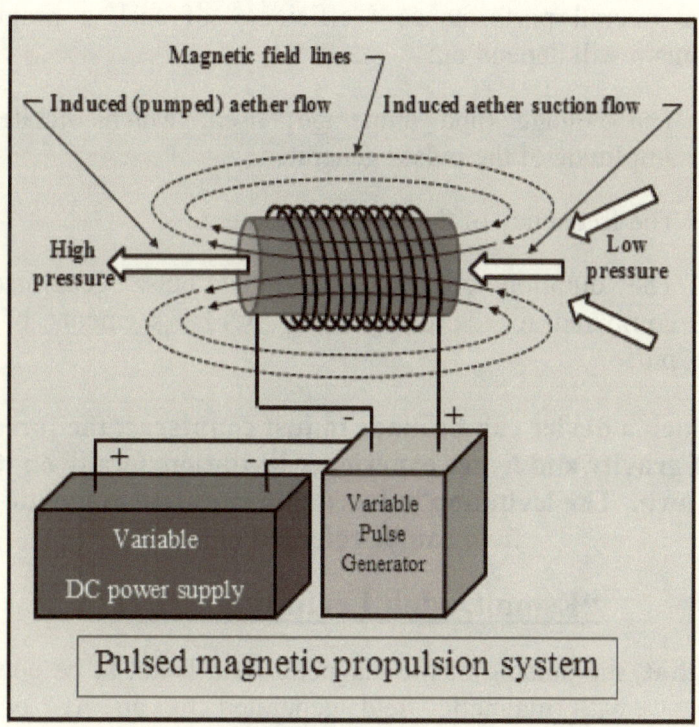

Magnetic field lines

Induced (pumped) aether flow   Induced aether suction flow

High pressure   Low pressure

−   +

+   −

Variable DC power supply   Variable Pulse Generator

Pulsed magnetic propulsion system

The overall effectiveness of this type of propulsion system depends on the strength of the magnetic field generated. Since the effects generated by such a system are accumulative, increasing the length of the magnetized region will automatically increase the overall speed of the aether flowing through. This is exactly like stacking up regular inline water pumps to increase the output pressure.

Such a propulsion mechanism generates its propulsive force by inducing a high pressure region in its back side, and inducing a low pressure region in its front side, just like a jet engine.

The proper operation of such a propulsion system depends on the way the magnetic field is induced. If the magnetic field is induced in a continuous manner, aether will form a complete loop. As a result, no net thrust of considerable magnitude will be generated. However, if it is generated in the form of short **unidirectional** pulses, it will demonstrate its ability in generating a fairly strong thrust force in the desired direction. The shorter duration of the pulse will ensure the formation of a parallel aether flow, since aether does not get the chance to complete a loop.

The overall performance / efficiency of such a propulsion mechanism will depend on:

- The voltage (and amperage) used, which dictates the amplitude of the pulses generated,

- The frequency of the pulses generated,

- The duration of each individual pulse generated (as compared to the duration of "OFF" segments of each pulse).

Such a device can be made to just counteract the force of gravity and hence experience levitation, totally on its own. The levitation induced due to pulsed magnetic field can be referred to as,

"Esmailzadeh Levitation Effect"

Note that, the pulsing of the magnetic field is not to be confused with magnetic field generated by an AC type of electricity. In the case of AC, the direction of the induced magnetic field is repeatedly changing direction. Whereas, in the case of pulsed magnetic field, the electrical power delivered to the electromagnet is DC which is chopped into narrow time segments.

To generate pulsed DC current, for such an application as pulsed magnetic field, one can either use a DC power supply with a current chopper or use an AC power supply with a rectifier. However, in case an AC power supply and a rectifier is used, either the top half or the bottom half of the sine wave should be used. However, the outputs corresponding to top half and bottom half portions of the sine waves can be fed separately into different capacitor banks, which can then be used in sequence to energize the electromagnet's coil, so that the magnetic poles stay the same when energized by different capacitor banks.

Near-Earth Probe Flyby Anomaly Explained
(Unexpected speed variations)

Over the last few decades, as different probes were directed to execute near earth flybys, to gain speed while changing direction towards their desired destinations, an anomaly was noticed. Starting with 'Galileo' spacecraft (flybys made during 1990 and 1992), and later on 'NEAR' (1998), 'Cassini' (1999), 'Messenger' (2005) and 'Rosetta' (2005, 2007 and 2009) some unexplained variations were detected in their respective speeds, as they had left earth behind.

The variations detected in the speeds of such probes were quite small in their magnitudes, the maximum being that of 'NEAR' at 13.48 mm/s. However, such variations were simply not expected, since all possible affecting forces were already accounted for. Also, the amounts of variations in the speeds of various probes were different. It should also be mentioned that, each flyby was executed at a different angle with respect to earth's equator and at a different altitude from earth's center of mass.

In this section, it will be shown that, once the effects of aether flow associated with earth's magnetic field are taken into account, not only such a variation in the speed of any probe performing a flyby routine around earth, or any other celestial body which possesses a magnetic field, becomes explicable but also their occurrences become quite expected and their magnitudes can even be predicted, in advance.

Any probe that performs a sling-shot maneuver around planet earth not only passes through the aether flow that is giving rise to earth's gravity, but also passes through a secondary aether flow which is associated with earth's magnetic field. Hence, as the probe follows a sling shot trajectory around earth, it gets dragged by this secondary aether flow, also.

Such a secondary aether flow is mainly from the Magnetic North Pole towards the Magnetic South Pole. However, there is also a minor aether flow associated with earth's magnetic field which is towards magnetic / geographic east, due to its rotation with earth, as a whole. Therefore,

"Any probe that executes a near earth flyby routine, regardless of its trajectory, experiences variations in its speed as well as its direction of motion. However, in most cases, depending on its specific trajectory, it may experience more of a change in its speed or more severe alteration in its direction of motion."

The effect will be most pronounced on the speed of the probe if the inclination angle of its trajectory, with respect to earth's magnetic equator, is close to either 90 or 270 degrees. In this case, while the speed of the probe will be affected towards the magnetic South Pole, its trajectory will be pushed towards the magnetic/geographic east.

Correspondingly, the effect will be most pronounced on the final direction of motion or heading of the probe (towards magnetic South Pole) if the inclination angle of its trajectory, with respect to earth's magnetic equator, is close to either 0 or 180 degrees. In this case, while the trajectory of the probe will be affected towards the magnetic South Pole, its speed will be affected towards magnetic/geographic east.

In other words, it shows the aether flow that is associated with the magnetic field of earth, as it exits the surface at the magnetic North Pole, flows towards the South and enters the surface at the magnetic South Pole.

Therefore, the effects of such an aether flow must be taken into account in all flyby trajectory computations, particularly for those probes that pass fairly close to planets such as earth with a fairly strong magnetic field.

The severity of the drag force exerted by such an aether flow on a given probe and the subsequent effects on the amount of speed gained or lost by that probe, during its flyby encounter with earth, depends on the exact trajectory of that probe relative to earth.

The following figure shows an overall view of the portion of earth's magnetic field that is fairly close to its surface.

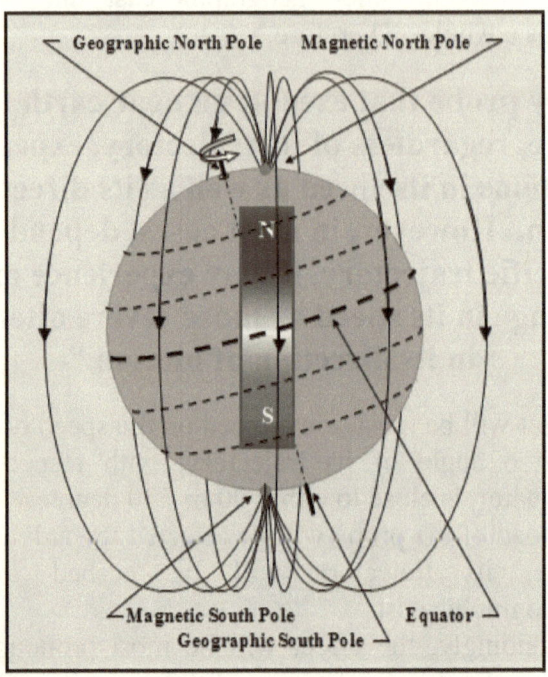

To properly analyze and quantify the expected effects of earth's magnetic field on both speed and trajectory of a given probe, during a near earth flyby routine, it is necessary to know the following information regarding that probe's motion as it approaches earth:

1- The initial speed of the probe along with its heading relative to earth's center of mass

These values will help determine the expected range of altitudes and corresponding speeds, the expected deflection trajectory due to earth's gravity, and hence the expected length of time that the probe will be within earth's magnetic field of varying intensities.

2- Probe's expected nearest distance to earth

The lowest altitude that the probe is expected to reach will help determine the maximum strength of the magnetic field to which the probe will be exposed.

3- The date that such a close encounter is expected to occur

The date of the flyby determines the general orientation of earth's rotational axis with respect to the tangent to its orbital path around the sun. This orientation varies by (23.45 × 2 = 46.90 degrees) during a full earth year.

4- The exact time that such a close encounter is expected to occur

The exact time of the flyby will narrow down the exact angle between earth's magnetic equator and the probe's trajectory, as it swings by.

5- The expected duration of such a close encounter

This piece of information allows calculation of the accumulated effect of earth's magnetic field on probe's speed, as well as its direction of motion, over the period probe is within range.

6- The longitude and latitude of the location on earth where the probe is expected to be the closest to earth's center of mass

This information will assist in pinpointing the location along its trajectory where the probe will encounter the maximum influence from such a drag force.

7- The angle between the trajectory of the probe and earth's equator

This angle will specify probe's passage route, the angle it will make with earth magnetic equator. Hence, the expected variations in its speed and its course can be readily calculated, as it will be expected to pass through certain zones of earth's magnetic field with known magnetic strengths.

8- The strength of the magnetic field present along the path which the probe is expected to follow

This information will allow proper calculation of the magnitude as well as the direction of the drag force that will be exerted on the probe, during its flyby encounter with earth.

It must be emphasized that, the aether flow that is associated with earth's magnetic field has a specific direction which is from the Magnetic North Pole towards the Magnetic South Pole. Therefore,

"As a probe executes a flyby routine around earth, not only its speed will be affected, but also its direction of motion can and will be affected, by the drag force of earth magnetic field."

As the trajectory of a given probe happens to be closely aligned with the direction of this aether drag force, the probe will receive either a push forward (hence gain speed) or a pull backward (hence lose speed), depending on its direction of motion being towards the Southern Magnetic Pole or towards the Northern Magnetic Pole, respectively.

Note that, as the trajectory of a given probe happens to be closer to the perpendicular to the direction of this aether drag force, the probe will receive more of a sideways alteration in its course, as compared to either a gain or a loss in its speed. Because, **The only in-line motion of aether in this case will be associated with the rotation of earth's magnetic field with earth, as a whole.**

Therefore, the course change will be either to the right or to the left, depending on the probe's direction of motion being towards the magnetic/geographic east or towards the magnetic west, respectively.

The exact amount of gain or loss in the speed of a given probe, as well as the severity of its course alteration, during any flyby experience, will depend on the above mentioned details / specifics of the probe's close encounter scenario with a celestial body that possesses a fairly strong magnetic field.

Note that,

Satellites that are already in earth orbit at various altitudes can readily assist in not only confirming the information presented in this section, but also gather precise information on the strength of earth's magnetic field at different altitudes.

Such a comprehensive information can be used to generate a 3-D map of earth's magnetic field strength so that more precise near earth flyby scenarios can be planned and executed, in the future.

It would be quite beneficial to perform separate experiments that would specifically confirm the inline and the cross-flow effects of such an aether flow on probes executing near earth flyby trajectories. Such experiments will allow the proper examination and hence verification of such effects in total isolation from each other. The following two experiments are designed to perform such specific tasks.

- First experiment:
  (The in-line effect of aether flow manifesting as earth's magnetic field on the speed of probes executing near-earth flyby routines)

As shown in the following figure, two probes are needed to perform this experiment.

One probe should be launched into space only to come back and perform a near earth flyby routine, just outside its atmospheric layer, in such a way that its trajectory would take it along the North-South Magnetic Polar route. In this case, the probe will experience a force dragging it in the forward direction, leading to the maximum gain in its speed and minimum deflection of its course trajectory, both at that particular altitude from earth's center of mass (earth's magnetic core, to be more precise).

The only cross flow of aether in such a flyby scenario will be associated with the rotation of earth's magnetic field with earth, as a whole. Therefore, <u>a minimal deflection in the course trajectory towards</u> **left** <u>(towards magnetic/geographic east) should be detected.</u>

The very same probe can be guided to repeat such a pass at different altitudes, allowing the collection of very crucial data regarding the effectiveness of earth's magnetic field at different distances from its center of mass (earth's magnetic core, to be more precise).

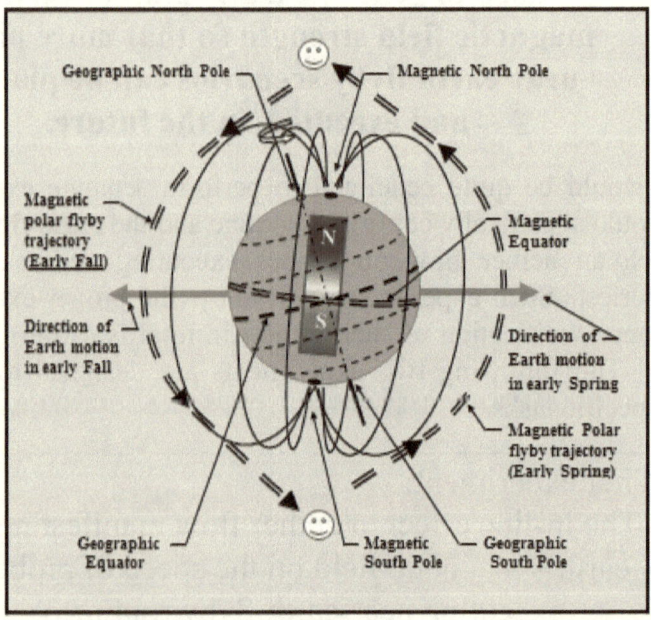

Another probe can repeat the very same experiment but in the opposite direction. The drag force experienced by such a probe will also be the strongest but in the backward direction, leading to the maximum loss of speed at the altitude from earth's center of mass (earth's magnetic core, to be more precise) its trajectory takes it.

Correspondingly, the only cross flow of aether in such a flyby scenario will be associated with the rotation of earth's magnetic field with earth, as a whole. Therefore, <u>a minimal deflection in the course trajectory towards</u> **right** <u>(towards magnetic/geographic east) should be detected.</u>

The very same probe can be guided to repeat such a pass at different altitudes, allowing the collection of very crucial data regarding the effectiveness of earth's magnetic field at different distances from its center of mass (earth's magnetic core, to be more precise).

> • Second experiment:
> (The sideways effect of aether flow manifesting as earth's magnetic field on the direction of motion of probes executing near-earth flyby routines)

As shown in the following figure, two probes are needed to perform this experiment.

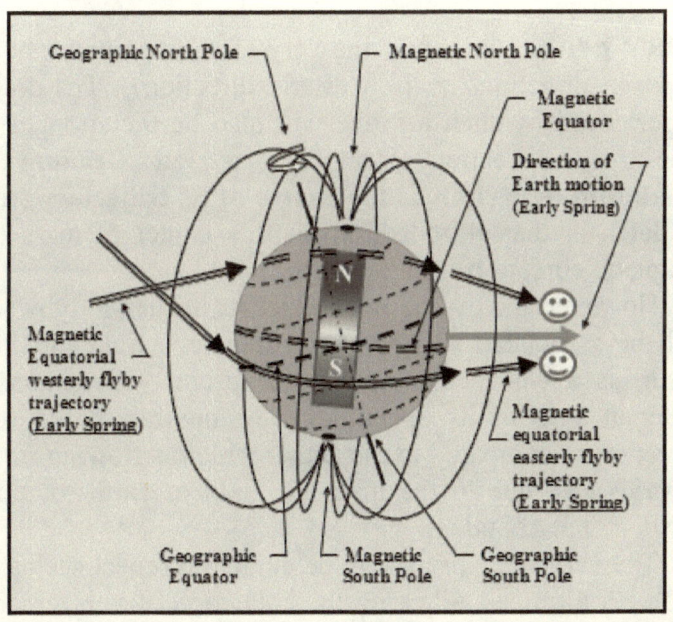

One probe can be launched into space only to come back and perform a near earth flyby routine, just outside its atmospheric layer, in such a way that its trajectory would take it along the magnetic equator, in the easterly direction.

The drag force experienced by such a probe will be the strongest in the side-to-side direction (from left to right), **leading to the maximum deflection of the course of its** trajectory (towards its right), at that altitude from earth's center of mass (earth's magnetic core, to be more precise).

However, the in-line flow of aether in such a flyby scenario will be associated with the local magnetic field rotating with earth, as a whole. Therefore, the probe will demonstrate a minimal gain in its speed, due to moving in the very same direction as the aether which is flowing from west towards east due to the magnetic field of earth rotating with earth, at that altitude.

The very same probe can be guided to repeat such a flyby routine at different altitudes, allowing the collection of very crucial data regarding the effectiveness of earth's magnetic field at different distances from its center of mass (earth's magnetic core, to be more precise).

Another probe can repeat the very same experiment but in the opposite direction (in the westerly direction). The drag force experienced by such a probe will also be the strongest in the side-to-side direction (from right to left), **leading to the maximum deflection of the course of its trajectory (towards its left),** at that altitude from earth's center of mass (earth's magnetic core, to be more precise).

However, the in-line flow of aether in such a flyby scenario will be associated with the local magnetic field rotating with earth, as a whole. Therefore, the probe will demonstrate a minimal loss in its speed, due to moving in the opposite direction as compared to the aether which is flowing from west towards east due to the magnetic field of earth rotating with earth, at that altitude.

The very same probe can be guided to repeat such a pass at different altitudes, allowing the collection of very crucial data regarding the effectiveness of earth's magnetic field at different distances from its center of mass (earth's magnetic core, to be more precise).

Using Earth's Magnetic Field to Maneuver Satellites

Depending on its orbital path, any satellite in orbit around a planet such as earth which has a magnetic field can use that field as its power source to perform certain maneuvers such as temporary orbital trajectory changes. It can even use the local magnetic field to slow down and eventually plunge towards the surface of the planet and/or get burned up as it hits the atmosphere of that planet.

A variety of experiments can be designed to test this proposal. One version of such an experiment is described below.

- Experiment:
  (Causing a satellite to alter its altitude or even fall)

This experiment is intended to demonstrate the feasibility of maneuvering a satellite by using the induced electrical current in a specifically designed major component of the satellite, due to crossing the magnetic field that is around earth. The extracted electrical power can be applied to generate heat or perform certain needed tasks onboard, or even be stored in batteries. The following figure shows the general details of such a setup.

As shown in the figure, a sail like component can be deployed to extract electrical power from the local magnetic field. Even though the above figure shows a sail that is deployed behind the satellite and its orientation is vertical, it can also be in the horizontal position, or at any angle in between, so long as the electrical conductors imbedded in the sail are horizontal and perpendicular to the direction satellite is moving. In fact, the electrical conductors can be organized in a non-magnetic container, so that many layers can be fitted next to each other, just like the wire windings inside an electric motor.

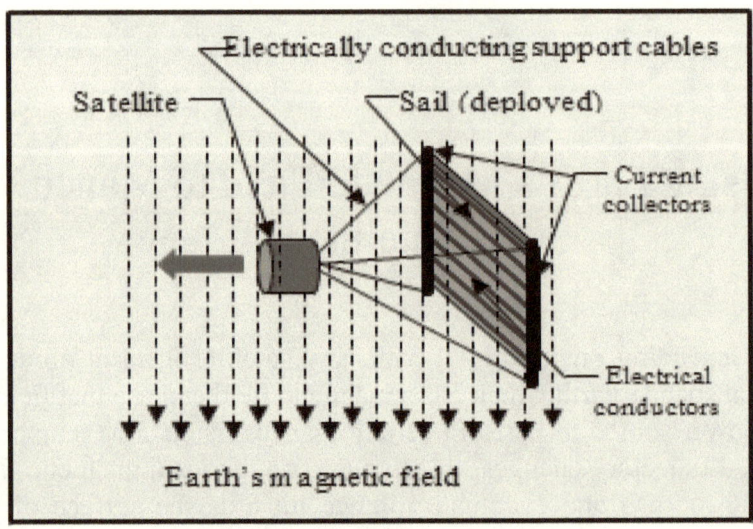

The orbital speed of the satellite will gradually decrease as electrical power is extracted from earth's local magnetic field. The action taken is basically like operating an electric generator, but using earth's local magnetic field.

To maximize the draw from the conductors, hence maximize the braking power exerted on the satellite, the ends of the support cables which are used to transfer the induced electrical power from the sail to the satellite can be shorted together, through a switch.

Note that, such a setup can even be used to save on maneuvering thruster propellants, since it can readily reduce the altitude of the satellite by any desired amount. Also, by storing the electrical power generated, the reverse of the maneuver back to a higher altitude can be performed, using an ion engine. The stored electrical power can even be fed into the sail itself which would then act just like the rotor of an electric motor and promote motion of the satellite as a whole towards a certain direction, even gaining altitude.

The overall size of the sail can be designed so that it can perform the needed maneuvers at the desired rate.

As needed, either by closing one switch the induced electrical power from the sail can be stored, or by throwing another switch and shorting the two main electrical leads, the generated electrical power can be pushed to its limit, as stronger braking / faster altitude reduction rate is desired. However, shorting the circuitry would require that the overall system should have sufficient radiative capacity to dump the generated heat into space to prevent any potential damage due to excessive increase in the temperature of the components involved.

Measuring the Propagation Speed of the Magnetic Field of a Magnet in space and Through a Matter Medium

To measure the speed at which magnetic field of a given magnet acts on an object such as a compass, the magnet and the compass must be moving relative to each other. However, they cannot be moving passed each other, along two parallel paths. Since, as shown below, each of the magnetic field lines that are offset from the magnet's main axis follows a different route from the North Pole to the South Pole.

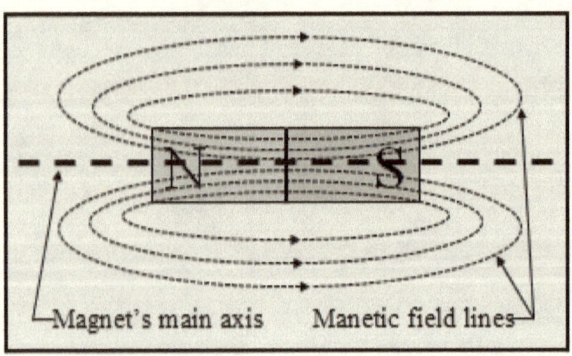

Magnet's main axis   Manetic field lines

The only consistent positions which can be used to perform such an experiment are located along the magnet's main axis. Therefore, the compass must be continuously positioned, at a fixed distance, along the main axis of the magnet, and not offset from it. In other words,

The magnet has to move relative to the compass, but its main axis must always be pointing towards the stationary position of the compass.

Therefore, the magnet must follow a circular path centered at the compass's pivot point, as shown below. The essential components include:

1- A strong electromagnet.

2- An electric power supply (Direct Current, DC) with a wide range of output voltages.

3- A circular track on which the magnet can move. The electrical power is supplied to the electromagnet through the track.

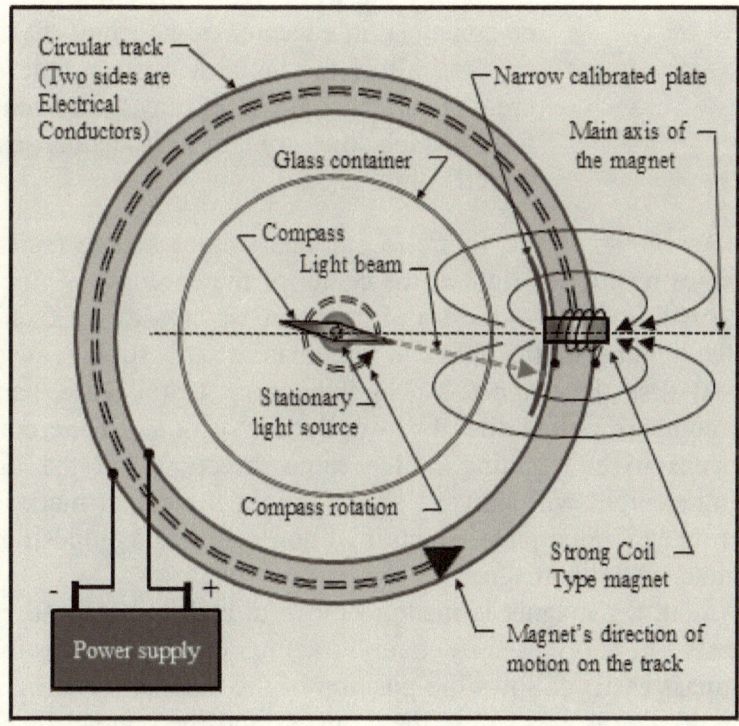

4- A compass with an opening at its tip (only in the direction that is pointing towards the Pole of the electromagnet which is facing it).

5- A light source, such as a laser, positioned independent of the compass, but right at compass's pivot point. The direction it points towards can be set to any desired angle (along the 360 degrees).

6- A narrow calibrated non-magnetic plate, with (0,0) mark positioned right at the main axis of the magnet. This plate is mounted directly on the magnet and will be moving with it.

7- A glass container in which the compass and the light source are housed. Once assembled, the air inside this glass container must be vacuumed out so that the compass can rotate freely, without facing any resistance due to the air molecules.

    Note that, preferably the whole experimental setup should be contained in a vacuumed housing. That way, any and all of the unknown effects due to the presence of the glass, or any other material, from which the container is made, can be directly eliminated.

As shown in the figure, a stationary light source (such as a laser) is positioned right at the center of the compass. Only when the small opening at the tip of the compass crosses the fixed path of the light beam, the light can illuminate what is straight ahead.

At first, the magnet is held stationary. In this case, naturally the compass points directly towards its position. If the compass happens to be pointing in the same direction that the light is pointing, light will shine at and highlight the center mark on the narrow calibrated plate which is of non-magnetic composition and is attached to the magnet.

Next, the magnet is made to move at a slow but fixed rate of speed. In this case, by freely rotating on its pivot point, the compass easily follows the position of the magnet. The light will shine through, every time the compass and the magnet line up in the same direction as the light beam.

In other words, the light beam used and the opening at the tip of the compass work as a union and generate flashes of light only towards a certain preset position, along the track. Together, they act just like a strobe light used in visualizing the timing at which cylinders fire in a gasoline engine relative to the Top-Dead-Center mark on the crankshaft pulley.

As long as the magnet moves relatively slowly and the compass can keep up, the light flash will be reflected off of the calibrated plate, and it will highlight the exact position of the magnet's main axis.

However, as the magnet is encouraged to move faster and faster and stabilize its speed, the light passing through the opening of the compass will indicate that the compass is falling behind with respect to the magnet.

The rotational speed of the magnet around the compass and therefore the rotational speed of the compass around its pivot point are expected to be quite high before noticeable deviations in the timing of the light beam and the main axis of the magnet can be observed. This is why it is necessary that at least the compass along with the light source must be housed inside a non-magnetic vacuumed container.

The orbital speed of the magnet around the compass must be increased in increments and it must be stabilized at a fixed rate of speed so that the compass can catch up and stabilize in its amount of lag time, the best that it can. Then, the amount of deviation observed on the calibrated plate must be recorded. Such procedures must be repeated for several speeds.

It should become readily evident that, as the speed of the magnet on the track is increased the amount of deviation highlighted on the calibrated plate by the light beam will be more pronounced. This is due to the fact that, as the compass responds to the flow of aether which is initiated by presence of the magnet, in its vicinity, the magnet has moved away and so the light beam will highlight a position on the calibrated plate that is offset from the main axis of the magnet.

The very same overall procedures should also be repeated as the electromagnet is energized by different voltages.

During such experiments, what is not readily expected but will be observed is that,

"For any given orbital speed of the magnet around the compass, as the strength of the electromagnet is increased, the response time of the compass and hence the amount of deviation highlighted on the calibrated plate will decrease. This is due to the fact that, the stronger magnetic fields encourage faster motions of the local aether which in turn shorten the delays in the timing of the compass adjusting its direction towards the ever-changing position of the magnet."

Once the amount of delay time is recorded for any given magnetic field strength, at a given orbital speed of the magnet, the speed of the aether flow which is manifesting as the magnetic field can be calculated. Since, the spot on the calibrated plate is highlighted by a light beam, the propagation speed of which is known in that medium. Also, the other needed pieces of information such as the orbital speed of the magnet and its magnetic strength at the pivot point of the compass are known, as well.

The amount of delay highlighted on the calibrated plate (when the orbital speed of the magnet is known) indicates how long it has taken for the magnetic effect of certain strength to be sensed by the compass.

By tabulating such findings, one can readily draw charts that would clearly demonstrate the relation between the orbital speed of the magnet, its strength and the amount of delay time expected for the magnetic field to be noticed at a certain distance.

- Measuring the propagation speed of magnetic field through a matter medium

The same procedures as mentioned earlier can be followed to measure the speed at which magnetic field travels through a given matter medium. But, **only non-magnetic materials can be used.**

As shown below, a doughnut shaped section made of the matter medium of interest can be installed in the gap between the moving magnet and the glass container housing the central compass.

By reading the amount of deviation in the direction pointed by the laser light, and comparing it with when the matter medium is not used, the speed at which magnetic field propagates through the matter medium of interest can be readily calculated.

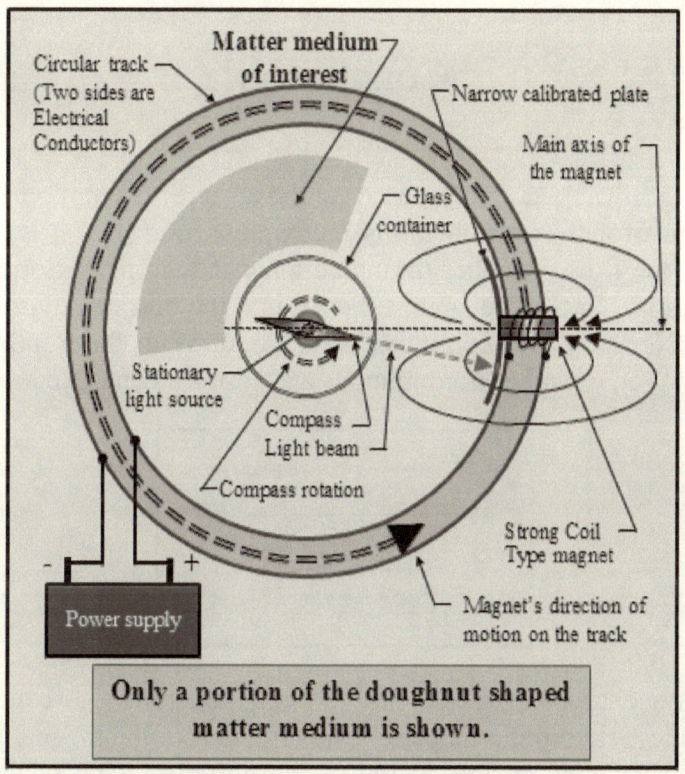

Only a portion of the doughnut shaped matter medium is shown.

Note that, if the matter medium of interest can be formed into a cylindrical shape it can be readily moved up and down, so that it can be inserted in and taken out of the path of the magnetic field lines. Hence, the experiment can be readily repeated for any given rotational speed of the magnet and also any given magnetic strength, without stopping and starting, or waiting for the chamber to be re-vacuumed.

Conclusion

Magnetic field is proposed to be a type of motion in the medium of aether which forms a complete loop. As it is shown below, the flow of aether through a magnet is from its South Pole towards its North Pole, and as it exits into the outside environment at the North Pole it continues until it returns to the South Pole. This motion of aether is continuous and forms a complete loop.

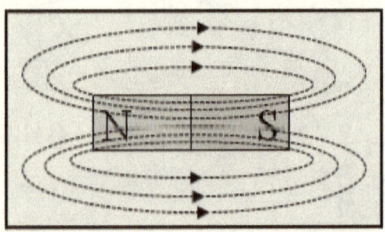

Any charged particle in motion generates a concentric magnetic field around its path. The direction of the magnetic field induced depends on the charge of the particle, being positive or negative. In other words,

**By their very motion in the aether medium,
charged particles encourage the local aether to
be stirred in one direction or another, depending
on their charges being positive or negative.**

By demonstrating the effects of magnetic field on a beam of light, as shown in both cases of cross flow as well as direct in-line flow, as well as the pace at which 'Time' is experienced, or the rate at which falling objects accelerate towards earth, it is shown that magnetic field is in fact a form of motion induced in the local aether. This type of motion in aether forms a complete loop, just like the air flow that is generated by a regular household fan. Such a flow generated in the local aether medium can be quite small (as in the case of magnetic fields generated by small regular magnets) or quite large in scale (as in the case of magnetic fields generated by individual planets and stars).

It is also shown that, it is the aether flow associated with earth's magnetic field that gives rise to such mysterious phenomenon known as **'magnetic hill'** or **'gravity hill'** effects. It is proposed that, the sideways flow of aether that is associated with earth's local magnetic field literally deflects the path of aether which is flowing vertically downward towards earth's center of mass, the same accelerated flow of aether that is giving rise to earth's force of gravity.

As it is proposed in these pages, magnetic field also affects non-magnetic materials and even neutral matter particles such as neutrons. Such effects can be likened to the way gravity affects each and every object and matter particle within its reach.

Note that, the stronger interactions manifested between magnetic fields of magnets and magnetic materials that happen to be nearby are due to aether flow patterns generated by them both. The following figure shows samples of such scenarios in a simple fashion.

- If the two magnets complement each other's induced aether flow, they will attract each other and hence generate a stronger overall aether flow. Such cases happen as opposite poles of two magnets are brought close to each other.

    Even as a magnetic (but not magnetized) material is brought close to a magnet, first it becomes magnetized by the magnet in such a way that the part closest to (and possibly touching) the magnet becomes a continuation to that magnet, as a whole.

    In both cases, they attract each other, since the induced aether flow by both objects are in the same direction. Therefore aether simply follows a longer path.

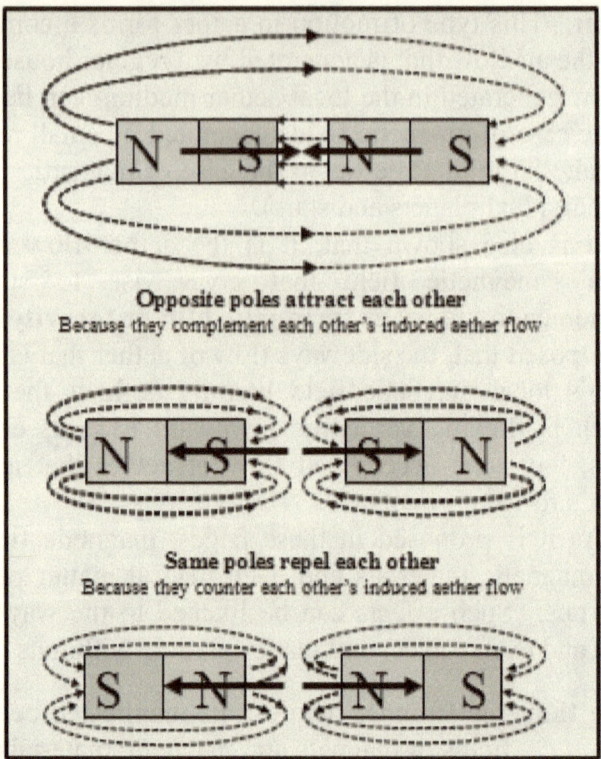

Opposite poles attract each other
Because they complement each other's induced aether flow

Same poles repel each other
Because they counter each other's induced aether flow

- If their induced aether flows counteract each other, they will repel one another. Since, the two locally induced aether flows will act just like two streams of water coming out of two hoses which are placed face-to-face.

  Also, if the same poles of two magnets are forced to touch each other, their overall strength will be reduced. The fractional decrease in their overall strength will be equivalent to the fractional decrease in the induced aether flow that was associated with the surface areas on both magnets that are jointly covered. Since, the two magnets will partially block each other from inducing a flow in the local aether medium.

Any probe experiencing a close encounter with a planet such as earth not only passes through the aether flow that is giving rise to that planet's gravity, but also passes through a secondary aether flow which manifests as that planet's magnetic field. Hence, as the probe follows a sling shot trajectory around such a planet, for example earth, it is expected to get literally dragged by this secondary aether flow, as well.

Once the effects of aether flow manifesting as earth's magnetic field are taken into account not only the variations in the speed of probes performing near-earth flyby routines become explicable, but also their occurrences become quite expected. In fact, the magnitude of the effect of earth's magnetic field on the motion of such probes can be predicted in advance.

It must be emphasized that, the aether flow that is associated with earth's magnetic field has a specific direction. The portion that is outside of earth, as a whole, is from the Northern Magnetic Pole towards the Southern Magnetic Pole. Therefore,

"As a probe executes a near-earth flyby routine, not only its speed will be affected, but also its direction of motion will be affected, as well."

The 'Time' anomaly predicted to be experienced by probes in polar orbits around earth is also explained to be due to the very same effect, as those experienced by probes executing near-earth flyby routines.

Also, it was proposed in the chapter titled "What is Light?" that, the efficiency of a LASER equipment can be improved upon, simply by exposing the equipment to an external magnetic field which is oriented in a specific direction with respect to the expected propagation direction of the amplified light waves.

A few of the experiments presented in this chapter demonstrate how the very presence of a magnetic field affects the rate at which the passage of 'Time' is experienced. In other words, such experiments provide the needed proof that,

"Presence of magnetic field gives rise to 'Time dilatation'."

Such an effect of magnetic field on 'Time' is a clear indication that magnetic field must be a type of motion induced in the aether medium, since it is giving rise to a relative motion between the object of interest and aether that is in its immediate vicinity.

It is also proposed that, artificially induced aether flow, using a strong pulsating magnetic field, can be used to generate the type of propulsive force which would be of great demand for spaceships.

The flow speed of aether manifesting as magnetic field through any medium, including any type of matter medium, can be calculated using a setup such as the one shown in the last experiment in this chapter. In short, according to the information presented in this chapter:

**"Magnetic field is due to a type of aether flow
which forms a complete loop."**

Note that, it is proposed in this chapter that, magnetic field is a form of motion induced in the medium of aether. Therefore, its strength is dependent on the density of the local aether. Hence,

**"As aether's density is gradually
decreasing in this universe, the strength
of the magnetic field induced by the
motion of any kind of charged particle is
correspondingly weakening with time."**

What
is
Electric Field?

Introduction

In 1865, Mr. Maxwell proposed his theory on electromagnetism which was based on the existence of a medium through which light and other electromagnetic waves were assumed to propagate. At that time, this medium was referred to as aether. However, what aether was made of and what kinds of properties it had were not known. Aether was accepted as a needed medium through which electromagnetic waves propagate. To electromagnetic waves the stationary aether medium was thought to be just as a medium such as air is to sound waves.

However, contrary to what was accepted back then, aether is not a stationary medium. The varieties of motions that exist in that medium can be likened to the motions of air molecules in the atmosphere or the motions of water molecules in the oceans. The varieties of motions in the aether medium give rise to and manifest as such phenomena as gravity, magnetic field and electric field, among many others.

Electric field and magnetic field are very closely related to each other. The most important common point between these two fields is that, both owe their existences to the existence of charged particles in this universe. Since, the very existence of a charged particle at any location induces an electric field in its surrounding environment. Also, if a charged particle happens to pass by a given location, it induces a magnetic field at that particular location. In other words,

"The historical existence of electric field and magnetic field goes back to when the first charged particles were formed, in this universe."

Electric field is one of many kinds of flow patterns that can exist in the aether medium. Electric field is stated to be an aether flow of some sort, because it exhibits a preferred direction which is away from the negatively charged particles (as if aether is pushed away due to being under higher pressure near such particles) and flows towards the nearest positively charged particles (as if there is a pressure gradient encouraging aether to do so), as shown below.

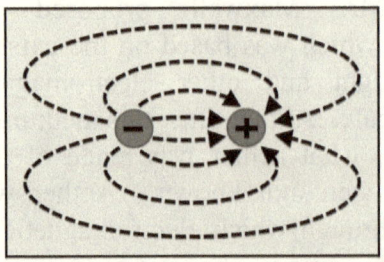

Such an aether flow pattern can be readily likened to the flow pattern of a **medium** such as air away from the front side of a household fan in operation, and towards its back side, as shown below.

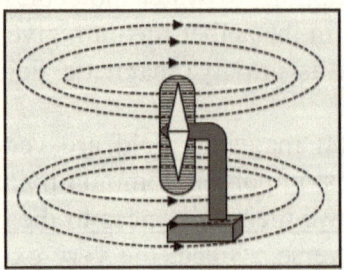

In other words, charged particles act as if they are the cutoff faces of a hose through which aether is flowing either out of or into, depending on the charged particle being a negatively charged particle or a positively charged particle, respectively. To better visualize this hose analogy one can consider a hose which is used to connect the outlet of an air pump directly to its inlet, as shown below.

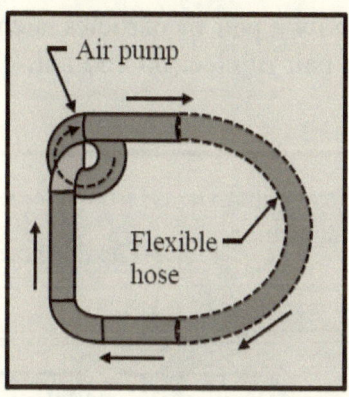

If the hose is cut off by a saw, at a certain point along its length, while the air pump is operating, as shown in the following figure, even though the two cutoff faces of the hose will be of exact same size and shape, they will act opposite of each other in one very important and distinctive way. Since, as soon as the hose is cut off, the cutoff face of the section of the hose that is connected to the inlet of the air pump will be sucking air from its surrounding, while the cutoff face of the section of the hose that is connected to the outlet of the air pump will be blowing air into its surrounding. The two cutoff faces of the hose will basically act like a positively charged particle and a negatively charged particle, respectively.

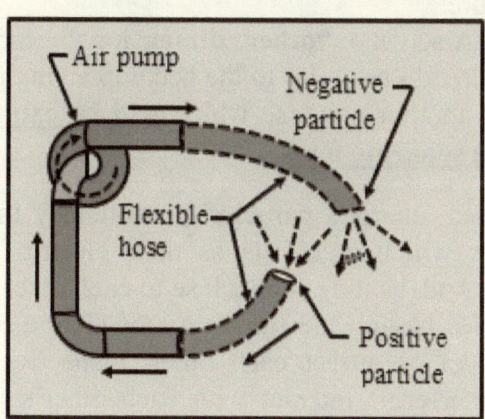

In fact, if the plane of the saw blade is considered to be this universe and the two cutoff faces of the hose were allowed to move in different directions but stay in their initial common plane, as shown in the following figure, the above mentioned scenario describes precisely how a pair of particles such as a pair of proton and anti-proton or a pair of electron and anti-electron (positron) is produced in nature.

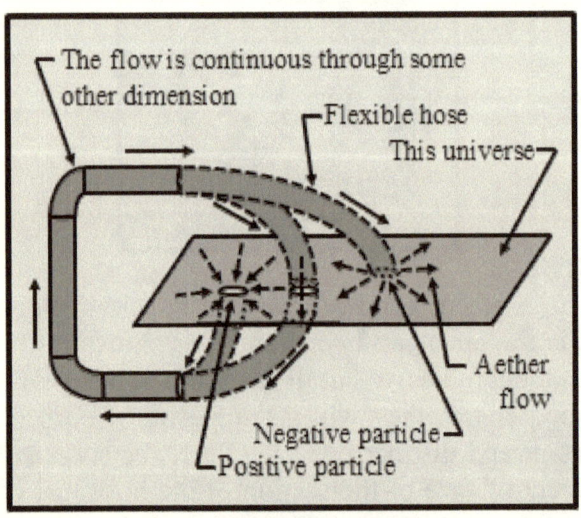

Note that, the two-dimensional cutoff faces of the hose in the above example resemble the three-dimensional bubbles / voids which are the charged particles in this universe which is shown as a two-dimensional plane in the figure.

Also, the **"other dimensions"**, indicated in the figure, do not refer to the accompanying universe but to yet another universe which **will be explored in detail in a separate book.**

Once formed, charged particles continuously induce a steady flow of aether which manifests as their electric fields in their surroundings. And, as they come close to each other, depending on the two particles having the same type of charge or the opposite types, they repel or attract each other, respectively, since they mandatorily interfere / interact with each other's induced aether flow patterns.

The induced aether flow by any given charged particle is at a steady pace. Therefore, the rate at which aether is flowing towards a positively charged particle (or away from a negatively charged particle) is inversely proportional to the square of the distance between the location of interest and the charged particle.

If a charged particle happens to be stationary at a given location, its induced aether flow at any other location in its surrounding will be at a fixed rate. However, if the charged particle happens to be passing by a given location its induced aether flow rate at that particular location will gradually increase as it approaches that location and then will gradually decrease as it goes away from that location. Also, if a particle approaches a relatively stationary charged particle and then recedes from its location, that particle will experience an electric field which becomes gradually stronger during its approach and then becomes gradually weaker as it goes away from that charged particle's location.

Note that, it is proposed in this chapter that, electric field is a form of motion induced in the medium of aether. Therefore, its strength is dependent on the density of the local aether. Hence, as aether's density is gradually decreasing in this universe, the strength of the electric field induced by the presence of any kind of charged particle is correspondingly weakening with time.

The most effective way of demonstrating the fact that electric field is a type of aether flow is to experimentally show its various effects. The following experiments are specifically designed to explore different effects of a certain type of aether flow which manifests as electric field.

Effects of Electric Field on the Passage of 'Time'

A variety of experiments can be designed and conducted that will demonstrate how the flow of the local aether relative to objects (or living beings), manifesting as the local electric field, affects the rate at which they experience the passage of time. Details of two experiments are provided in the chapter titled "What is Time?".

- First experiment: (Using electric field to induce 'Time Dilation')

- Second experiment: (Electric field powered time dilation machine)

These experiments are proposed to confirm the validity of the aether-based theories of 'Electric field' and 'Time'.

Effects of Electric Field
on the Speed of Light
and its Direction of Propagation

Electric field is basically one type of aether flow which starts from the negatively charged particles and ends at positively charged particles. The effects of electric field on the propagation speed of light and its direction of propagation are described in great detail in the chapter titled "What is Light?".

The following two experiments are also described in detail in that chapter, experiments which are specifically designed to demonstrate the effect of electric field on the speed of light or any other electromagnetic wave and the direction of their propagation, respectively.

- <u>First experiment:</u> (Effect of electric field on the speed of light)

- <u>Second experiment:</u> (Effect of electric field on the direction of light's propagation)

These experiments are proposed to confirm the validity of the aether-based theories of 'Electric field' and 'Light'.

Effect of Electric Field on the Strength of the Force of Gravity

---
- **Experiment:**
  (Effect of electric field on Mr. Galileo's experiment)

Suppose two electrically conducting plates (each having a slightly oval shaped hole) are positioned nearly horizontally, one above the other. Their common tilt angle should be just enough so that their holes would line up vertically above each other. A glass tube can be passed through their holes and be secured in a fixed frame, as shown below.

Two plugs or caps are used to close to two ends of the glass tube. The top cap should be equipped with three specific features:

- **A release mechanism** so that different small objects can be released to pursue a free fall journey downward towards earth.

- **A tube connected to a vacuum pump** to allow the air inside the glass tube to be vacuumed out to nearly eliminate air resistance.

- **An air intake valve** to allow the tube to be refilled with air so that the lower cap can be removed to retrieve the sample after it has fallen. This feature will also allow the top cap to be removed and reloaded with a new sample object.

The two plates should be connected to an electrical power supply capable of automatically delivering DC electricity at gradually increasing voltages, over a preselected time period, from zero voltage to variably adjustable very high voltages. A trigger mechanism such as a laser beam and a light sensor connected to a relay is needed to activate the variable power supply to the plates, right after the instant each of the objects pass through the hole in the top plate. Also, a vacuum system is required to generate a nearly vacuum condition inside the tube. Such a setup will allow the famous Galileo's experiments to be repeated, but under different conditions.

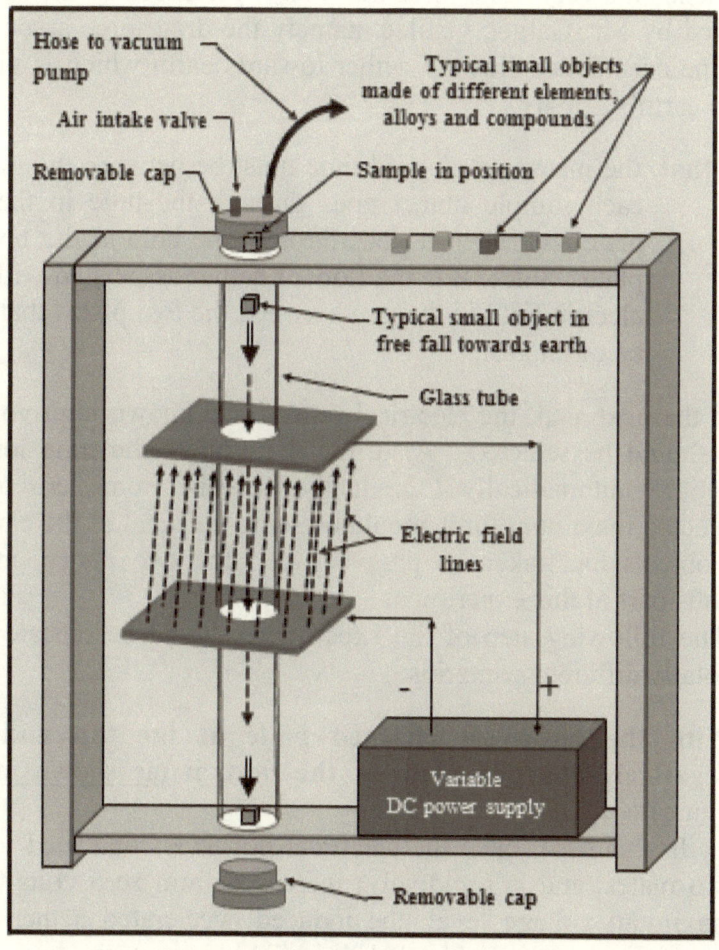

At first, before connecting the power supply to the plates, one by one a few of the small non-magnetic, not-electrically charged objects should be placed inside the top cap. Then, after the air inside the tube is vacuumed out, they should be released to experience their respective free falls towards earth. The amount of elapsed time as they pass through the two plate holes must be measured and recorded. It will be shown that, in the absence of air resistance (and, without the presence of an electrical field), all objects regardless of their densities and/or compositions will experience free fall towards earth at an identical rate of acceleration.

During this part of the experiment, the small objects used will experience the very same force that was experienced by the objects dropped by Mr. Galileo Galilee, namely the drag force associated with the accelerated flow of aether towards earth which is giving rise to earth's gravity.

Note that, the measured elapsed time must be between the instant each sample object goes through the hole in the top plate and when it goes through the hole in the bottom plate. Since, it is the flow of aether associated with the electric field in the gap between the two plates that will be of interest.

In the next step, the electrical power of a known high voltage level should be selected. Also, the duration of the time for the voltage to automatically / gradually increase from zero to its preselected maximum limit should be set to be equal to the elapsed time objects had taken to pass between the two plates, in the previous part of this experiment.

The following step of the experiment should be repeated for two totally different scenarios:

1- **With the positively charged plate at the top and the negatively charged plate at the bottom** (as shown in the figure)

In this case, once the electrical power is connected to the two plates, and is (gradually) increased from zero volts to its maximum voltage level, the induced accelerated aether flow manifesting as the induced electric field between the two plates will be upwards. Therefore, it will partially counteract the downward accelerated flow of aether which is giving rise to earth's force of gravity. Hence, regardless of their compositions, as sample objects are one by one dropped from the top cap, all of them will accelerate towards earth at a **slower pace**, as compared to when the two plates were not electrically charged. The amount of elapsed time as each object passes through the gap in between the two plates should be measured and recorded.

This procedure can be repeated as the two plates are charged with different maximum voltage settings. The higher maximum voltage settings on the power supply will encourage the local aether between the two plates to accelerate upwards at higher rates, as every time the voltage level is raised from zero to its higher maximum voltage levels, within the fixed time period, measured during the first part of this experiment.

Note that, if a sufficiently high voltage setting is selected as maximum on the power supply, and it is applied to the plates, the induced upwardly accelerating aether flow will cancel out the downwardly accelerating aether flow which is responsible for inducing the drag force called gravity. Hence, **the dropped sample objects momentarily will not experience any acceleration towards earth. And, if they were positioned in between the two plates, while hanging from a string attached to some sort of elastic acting as a scale, their weight would temporarily drop to zero, as the elastic would temporarily contract, indicating that they are experiencing weightlessness / levitation.**

On the surface of earth, the rate at which aether is accelerating towards earth is about 9.8 meters per second squared. Therefore,

"Varying the electrical voltage delivered to the two plates at a rate that corresponds to an acceleration rate of the local aether of about 9.8 meter per second squared in the upward direction, will encourage any sample object that may be

hanging from a string to momentarily experience a state of weightlessness, as it will float inside the tube between the two plates."

2- **With the negatively charged plate at the top and the positively charged plate at the bottom** (opposite of what is shown in the figure)

In this case, once the electrical power is connected to the two plates, and is (gradually) increased to its maximum level, the induced accelerated aether flow manifesting as the induced electric field between the two plates will be downwards. Therefore, it will complement the downward accelerated flow of aether which is giving rise to earth's force of gravity. Hence, regardless of their compositions, as sample objects are one by one dropped from the top cap, all of them will accelerate towards earth at a **faster pace**, as compared to when the two plates were not electrically charged. The amount of elapsed time as each object passes through the gap in between the two plates should be measured and recorded.

This procedure can be repeated as the two plates are charged with different maximum voltage settings. The higher maximum voltage settings on the power supply will encourage the local aether between the two plates to accelerate downwards at higher rates, as every time the voltage level is raised from zero to its higher maximum voltage levels, within the fixed time period, measured during the first part of this experiment.

If the sample objects were positioned in between the two plates, while hanging from a string attached to some sort of elastic acting as a scale, their weight would temporarily increase every time power is connected to the plates, as the elastic would temporarily stretch, indicating that they are experiencing having excess weight.

The results obtained during either parts of such an experiment will indicate that, wherever the acceleration of the induced flow in the local aether (manifesting as the electric field) is comparable to the acceleration of the local downward aether flow (manifesting as earth's force of gravity), it can directly affect the strength of the local force of gravity.

Note that, if the induced accelerated aether flow (manifesting as the electric field) and the accelerated aether flow associated with the force of gravity cross each other's paths, the local force of gravity will act at an angle other than directly perpendicular to the horizon (towards the center of mass of earth). The severity of such a deviation or the magnitude of such a tilt angle will depend on:

- The relative magnitude of the two induced accelerations in the local aether, giving rise to the electric field and the force of gravity, and

- The relative angle between these two induced aether flows.

It should be emphasized that, if the electrical power of a constant voltage be connected to the two plates, during the whole period each of the samples are released by the top cap and when they are stopped by the bottom cap, the induced flow of aether manifesting as the **constant electric field would not have any effect, whatsoever, on the rate at which those sample objects will accelerate towards earth. Since,**

"Aether flow generates a drag force only when it is accelerating."

Electric Field Propulsion System

The main components in this type of propulsion system are two electrically conducting plates and a high voltage DC electricity power supply. The two plates are positioned so that one is facing towards the front end of the unit and the other facing towards its back. The power supply can be positioned anywhere in between the two plates. A general setup of such a propulsion system is shown in the following figure.

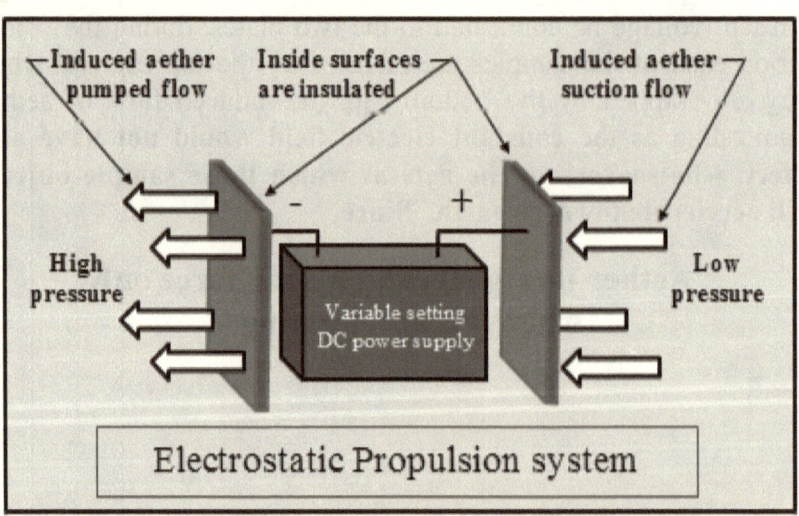

The overall setup used in such a propulsion system can be likened to a water pump which is part of a piping system and the whole network, including its extended inlet and outlet pipes, is submerged in water. In such a case, the length of the piping used between the inlet and pump, as well as between the pump and the outlet can be of a variety of sizes and lengths. The inflow of water into the inlet will exert a suction force (low pressure) and the outflow of water from the outlet will exert a repelling force (high pressure) on their respective surrounding water mediums.

Such a propulsion mechanism generates its propulsive force by absorbing aether (generating a low pressure region) towards its front and by releasing aether (generating a high pressure region) towards its rear. The following figure shows the general shape of a UFO which can be benefiting from such a propulsion mechanism.

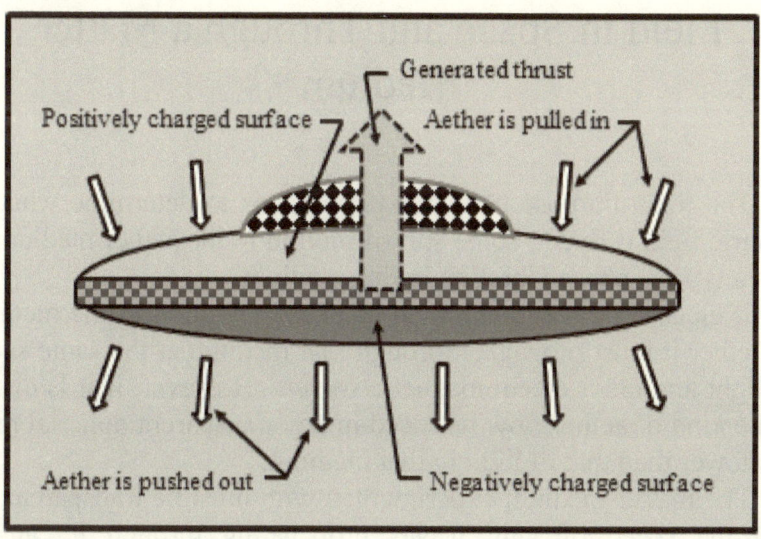

Note that, in such applications, just by controlling the voltage delivered to different segments of the upper and lower surfaces, the craft can be tilted and propelled towards any desired direction.

The overall performance of such a propulsion system depends on the voltage used.

Such a device can be made to just counteract the force of gravity and hence generate stationary levitation.

Measuring the Propagation Speed of Electric Field in Space and Through a Matter Medium

The main purpose of this experiment is to determine whether electric field is due to some sort of motion in the aether medium or it is a type of phase vibration in that medium.

If electric field is some kind of phase vibration in the medium of aether it must propagate through that medium at the same speed as light and other electromagnetic waves. However, if it is due to some kind of aether flow its speed in any transparent material must be slower than that of light in that medium.

The matter medium experimented with must be transparent, so that the speed of light waves propagating through it can be measured, simultaneously. In other words, any solid, liquid or gaseous material which satisfies both conditions, namely being a conductor of electric field and being transparent to light waves, can be used for this experiment. The following figure shows the experimental setup used.

This experimental setup consists of a test segment of a specific length. If need be this segment can be replaced by a canister which can be filled with a liquid or a gaseous material.

Once the electrical switch is closed, the current will simultaneously reach the plate and the light source. If electric field is a form of phase vibration in the medium of aether it will reach the other end of the matter medium at the same time as the light waves generated by a light source and the two light bulbs will turn on simultaneously. However, if electric field is some kind of motion induced in the aether medium, in between the two plates, it will have to have a slower speed as compared to that of light, and the difference should be demonstrated by the two light bulbs turning on at different times.

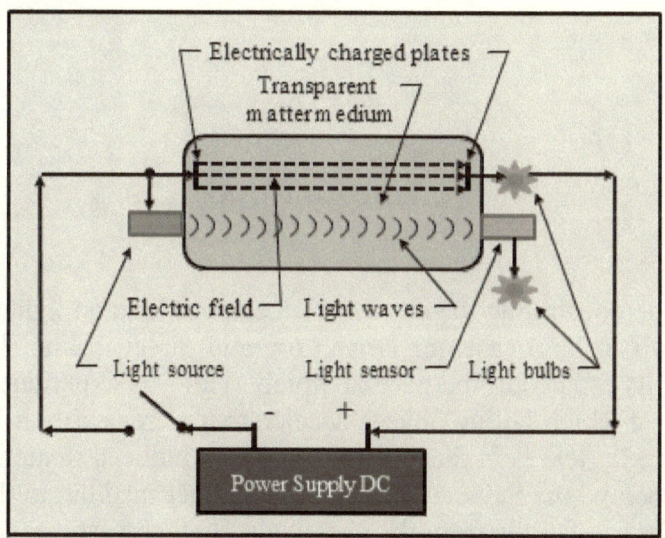

A high speed camera can be used to record the timing of the two light bulbs turning on.

Note that, if electric field is proven to be a form of motion induced in the local aether medium, by repeating the same experiment with different voltage settings delivered to the plates one can show that the speed at which aether is encouraged to flow is dependent on the strength of the electric field generated.

The very same experiment can be conducted after the canister has been vacuumed out. This way the speed of the induced aether flow in free space can be measured. Again, such an experiment will readily prove that electric field is either another type of phase vibration in the medium of aether or an induced flow in that medium.

Conclusions

By demonstrating the effects of an electric field on light waves, as shown in both cases of cross flow and direct in-line flow, as well as its effects on the pace at which 'Time' is experienced and the rate at which falling objects accelerate towards earth, it is clear that electric field is in fact a form of motion induced in aether. As shown below, the induced motion in the aether medium in the case of an electric field is from the negatively charged particles towards the positively charged particles.

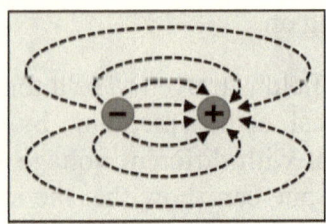

Such an aether flow pattern can be readily likened to the flow pattern of a medium such as air away from the front side of a household fan in operation, and towards its back side, as shown below.

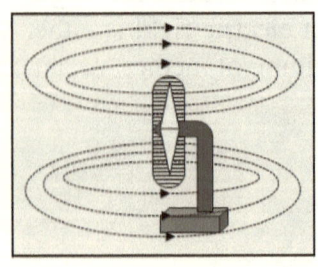

It is also proposed that, electric field affects neutral matter particles such as neutrons, as well. Such effects can be likened to the way gravity affects each and every object and all matter particles within its reach.

Note that, the stronger interactions manifested between electric fields and charged particles can be likened to the way magnets act differently on magnetically inclined materials, since in both cases the induced effects are due to the aether flow patterns generated by them both.

Two oppositely charged plates complement each other's induced aether flow directions, hence they attract each other. However, if their induced aether flows counteract each other, they will repel one another, since the two locally induced aether flows will act just like two streams of water coming out of two hoses which are placed face-to-face.

The following figure shows samples of such scenarios in a simple fashion.

Also, it was proposed in the chapter titled "What is Light?" that, the efficiency of a LASER equipment can be improved upon, simply by exposing the equipment to an external electric field which is oriented in a specific direction with respect to the expected propagation direction of the amplified light waves.

One of the experiments presented in this chapter demonstrates how the very presence of an electric field affects the rate at which the passage of 'Time' is experienced. In other words, such an experiment provides the needed proof that,

"Presence of electric field gives rise to 'Time dilatation'."

Such an effect of electric field on 'Time' is a clear indication that electric field must be a type of motion in the aether medium, since it is giving rise to a relative motion between the object of interest and aether that is in its immediate vicinity.

The flow speed of aether manifesting as electric field in any medium, including any type of matter medium, can be calculated, as shown in this chapter.

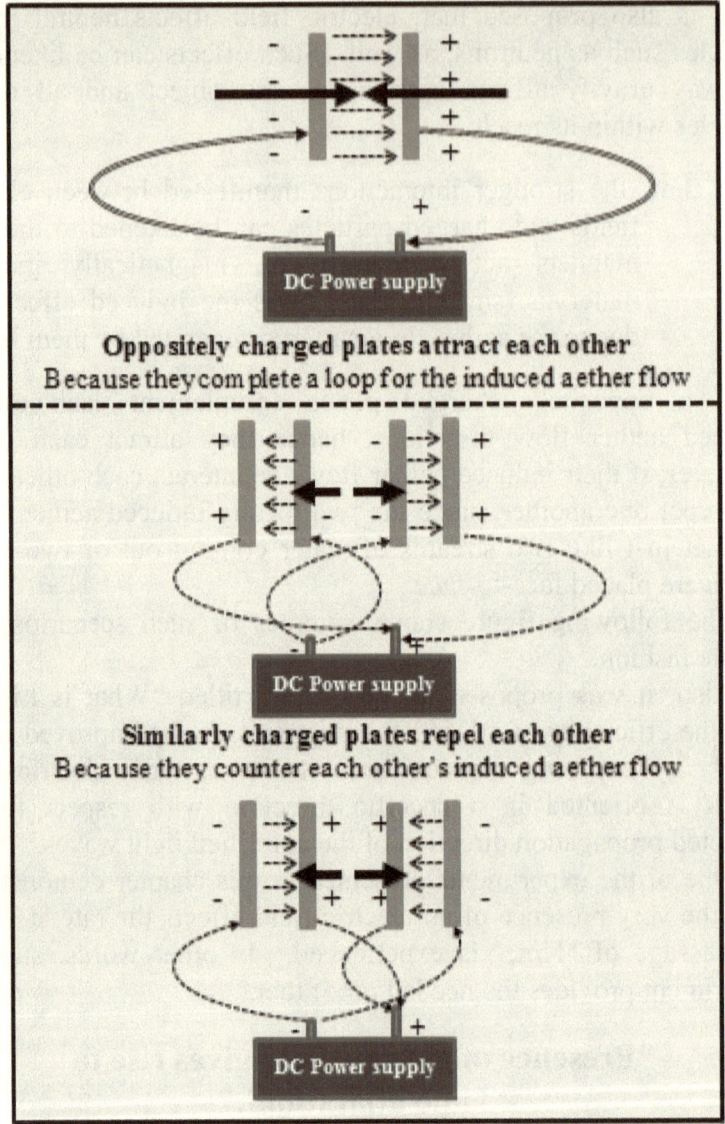

In short, all of the experiments detailed in this chapter point to one common result, and that is,

"Electric field is due to the motion of aether, a type of one way motion which is due to the existence of charged particles."

Note that, it is proposed in this chapter that, electric field is a form of motion induced in the medium of aether. Therefore, its strength is dependent on the density of the local aether. Hence,

> **"As aether's density is gradually decreasing in this universe, the strength of the electric field induced by the presence of any kind of charged particle is correspondingly weakening with time."**

What
is
Electricity?

Introduction

There are two general types of electricity, namely static electricity and current (flowing) electricity. Both types are made of positively and negatively charged components. As the two components interact with each other they give rise to different effects in both types of electricity.

- **Static Electricity**

    Static electricity is due to either the buildup of excess number of electrons or the deficiency in the number of electrons in an object. Such an imbalance in the number of negatively charged electrons in one object as compared to another usually arises as different electrically non-conducting types of material come in direct contact with (or are rubbed against) each other.

    Objects with more electrons than usual are referred to as being "negatively charged", while objects with fewer electrons than usual are referred to as being "positively charged".

    The excess charges are relatively concentrated on and near the surfaces of objects, because like charges repel each other. Due to the very same reason, electrons also tend to be more focused in sections of the object that are relatively sharper in their geometries, such as sharp edges and particularly pointy tips. Higher electron concentrations in sharp or pointy parts of objects encourage the electrons to jump off unto another object which is starving for electrons and is close enough.

    This is why the sharper the very top end of a lightning rod (lightning arrester) is the more effective it becomes in performing its intended task. Since, a lightning rod is expected to act as a mediator allowing the exchange of the excess electrical charges that often build up between the ground and the air mass above it, in a safe and predictable manner.

When two oppositely charged objects come in direct contact, they have the tendency to share their charge imbalaces and reach a mutually neutral state. During such a process some of the electrons in the negatively charged object are literally pushed away, due to being repelled by the other electrons in that material medium, and unto the other object which is starving for electrons. As a result, when the physical contact is **nearly** established, a spark forms between the two oppositely charged objects. The size of the gap between the two objects across which a spark can form depends on the severity of their charge imbalance. The following figure shows the formation of such a spark between two oppositely charged objects, as they are pushed towards each other.

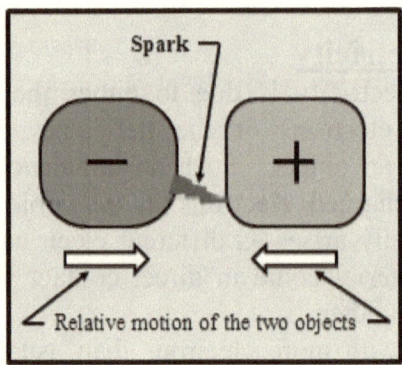

Once the two objects come into direct contact and stay connected they continue sharing their charge imbalance until they both possess the same charge, or totally neutralize each other's charges.

The very phenomenon of spark is due to phase vibrations induced in the local aether medium by electrons or other charged particles which are transferred from one region of space to another at a high rate of speed, particularly as they are transferred between objects with electrical charge imbalance. Such phase vibrations can cover a wide range of frequencies, since individual electrons (particles) move at different rates of speed and

follow quite a random path, as they are transferred between objects.

The length of a spark depends on the magnitude of charge imbalance between the two objects, as well as the presence of a medium in between which can possibly become ionized and allow electrons to literally cross the gap with multiple hops between its atoms.

Note that, sparks are electromagnetic waves in nature, whether they are in the visible range of the spectrum or not. Therefore, they can form in a pure aether medium (so-called vacuum), as well. However, in such cases, an observer can see such sparks only and only if he/she happens to be looking directly at their exact loactions, as they are formed.

Also, a spark formed in a pure aether medium will not have any kind of sound waves associated with it, since sound waves require a material medium to propagate through.

If a spark is formed within a liquid or a gaseous type of matter medium, the observer will become aware of the spark forming even if he/she does not look directly at the exact location where the spark occurs. This is due to the dispersion of the light waves by the matter content of the local medium. Such an effect is readily experienced by almost everyone, because at some point in time or another, they have noticed a flash of light in the air itself, as a lightning occurs at a distance.

Also, the very presence of objects nearby causes the light waves to become reflected towards the observer. That is how an observer can acknowledge the occurrence of a lightning just by receiving its reflected waves off of the nearby objects, such as buildings.

"It is very important to realize that, light waves themselves are not visible, when they are not directed at the observer's eyes."

As light waves get reflected by objects they make their presence known to the observer, since objects have the habit of reflecting light waves in literally all spatial directions. Even part of the light waves that hit the surface of any mirror, regardless of the incident angle, is reflected towards all spatial directions, other than those dictated by the incident angle.

To demonstrate the invisibility of light waves, one needs to shine his/her flashlight towards the sky and also shine it at nearby objects, both at night when the sky is moonless and dark and the air is clean. Objects will reflect the light waves and the existence of the light waves can be confirmed by the very visualization of those objects. Yet, the sky will not be iluminated since there is relatively speaking nothing there to reflect the light waves towards the observer. Hence, it seems like as if the flashlight is not even turned on when it is pointed towards the night sky.

Sparks generated due to the buildup of static electricity can be very small and weak in nature, such as the ones formed by walking on carpets and then touching a metal doorknob. They can also be quite enourmus in their sizes and strengths such as lightnings. Lightnings are formed as matter particles, by becoming ionized and hence acting as conductors, assist in providing a bridge or path for electrons to migrate between a negatively or a positively charged cloud mass and the ground. Such an exchange of electrical charge can also take place between two cloud masses which are oppositely charged.

The following figure shows diferent cases, as the water droplets in the atmosphere and a lightning rod or even a tall live tree assist in the formation of such a connection between highly charged clouds and ground, as well as between two oppositely charged clouds.

Thunder is due to the sudden expansion of the immediate air medium, as the local atoms / molecules along the length of the spark receive huge amounts of kinetic energy from the stampede of electrons. The sudden increase in their kinetic energies generates a compression wave-front in the surrounding air medium which in turn carries/spreads the generated waves (phase vibrations) in all directions. The duration of the sound or thunder heard by an observer at a distance depends on the distance between the observer and different parts of the lightning, from one end to the other end.

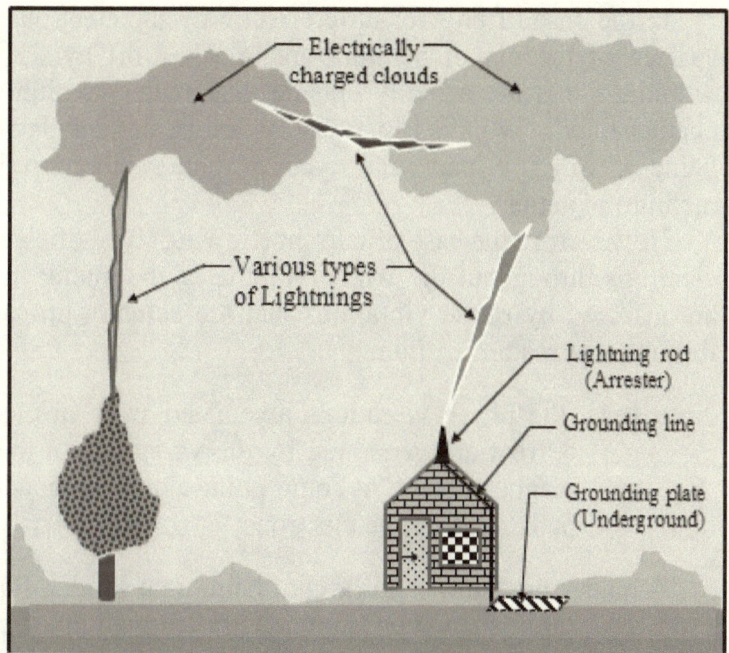

- **<u>Current Electricity</u>**

It is proposed in these pages that, **current type of electricity is due to the propagation of a certain type of phase vibrations which is induced in (or transferred to) the medium of aether within the volume of a matter medium.**

Such phase vibrations encourage the electrons to jump from one atom to another and hence form a current. The effect experienced by electrons in such a case is basically the same as photoelectric effect. In other words, <u>this newly proposed theory for the formation of current type of electricity</u> can be summarized as follows,

"It is the propagating phase vibrations (of a certain type) through the medium of aether which is superimposed by the conducting matter medium that encourage the electrons to let go of their atomic obligations and jump to other atoms downstream, hence form a current."

In the case of photoelectric effect only the electrons at the surface of the matter medium are knocked off by the phase vibrations that are actually outside that matter medium. The chapter titled, "What is Light?" provides greater details on light being a wave, a particular type of phase vibration in the medium of aether.

However, in the case of current (flowing) type of electricity electrons throughout the whole volume of the matter medium are affected by phase vibrations that are actually propagating through that matter medium.

Note that, the phase vibrations associated with an electrical current are generated by the variations in the local magnetic field, at some point along the conducting path taken by the electrons.

A matter medium, regardless of being in a solid, a liquid or even a gaseous state, that supports the formation of such a current of electrons is referred to as an **electrical conductor**. Correspondingly, a matter medium which does not support the formation of a current of electrons is referred to as an **electrical insulator.**

Furthermore, it is proposed that,

"The speed at which phase vibrations associated with an electrical current propagate in a given matter medium is the same as the speed of light in that particular medium. Since, both are phase vibrations in the medium of aether."

However, **the speed of individual electrons in a given conductor of electricity is only a very small fraction of the speed of phase vibrations in that medium.** This is due to the fact that, electrons are repeatedly captured and released by various atoms (nuclei) that make up the matter medium.

The propagation of phase vibrations associated with an electrical current, within a conductor, and the confinement of such waves to that conductor's boundaries can be readily likened to the confinement of light waves (which are also phase vibrations in the medium of aether) propagating through optical cables. In both cases, phase vibrations follow the contour of their respective conductive mediums by getting reflected, repeatedly, off of their surface boundaries.

As shown in the following figure, the inner transparent crystaline medium (inner core) in an optical cable and the transparent medium immediately surrounding it (cladding) have different refractive indexes.

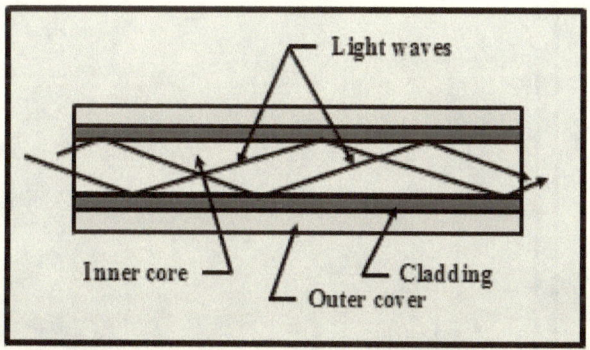

As light waves propagte through the inner medium, they get reflected repeatedly at the boundary between the two mediums. Hence, they stay confined to the core medium, over long distances.

The phase vibrations associated with electrical currents can also be likened to sound waves propagating through a solid material placed in a vacuum. Even if the wave carrying material may be bent into different shapes/directions, even at angles that are greater than 90 degrees, the sound waves will still manage to travel its full length. Once they reach the end of the conducting medium they will echo back and generate a humm inside that medium. But, they cannot propagate passed the end of the solid material, since the outside is a vacuum. However, if another material (soild, liquid or even gaseous) is brought in direct contact with the first material medium the sound waves can and will continue to propagate through that medium.

The very same applies to phase vibrations associated with an electrical current, since they cannot readily leave the conducting medium in which they are propagating. They can do so only as their current medium comes in direct contact with another conducting medium so that the transfer of phase vibrations becomes allowable.

The case of sound waves being transferred between two objects inside a vacuumed chamber is shown in the following figure.

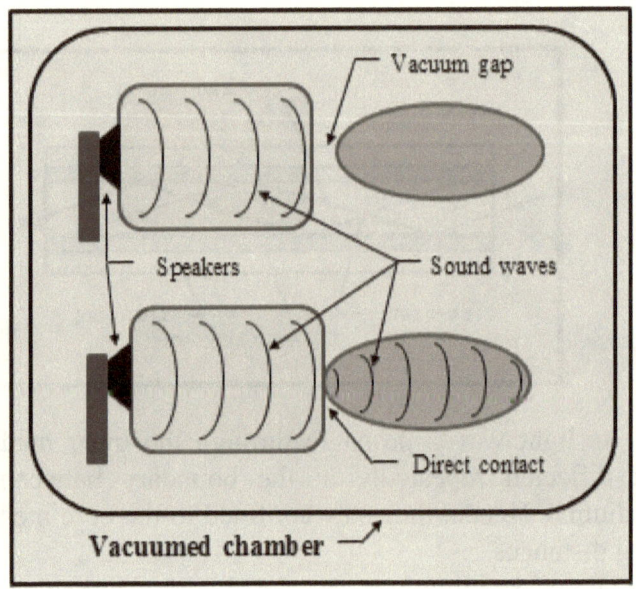

If there is a gap in between the two objects it will act as a barrier for the propagation of the sound waves. As it is shown, phase vibrations induced in one object can be transferred to the other object only and only if the two objects come in direct contact with each other.

This is what is taking place between two electrical conductors along a circuit, particularly within an electrical switch, since when the two conductors directly touch each other the phase vibrations induced within one gets transferred into the other. Otherwise, the gap in between acts as a barrier for the propagation of the phase vibrations which are associated with the electrical current.

Note that, if the phase vibrations associated with an electrical current are very intense, as they encounter a gap between two electrical conductors they saturate their immediate matter medium with the type of charge (positive or negative) that they are promoting. Hence, they build up a local static charge, at the end of that medium, as well as at any existing sharp edges or pointed sections.

If the amount of charge is sufficient for the size of the gap, the electrons (or ions, as it may be the case) will be made to take the leap across the gap unto the other conductor. Since, they will be literally pushed off board by the other electrons (or ions), due to having the same charge.

Of course, the jump (**spark**) will occur more readily if the gap is filled with some kind of solid, liquid or even gaseous material which is even slightly conductive.

As shown in the following figure, very sharp or pointy edges encourage phase vibrations to become more focussed / concentrated / compressed.

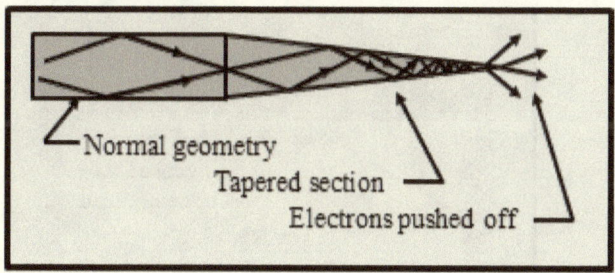

Therefore, as phase vibrations encounter the tip they are propagating nearly perpendicular to its surface and hence are encouraged to pass through and escape the conductor. This is why sparks first form at parts of the conductor that are sharper in their geometries.

This is also why the sharper the very top end of a lightning rod (lightning arrester) is the more effective it becomes in performing its intended task of exchanging its electrical charge with the oppositely charged surrounding medium.

The propagation of phase vibrations through an electrical conductor that has variable thicknesses at different sections of its length can be readily likened to the propagation of sound waves through a tunnel with different cross sectional areas along its length. In both cases, phase vibrations simply occupy the whole cross section (volume) of their respective mediums, as they propagate through them.

Just as sound waves propagate through any and all available passageways in a complex, interconnected tunnel system, the phase vibrations associated with electrical currents also divide into different branches, as multiple passageways become accessible to them. This is how the current of electricity branches out into any and all available continuations of its immediate conducting matter medium.

The following figure shows an electrical circuit where the sizes of the wires used are exaggerated just to demonstrate the passage of phase vibrations through the medium of the conductor.

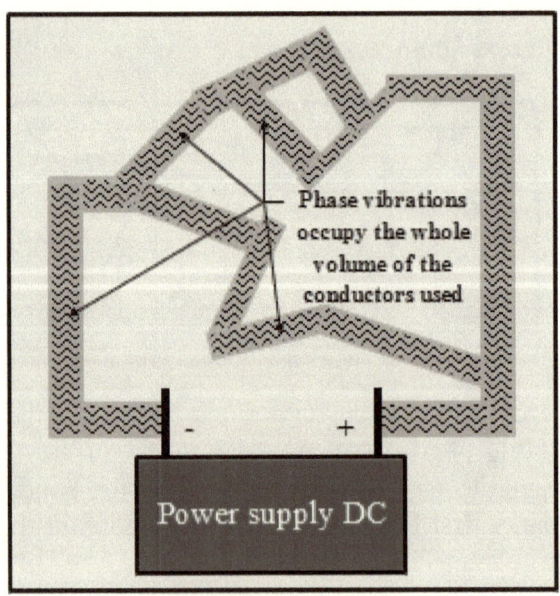

Phase vibrations occupy the whole volume of the conductors used

Power supply DC

Note that, just like in a wind tunnel's test section, as an electrical conductor's cross section is reduced the phase vibrations associated with the electrical flow in that conductor become more concentrated.

Hence, the individual electrons are encouraged to move along at a faster pace. However, <u>the propagation speed of phase vibrations remains unaffected by any changes in the cross sectional area of the conductor.</u>

It should be emphasized that, even though the phase vibrations associated with an electrical current propagate through a conductor spread in all directions and occupy the whole cross section of that conductor, the electrons jumping between atoms, due to having the same type of electrical charge, repel each other and get as far away from one another as they possibly can.

That is why electrons mainly travel closer to the surfaces of conductors rather than through their interiors. Of course, higher and higher intensities of the phase vibrations in a given matter medium will encourage the travelling electrons to hop between atoms that are deeper and deeper within the cross section of the conductor.

Note that, even at their lowest intensities, phase vibrations propagating along the length of a conductor keep on causing the electrons within the whole volume of that conductor to receive the very same treatment, as they are encouraged to let go of their immediate atomic obligations. Hence, there are always some electrons flowing through the whole interior portions of a conductor, as well. But, the relative magnitude of their current is very small and it is mainly directed towards the outer skin of the conductor, along which they will continue with their journey.

At this point, various terminology used in electrical circuitry can be defined.

Voltage is a measure of the relative intensity of the phase vibrations induced in the aether medium, phase vibrations which are confined within the volume of one or more conductors, between any two points along the length of an electrical circuit.

The intensity of phase vibrations (voltage) is weakened along the length of an electrical circuit, as phase vibrations are partly expended due to encouraging the local electrons to let go of their respective atomic commitments.

Amperage of an electrical circuit is a measure of the number of electrons that are successfully motivated to go through the cross section of the conductor per unit time.

Resistance of a conductor is an indication of how the flow of electrons is obstructed, as they pass through that conductor. In other words, resistance of a conductor is an indication of how effectively the atoms in that medium impede the flow of electrons which have been freed from their respective atomic bondages.

The resistance of a given conductor is increased either as its length is increased, or as its cross sectional area is decreased.

The intensity of the phase vibrations (voltage) is reduced particularly as the phase vibrations propagate through portions of the conducting circuit where either the cross section is noticeably reduced or the local electrons are more strongly attached to their corresponding nuclei.

Such portions can be and are in fact added to electrical circuits for a variety of reasons, including intentionally reducing the voltage delivered to the following components in the circuitry, or for the purpose of generating heat.

Since different parts of a given circuit can exhibit different resistances to the flow of electrons, meaning they allow different number of local electrons to join the parade downstream, the voltage drop measured across different points along the circuit will be different.

Note that, once the input voltage (the intensity of the phase vibrations) is specified, the overall magnitude of the resistance along a given electrical circuit dictates the amperage or electron flow density along the full length of that circuit.

If parts of the circuit happen to be in series, as shown below, each component reduces the intensity of the phase vibrations to some extent. The remaining intensity is what is delivered to the following components. The amount of reduction in the intensity of the phase vibrations passing through a particular component depends on the resistance of that component.

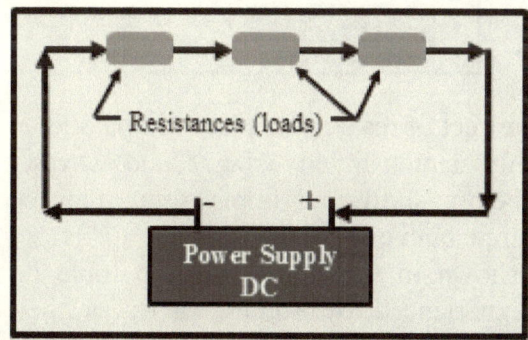

If parts of an electrical circuit happen to be in parallel to each other, as shown below, all of them will receive the very same initial phase vibration intensities (same initial voltages). Yet, due to having different resistances along their individual paths, each of them will allow different number density of electrons to pass through them.

As shown below, if a direct path with very low resistance is provided (by closing the switch provided), the majority of electrons will choose that route. Hence, the amperage through that path will be quite high, as compared to the amperages through the other paths.

The solid lines (arrows) in the figure indicate the routes taken by the electrons when the switch is open (off). The dashed lines indicate the routes followed by electrons after the switch is turned on.

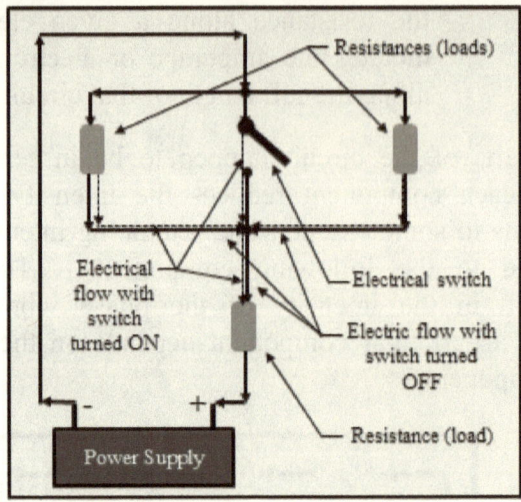

The effect of resistances placed in parallel and in series can be readily demonstrated using sound waves, as well. The figure below shows a simple setup that can be used to demonstrate both effects.

As shown in the figure, the top route demonstrates the effect experienced by sound waves as they are made to propagate through two narrow (restricted) passageways placed in series. Also, the top route as a whole can be considered as a route that is in parallel with the bottom route which has only one less severely restricted section. As shown, the sound waves reaching the end of the bottom route are much stronger as compared to those reaching the end of the top route, an effect which is expected, due to the difference in the resistances that those two paths impose onto the propagation of the sound waves.

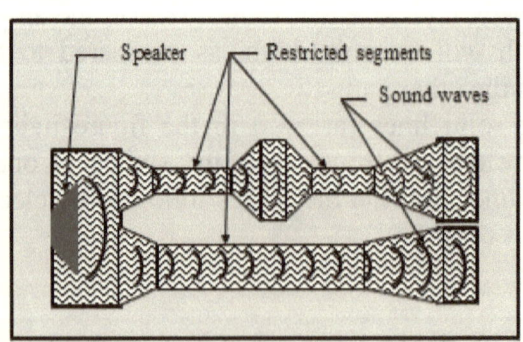

By manipulating the route taken by the exhaust gasses inside the muffler of vehicles, even though the gasses are still allowed to flow through, the sound waves are quite dampened to their desired levels. Such manipulation of routing of gasses can even alter the frequencies of sounds heard coming through the outlet, namely the tailpipe. The following figure shows a simple muffler design which benefits from such gas routing manipulation to dampen the engine noise. The internal segments act as obstacles (resistances) for the flow of exhaust gasses and by rerouting the flow they cause the sound waves to get trapped and blend in together, and nearly neutralize each other.

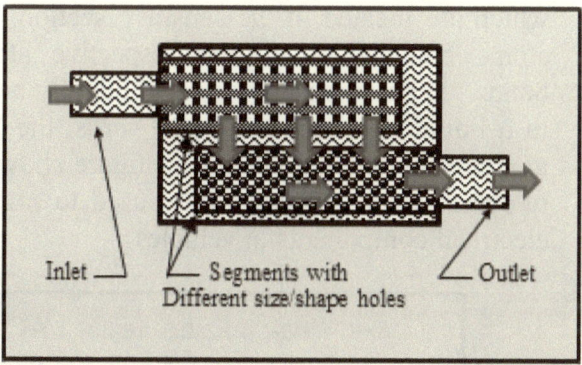

Inlet — Segments with Different size/shape holes — Outlet

The overloading of a conductor and its eventual meltdown is due to high intensity of phase vibrations propagating through the limited cross section of a given matter medium. Such high intensity phase vibrations encourage too many electrons to move through the limited available cross section. Therefore, the atoms which happen to be along the way are knocked on more frequently, as compared to the atoms that are located at other parts of the conductor which are larger in cross section.

Atoms of the conducting (matter) medium that are hit more frequently are encouraged to vibrate faster. The rate at which they vibrate is manifested as the temperature of that conductor, or the matter medium. Once the temperature of the matter medium reaches its melting point (when its atoms / molecules are energetic enough to break free from their bondages with their neighbors), the local matter medium has no choice but to melt at its hottest section. Hence, it forms a gap / barrier for the propagation of phase vibrations, as well as the flow of electrons. This is why certain parts of the circuit which are smaller in cross section heat up and eventually melt and disrupt the current of electricity.

Fuses are electrical components which are intentionally designed so that they include a narrower section along their respective conductive materials. The predesignated segments can withstand the passage of a certain amount of current (number of electrons per unit time).

As this limit is reached, excessive amount of momentum energy is transferred to the individual atoms which are located in the sensitive section. Hence, the atoms break free of their respective atomic lattice bonds. In so doing, they start to literally act like atoms in a liquid medium. In other words, the local matter medium melts. The following figure shows two types of fuses which are commonly used to protect various electrical components in vehicles.

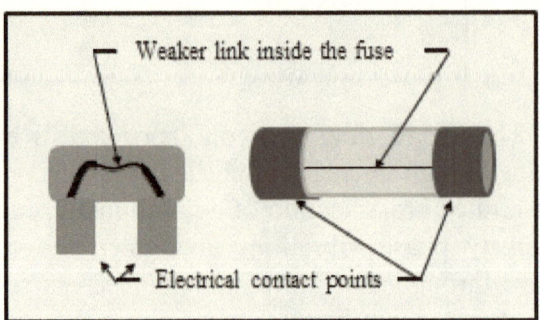

Of course, if the amount of increase in the current of electrons experienced happens to be exceedingly high, due to the excessive amount of kinetic energy received by the individual atoms over a very short period of time, the conductive material in the predesignated section can and will literally vaporize.

Superconductors are materials in which, below certain temperature level, the electrons are provided with an easy passage through. At such low temperatures, some of the electrons are no longer quite bound to specific nuclei. In other words, in a superconductor certain electrons are free to float, as if they were dust particles in the air. Just as a gentle breeze can move dust particles in any desired direction, the phase vibrations in the medium of aether can also readily encourage the movement of these free electrons in such materials.

The reason for the nuclei exhibiting such a behavior is that, below certain temperatures the atomic vibrations are reduced to a minimal amount and even the vibrations of the nuclei are quite limited in such states. As electrons encounter such a medium either they are allowed to pass through or they are not.

If electrons are allowed to pass through, it simply means that the atomic nuclei are organized in a crystalline lattice structure with limited randomness in their local motions. In other words, the nuclei are relatively fixed in their respective positions. In such a case, the free electrons follow paths which are open in between the rows of nuclei and can also be easily pushed along and guided by the phase vibrations. Such a medium is referred to as a superconductor.

The following figure shows the atomic structure of regular conductors, at room temperature. The top section shows the atoms in liquids, gasses and even some of the solids (as their temperatures are raised) in which atoms / molecules as a whole have certain amount of vibrational motions as well as translational motions associated with them. The bottom section shows the atomic structure of a solid conductor that is very organized in its lattice format. In this case, the atoms / molecules only possess local vibrational motions, but no translational motions.

The shaded areas around individual atoms indicate the spatial range of their vibrations. In both cases, the electrons are literally tossed between the nuclei, as they are pushed along by the phase vibrations.

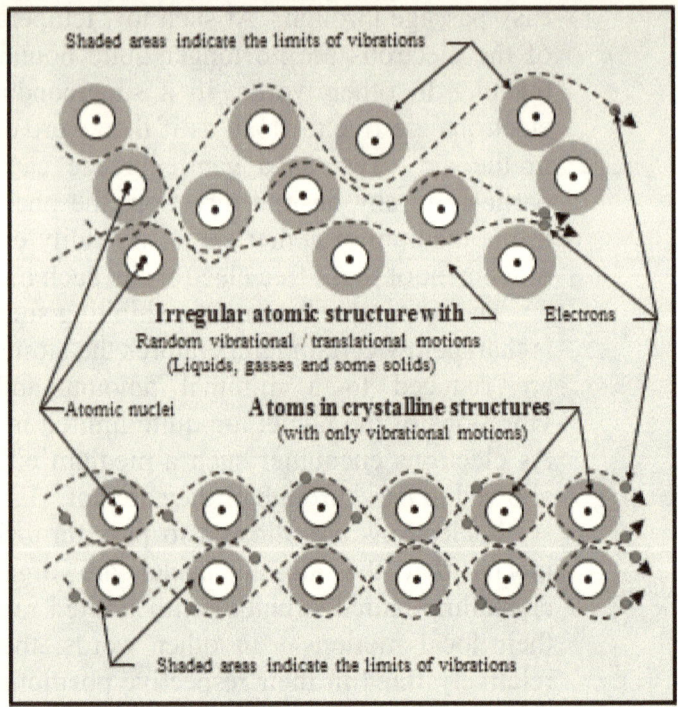

In the case of liquids and gasses, shown at the top, electrons follow fairly random paths. However, in the case of solids with atoms organized in crystalline structures, shown at the bottom, electrons follow a fairly well-defined path.

The following figure shows the atomic structure in a superconducting material in which atoms and hence the nuclei are nearly fixed in their positions, in a lattice structure. Therefore, electrons can freely pass through, in between the nuclei, without the nuclei even noticing their presence.

The speed of electrons moving through any given superconducting material is quite high, as compared to when they pass through a regular conductor. Since, in the case of a superconducting medium, electrons no longer follow a zigzag path, due to being tossed from one nucleus to another, but rather float in between the nuclei, following a fairly straight path.

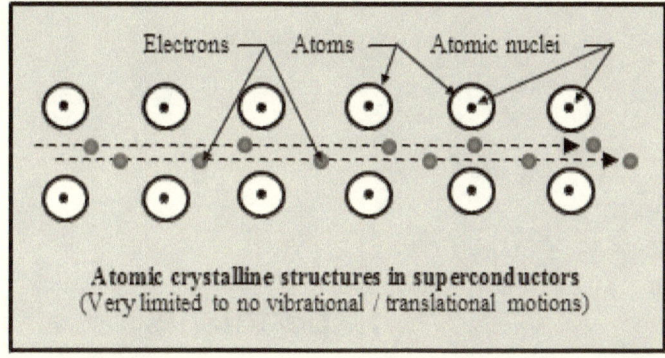

Atomic crystalline structures in superconductors
(Very limited to no vibrational / translational motions)

As a superconducting material is gradually warmed up, its atoms / molecules start to vibrate and hence the individual atomic nuclei cover wider spatial ranges. As they randomly move about their lattice positions, due to their electric charges, the nuclei automatically generate local, variable and independent magnetic fields. As electrons pass through such an environment, they get affected by these randomly oriented, local magnetic fields and therefore get pushed at right angles to their directions of motion. In other words, the nuclei basically generate resistance to the very motion of electrons by literally tossing them sideways. At such temperature levels, the material is no longer in its superconducting state.

As the temperature of the matter medium is raised even further, depending on how more randomly the atoms / molecules move or if they become semi organized at their local lattice positions, the overall resistance of that medium will increase or decrease, respectively.

Experiments

The following experiments will demonstrate that the propagation speed of phase vibrations associated with an electrical current in a given matter medium is:

1- Dependent on the matter medium used as the conductor.
2- Independent of that medium's thickness.
3- Equivalent to the speed of light waves through that matter medium.

1- Propagation speed of phase vibrations associated with electricity in any given conducting matter medium is independent of the size of that medium's cross sectional area or its shape

The speed of phase vibrations associated with electricity along a given conductor is independent of the thickness of that conductor. This fact can be readily demonstrated by using wires (conductors) which over a specified length, used as a test section, are only different in their thicknesses. Such a setup is shown in the figure below.

All of the wires / conductors, including test segments, used in this experiment are made of the very same material, and the test segments are different only in their thicknesses but not their lengths.

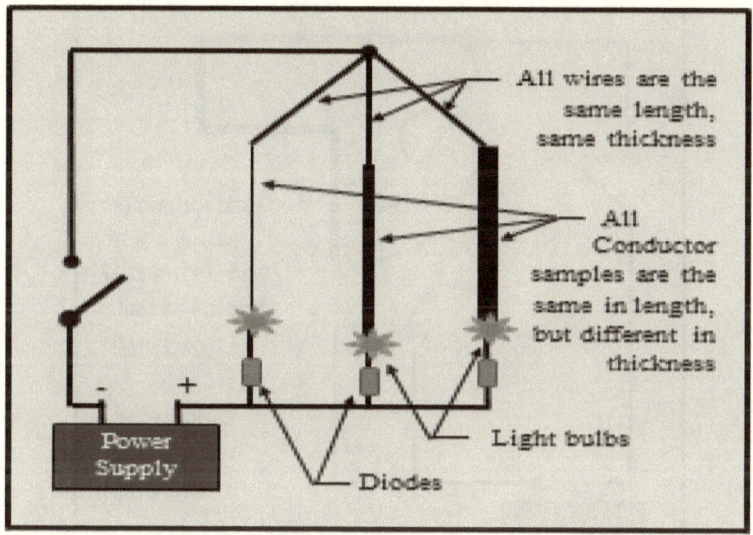

Even though each light bulb used in this setup receives its electrical current independently of the others, due to diodes used in all branches of the circuit, once the electrical switch is closed, all of the light bulbs will light up simultaneously. Their timing can be easily verified with the help of a high speed camera which can be used to record the event.

Note that, in the above figure it is assumed that electricity flows from the negative post of the battery towards its positive post. Just to clarify any doubts, the very same experiment can be repeated by switching the wires at the battery terminals.

The following figure shows a setup which can be used to demonstrate that the speed of phase vibrations associated with an electrical current in a conductor is independent of the shape of the path taken.

In this case, once the electrical switch is closed both light bulbs will light up at the very same instant, indicating that phase vibrations associated with electrical currents propagate (through a given matter medium) at the very same speed, regardless of the shape of the route taken.

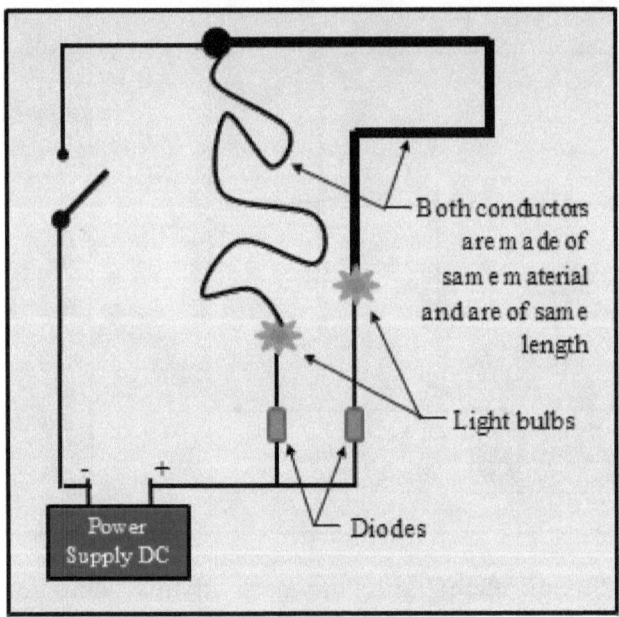

2- Measuring the propagation speed of phase vibrations associated with electricity in a given conducting matter medium

The propagation speed of phase vibrations associated with an electrical current along a given conducting material can be meassured by using wires (made of the conducting material of interest) which are only different in their lengths. Such a setup is shown in the following figure.

In this case, once the electrical switch is closed the light bulbs will turn on in sequence, starting with the one connected to the shortest wire used. Their timing can be easily verified by a high speed camera or another type of timing device that can record the event. Due to the presence of diodes in each branch of the circuit, each light bulb is independent of the other.

The difference in the timing of the two light bulbs turning on and the difference in the length of the conductors used in each branch will allow the calculation of the speed at which phase vibrations associated with electrical current propagate in the matter medium used as the conductor.

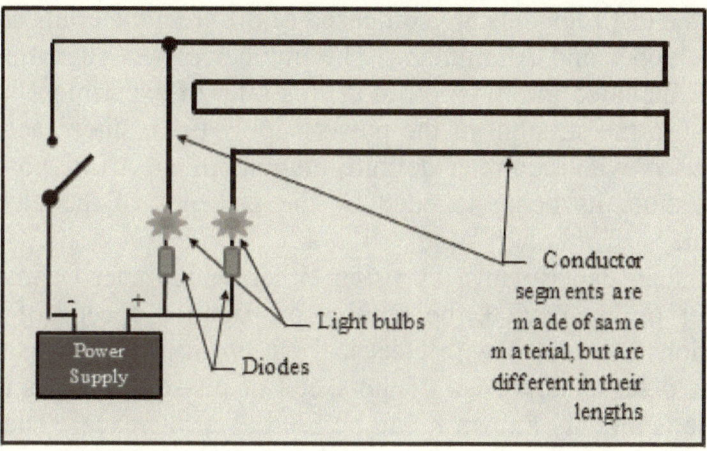

3- Directly measuring the relative speed at which phase vibrations associated with electricity propagate through different conducting matter mediums

The relative propagation speeds of phase vibrations associated with electrical current in various matter conductors can be determined using a setup like the one shown below.

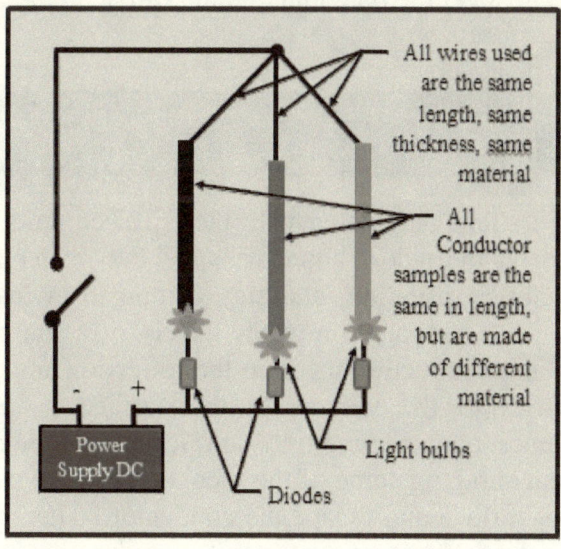

In such a setup, all of the wires used are identical. However, the three test segments are composed of different materials such as Iron, Copper and Aluminum. The number of test segments that can be included in any one run of this kind of experiment is only limited to the strength of the power supply used, since each light bulb used will require a definite amount of electrical power to demonstrate its being touched by the presence of the electrical current.

In such experiments, by using either a high speed camera or another timing device, the relative propagation speeds of phase vibrations associated with electrical current in different materials can be determined. Also, if the speed of phase vibrations in one sample is already known it can be used as a reference. Then, the speed of phase vibrations associated with electrical current in other materials can be determined.

Note that, the matter mediums used for the test segments do not have to be in a solid form/state. They can be in a liquid or even in a gaseous state, as well.

However, when liquids are considered, they must be well contained (with sufficient air gap to allow for their expansion, as they warm up). Also, if gaseous mediums are considered, the container must be able to withstand the potential amount of rise in their pressures.

4- Loss of electrons due to sharp ends/edges in a conducting matter medium

Phase vibrations associated with electrical currents repeatedly get reflected by the sides (boundaries) of the matter medium in which they are propagating, and stay contained in that medium. However, as they propagate towards the tip of a sharp edge or a tapered end in a conducting medium the reflection angle gradually becomes closer to the vertical to the surface. Also, due to becoming more concentrated in such locations, they are quite capable of encouraging some of the local electrons to pass through the boundary surface and escape the conductor. These two effects are shown in the following figure.

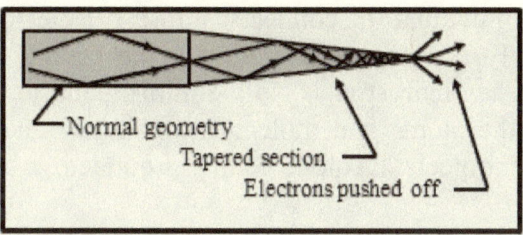

That is how and why sparks first form at parts of the conductor which are sharper in their geometries. The following figure shows a simple setup which can be used to detect such leakages of electrons from electrically charged sharp edges along a regular circuit carrying DC electricity.

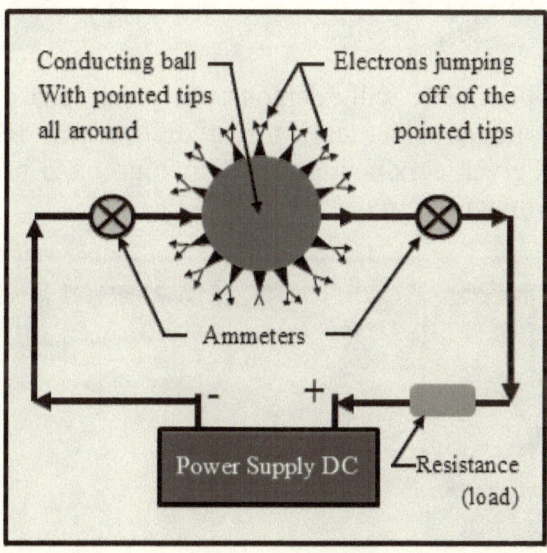

The sharply pointed segments shown in the figure are just like the hair strands that stand up and form a hair ball on the person's head, as he/she touches the surface of an electrostatically charged ball in a science lab. In such a case, the head of the individual acts as the conducting ball and his/her hair strands act as the pointed tips. As the electrons become more concentrated at the tips of the hair strands, they repel each other and hence try to get as far away from each other as they possibly can. The only way they can achieve their mutual objective is by making the person's hair to stand upright forming a hair ball.

As the individual disconnects himself / herself from the electrostatically charged ball, gradually the charge received and stored in his/her hair strands will diminish due to exchange of charge with the atoms and molecules in the air and also due to touching other objects, let alone getting grounded, as the individual steps away.

The amount of electricity that is lost due to electrons being pushed off of the surface at the pointed tips should be measurable. The difference between the readings of the two ammeters used will indicate such losses.

5- Presence of electricity throughout the whole volume of a given conducting matter medium

This experiment will demonstrate that phase vibrations associated with an electrical current are present in the whole volume of a given conducting matter medium, and not just along its contact points in a circuit.

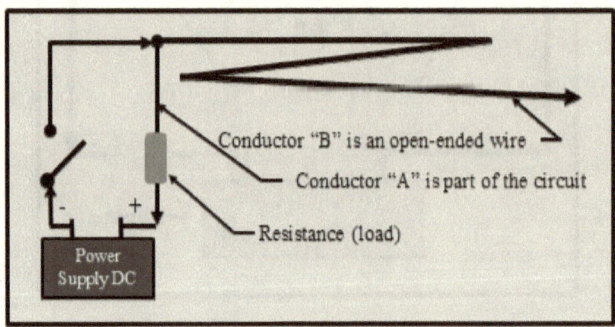

Conductor "B" is an open-ended wire

Conductor "A" is part of the circuit

Resistance (load)

Power Supply DC

Using a voltmeter, it can be readily shown that the voltage across the resistor (load) and across the loose end of wire "B" and the positive terminal on the power supply are the same. This is due to the fact that, the phase vibrations associated with electrical current in a conducting medium occupy the whole volume of that conductor and not just the portions that are forming a complete loop between the negative and the positive terminals of the power supply. Even the open end of the electrical switch is actually negatively charged and when it is set in the closed position it conveys its charge to the rest of the circuit.

Note that, that is why it is not safe to touch a hydro wire which is hanging down from a hydro pole, since if the power is not disconnected, the wire will be carrying the full local voltage used. If it is touched by any object including a living being, the charge will build up at all farthest points, causing the object / living being to get electricuted.

6- Proof that the flow of electricity is generated by a certain type of phase vibrations in the medium of aether that is occupied by a conducting matter medium

An electric circuit has to form a complete loop, otherwise electricity inducing phase vibrations propagating through the matter medium come to a complete halt. Once such waves stop propagating they can no longer encourage the electrons in their path to form a current. Once the loop is broken (the switch is turned off) the phase vibrations come to a halt nearly instantly, due to their propagation speed being that of light in that matter medium.

An experiment such as the one described below can be conducted to demonstrate that, it is the propagation of electricity inducing phase vibrations in a conducting medium that induces the flow of electrons.

A light bulb can be installed along a broken circuit which is quite long, as shown in the following figure. First, switch "A" is temporarily turned on to let the whole volume of the long wire to become positively charged, as in becoming desperate to receive any available electrons.

Next, switch "A" is turned off, and then switch "B" is turned on. Even though the long wire part of the circuit stays open-ended, the very temporary propagation of phase vibrations along its length that is just made available to such phase vibrations will cause the light bulb to show a correspondingly very temporary flash of light indicating the temporary flow of electrons moving passed its position. The electrical current is momentarily established by the temporary propagation of phase vibrations through the position of the light bulb, before they hit the dead end along their propagation path.

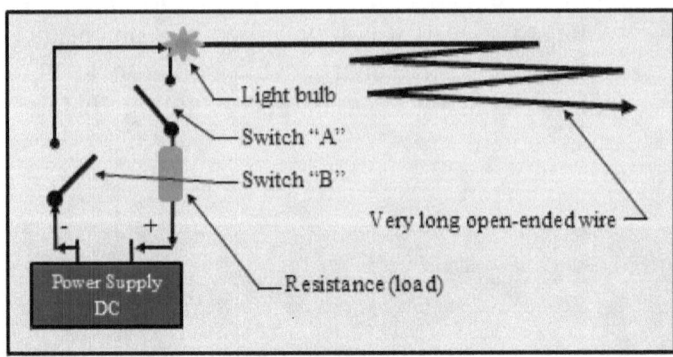

The propagation of electricity inducing phase vibrations in such a scenario can be readily likened to the flow of air from a highly pressurized air tank into a smaller air tank with atmospheric pressure air inside. As the air that is inside the larger tank flows passed the position of the valve and fills the smaller air tank, it can perform a certain task such as blowing a whistle. The amplitude of the sound waves generated by the whistle will be at its maximum in the beginning and it will gradually reduce to nothing, as the air pressure inside the smaller air tank gradually becomes the same as that of the air inside the compressor's tank. The following figure shows such a scenario.

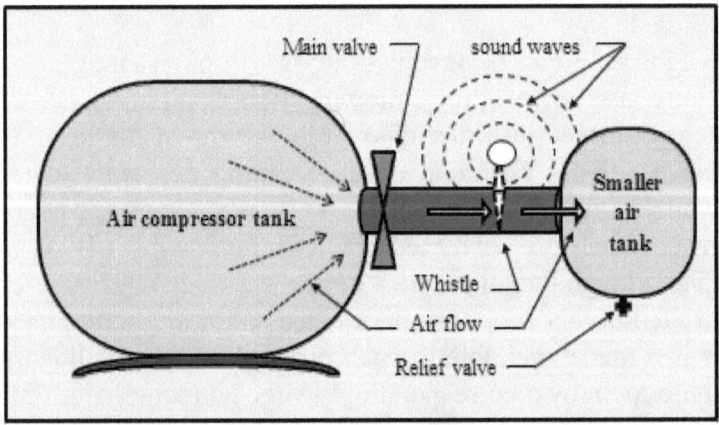

Correspondingly, in the case of the electrical circuit, the light bulb will shine the strongest at the instant the switch is turned on but it will quickly dim down and turn off due to electricity inducing phase vibrations, hence number density of electrons passing through its position weakening / decreasing down to zero.

Note that, the length of the time that the light bulb will shine will be equivalent to the length of time it takes for light waves to travel that conductor's full length.

7- Propagation speed of phase vibrations associated with electricity through a given conducting matter medium is equivalent to the speed of light in that medium

To perform this experiment the conducting matter medium must also be transparent, so that the speed of light waves propagating through it can be measured, simultaneously. In other words, any solid, liquid or gaseous material which satisfies both conditions, namely being a conductor of electricity and being transparent to light waves, can be used for this experiment.

The following figure shows a simple electrical circuit consisting of a test segment which is of a specific length. If need be this segment can be replaced by a canister which can be filled with a liquid or a gas.

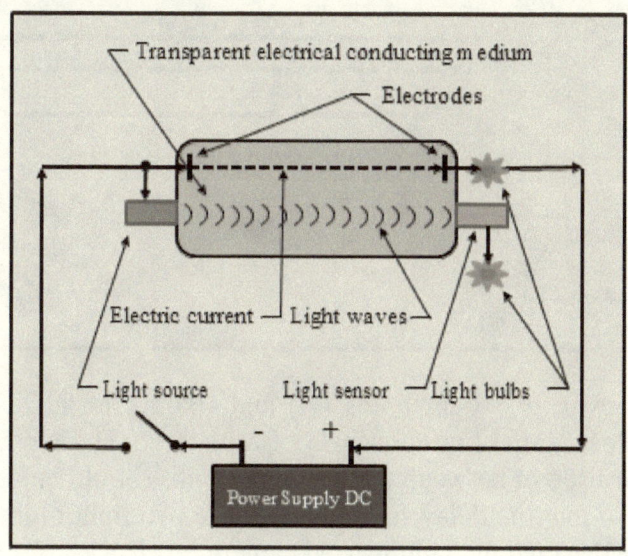

Once the electrical switch is closed, the current will simultaneously pass through the transparent conducting medium in two different forms, one as electrical phase vibrations and the other as light waves. Since both light waves and phase vibrations associated with electrical currents are phase vibrations in the aether medium which happens to be superimposed by a transparent conducting matter medium, both signals will propagate through the medium at the very exact same speed. Therefore, the two light bulbs are expected to light up at the very same instant. To confirm the simultanity of the two light bulbs turning on, a high speed camera can be used to record the whole experiment.

8- Do phase vibrations associated with electricity propagate through conducting matter mediums from the negative terminal of the battery towards its positive terminal or the other way around?

The following figure shows a simple setup to determine the propagation direction of phase vibrations associated with electricity.

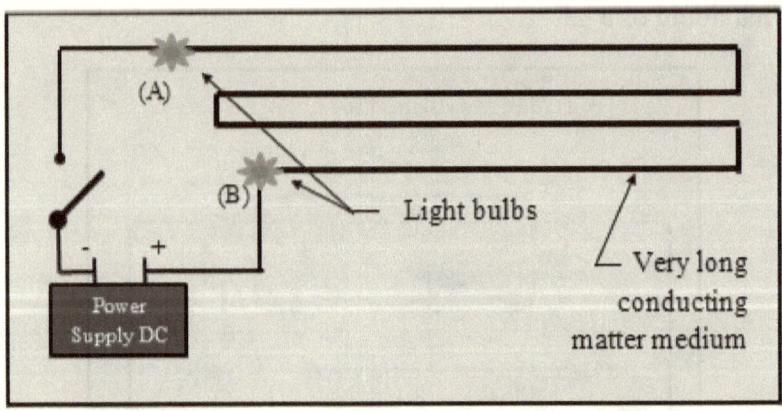

As shown, two light bulbs (A) and (B) are installed along a circuit, close to the two terminals of a battery. The length of the conductor used in between the two light bulbs should be as long as possible so that the delay time between the two light bulbs turning on can be detected by a high speed camera.

Once the switch is turned on, phase vibrations associated with the electricity will be propagating in the whole loop of the circuit. As the phase vibrations propagate through the medium of the conductor they will encourage the local electrons to form a current by joining the stampede along the conductor.

Therefore, as the leading edge of the phase vibrations associated with electricity that is initiated in the conductor reaches the position of each of the light bulbs, that light bulb will demonstrate its reaction by lighting up. Phase vibrations have a finite speed in the medium of the conductor. Hence, depending on the direction of the propagation of these phase vibrations, either light bulb (A) will turn on first or light bulb (B).

If the propagation direction of phase vibrations associated with electricity is from the negative terminal of the battery towards its positive terminal, light bulb (A) will turn on first. Otherwise, light bulb (B) will turns on first.

Conclusion

In this chapter, it is proposed that **current (flowing) electricity** is due to certain type of phase vibrations in the medium of aether that are induced in (or transferred to) a conducting matter medium. Furthermore, it is proposed that the speed of electrical signal (not the speed of electrons) through any given matter medium is exactly equal to the speed of light (electromagnetic waves) in that medium. In other words,

"It is the propagation of a certain type of phase vibrations in the aether medium which is superimposed by the conducting matter medium, that encourages the electrons within the volume of that matter medium to let go of their atomic responsibilities and by progressively hopping unto atoms downstream, form a current."

This process is the very same as the photoelectric effect in which case light waves enable the electrons to become free of their corresponding nuclei and start wandering in the conducting medium.

The phase vibrations associated with an electrical current propagate through the whole volume of the conductor. And, they motivate the electrons throughout the whole volume of the conductor the very same way. Electrons that receive sufficient amount of energy break free from their atomic families, regardless of their position within the conductor.

Due to possessing the same electrical charge and hence repelling each other, electrons instinctively get as far away from each other within the conducting medium as they possibly can. That is why they move towards the surface of the conducting medium and continue travelling along/near the surface of that medium.

The propagation speed of phase vibrations within a matter medium is experimentally proposed to be independent of the thickness of the conducting medium or its shape or irregularities in its cross-sectional area.

Note that, the speed of phase vibrations in the medium of aether is dictated by the density of the local aether and its pressure. Therefore, as the density of aether and that medium's internal pressure in this universe is gradually decreasing, due to the overall expansion of that medium and leakage of aether into the accompanying universe, the propagation speed of all forms of phase vibrations, including those associated with the generation of electrical currents in various matter mediums is gradually increasing with time. In other words,

"The speed at which phase vibrations associated with the generation of electrical currents propagate through any and all matter mediums is gradually increasing."

In short,

"The speed of electrical signals is gradually increasing, in this universe."

Reducing
the Frequency and Intensity
of
Natural Disasters

Droughts, Earthquakes, Floods, Hurricanes
Tornadoes, Tsunamis and Volcanic Activities

The contents of this particular chapter may be reproduced and distributed, in every shape and form, as is necessary, by anyone who is doing so for the sole purpose of getting the information herein to the right individuals who can implement them, so that the public at large may benefit from their applications.

Bahram Esmailzadeh
bahram965@gmail.com

Foreword
(To this particular chapter)

This research was undertaken by an independent individual who has been and is quite concerned about the future of not just the human race but the future of all life forms on this planet.

Man has managed to transform the face of this planet due to such activities as farming, mining, power generation, housing, manufacturing and so on. Some changes have had temporary effects on the environment as a whole, while some others have left scars all over this planet. Over the last few decades, human race has woken up and is gradually starting to realize the consequences of at least some of his actions. He can try to counter the adverse effects of some of his doings, while the other more severe ones have to be left to Mother Nature, to take care of itself over millions of years.

At the present time, one of the most important issues that human race, along with all other life forms on this planet, are facing is the rapid increase in the frequency and the intensity of various types of natural disasters occurring on a global scale. Each and every one of the varieties of natural disasters has been studied by many local as well as multi-national agencies. They all have confirmed that, as time goes on, natural disasters are becoming increasingly more frequent and more intense.

This particular research was initiated in 2004 and was finished in 2010. It was conducted without any kind of request, assistance and/or support from any local or international either private or government related agencies / departments, on any level. It was made possible purely due to a personal goal and desire to solve this global problem, which is affecting the lives of millions of individuals, animals of all kinds and all plant life forms on this planet, every year.

This research has led to the identification of the cause of the observed rapid increase in the number as well as the intensity of various types of natural disasters, in recent decades. It also provides a practical and readily doable solution for it.

During February of 2011, printed copies (and CDs) of this report were personally hand delivered by the author to the Consulates of Brazil, China, France, Germany, India, Japan, Mexico, Norway, South Korea, Spain and Sweden, in Toronto, Canada. And, printed copies (and CDs) were mailed to the Embassies of Australia, China, Egypt, France, India, Iran, Japan, Turkey, Russia, Ukraine and United Kingdom, in Ottawa, Canada. Also, printed copies (and CDs) were mailed to Ontario ministry of environment and various United Nations Environmental / Meteorological Departments based in Switzerland and Mr. Al Gore's office in Arkansas, USA.

At the same time, printed copies (and CDs) of this report were also mailed to over 30 university professors globally, the list is given below.

This report was published in the Iranian monthly science magazine "Daneshmand" based in Tehran, Iran in May of 2011. A few months later, it was also published in the Iranian weekly newspaper "Daneshmand" which is based in Vancouver, Canada. Later on, during 2012, this report was also shared with the general public on the internet via face book.

The author is hereby respectfully submitting the results of his findings to you the reader, as well, not with the hope of receiving any kind of assistance in return, but with the hope of seeing its contents be guided through proper channels and be implemented, as recommended at the end of this chapter. Also, as it was stated in the cover page of this chapter,

The contents of this particular chapter may be reproduced and distributed, in every shape and form, as is necessary, by anyone who is doing so for the sole purpose of getting the information herein to the right

individuals who can implement them, so that the public at large may benefit from their applications.

Thank you, and may we all have many pleasant years ahead.

Bahram Esmailzadeh
bahram965@gmail.com

The following is a list of the university professors in the USA, Russia and Japan, as well as the branches of the United Nations and Mr. Al Gore's offices, to whom printed copies (along with CD's) of the results of this research were mailed, during February of 2011.

Missouri University of Science and Technology
Physics Department
Dr. George D. Waddill, Chairman and Professor
Nuclear Engineering
Dr. Arvind S. Kumar, Chairman and Professor
Dr. Gary E. Mueller
Dr. Nicholas Tsoulfanidis

University of California, Berkley
Department of Physics
Dr. Frances Hellman, department chair

Massachusetts Institute of Technology
School of Science's Dean's Office
Dr. Marc A. Kastner

Harvard University
School of Engineering and Applied Sciences, Dean's Office
Dr. Cherry A. Murray
Area Dean for Environmental Science & Engineering
Dr. Steven C. Wofsy

Yale University
Professor of Chemical & Environmental Engineering
Dr. Paul Van Tassel, Department Chair

North Western University
Environmental Science, Engineering, and Policy Program
Dr. Neal Blair, Department Head

Purdue University
Physics Department
Dr. Nicholas J. Giordano, Department Head

Cornell University
Earth and Atmospheric Sciences
Dr. Richard Waldron Allmendinger
Department of Earth and Atmospheric Sciences
Dr. Larry Brown, Chair
Dr. Arthur DeGaetano, Associate Co-Chair:
Director of Graduate Studies, Geological Sciences
Dr. Teresa Jordan
Director of Graduate Studies, Atmospheric Sciences:
Dr. Dan Wilks
Director of Undergraduate Studies, Science of Earth Systems:
Dr. Natalie Mahowald
Director of Undergraduate Studies, Atmospheric Sciences:
Dr. Mark Wysocki

Columbia University
Department of Earth and Environmental Engineering
Dr. Klaus S. Lackner, Department Chair

Iowa State University
Environmental Sciences, Graduate Program Office
Dr. Charles R. Sauer

Ohio University
Department of Geological Sciences, Graduate Chair
Dr. Douglas Green

Moscow State University
Faculty of Physics
Dr. Vladimir I. Trukhin, Dean
Dr. Viktor Antonovich Sadovnichy, Rector
Faculty of Geology
Dmitry Y. Pushcharovsky, Dean

Stanford University
W.M. Keck Professor, Environmental Earth System Science
Dr. Robert Dunbar
Dr. Scott Fendorf
Dr. Page Chamberlain
Professor and Department Chair, Geological & Env. Sciences
Dr. Jonathan Stebbins

University of Tokyo,
Department of Urban Engineering, Graduate School of Eng.
fso@ue.t.u-tokyo.ac.jp

United Nations
Global Climate Observing System
Switzerland
World Meteorological Organization
Switzerland

Mr. Al Gore's offices
Arkansas, USA

Introduction

The occurrences of various types of natural disasters such as Droughts, Earthquakes, Floods, Hurricanes, Tornadoes, Tsunamis and Volcanic activities, have been recorded over the years. The frequency of these unpleasant events has been increasing dramatically. Particularly, <u>since the 1940s, the occurrences of various types of natural disasters on a global scale have been nearly doubling every 10 years.</u>

But, why would the frequency of natural disasters increase, so drastically?

<u>What could be the cause of this rapid growth in the occurrences of natural disasters, on a global scale, in recent decades?</u>

According to the official reports by agencies directly related to the United Nations, during the same time period, natural disasters have been increasing both in numbers and in intensity. Hence, the damage caused by natural disasters has also been quite extensive in economic terms, as well as loss of life. As a result, every year, insurance companies have been compensating their clients with larger sums of money, to at least cover their property damages.

According to the Environmental Protection Agency affiliated with the United Nations, over the last several years, the total amount of money handed out by the insurance agencies has exceeded 100 Billion U.S.D., on yearly bases, which is a substantial amount.

Between 1950 and 2009, over 2 Million people lost their lives due to major natural disasters alone. During the very same period, the monetary value of the damage incurred to properties was over 2,000 Billion U.S.D. Tens of millions of people have also been injured due to these natural catastrophes.

It is commonly assumed that, all of these disasters have been happening naturally. So far, some of the scientists and related specialists have tried to connect the occurrences and growth in severity of natural disasters to the global warming phenomenon. However, over the last 100 years, the increase in the global temperature has been about 0.74 degrees Celsius.

Now, again, the main question remains:

What could have caused such an exponential increase in the occurrences of various types of natural disasters, including volcanic activities, on a global scale, in recent decades?

Solar Storms could be a major contributor to these occurrences on earth. Over the past, these storms have managed to cause serious problems for communication satellites, interrupt all kinds of radio communication systems and even shot down major national power grids.

Solar storms are in fact like sand storms except that instead of sand particles they are made of charged subatomic particles and ions. These particles are thrown into space by the sun as it experiences its regular cyclic magnetic indigestions. The strength of these storms goes through a peak once every 11 years and then goes back to its normal levels. These cycles are well recorded and the seriousness and extent of their effects are well expected as the new peak periods approach.

Van Allen magnetic belts which are around the earth protect it from these storms by deflecting these particles back into the space. Some of the more energetic particles manage to infiltrate these belts and enter the upper layers of earth's atmosphere. They get trapped in earth's magnetic field and by reacting with ions in the ionosphere and various molecules in those layers they manage to create the amazing Northern Lights (the Aurora Borealis) and the Southern Lights (the Aurora Australis).

Solar Storms are one of the most regular and natural rhythms which have been repeating their cycles every 11 years, for millions of years. The last solar storm peak happened during 2011. These cyclic effects show themselves as an increase of about 30% in the total number of natural disasters happening during the peak years as compared to the average of these events, over a period of a few years, immediately before.

Solar storms have been affecting the frequency of occurrence of various types of natural disasters over a lot longer period of time than just the past 70 years or so. Therefore, there has to be a different and most definitely newly created cause for the rapid increase in both number and intensity of various types of natural disasters. By studying the temporary and cyclic effects of solar storms (of charged particles and ions) on the frequency of various types of natural disasters occurring on earth, one very important point stands out. That important point is the effect that the electromagnetic fields generated by these moving charged particles have on the magnetic field of earth.

Lightning is another type of natural phenomenon which most people are familiar with. The reason for the generation of the extremely intense spark of light and the following thunder is due to the existence of electric charge in the ions, atoms and molecules floating in the air mass that makes up the clouds. Based on known laws of physics, particularly related to electricity and magnetism, clouds which carry electrical charges are automatically affected by the local magnetic field of earth.

Movements of air in general, and clouds in particular, are mainly caused by differences in atmospheric pressures in adjacent regions. However, the existence of electrical charges in the air, and particularly in the clouds, and hence their interactions with earth's local magnetic field encourage extra movements of air masses. In other words, earth's magnetic field has direct effect on the movements of clouds, among other things, as it is explained below.

The following is a brief description of how variations in the strength of earth's local magnetic field directly affect the occurrences of various types of natural disasters.

1- Tornadoes and Hurricanes

Tornadoes and hurricanes are not caused by linear motions of the air masses which are usually due to atmospheric pressure differences in adjacent regions, but rather by the effect that the local magnetic field of earth has on charged particles in the air and the circular motions generated in the air masses due to the Coriolis force. Since, as charged particles in the air interact with earth's local magnetic field, the reaction is an upward movement of the air mass which is usually trapped underneath a thick layer of wide spread cloud.

These upwardly directed induced pressures, which are directly dependent on the strength of earth's local magnetic field, encourage the clouds to give way and eventually a hole is punctured through. The more frequently any variations are introduced into the strength of earth's local magnetic field, the faster it will cause such holes to be punctured through. This is due to the variable nature of the force exerted, since it creates a dynamic force which automatically induces a **Jack hammer effect**. Subsequently, the warmer air mass that is trapped below is allowed to rise up and go through the cloud layer.

The free upward movement of the warmer air mass automatically is affected by the Coriolis force and hence starts to rotate. The direction of rotation induced is counter clockwise in the Northern hemisphere and clockwise in the Southern hemisphere. However, in order for the magnetic field of earth to be able to contribute its effect on the air mass, the air itself has to be electrically charged.

2- Floods and Droughts

Floods and droughts are caused by opposite types of cloud movements. In the case of floods, too much cloud is guided towards certain regions and their eventual natural seeding causes the formation of droplets, in fact too much of it. However, in the case of droughts, not enough clouds are allowed to follow their natural routes which they had been pursuing over millions of years.

Again, both of these abnormal movements of clouds are the result of changes in earth's local magnetic field and its interaction with the charged atoms / molecules in the respective air masses.

3- Earthquakes

Earthquakes happen because of fluctuations induced in earth's local magnetic field. Since, earthquakes are caused by the movements of different layers of earth's crust as they go over and under each other, particularly at the fault lines. Earth's magnetic field is causing the overall movement of these continental plates, since they are mainly composed of Iron and other magnetically inclined elements.

Normally, the relatively speaking steady-state force of earth's local magnetic field exerts a steady push on these continental plates, towards one direction or another. However, as the strength of earth's magnetic field fluctuates due to let's say its interaction with the variable electromagnetic fields generated by the solar storms, the effect becomes as though the same pushing force has now changed to a variable type of a force, which runs into harmonics and <u>creates a Jack Hammer type of an effect</u>. This type of force is no longer a Static force but rather a Dynamic force, which is more effective in initiating a motion in the tectonic plates.

4- Volcanic Activities:

Volcanic activities also happen due to the fluctuations induced in earth's local magnetic field. Since, volcanic eruptions are encouraged by movement and concentration of the molten magma just under earth's crust, particularly at locations where the upper layers are not capable of handling the induced pressures. The movements of these magma masses are affected by the fluctuations in earth's magnetic field, because magma masses are mainly composed of iron and other magnetically inclined elements. Therefore, any normally or abnormally induced variations in earth's local magnetic field would in turn affect the movements of these magma pockets, accordingly. Also, as the force exerted becomes variable, it changes from a Static force to a Dynamic force, which is more effective in initiating serious movements in the magma masses.

Note that, lava deposits, all over the globe, are basically chilled out magma masses. They have been quite helpful to scientists in their research on the changes which have occurred in earth's local magnetic field directions, as well as shifts in the magnetic poles of the planet. This is due to the fact that, after these iron based masses make their way to the surface of earth, but before solidifying, they orient themselves to the local magnetic field lines. Hence, they create a solid record of the local magnetic field lines, corresponding to that particular point in time.

All facts and scientific observations lead to the conclusion that earth's overall magnetic field, and its local variations / fluctuations, directly affect all types of natural disasters occurring on this planet. **Now, again:**

What phenomenon could have had such an ever increasing effect on the occurrences of various types of natural disasters, on a global scale, especially in the last 70 years?

Solar storms directly affect the total number of various types of natural disasters happening all over the globe through the interactions between the electromagnetic fields generated by the motion of their charged particles and earth's magnetic field. These interactions lead to fluctuations in earth's local magnetic field. Such induced fluctuations in earth's local magnetic field in turn encourage the occurrences of various types of natural disasters. Therefore, one has to concentrate his/her research on other phenomena which are also capable of affecting earth's local magnetic field. These phenomena must have the following specific characteristics:

- They have to be able to affect earth's local magnetic field,
- They have to be of a more recent nature / short history of existence,
- They have to be consistent in their ever-increasing strengths with the observed / recorded data on the occurrences of all types of natural disasters.

The most effective of all industries developed by man on this planet that could have had such an adverse effect on earth's magnetic field, which has also experienced an exponential growth rate over the last 70 years or so, has been the electricity industry and all of its related technologies.

The two most commonly used types of electricity are Direct Current (DC) and Alternating Current (AC). Direct current type of electricity is the type of electricity used in such industries as automotive and electronics. Direct current electricity which operates at a constant voltage induces a constant magnetic field around its conducting medium such as a copper wire. And, as its flow starts, stops or as its voltage level is altered, the strength of its generated magnetic field varies, accordingly. However, once its voltage is stabilized at a certain level, its generated magnetic field becomes of a fixed strength.

Note that, constant strength magnetic fields which are artificially induced by constant voltage DC type electricity **do not** generate any kind of cyclic fluctuations in earth's local magnetic field.

An alternating current type of electricity can be viewed as being a DC type of electricity with a continuously changing voltage level, hence the name Alternating Current (AC). In fact, an alternating current repeatedly changes its flow direction many times during each second. Hence, its voltage level follows a sine wave. As a result, this type of electricity, as it passes through any conducting medium such as a copper wire, induces a cyclically varying magnetic field around that medium.

Note that, variable strength magnetic fields which are artificially induced by AC type electricity **generate** cyclic fluctuations in earth's local magnetic field.

This very characteristic of AC type electricity, namely inducing a continuously varying magnetic field, has made the use of such devices as transformers possible. The possibility of transforming the voltage of the electricity generated at the power stations to much higher levels, has allowed the economical transfer of electrical power over very long distances. That was why the alternating current electricity replaced the direct current electricity as the main type of electricity used in power generation, transmission through nationwide power grids, as well as the eventual usage by the variety of electricity driven appliances / equipment.

Note that, <u>scientists have always been aware of the varying magnetic field induced by the alternating current type of electricity. But, they have chosen to ignore its potential short-term and long-term side effects on various phenomena such as earth's local magnetic field, as well as living tissues/cells in animals and plants.</u>

The growth history of electricity industry on this planet can be summarized as follows. The electricity industry started its existence just over a hundred years ago. In the beginning, its development and expansion were very limited. However, over time, the development of more and more applications encouraged faster growth of electricity industry, on a global scale. This growth burst has been particularly noticeable after the 2nd World War.

In the beginning, individual power stations were built to serve small communities / areas. As time went on and power plants with ever larger capacities were developed and built, larger areas were connected to the same power plants. Also, as large hydroelectric dams were constructed in remote places, due to topographical / geographical necessity, there was a need for construction of high voltage power transmission lines to transfer their generated electricity to where they were needed.

As more power plants of different kinds like hydroelectric dams or Coal, Oil or Gas burning power stations or even Nuclear power plants were built, having a State, Province or district wide power transmission network was unavoidable. Therefore, every single country that had industrial expansion in mind had to start with constructing its own national power grid. Designers of such power grids had to make sure that they had plenty of reserve in generating capacity, as well as redundancy in their delivery routes.

Over time, neighboring countries connected their respective national power grids together. Since, by being connected together they could help each other in both generation and consumption of electric power. They could also help each other in emergency situations, whenever one or more transmission lines get damaged, particularly due to extreme weather conditions.

At the present time, nearly all of the countries, on each continent, have their national power grids interconnected. These interconnected power grids have grown to colossal sizes. For example, the interconnected power grid connecting Canada, United States and Mexico literally covers most of the North American Continent and includes over 350,000 Kilometers of high voltage power transmission lines alone. Even though this huge network is divided into several major power districts, they are all literally interconnected and power can be transferred from anywhere to anywhere, across the continent, as needed.

Another example is the European continental power transmission grid. This power grid network not only connects literally all of the countries in Europe, but also itself, as a whole, is also connected to the power grids in Ukraine and Russia which in turn are connected to other countries. Countries in the west-central Asia, which used to be parts of the old USSR power block, are now connected to both the new Russian power grid, as well as the power grids of their other neighbors such as Iran, Afghanistan and Turkey. Arabic countries also have their own interconnected power grid, which in turn is connected to many other countries in Northern Africa.

South East Asian countries are also connected together via a common interconnected power grid. Also, China and India have their own huge power grid systems and currently are growing at a great pace, due to extensive industrial expansions in both countries. Other Asian countries such as Japan, North and South Korea, Pakistan, Taiwan and South American countries such as Brazil, Argentina, Venezuela, Peru, Chile, as well as countries in Central America and Australia are covered with high voltage power transmission lines of their own.

At the present time, five regional power transmission grids connect the African nations in their respective regions. These five grids are already planned to become interconnected, in the near future.

It should be emphasized that, before 2nd World War the electricity delivering power grids were of a more or less isolated local nature. That is, only relatively small areas were connected together and each had its own independent power generating station. However, after 2nd World War, due to industrial expansion, networks of national power grids were created in literally all industrialized countries. These high voltage power grids were a necessity in order to deliver the needed power to newly installed light and heavy industries, as well as homes and office buildings.

The main common feature between all of these power grids installed all over the globe is that, the electricity carried by almost 100% of them is of the Alternating Current or AC type. In the last 50 years or so, a few major power transmission lines were specifically built to carry Direct Current or DC type electricity. These lines carry DC type electricity which is in fact rectified high voltage Alternating Current type of electricity. Power lines carrying high voltage DC type electricity were built mainly because they were more economical over long run, particularly when used to transfer electrical power over long distances without interruptions. However, even the electrical power carried by DC lines were and still are transformed into AC type electricity before they are fed into the local lower voltage power networks, which in turn deliver the electricity to consumers of all sorts.

The AC electricity generated, transferred through national and international grids and used by various consumers, in all of the countries, are of either 50 or 60 cycles per second in frequency. While the European and the Asian interconnected power grids operate on 50 cycles per second, the North American power grid operates on 60 cycles per second. Other countries, depending on where from they get their electrical equipment and their related supplies, operate their power grids at either 50 or 60 (or both) cycles per second, accordingly.

Again, it is of great importance to understand that nearly all of the electricity used by man, on the global scale, is of the alternating current type. The AC electricity is carried at high voltages through huge network of power transmission lines. Then, step by step, its voltage is reduced, as it is fed through the local systems. And, eventually it is delivered to its end users. These colossal networks of power grid lines are nothing but huge electrical coils (windings) which not by design, but by nature are creating super strong, super wide-spread, pulsating electromagnetic fields in their respective vicinities.

The interconnected power grid system in the North America which includes Canada, United States and Mexico, is one of the most extensive power grids built on the planet. As the alternating current type of electricity is carried through this grid, the resulting electromagnetic field literally covers most of the North American Continent. AND, **because it is a synchronized network**, its pulsating action at 60 cycles per second, literally gives continental-wide, super strong, Jack Hammer type of shocks to both, earth's local magnetic field and the electrically charged atoms and molecules in the air. And, as its response,

Earth's locally manipulated / variable magnetic field demonstrates its complaints by blowing, splashing, shaking, spitting or even violently shouting at the ignorant creatures who do not wish to demonstrate a two-way respect.

The power grid in North America is just an example. When considering the whole planet, the power grids in Europe, China and Russia which are just as extensive and elaborate, as well as all of the other interconnected and/or isolated power grids on other continents, also create their own respective pulsating electromagnetic fields around themselves.

These pulsating electromagnetic fields, generated at either 50 or 60 cycles per second, directly affect earth's local magnetic field. Hence,

"The increases observed in the frequencies and intensities of various types of natural disasters, occurring on a global scale, are simply adverse consequences of man's actions."

Since,

"Especially after 2nd World War, the growth in the amount of electrical power generated, the expansion of the national power grids and the synchronization of ever larger areas are clearly in direct relation with the increase in the number and intensity of

various types of natural disasters,
documented on the global scale."

The global maps of the annual worldwide occurrences of various types of natural disasters provide sufficient proof for the overall message delivered through these pages, since they clearly indicate the following important points:

- More than 80% of the natural disasters that had occurred during 2004-2009, had taken place in areas with the strongest and/or the most widespread national synchronized power grids. Places such as North America, Europe (including western Russia), China, India and Japan had experienced the most in numbers and the worst in severity of the natural disasters during those years.

- Vast expanses like the northern regions of Canada, the whole of Alaska, the Amazons, Central Africa, central region of Australia, the whole of central and northern Asia which includes all of Siberia, Mongolia, Tibet and China's north central desert, all of which basically have minimal widespread synchronized power grids, rarely had experienced any kind of serious natural disasters, during those years.

Conclusions

Solar storms affect the occurrences of various types of natural disasters on a global scale. However, their effect is cyclic in nature, repeating every 11 years, due to the intensity of sun's activities going through a regular cyclic variation. Such cyclic variations in solar activities have been repeating over millions of years. Therefore, solar storms could not have been and cannot be the reason for the unusual and consistent increase in the frequency and intensity of natural disasters occurring on this planet, over the last 70 years, or so. Solar storms, however, point towards what is actually taking place.

The variable electromagnetic fields generated by the motion of charged particles that are parts of solar storms introduce fluctuations in the strength of earth's magnetic field. Varying earth's magnetic field in turn, due to becoming a dynamic rather than a static force, encourages more frequent occurrences of various types of natural disasters.

Therefore, the unusual growth in the number of natural disasters and their intensities, over the last seven decades have to be due to some new phenomenon which affects earth's magnetic field in a similar way.

This phenomenon is shown to be the type of electrical power used by man, namely the Alternating Current (AC) electricity. The continental-wide power grids deliver electricity from different power stations to the end users, in the form of alternating current (at either 50 or 60 cycles per second). **AC electricity automatically generates a varying magnetic field matching its frequency. Such variable magnetic fields interact with earth's local magnetic field and cause it to fluctuate.**

The variable earth local magnetic field in turn, by acting as a dynamic force rather than a static force, delivers the needed encouragements to the charged particles floating in the air, as well as the magnetically inclined layers of earth's crust, including the magma. Hence, the occurrences of various types of natural disasters have become ever more frequent than before. In short:

"It is the alternating current type of electricity that is inducing the fluctuations in earth's magnetic field, hence leading to more frequent occurrences of various types of natural disasters on a global scale."

Man can still continue with his current trend of mechanizing every aspect of his life and become as dependent on electricity as he pleases. However, he has to understand 'why' he needs to abide by certain rules, as he utilizes the generated electrical power. He must also be willing to change the type of electricity he uses. He needs to alter his machinery and so on, so that they can use Direct Current (DC) type of electricity rather than Alternating Current (AC) type. Because, Direct Current type of electricity has no adverse (cyclic) effect on earth's local magnetic field. In other words,

"If man is expecting to live in peace with the forces that are strongly protecting the nature, he has to abide by certain rules and learn to have respect for the Mother Nature and its health."

Recommendations

The needed steps to implement the necessary changes, from AC to DC type of electricity, are briefly outlined below.

Step one:

The electricity carried through all of the power transmission lines **(starting with the ones carrying the highest voltages and most currents)** should be changed from AC to DC type. The DC type electricity carried through the main grid lines should be transformed into AC type at multiple sub-stations which would in turn feed their locally **isolated** grid networks covering only relatively small regions (the smaller the size of these regions will ensure more effective results). In other words, **the AC electricity fed into adjacent regions must not be in synch with each other.**

> **"The unhealthy effect of using AC type of electricity imposed on earth's local magnetic field will be reduced drastically just by making sure that the alternating currents fed into small neighboring regions are not synchronized, in their cycles."**

Note that, the difference between vast interconnected grid systems which are not synchronized as compared to those that are synchronized, and their respective overall effects on earth's local magnetic field, is like the difference between hearing 100 members of a choir group singing a song in total random timing as compared to the same members singing the very same song in total harmony.

By being out of phase with each other, small neighboring regions will actually partially cancel each other's effects on earth's local magnetic field. Whereas, by being synchronized, they give rise to resonances and wider spread and stronger harmonics which must be avoided.

The technology related to high voltage DC type power transmission grids has been around for several decades. In fact, even though it has been incorporated in a relatively small number of projects, there are quite a few countries which are already using this technology. This technology has also been applied in situations where electrical power was to be carried across water ways, as it has obviously been the case for interconnections between Sweden and its neighbors.

So far, power economy has been the major factor considered, by the power companies, as such high voltage power transmission lines carrying DC type electricity were installed along certain long distance routes. However, at the present time, man should seriously consider using DC type power lines on all routes. They are more economical, and their usage will drastically reduce the potential damages inflicted through various types of natural disasters on lives, the environment and the overall eco-system, as a whole.

Note that, <u>by implementing these recommended changes,</u> **man can expect to see the frequency and intensity of various types of natural disasters, occurring on the global scale, to be reduced to the levels recorded back in the 1940s.**

Step two:

Eventually, man should try to use only DC generators and DC equipment so that there would be no AC type electricity in use anywhere on the planet.

Note that, <u>by implementing these recommended changes,</u> **man can continue his modern way of living but with the number of natural disasters reduced to the levels experienced back in the early 1900s.**

By properly implementing these required changes in his electrical power generation as well as the transmission lines, on a global scale, man will reduce the unpleasant effects of AC type electricity on earth's local magnetic field to a minimum.

It is expected that, implementing the required changes on the overall existing power transmission lines, on the global scale, will have an astronomical price tag associated with it. However, by reducing the number of natural disasters which would have otherwise occurred in the future, the costs associated with replacement of damaged properties, let alone lost lives, will also be reduced drastically.

At the present time, every year insurance companies, on the global scale, are paying over 100 Billion U.S. Dollars to their clients who happen to have insurance on their damaged properties. But, most others are helped by their local governments, even foreign governments or in many cases are simply left alone to start all over totally on their own.

Since the insurance industry is the only industry which is going to directly benefit financially from these implemented changes to the power grid systems, they should be held responsible for covering a major portion of these expenses. That is, while they keep their charged rates at their current levels, they can forward the excess to the local governments, hydro companies as well as the manufacturers of the required components.

As time goes on and the required expenses are reduced, the insurance companies can also lower the amount they charge their clients and also reduce their contributions towards these power grid modification projects. As for the governments, they can use part of their future budgets, which would have been allocated to deal with such compensations, to improve the accommodations they provide for their citizens. These applications could include huge green spaces, a variety of recreational facilities, as well as hospitals, schools and so on.

A Friendly advice

The ever increasing frequency and intensity of various types of natural disasters, on the global scale, is no longer a problem that just the human race is faced with. Literally all animal life-forms and all plant life-forms on this planet have been and are being affected by these occurrences. Every year, literally millions of people and animals lose their lives, get injured or at least lose their habitats due to the occurrence of various types of natural disasters.

This problem has been solely due to the ignorance of human race, not thinking through all of the potentially negative consequences of his actions.

If we, as the current members of the human race, value our own existence, our future generation's existence or even the existence of any particular kind of animal or plant life-form on this planet of ours, we must take the required steps to alleviate this man-made problem.

**<u>May we all act more responsibly,
before it is too late.</u>**

A Short Note on the Global Warming

There are two relatively new major man-made/man-caused contributors to the global warming on this planet which need to be taken seriously:

- **The overall amount of heat energy generated by all of the power plants (Nuclear, Hydro, Coal, Gas and Oil) on a global scale.**

  According to the United Nations' affiliated agency, The Global Energy Resources and Consumption, the amount of energy generated by man during 2008 has been equivalent to the energy produced by a power plant generating 15 Million MW (of heat) operating none stop 24 hours a day for 365 days of the year. **This is a substantial amount of additional heat which simply was not there, before.**

  Even though the amount of power usage in developed countries has reached its saturation level, but due to the rapid growth of various types of industries in countries such as China, India, Brazil, Argentina and other 2nd and 3rd world countries, **the amount of this excess heat is rapidly increasing, every year.**

  Note that, the heat generated due to the consumption of millions of tons of various types of fuels on a grand scale is the reason why during winter the temperature level in the downtown areas of major cities such as Toronto is reported to be a few degrees higher than their respective surroundings, including their own major airports.

- **The overall excess amount of heat energy generated by the ever increasing number of underwater volcanic activities.**

  By literally dumping all of their heat energy directly into the water that is forming the oceans, underwater volcanic activities which have increased drastically over the last few decades, have been a major contributor in raising the oceans' water temperature and causing the melting of the ice caps near the poles and hence have directly contributed to the rise in the overall global temperature.

  ===

  =================================

  ===============

Locating the Birthplace
of
the Universe

Introduction

Man has always tried to understand how this universe has come into existence and how it has evolved. Particularly over the last 200 years or so, using a variety of powerful telescopes, he has managed to collect and catalog quite a comprehensive amount of data about thousands of galaxies. Currently, the universe is estimated to be about 13.7 billion years old, and it is also estimated to be over 93 billion light years in diameter. According to the observations performed by such scientists as Mr. Hubble, all galaxies are getting away from each other. The speed at which galaxies are receding is directly proportional to their distances. In other words, the farther two galaxies are from each other, the faster they are getting away from one another.

By hypothetically reversing the expansion process of the universe and following the motion of the galaxies backwards in time one can visualize the contents of the universe approaching each other and get squeezed into an ever smaller region of space, a region with a specific set of spatial co-ordinates. In other words,

"The ongoing expansion process of the universe must have started from a specific location in space."

The aim of this chapter is to demonstrate that by using the physical evidences made available through the use of telescopes the birthplace of this universe can be identified. In fact, five independent methods of identifying the physical location of that particular region of space are introduced in this chapter. Since these five methods are independent of each other, they can also be used to confirm each other's results.

Necessary background information

The first thing that catches one's attention about the contents of this universe, on literally every scale, is that planets, stars and even galaxies are demonstrating two types of motions, namely a rotational motion and a translational motion.

These two types of motions, namely the rotational motion around their own axis and the translational motion around something else, in most cases, are in nearly the same plane. Such occurrences have not happened by chance.

The origins of all spins associated with any of the universe's contents, on various levels of complexity, have started from the time the simplest of all unions were formed between particles. As it is shown below, the rotational motions have been due to the differences between the speeds of the individual particles uniting with each other while they were running away from the center of expansion.

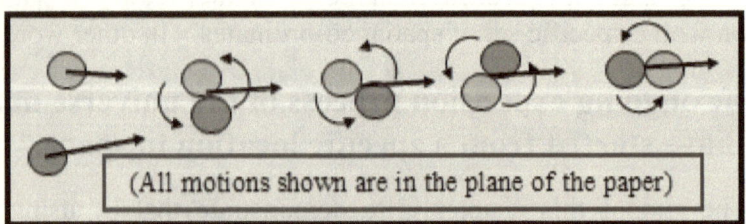

(All motions shown are in the plane of the paper)

Note that, in the above figure, the individual constituents are moving away from a common region of space (a common center), when considered in 3 dimensions.

In time, as more and more particles (pieces) were drawn together, due to their mutual gravitational attraction towards each other, they made up for the differences in each other's speeds by forming a rotational motion around each other. All of the planets, stars, solar systems and galaxies clearly demonstrate such rotational and orbital motions. Again, in most cases, the two types of motions for each planet, star, or galaxy are in nearly the very same plane.

Based on the data collected through direct observations, majority of the planets in the solar system (with the exception of Uranus and Pluto) spin around their own axis in a plane which makes less than 30 degrees with their respective orbital planes around the sun. Their orbital planes are important because they literally pass through the sun. Also, the orbital planes of all planets are offset from sun's equatorial plane by less than 7.16 degrees. For details please refer to the table below.

Planets in the Solar System	Inclination of Rotational Axis, Degrees	Inclination to ecliptic Plane, (Earth=0) Degrees	Inclination to Sun's Equatorial Plane, Degrees
Mercury	0.00	7.00	3.38
Venus	-2.70	3.39	3.86
Earth	23.45	0.00	7.16
Mars	23.98	1.85	5.65
Jupiter	3.12	1.30	6.09
Saturn	26.73	2.49	5.51
Uranus	97.90	0.77	6.48
Neptune	29.56	1.77	6.43
Pluto	122.00	17.20	

The rotation of all planets is due to the differences in the speeds/momentums of their constituents, namely the particles that were thrown into space by the sun. As these particles kept on orbiting the sun, they either united with other particles or were absorbed by existing planets. Either way, the difference in their linear momentums caused their unions to manifest a rotational motion which is in a plane that is close to their orbital planes around the sun.

Of course, in the case of the solar system all of the planets are imbedded in a relatively thick pancake like volume of space. In other words, the case of the solar system can be considered as being a semi-two-dimensional scenario. However, in this universe, galaxies are spread in a true three-dimensional volume of space.

As it is explained in the chapter titled "Formation and Development of the Universe", matter and anti-matter particles were formed almost simultaneously across the region of space which was flooded with phase vibrations, the same phase vibrations that were spreading in all spatial directions, away from their common point of origin, namely the birthplace of this universe. Therefore,

Upon their formation from phase vibrations, matter and anti-matter particles already possessed a certain amount of momentum in the general direction that was away from the birthplace of the universe, from where phase vibrations had started to spread in all spatial directions.

Over time, due to the gravitational forces that matter particles exerted on each other, they formed giant cloud-like gatherings, throughout space. The force of gravity also encouraged the formation of numerous even more localized concentrations of matter particles within each of these giant cloud-like gatherings. The localized concentrations of matter particles gradually became denser and denser and eventually led to the formation of star systems. As stars became organized in their motions relative to each other, the giant cloud-like gatherings of matter particles were transformed into galaxies. The following figure shows such a transformation. For more details, please refer to the chapter titled "Formation and Development of the Universe".

Initially, the newly formed galaxies were generally spherical / elliptical in their geometries. However, over time, as their stars compromised in their momentums, as they passed by each other, spherical galaxies became flattened and eventually formed spiral galaxies.

Again, it must be emphasized that,

All of the matter particles were formed by phase vibrations that were already propagating away from the birthplace of this universe. Therefore, the initial momentum of all of the matter particles had to have been away from the center of expansion, also.

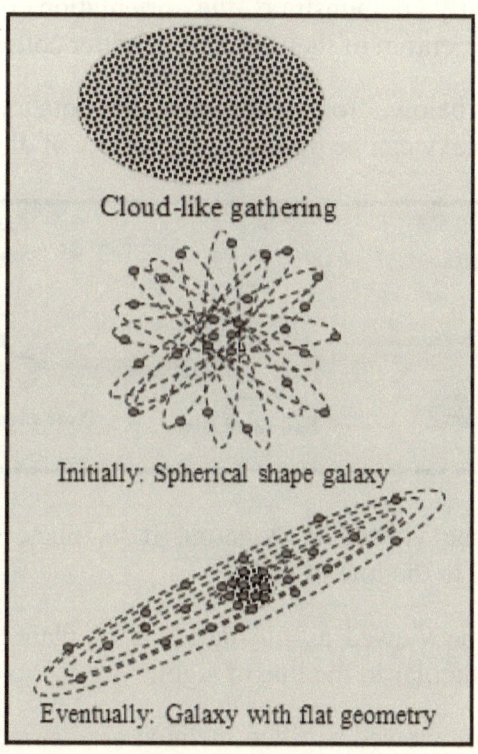

Cloud-like gathering

Initially: Spherical shape galaxy

Eventually: Galaxy with flat geometry

Consequently, as the large gatherings that they had formed evolved into spiral galaxies, the orientation of their rotational planes which is the representation of the overall momentum of their constituents must have been parallel to the line of sight from the center, the birthplace of the universe, as well. In other words, it is expected that:

- The general motion of individual galaxies, as a whole, must be away from the center, from where the phase vibrations had started their propagation in the aether medium.

- The extensions of the rotational planes of the spiral type galaxies (which are readily verifiable in their geometries) must pass through the common center or the birthplace of the universe.

Note that, the spiral galaxies are of particular interest in this study, since they are well developed in their geometries and clearly demonstrate the orientation of the overall momentum of their localized matter collectives.

As shown below, from a distance, the rotational plane of a given spiral galaxy can be oriented in a variety of directions:

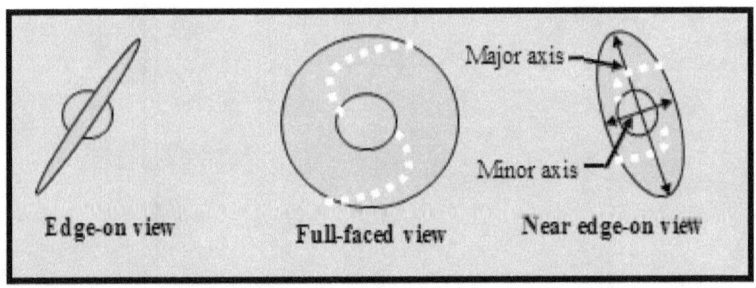

- It can be viewed as edge-on, if its plane happens to be parallel to the line of sight.

- It can be viewed as full-faced, if its plane happens to be perpendicular to the line of sight.

- It can be viewed as tilted/inclined at a variety of angles, as its plane happens to be diagonal to the line of sight.

Methods of locating
the Birthplace of the Universe

Five independent methods of locating the birthplace of this universe are introduced / presented with sufficient details in the following pages.

The first four methods use the orientations of the rotational planes of galaxies, which are well-developed into their planar / spiral geometries, to achieve their common task. While, the last method, uses the temperature distribution in the cosmic microwave background radiation, the phase vibrations existing in this universe, to locate where this universe was born.

- First Method:
(Using the intersection of the rotational planes of spiral galaxies)

This method is based on the fact that, the extension of the rotational planes of most spiral galaxies must pass through a common region of space, the same region from where this universe has started its expansion. The following figure shows the basic idea behind this particular method.

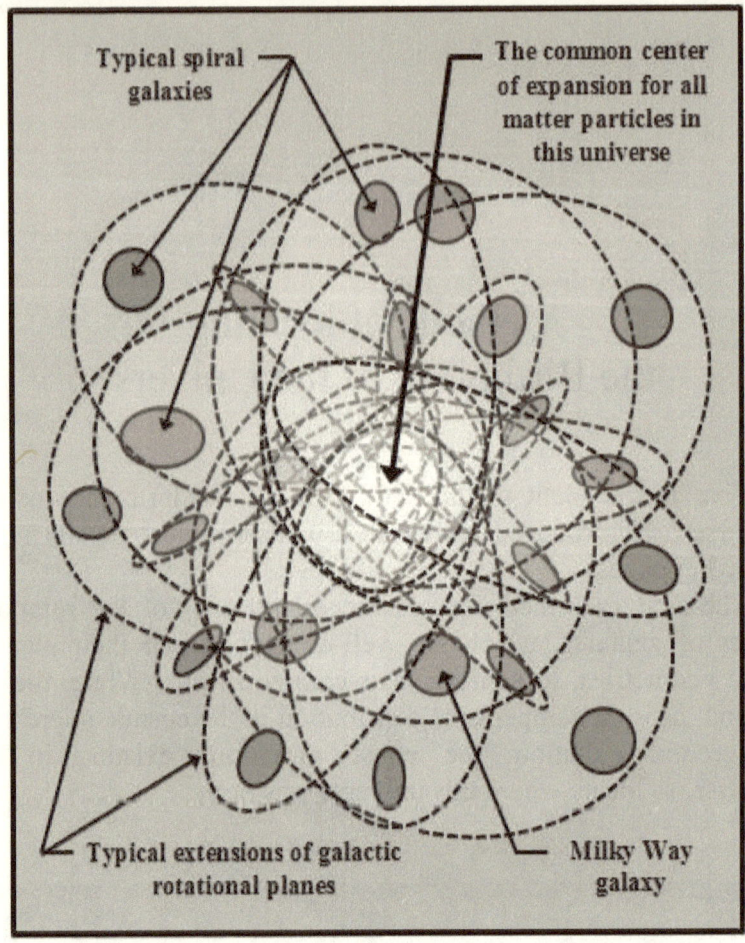

In this method one needs to follow the steps outlined below:

Step one:

He/she needs to make a simple, small model of a planar galaxy or a disk with two pivotal axes, which are at 90 degrees to each other, as shown below.

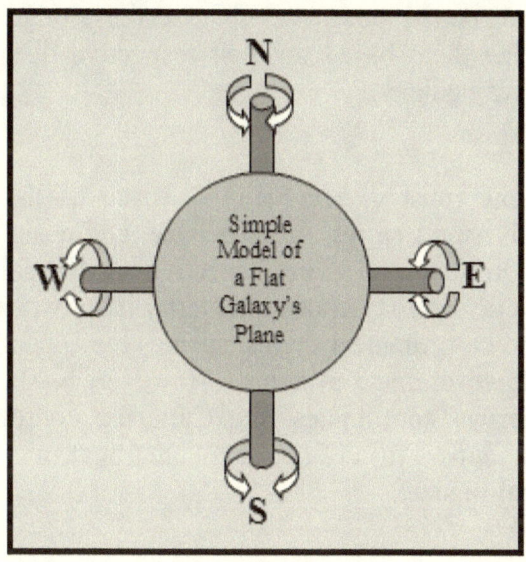

Step two:

He/she needs to hold this model in such a way that one axis points in the north-south direction and the other in the east-west, as seen directly face-on. Then, by turning it on its two axes (one at a time) through different angles, he/she can duplicate the true angles at which the rotational plane of any given galaxy in mind is rotated with respect to the line of sight from earth.

Step three:

He/she needs to go through all of the available galaxy pictures and sort out the ones which clearly exhibit an overall disk shaped structure.

Then, he/she must tilt the disk model about its two axes and determine the respective apparent tilt angles of each and every galaxy with respect to the line of sight from earth.

Next, using the degrees at which their respective planes are tilted, as well as the longitude and latitude angles at which each galaxy is located with respect to earth using a universally accepted coordinate system, he/she needs to write the equation describing the rotational planes for each and every one of these galaxies, (algebra, the section on equations of planes and their perpendicular distances to any given point).

Then, he/she needs to organize these data in a table and categorize them based on their respective distances from the Milky Way galaxy.

Step four:

He/she must divide the 3-D space of the universe into relatively small **cubes**, of equal size, and designate them with the co-ordinates of their respective center points. For this purpose he/she may choose to start with any of the universally accepted co-ordinate systems, such as the Galactic or the Super Galactic co-ordinate systems. However, he/she must convert all spherical coordinates into Cartesian coordinates and use straight cubes to define small regions of space that are **identical in size,**.

Step five:

Using the formulas given in algebra, he/she must calculate the shortest distances (perpendiculars) from center of each and every one of the cubes to the plane of each and every one of the galaxies considered.

If the calculated distance is less than one half of the cube's dimensions, it would mean that, if that particular galaxy's rotational plane were to be stretched/extended, it would pass through that particular cube.

This procedure needs to be repeated for every single cube and with respect to every single galaxy that is considered.

These calculations may be performed using a personal computer. The only limitation for performing such calculations would be having access to various comprehensive galaxy catalogs which include pictures of individual galaxies, as well as detailed information about their locations with respect to the Milky Way galaxy.

Note that, to obtain better precision, one needs to use pictures of as many galaxies as possible. The larger the number of galaxies considered, will automatically result in a more definite indication of the cube(s) of space from where this universe has started its existence.

If need be, one can start with fairly large size cubes to limit the computation time required. Then, step by step, by ignoring the cubes through which nearly none of the galactic planes pass, he/she can use finer size cubes to obtain as accurate a final result as desired.

"The location of the birthplace of this universe is simply within or quite near the cube(s), region of space, through which the extensions of most of the galactic rotational planes pass."

Note that, due to the long term effects of gravitational forces between neighboring galaxies, the general orientation of most galaxies rotational planes are no longer what they used to be, when they were first formed.

Therefore, one cannot expect the difference between the percentage of galaxies with rotational planes that pass through a certain region of space to be drastically higher than those of the others regions in space. However, its magnitude should still be readily distinguishable among those percentages detected / calculated for other regions of space.

By only considering galaxies that are farther away from earth, one can expect to obtain higher precision in locating the region of space where this universe was born. Since, the gravitational forces of such distant galaxies have not had sufficient time to randomize the orientation of their rotational planes.

- Second Method:
(Using edge-on view of the rotational planes of spiral galaxies)

In this method, one only needs to pay close attention to the apparent orientation of the rotational planes of spiral galaxies, regardless of their distances from earth.

As it was explained earlier, in the section "Necessary Background Information", it was the differences between the speeds/momentums of the particles that caused each of the cloud-like gatherings gradually evolve into galaxies with a more or less flat geometry.

Therefore, one can expect to see most of the galaxies which are flat or planar in their geometries have their rotational planes pointing roughly towards the center, where universe was born. That is, if their rotational planes were extended long enough, they would definitely pass through (or at least pass nearby) the region of space from where the universal expansion had begun. In other words,

"The rotational planes of most of the flat/planar galaxies located along a straight line, between the birthplace of the universe and its outer perimeter (in any radial direction), should be viewed by each other as a narrow strip or edge-on (or close to it) rather than as a wide plane or full-faced view."

The following figure shows **the most likely orientation** of galactic planes along a typical straight line between the birthplace of this universe and its outer perimeter.

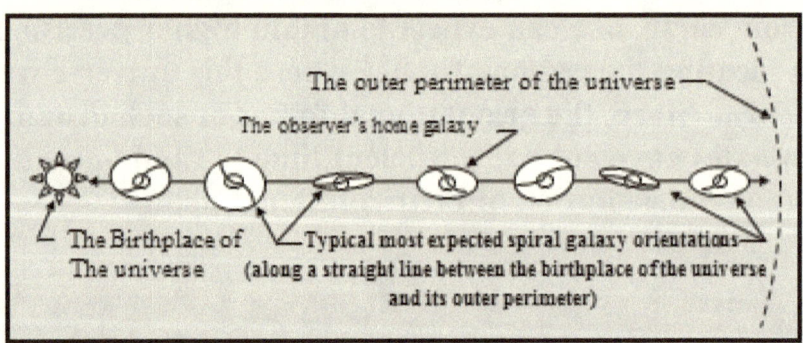

Some standard catalogs containing the observational data on galaxies, also include data on the planar tilts of these galaxies. They present such data in the form of a ratio of the longer and the shorter dimensions of their apparent planes (ratio of Major axis to Minor axis). This information is exactly what is needed to perform the task at hand using this method. The figure below shows the variety of shapes that a typical galaxy with a relatively flat geometry would appear along the line of sight.

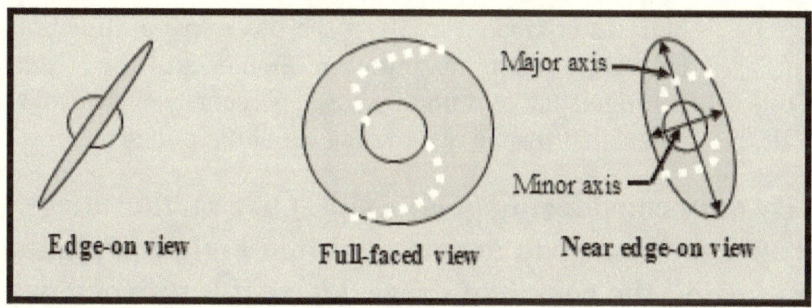

Notes:

- **The orientations of these edge-on views (the orientation of galaxies rotational plane's major axis) are irrelevant.** That is, their major axis can be oriented in the north-south, east-west or any other direction, as long as their rotational planes are viewed as being edge-on or close to it.

- **The directions of rotation of the galaxies, as in being clockwise or counter clockwise, are also irrelevant.**

As one examines the data regarding galaxies which have evolved into their planar shapes, he/she should be able to distinguish <u>two opposing directions in space that would have the</u> **highest percentages** <u>of the edge-on galaxies in view</u>, as compared to all of the other directions.

Of these two distinct opposing directions, one will be pointing directly towards the birthplace of the universe, while the other will be pointing directly towards its outer perimeter.

Note that, due to the long term effects of gravitational forces between neighboring galaxies, the general orientation of most galaxies rotational planes are no longer what they used to be, when they were first formed.

Therefore, one cannot expect the differences between the percentages of galaxies with rotational planes that are close to edge-on relative to his/her line of sight, in the two opposing directions to be drastically high, as compared to all of the other spatial directions. However, their magnitudes should still be readily distinguishable among those percentages detected / calculated for the rest of the spatial directions.

By only considering galaxies that are farther away from earth, one can expect to obtain higher precision in locating the region of space where this universe was born. Since, the gravitational forces of such distant galaxies have not had sufficient time to randomize the orientation of their rotational planes.

- Third Method:
  (Using full-faced view of the rotational planes of spiral galaxies)

This method is closely related to the "second method" which expected that,

"The rotational planes of most of the flat/planar galaxies located along a straight line, between the birthplace of the universe and its outer perimeter (in any radial direction), should be viewed by each other as a narrow strip or edge-on (or close to it) rather than as a wide plane or full-faced view."

The method presented in this section, takes advantage of the fact that the opposite of the above statement must be true, also. In other words,

"The rotational planes of nearly none of the flat/planar galaxies located along a straight line,

between the birthplace of the universe and its outer perimeter (in any radial direction), should be viewed by each other as being full-faced or nearly full-faced."

The following figure shows **this least likely orientation** of galactic planes along a straight line between the birthplace of this universe and its outer perimeter.

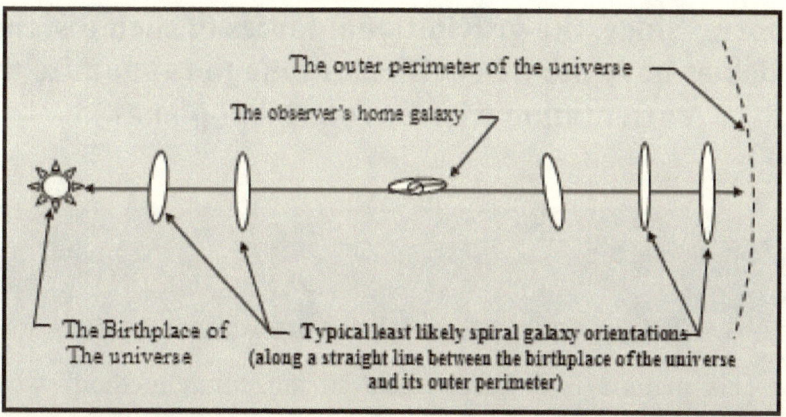

As one examines the data regarding galaxies which have evolved into their planar shapes, he/she should be able to distinguish <u>two opposing directions in space that would have the</u> **least percentages** <u>of the full-faced galaxies in view</u>, as compared to all of the other directions.

Of these two distinct opposing directions one will be pointing directly towards the birthplace of the universe, while the other will be pointing directly towards its outer perimeter.

Note that, due to the long term effects of gravitational forces between neighboring galaxies, the general orientation of most galaxies rotational planes are no longer what they used to be, when they were first formed.

Therefore, one cannot expect the differences between the percentages of galaxies with rotational planes that are close to full-faced relative to his/her line of sight, in the two opposing directions to be drastically low, as compared to all of the other spatial directions. However, their magnitudes should still be readily distinguishable among those percentages detected / calculated for the rest of the spatial directions.

By only considering galaxies that are farther away from earth, one can expect to obtain higher precision in locating the region of space where this universe was born. Since, the gravitational forces of such distant galaxies have not had sufficient time to randomize the orientation of their rotational planes.

- Fourth Method:
(Using horizontal edge-on view of the rotational planes of spiral galaxies –forming a 360 degree belt)

This method is closely related to the "third method" which expected,

"The rotational planes of <u>nearly none of the flat/planar galaxies</u> located along a straight line, between the birthplace of the universe and its outer perimeter (in any radial direction), <u>should be viewed by each other as being full-faced</u> or nearly full-faced."

To be more specific, the method presented in this section takes advantage of the fact that the above statement must also imply that, as shown below, the rotational planes of nearly none of the galaxies located to the sides of any given galaxy should be perpendicular to the lines connecting their respective locations to the birthplace of the universe.

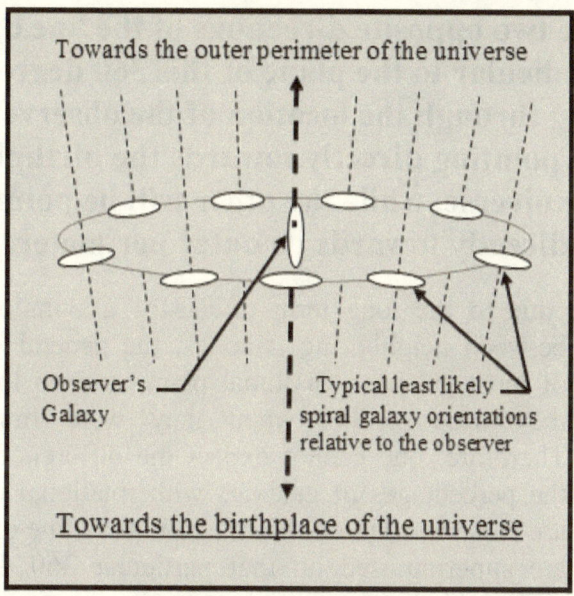

Therefore, **in this method, the orientations of these edge-on views of galaxies are quite important.** In other words, as shown in the figure,

The rotational planes of nearly none of the flat/planar galaxies located in a region of space forming a 360 degree belt around the observer (the plane of which is perpendicular to the line connecting the observer's location to the birthplace of the universe), should be viewed as being edge-on and also superimposed on the plane of that belt.

Therefore, as one examines the data regarding the galaxies which have evolved into their planar shapes, he/she can expect to distinguish a belt-shaped region in space surrounding his/her position that would have the **least percentages** of the edge-on galaxies with rotational planes that are superimposed (or nearly so) on the plane of that belt. The plane of that particular belt is perpendicular to the line which connects the observer's position to the birthplace of the universe.

Of the two opposite directions of the line drawn perpendicular to the plane of this 360 degree belt, passing through the location of the observer, one will be pointing directly towards the birthplace of the universe, while the other will be pointing directly towards its outer perimeter.

Note that, due to the long term effects of gravitational forces between neighboring galaxies, the general orientation of most galaxies rotational planes are no longer what they used to be, when they were first formed. Therefore, one cannot expect the differences between the percentages of galaxies with rotational planes that are close to edge-on relative to his/her line of sight and are superimposed on that particular 360 degree belt around his/her home galaxy to be drastically low, as compared to belts forming in all of the other spatial directions. However, their magnitudes should still be readily distinguishable among those percentages detected / calculated for the rest of the spatial directions.

By only considering galaxies that are farther away from earth, one can expect to obtain higher precision in locating the region of space where this universe was born. Since, the gravitational forces of such distant galaxies have not had sufficient time to randomize the orientation of their rotational planes.

- Fifth Method:
  (Using the temperature distribution in the cosmic microwave background radiation, CMBR)

There are billions of galaxies in this universe which are fairly randomly positioned across its entire volume. Each galaxy, with the help of its stars, is basically acting as a huge heating element.

And, the overall region of space in this universe that is hosting matter (as well as anti-matter) particles can be likened to a giant spherical furnace. As these heating elements (stars in the galaxies) operate and generate heat, in the form of various types of radiation, the entire volume of this furnace will eventually experience nearly the same temperature.

However, the local energy levels (temperatures) are continuously affected by the ongoing nuclear activities within local stars. In other words,

Regions of this universe that host higher number densities of galaxies (active stars, to be more precise), automatically contribute more effectively to the overall energy level (temperature) of the background radiation in their respective vicinities. Hence, they cause the formation of hot spots in their localities.

Suppose a heat sensing satellite is placed inside this universe (the universal furnace) so that it is somewhere between its center and its outer limits.

If sensitive enough, the satellite will detect the direction pointing roughly towards the outer perimeter of the universe that is nearest to it. It will do so by detecting ONE major region of space which is the coldest as compared to all of the other regions. This is expected due to <u>net energy loss</u> through the exterior boundaries of the region of space which is hosting galaxies. In other words, such a satellite will detect the direction pointing towards the region of space that is void of any heating sources, namely galaxies or even stars, and does not radiate any energy back.

"Once the direction pointing towards the coldest region of space is found, the opposite direction will be pointing towards <u>the general direction</u> of the birthplace of the universe."

The opposite direction of the coldest region of space detected is stated as being **<u>the general direction</u>** pointing towards the birthplace of the universe because galaxies located near the outer limits of the region of space which is hosting matter particles, are not spaced uniformly. They are more or less randomly positioned, due to the availability of matter particles that gravitated towards each other and eventually formed those galaxies. The following figure shows this concept.

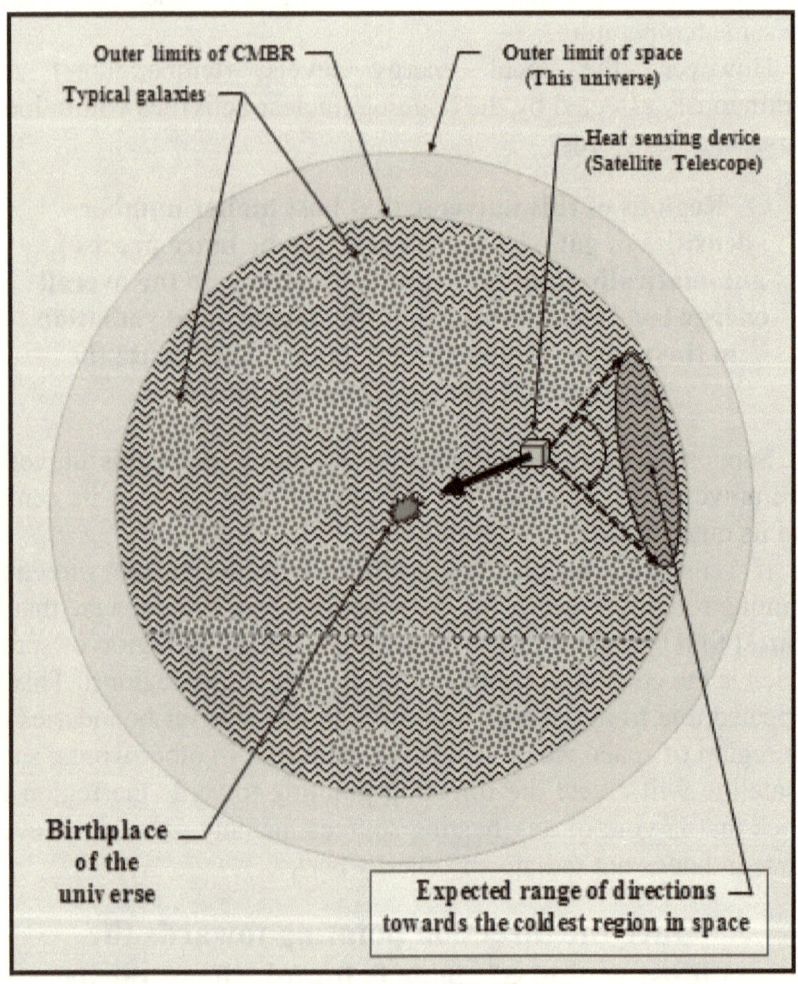

Note that, even though the direction indicated by this method can only be a general one, it still has to be close to the directions indicated by the other methods presented in this chapter.

The Cosmic Microwave Background Radiation (CMBR) actually represents this background temperature of the whole universe. The underlying waves in the cosmic microwave background radiation are due to:

- The initial phase vibrations which were introduced into this space and prompted the birth of this universe, and

- The annihilation of the matter and anti-matter particles, nearly at the end of the rapid expansion era.

However, as it was stated earlier, each and every active star contributes to and enhances (affects) the energy content (temperature level) of the background radiation in its immediate vicinity. This is how galaxies and clusters of galaxies, as a whole, promote the formation of hot spots in this universe.

In short,

"To identify the direction pointing towards the birthplace of this universe, one needs to use a satellite that is capable of sensing the temperature of the background radiation in all spatial directions. Since, the direction pointing towards the birthplace of the universe is in the general opposite direction of where the coldest spot is located."

Conclusion

So far, it has been taught in the universities that the birthplace of this universe cannot be located, or there is no such a unique place that can be referred to as (or rightfully hold the title of) "The Birthplace of the Universe". Since, <u>according to currently accepted theories, the expansion of the universe is **solely** due to space itself expanding and not due to galaxies moving **in** space.</u> The common analogies used by astronomers and astrophysicists to describe such motions of galaxies **with space** rather than **in space** are:

1- The increasing distances between dots on the surface of an ever-inflating balloon, and

2- The increasing distances between raisins in a raisin cake/dough as it is baking.

The aim of this chapter has been to demonstrate that by using the physical evidences made available through the use of various types of telescopes the birthplace of this universe can be identified. In fact, five independent methods of identifying the physical location of that particular region of space are proposed, presented in detail, in this chapter.

The first four methods perform their tasks by using the already collected and hence available data regarding the orientations of the rotational planes of individual planar / spiral galaxies. These data include galaxies' locations, apparent shapes as well as the major and the minor axes of their rotational planes, as seen from earth (or the Milky Way galaxy). The fifth method uses the background temperature distribution, namely the variations in the cosmic microwave background radiation to perform its task.

Note that, since these five methods are independent of each other, they can also be used to confirm each other's findings.

The results obtained from the applications of three of the methods (the second, the third and the fifth methods) proposed in this chapter, to locate the birthplace of this universe, are presented in the following chapter titled "The Birthplace of the Universe (Discovered)".

The Birthplace
of
the Universe
(Discovered)

The contents of this particular chapter may be reproduced and distributed, in every shape and form, as is necessary, by anyone who is doing so for the sole purpose of getting the information herein to the right individuals who can implement them into the educational system on various levels, so that the new generations of astronomers and astrophysicists may benefit from their applications, as they pursue their research in furthering the proper understanding of the universe in which we all are living.

Bahram Esmailzadeh
bahram965@gmail.com

Foreword
(To this particular chapter)

This research was undertaken by an individual who has been and is quite interested in finding answers for apparently inexplicable questions and hence provide a better understanding of this universe, as a whole.

One such puzzle has been the physical location of the birthplace of this universe, even though thousands and thousands of scientists, professors as well as graduate students, globally, have researched the structure of this universe from a variety of viewpoints and with the help of a variety of sophisticated equipment.

The overall conclusion arrived early on by astronomers and astrophysicists has been that, there is no such a unique place in this universe that can be referred to as (or rightfully hold the title of) "The Birthplace of the Universe". Since, according to their accepted theories, the expansion of the universe is solely due to space itself expanding and not due to galaxies moving in space. The common analogies used to describe such motions of galaxies **with space** rather than **in space** are:

1- The increasing distances between dots on the surface of an ever-inflating balloon, and

2- The increasing distances between raisins in a raisin cake/dough as it is baking.

Here in this chapter, using the visual aspects of the spiral galaxies, as well as the temperature distribution in the cosmic microwave background radiation, it will be clearly demonstrated that not only the direction pointing towards the birthplace of this universe can be specified, but also its distance from earth (or the Milky Way galaxy) can be calculated.

In fact, **five independent methods are described in sufficient details in the previous chapter, titled "Locating the Birthplace of the Universe".** In this chapter the **second method** (Using edge-on view of the rotational planes of spiral galaxies), the **third method** (Using full-faced view of the rotational planes of spiral galaxies) and the **fifth method** (Using the temperature distribution in the cosmic microwave background radiation) are applied and their respective findings are presented.

This research was started back in 1987, but due to the lack of access to the needed data it was put on hold, until 1989. During 1989, the first method presented in the previous chapter was applied using a limited number of data points (actual pictures of about 300 spiral galaxies), from the Catalogs which were available at the University of Toronto's astronomy department library. It was found to be quite feasible and promising to perform such a task of finding the birthplace of the universe. However, again, this research had to be put on hold due to not having access to extensive data bases, gathered by various international groups, to perform a proper study and analysis.

Recently, the needed data to perform the second and third methods, presented in the previous chapter, have become available to the general public through the World Wide Web system. Therefore, the research for locating the birthplace of the universe was given a proper chance to be completed.

The results of the second method was written in the form of an article and was officially published in English, in the Iranian weekly magazine, "Daneshmand", based in Vancouver, Canada, in January 2013. On March 21st, **the same article was <u>emailed directly</u> to 42 astronomy professors** affiliated with different universities, in Canada and in the USA. The same article was also summarized and translated by the author and was officially published in the Iranian monthly science magazine "Daneshmand", based in Tehran, Iran, in June of 2013. The said article was later on shared with the general public on the facebook and also through various internet groups.

The methods of locating the birthplace of the universe were also presented as a chapter in the author's book "Aether: Past, Present and Future of the Universe", Xlibris publishing company, Indiana USA, in 2012. That book was translated by the author into Farsi and with an additional chapter that presented the full length results of the second method was officially published in Iran in 2013.

The data from the Plank Space Telescope, released on March 23rd, 2013 provided the information needed to perform the task of locating the birthplace of the universe using the fifth method, presented in the previous chapter. Also, during 2014, this study was furthered again, by performing the task using the third method, presented in the previous chapter. <u>All three methods agree in their results, as in where this universe was born.</u>

This chapter provides the overall results obtained from the three methods performed / completed, so far. The obtained results from these three independent methods clearly demonstrate that, in fact,

<u>"There is such a unique place in this universe that can be rightfully referred to as 'the birthplace of the universe'"</u>

In fact, as it is presented in the following pages,

"The location of the birthplace of this universe has been discovered, by <u>the author of this book.</u>"

Note that, the data required to perform the remaining two methods have already been gathered and have been cataloged by various astronomical groups, but they are not readily available to the general public through the internet, yet.

The author is hereby respectfully submitting the results of his findings to you the reader, as well, not with the hope of receiving any kind of assistance in return, but with the hope of seeing its contents be guided through proper educational channels and be put into proper use. And, as it was stated in the cover page of this chapter,

The contents of this particular chapter may be reproduced and distributed, in every shape and form, as is necessary, by anyone who is doing so for the sole purpose of getting the information herein to <u>the right individuals who can implement them into the educational system, on various levels</u>, so that the new generations of astronomers and astrophysicists may benefit from their applications, as they pursue their research in furthering <u>the proper understanding</u> of the universe in which we all are living.

Thank you, and may we all have many pleasant years ahead.

Bahram Esmailzadeh
bahram965@gmail.com

The following list provides the names of the university professors affiliated with different universities in Canada and in the USA, who have received a direct email along with an attached file containing a copy of the results of the second method, on March 21st, 2013.

Univ. of Arizona Astronomy, USA

Dr. Xiaohui Fan,
Dr. Edward W. Olszewski,
Dr. Dennis Zaritsky,
Dr. Peter A. Strittmatter

Dr. Dan Marrone,
Dr. Marcia J. Rieke,
Dr. Ann Zabludiff,

University of California Berkley, USA

Dr. Uros Seljak,

Dr. Gibor Basri

Univ. of Calif. Santa Barbara, USA

Dr. Crystal L. Martin

University of Toronto, Canada

Dr. Ray Carlberg,
Dr. James Graham,
Dr. Keith Vanerlinde

Dr. Roberto Abraham,
Dr. John R. Percy,

Univ. of British Columbia, Canada

Dr. Mark Halpern,   Dr. Ludovic Van Waerbeke,
Dr. Douglas Scott,   Dr. Kriss Sigurdson

Perimeter Institute, Canada

Dr. Neil Turok,   Dr. Niayesh Afshordi,
Dr. James E. Taylor,   Dr. Avery E Broderick,
Dr. Mike Hudson

McMaster University, Canada

Dr. Laura Parker,   Dr. William Harris,
Dr. Cliff Burgess,   Dr. Sarah Symons

Arizona State University, USA

Dr. Judd D. Bowman,   Dr. Sangeeta Malhotra,
Dr. James Rhoads,   Dr. Evan Scannapieco,
Dr. Rogier Windhorst

Princeton University Astrophysics, USA

Dr. David N. Spergel (Chair),   Dr. Neta A. Bahcall,
Dr. Adam Burrows,   Dr. Renyue Cen,
Dr. Christopher Chyba,   Dr. Michael A. Strauss,
Dr. Robert Lupton,   Dr. Gillian R. Knapp

Princeton Institute for Advanced Studies, USA

Dr. Scott D. Tremaine,   Michele Turansick (Academic Assistant)

NASA's galactic database website staff, USA

Tom.McGlynn@nasa.gov,
stephen.a.drake@nasa.gov

Introduction

All along history, man has always tried to understand how this universe has come into existence and how its contents have evolved. Using a variety of powerful telescopes, he has collected quite a comprehensive amount of data regarding thousands of galaxies. It is estimated that there are billions of galaxies in this universe which is billions of light years across.

Based on direct observations, performed by such scientists as Mr. Hubble, galaxies are receding from each other. Their relative speeds are directly proportional to their respective distances which means the farther two galaxies are from each other, the faster they are getting away from one another. In other words,

The whole universe is expanding.

The observed receding motion of galaxies from each other and hence the overall expansion process of the whole universe, is the greatest indication that the universe has started its existence from a much, much smaller region of space, and over time it has expanded to its current size. The following is one of the most basic questions that can be raised in regards to the beginning of this universe.

Is it possible to locate the birthplace of this universe?

So far, it has been taught in the universities that such a place cannot be located, or there is no such a unique place that can be referred to as (or rightfully hold the title of) "The Birthplace of the Universe". Since, according to currently accepted theories, the expansion of the universe is solely due to space itself expanding and not due to galaxies moving in space. The common analogies used by astronomers and astrophysicists to describe such motions of galaxies **with space** rather than **in space** are:

1- The increasing distances between dots on the surface of an ever-inflating balloon, and
2- The increasing distances between raisins in a raisin cake/dough as it is baking.

Here in this chapter, using the visual aspects of the spiral galaxies, as well as the cosmic microwave background radiation, it will be clearly demonstrated that not only the direction pointing towards the birthplace of this universe can be specified, but also its distance from earth (or the Milky Way galaxy) can be calculated.

To perform this task, five methods were described in details in the previous chapter, titled "Locating the Birthplace of the Universe". In this chapter the **second method** (Using edge-on view of the rotational planes of spiral galaxies), the **third method** (Using full-faced view of the rotational planes of spiral galaxies) and the **fifth method** (Using the temperature distribution in the cosmic microwave background radiation) are applied and their respective results are presented.

Using the Orientation
of the
Rotational Planes of Spiral Galaxies

• Necessary background information

The first thing that catches one's attention about the contents of the universe, on literally every scale, is that moons, planets, stars and even galaxies are demonstrating two types of motions, namely a rotational motion around their own axis and a translational motion around something else. For example, earth's moon is both spinning on its own axis and is following an orbital path around the earth. Also, earth is both spinning on its own axis and orbiting around the sun. Even sun is both spinning on its own axis and also following its set trajectory around the center of mass of the Milky Way galaxy, and so on. All of the rotational and translational motions observed in this universe are due to a common phenomenon.

Billions of years ago, matter and anti-matter particles were formed due to resonances (spikes) in the phase vibrations which were propagating in space. Since the direction of the propagation of these phase vibrations was away from the region of space where they were first formed, as they gave birth to matter and anti-matter particles across this universe (nearly at the end of the rapid expansion era), those particles possessed an initial momentum (motion in space), which in general, was carrying them away from the center of expansion, as if they were individual fragments of an explosion process.

As it is shown in the figure below, the differences between the speeds/momentums of individual particles (objects) that step by step joined together caused their respective collectives, on all levels of complexity, namely planets, stars, and galaxies, to possess rotational motions.

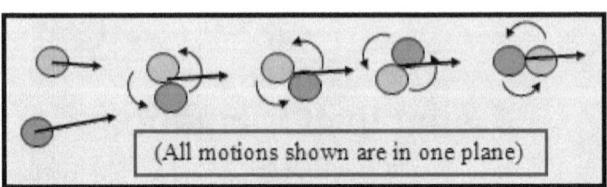

(All motions shown are in one plane)

Note that, in the above figure, the individual constituents are moving away from a common region of space (a common center), when considered in 3 dimensions.

Over time, due to the gravitational forces that matter particles exerted on each other, they formed giant cloud-like gatherings, throughout the region of space that was occupied by phase vibrations.

The force of gravity also encouraged the formation of numerous even more localized concentrations of matter particles within each of these giant cloud-like gatherings. The localized concentrations of matter particles gradually became denser and denser and led to the formation of star systems. As stars became organized in their motions relative to each other, the giant cloud-like gatherings of matter particles were eventually transformed into galaxies. The following figure shows such a process.

Again, **since all of the matter particles were formed by phase vibrations which were propagating away from the birthplace of this universe, the initial momentum of all of the matter particles had to have been away from the center of expansion, also.** Therefore, as they were forming large gatherings, the orientation of the rotational plane of their collectives must have been parallel to the line of sight from the center, as well. In other words, it is expected that:

- The general motion of individual galaxies, as a whole, must be away from the center, from where the phase vibrations had started their propagation in this universe. Also,

- The extensions of the rotational planes of the spiral type galaxies (which are readily verifiable in their geometries) must pass through the common center or the birthplace of the universe.

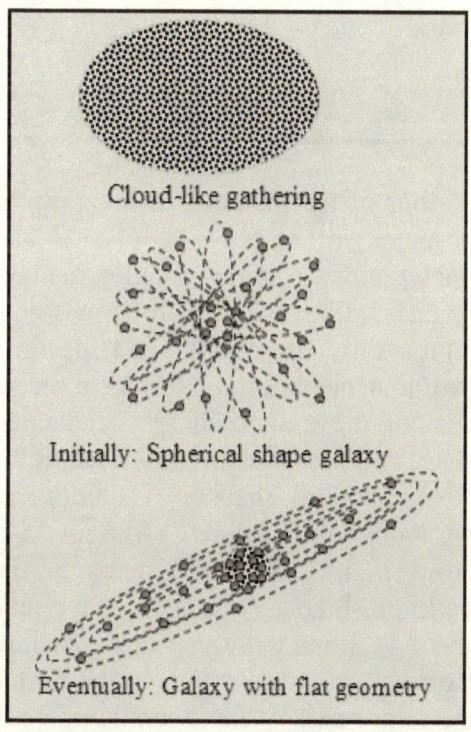

Cloud-like gathering

Initially: Spherical shape galaxy

Eventually: Galaxy with flat geometry

Note that, the spiral galaxies are of particular interest in this study, since they are well developed in their geometries and clearly demonstrate the orientation of the overall momentum of their localized matter collectives.

As shown below, from a distance, the rotational plane of a spiral galaxy can be oriented in a variety of directions:

- It can be viewed as edge-on, if its plane happens to be parallel to the line of sight.

- It can be viewed as full-faced, if its plane happens to be perpendicular to the line of sight.

- It can be viewed as tilted/inclined at a variety of angles, as its plane happens to be diagonal to the line of sight.

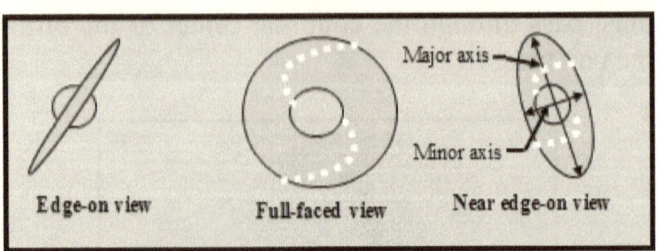

Even though four of the methods of locating the birthplace of the universe presented in the previous chapter are based on the very same characteristics of spiral galaxies, namely the angle of their rotational planes with respect to observer's line of sight (from earth, for example), only the second and the third methods are readily doable without needing any further direct examination of the pictures taken of the individual spiral galaxies. The results obtained from these two methods are described in this chapter.

Most of the standard galaxy Catalogs such as "The Morphological Catalog of Galaxies" (Moscow State University, USSR) and "The Principal General Catalog of Bright Galaxies" (Observatoire de Lyon, France) also include the inclinations of the rotational planes of galaxies with respect to the line of sight from earth. They report galactic inclinations in the form of ratios of the major axis to the minor axis of the overall apparent plane of each spiral galaxy. They also provide the sizes of the major and the minor axes, individually.

• Data / Catalogs used

The Morphological Catalog of Galaxies which contains detailed data regarding over 29,000 spiral galaxies was chosen for the first part of this study.

When plotted, according to their Longitudes and Latitudes, as shown below (using the galactic coordinate system), the spiral galaxies reported in this Catalog cover most of the spatial directions.

In this Catalog, the direction that corresponds to the Southern Equatorial Pole is left blank, since that portion of the sky is obscured by the horizon for telescopes that are located in the Northern hemisphere. Even though, The Morphological Catalog does not include any data on galaxies that are in the Southern most regions of the sky, yet it contains a major number of data points and proves to be quite helpful in visualizing the desired effect.

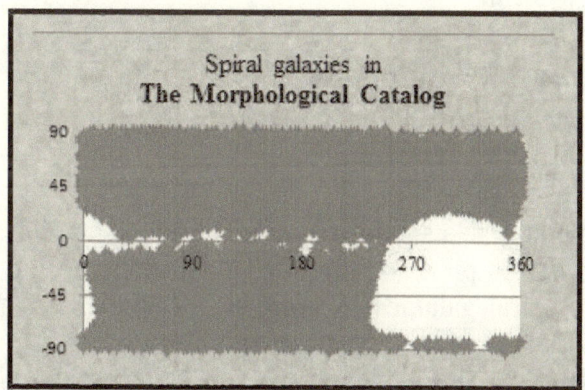

The Principal General Catalog of Bright Galaxies, which also contains detailed data regarding over 29,000 spiral galaxies, was chosen for the second part of this study.

When plotted, according to their Longitudes and Latitudes, as shown below (using the galactic coordinate system), the spiral galaxies reported in this Catalog cover almost all of the spatial directions.

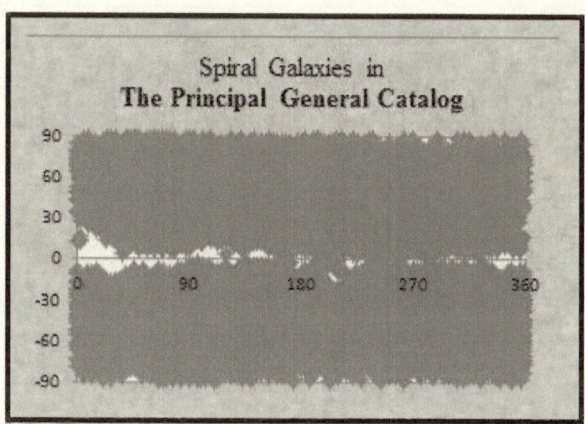

As it can be seen in the last two figures, there is a narrow longitudinal strip of space that is left nearly blank in both Catalogs. This narrow strip corresponds to the edge-on view of our own galaxy. The dust-like particles that are present in the Milky Way galaxy, by literally absorbing the incoming light, block the view to the galaxies that are located on the far side. The center of our galaxy marks the 0.0 Longitude and 0.0 Latitude in the galactic coordinate system.

The Second Method:
Using the edge-on view of the rotational planes of spiral galaxies

This method takes advantage of the fact that, the eventual orientation of the plane of rotation of galaxies, by the time they evolve into spiral galaxies, is more or less parallel to the line of sight from the birthplace of the universe.

Therefore, spiral galaxies which happen to be along any given straight line between the birthplace of the universe and its outer perimeters are expected to be viewed as edge-on or nearly edge-on to each other. In other words, they are likely to be oriented so that their planes of rotation are parallel to the line of sight from the birthplace of this universe.

The figure below shows **the most likely orientations of rotational planes of spiral galaxies** with respect to lines of sight from the birthplace, as well as from the outer perimeter, of this universe.

Using this method and the readily available data presented in standard galaxy Catalogs, one must be able to identify the direction (Longitude and Latitude coordinates) that points towards the birthplace of this universe. Because, according to the above figure, as one looks at all of the spiral type galaxies that are spread in different spatial directions,

He/she can expect to detect two opposing directions, hosting the highest percentages of galaxies with rotational planes that are

edge-on or nearly edge-on with respect to
his/her line of sight.

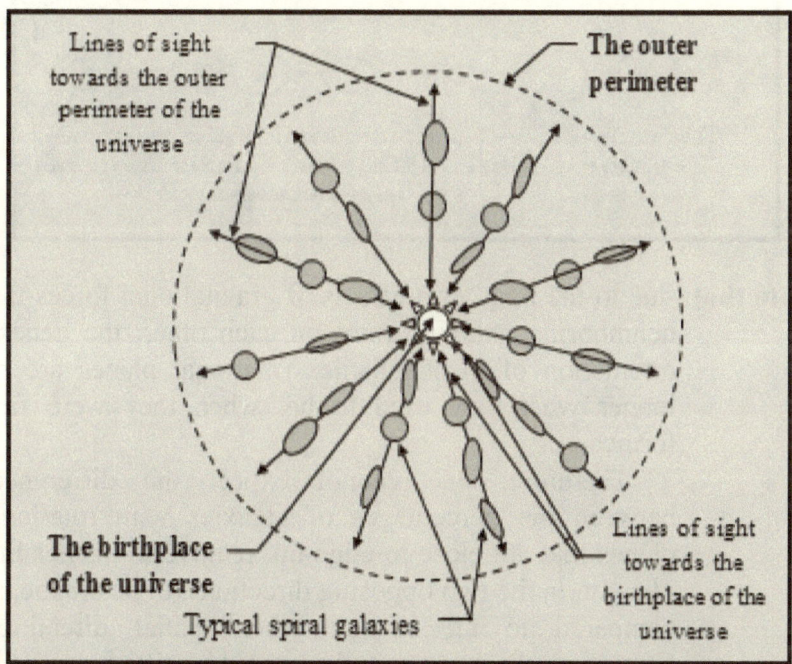

Notes,

- **The orientations of these edge-on views (the orientation of galaxies rotational plane's major axis) are irrelevant.** That is, their major axis can be oriented in the north-south, east-west or any other direction, as long as their rotational planes are viewed as being edge-on or close to it.

- **The directions of rotation of the galaxies, as in being clockwise or counter clockwise, are also irrelevant.**

The following figure shows several spiral galaxies that appear as edge-on or nearly edge-on to an observer, as he/she looks towards the birthplace of the universe or in the opposite direction, towards the outer perimeter of this universe.

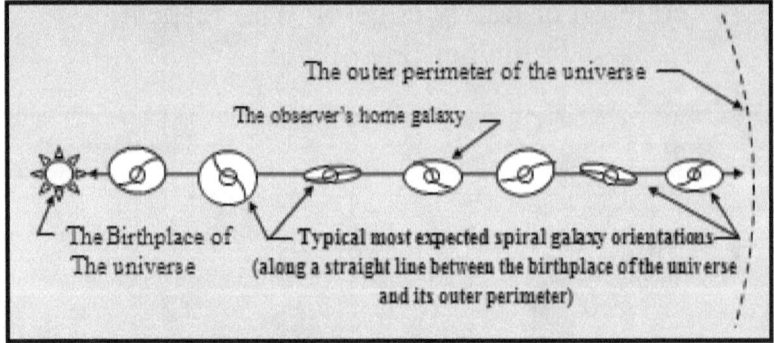

The outer perimeter of the universe

The observer's home galaxy

The Birthplace of The universe

Typical most expected spiral galaxy orientations (along a straight line between the birthplace of the universe and its outer perimeter)

Note that, due to the long term effects of gravitational forces that neighboring galaxies exert on each other, the general orientation of most galaxies rotational planes are no longer what they used to be, when they were first formed.

Therefore, one cannot expect the differences between the percentages of galaxies with rotational planes that are close to edge-on, relative to his/her line of sight, in the two opposing directions, to be drastic, as compared to all of the other spatial directions. However, their magnitudes should still be readily distinguishable among those percentages detected / calculated for the rest of the spatial directions.

By only considering galaxies that are farther away from earth, one can expect to obtain higher percentage differences between the desired directions and all of the other directions. Since, the gravitational forces of such distant galaxies have not had sufficient time to randomize the orientation of their rotational planes.

• Results

This method is based on identifying the two opposing directions in the sky that are hosting the highest percentages of spiral galaxies with rotational planes that are edge-on or nearly edge-on with respect to the line of sight, as they are viewed from earth. The percentages of spiral galaxies of interest are calculated and plotted for various Longitude and Latitude grid sizes, using an Excel spreadsheet program.

The two standard galaxy Catalogs used, namely The Morphological Catalog of Galaxies and The Principal General Catalog of Bright Galaxies, are examined individually and their respective results are presented separately.

- ## The Morphological Catalog of Galaxies

    This Catalog provides the galactic inclinations using a scale of 1 to 5, where 1 designates galaxies that are viewed as being flat-faced (full-faced) and 5 designates those that are viewed as edge-on with respect to the line of sight from earth. This Catalog also provides the size of the major and minor axes for individual spiral galaxies, as well.

    For this study, the inclination values of greater than 3 for the ratio of the major axis to the minor axis were chosen to isolate the galaxies which are viewed from earth as being edge-on or nearly edge-on.

    As shown below, when the sky around earth is divided into small Longitude and Latitude grid sizes (equivalent to 5 degrees by 5 degrees), the percentages of the spiral galaxies that are edge-on or nearly edge-on with respect to the line of sight from earth, show almost random distribution all over the plot. Only one portion of the sky shows an increased number density of peaks, representing higher percentages of the edge-on or nearly edge-on galactic planes in view.

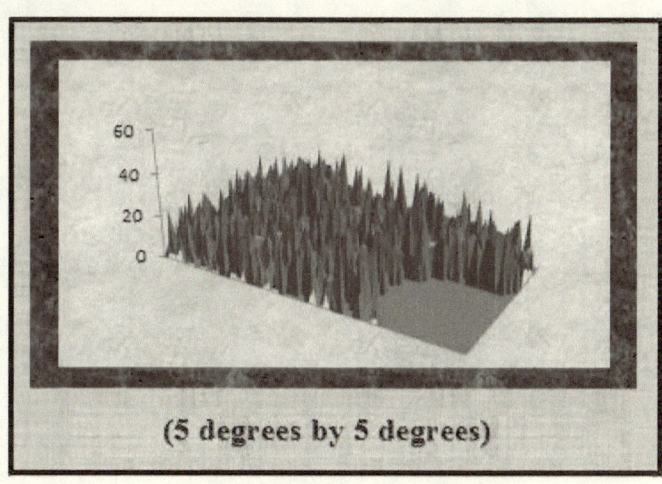

(5 degrees by 5 degrees)

However, as the data corresponding to the larger and larger grid sizes are plotted, such as the ones shown below for grid sizes of (10 by 10), (15 by 15), (22.5 by 22.5), (30 by 30) and (45 by 45) degrees, more and more of the peaks and valleys combine and average out and show as more or less flat sections. In the meantime, one major peak becomes more and more apparent and clearly stands out.

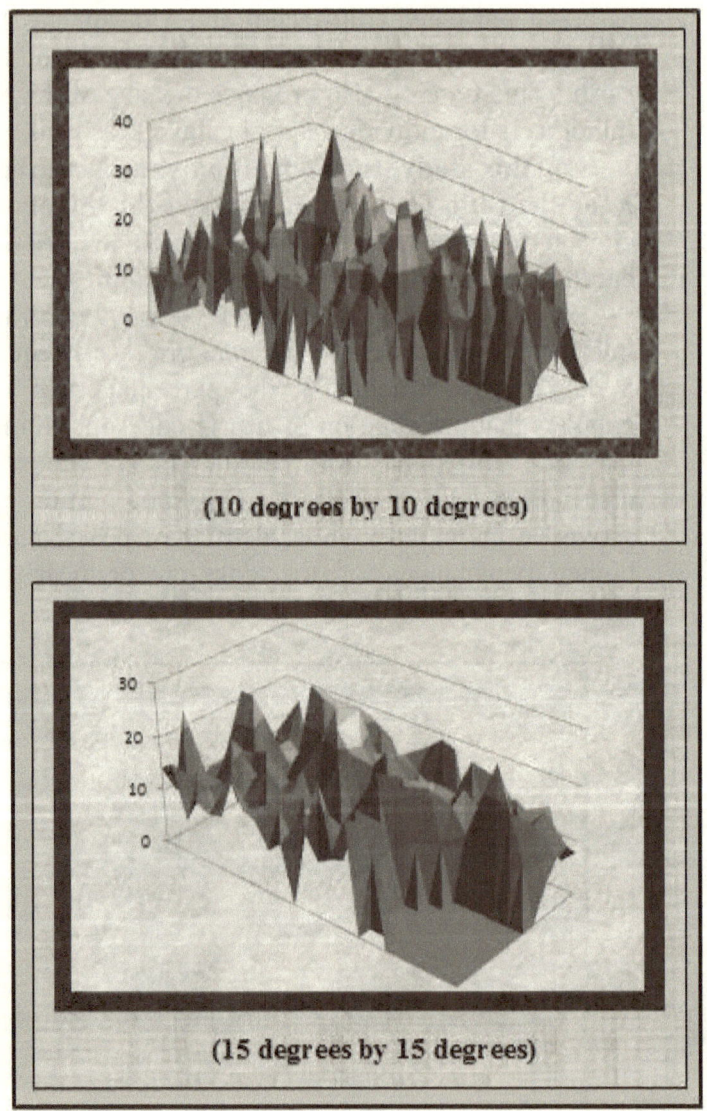

(10 degrees by 10 degrees)

(15 degrees by 15 degrees)

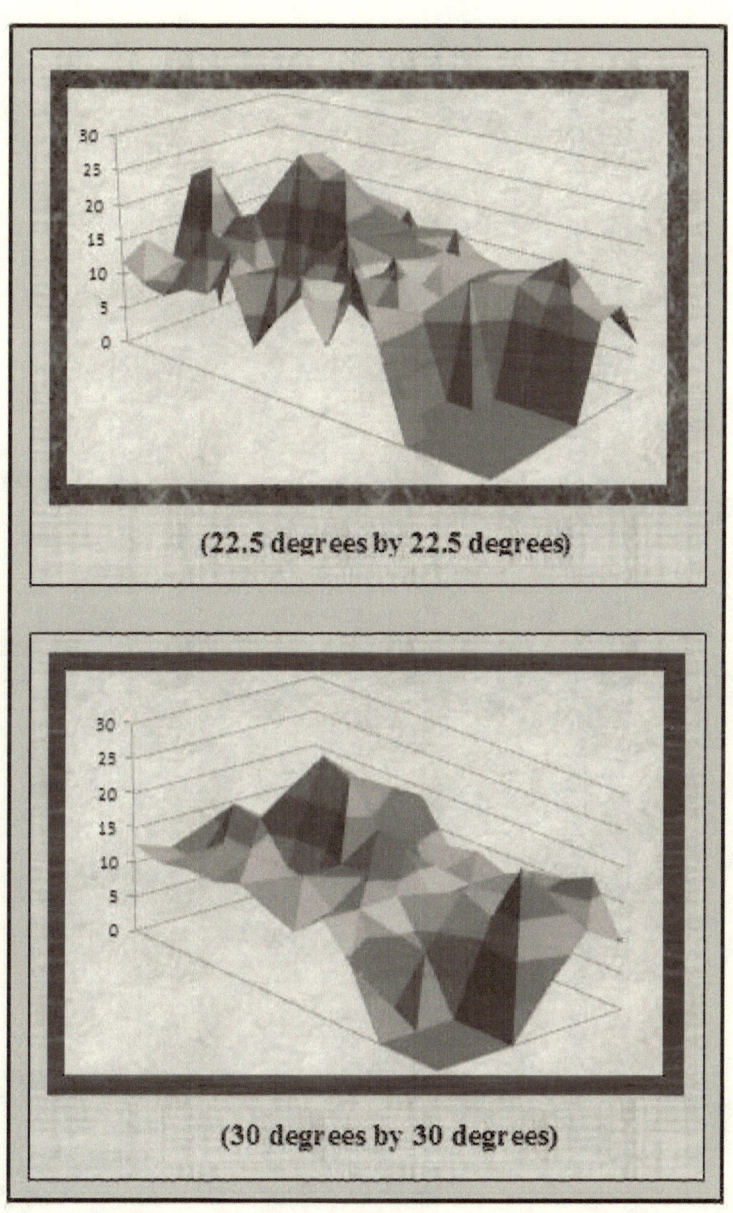

(22.5 degrees by 22.5 degrees)

(30 degrees by 30 degrees)

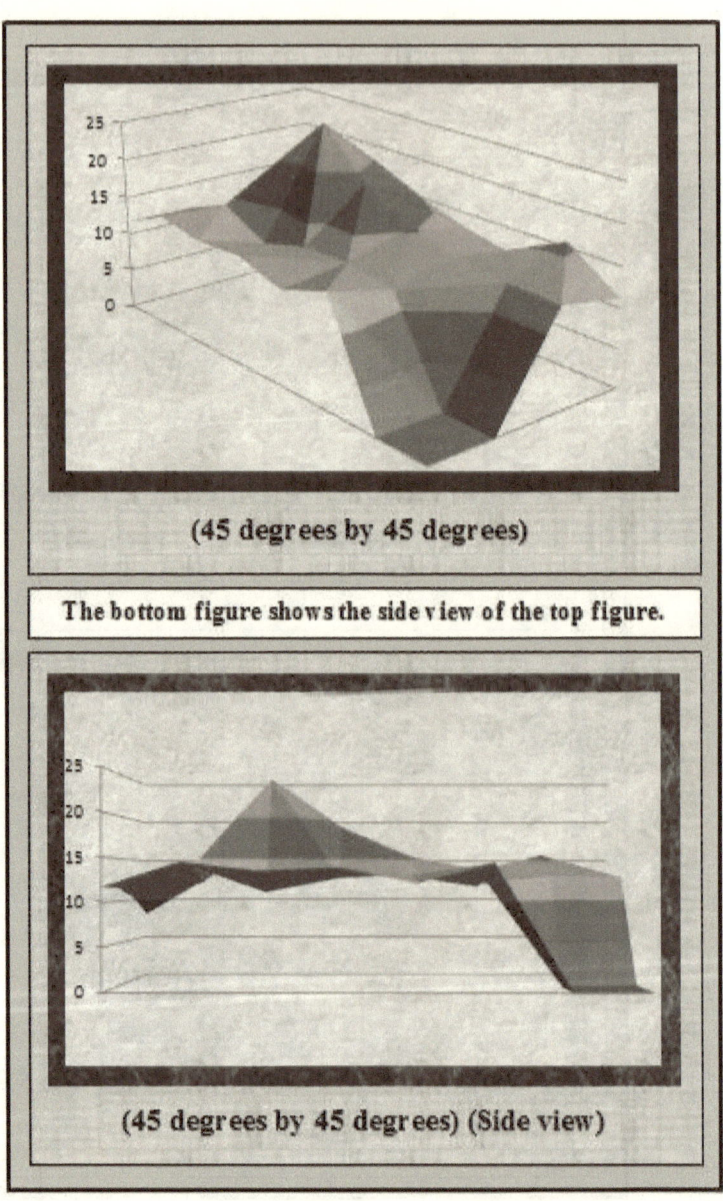

(45 degrees by 45 degrees)

The bottom figure shows the side view of the top figure.

(45 degrees by 45 degrees) (Side view)

As it is clearly shown in the last figure, for the data provided in The Morphological Catalog of Galaxies, generally the percentage of galaxies with rotational planes viewed as being nearly edge-on or parallel to the line of sight from earth, average at about 13 percent. Yet, the detected peak clearly demonstrates a height of about 25 percent, which is nearly double the average of the values observed in the other spatial directions.

The approximate location of the one peak detected in this section is at about 120 degrees Longitude East and about 30 degrees Latitude North (in the galactic coordinate system) which corresponds to the direction pointing towards the Northern Equatorial Pole.

The second major peak, which is expected to be in the opposite spatial direction, is missing. This is due to the fact that, the opposite direction to the first peak happens to correspond to the Southern Equatorial Pole where it is shown as a void, due to the lack of galactic data in The Morphological Catalog of Galaxies.

Therefore, by using a Catalog that includes the data on galaxies visible from the Southern hemisphere, the second peak is expected to be found at about 300 degrees Longitude East and about 30 degrees Latitude South which corresponds to the direction pointing towards the Southern Equatorial Pole.

- **The Principal General Catalog of Bright Galaxies**

This Catalog provides the galactic inclinations by listing the sizes of the major and the minor axes of individual spiral galaxies.

In this case, again values of greater than 3 for the ratio of the major axis to the minor axis were chosen to isolate the galaxies which are viewed as edge-on or nearly edge-on from earth.

In the previous part, one peak was detected by using the data provided in The Morphological Catalog of Galaxies. The data provided in The Principal General Catalog of Bright Galaxies is expected to confirm the existence and location of the first peak, and is also expected to show the second peak.

For this part of the study, the sky is divided into Longitude and Latitude grid sizes of (15 by 15), (22.5 by 22.5), (30 by 30) and (45 by 45) degrees.

Again, on the smaller grid sizes used, particularly for the grid size of (15 by 15) degrees (shown below), the percentages of spiral galaxies that are viewed as edge-on or nearly edge-on with respect to the line of sight from earth, show almost random distribution all over the plot.

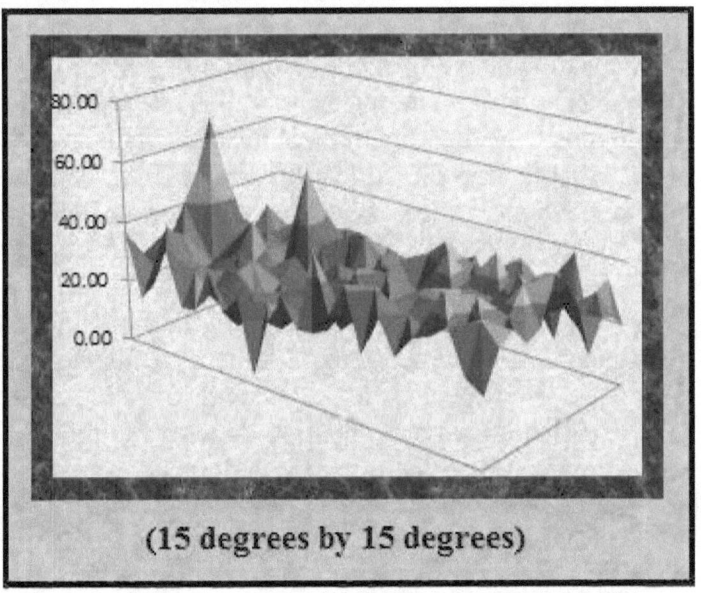

(15 degrees by 15 degrees)

However, as the data corresponding to the larger and larger grid sizes are plotted (shown in the following figures), more and more of the peaks and valleys combine and average out and show as more or less flat sections.

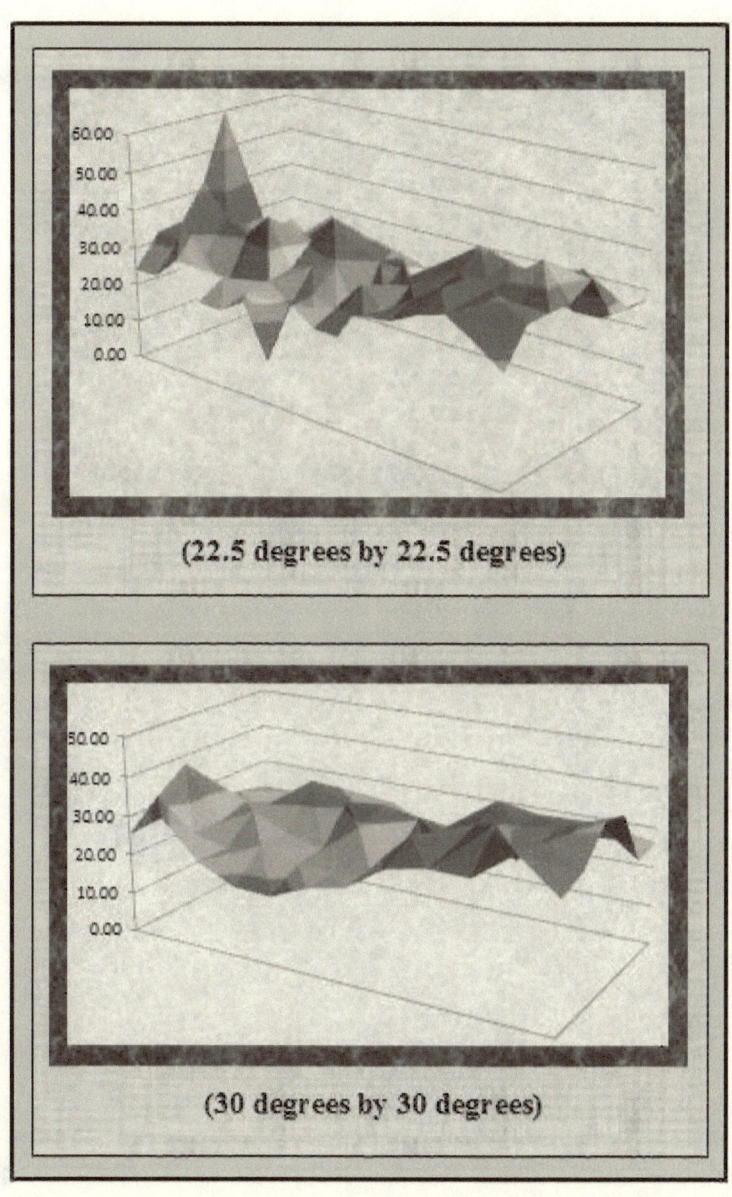

(22.5 degrees by 22.5 degrees)

(30 degrees by 30 degrees)

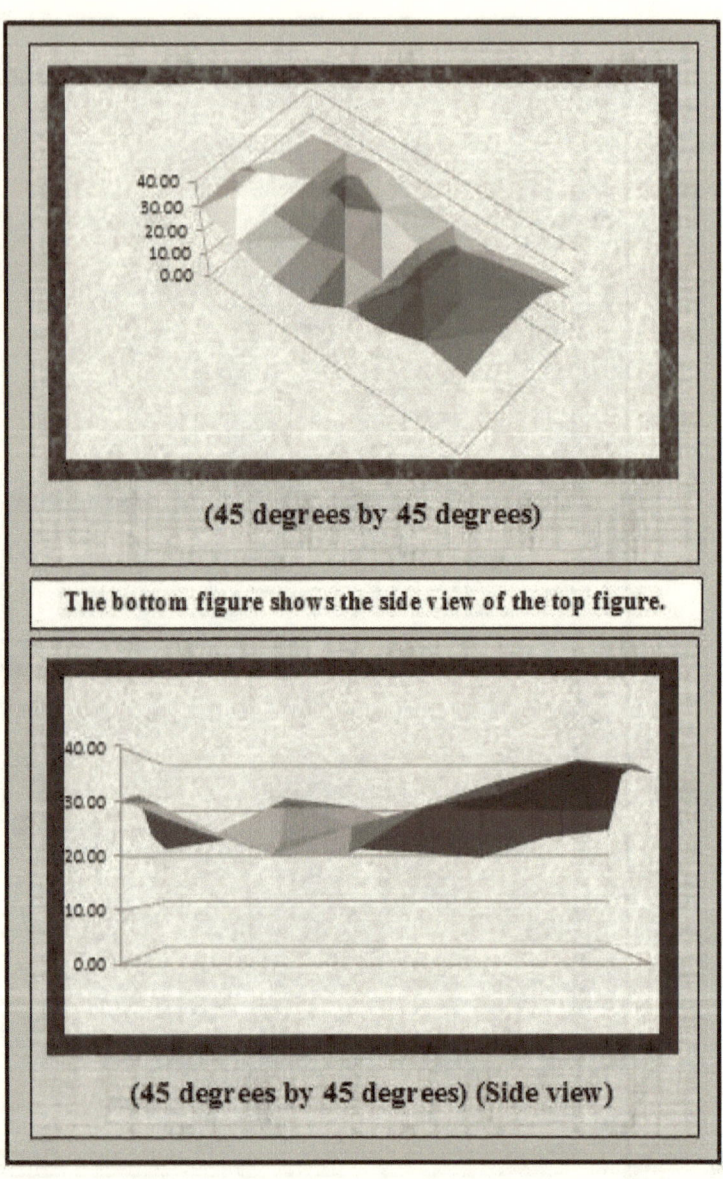

(45 degrees by 45 degrees)

The bottom figure shows the side view of the top figure.

(45 degrees by 45 degrees) (Side view)

During this part of the study, as it is shown in the last two figures, the existence and location of the first peak is confirmed. Also, the second peak is clearly demonstrated to exist where it was predicted to be. However, the second peak is found to be much wider spread than the first peak.

The size difference between the two peaks is due to the difference between the sizes of the view angles from earth's location looking towards this universe's birthplace as compared to looking directly towards its outer perimeter. The following drawing shows the reasoning for the existence of such size differences in the view angles.

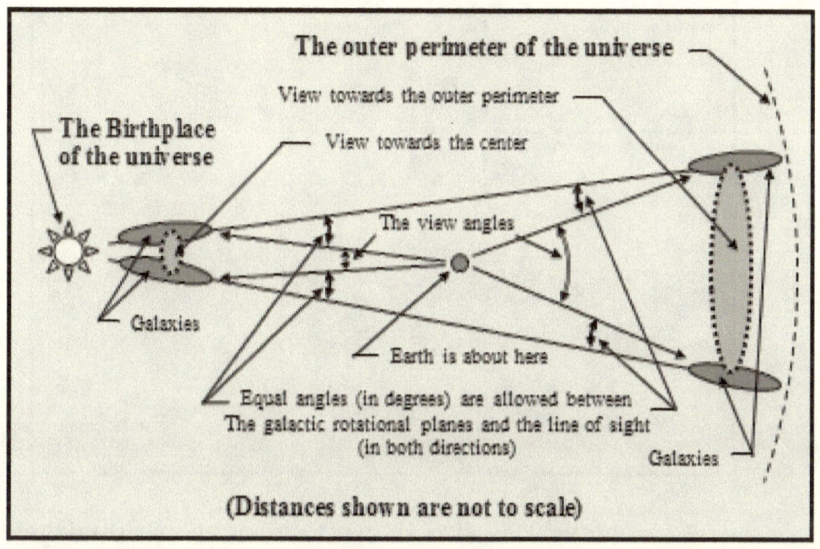

The outer perimeter of the universe

View towards the outer perimeter

The Birthplace of the universe

View towards the center

The view angles

Galaxies

Earth is about here

Equal angles (in degrees) are allowed between
The galactic rotational planes and the line of sight
(in both directions)

Galaxies

(Distances shown are not to scale)

Note that, as shown in the above figure, the maximum allowed angles that the rotational planes of galaxies can make with respect to the line of sight in both directions, namely towards the birthplace and towards the outer perimeter of this universe, are identical. Yet, the sizes of the view angles in those two directions are shown to be quite different.

This is the very effect that is detected in part 2 of this study and is clearly shown in figures corresponding to 45 degrees by 45 degrees grid size.

This method has indicated that the direction pointing towards the birthplace of the universe is at about 120 degrees Longitude East and about 30 degrees Latitude North, according to the galactic coordinate system. **These coordinates also correspond to the Northern Equatorial Pole.**

The following figure shows the direction pointing towards the birthplace of this universe, in relation to earth and sun.

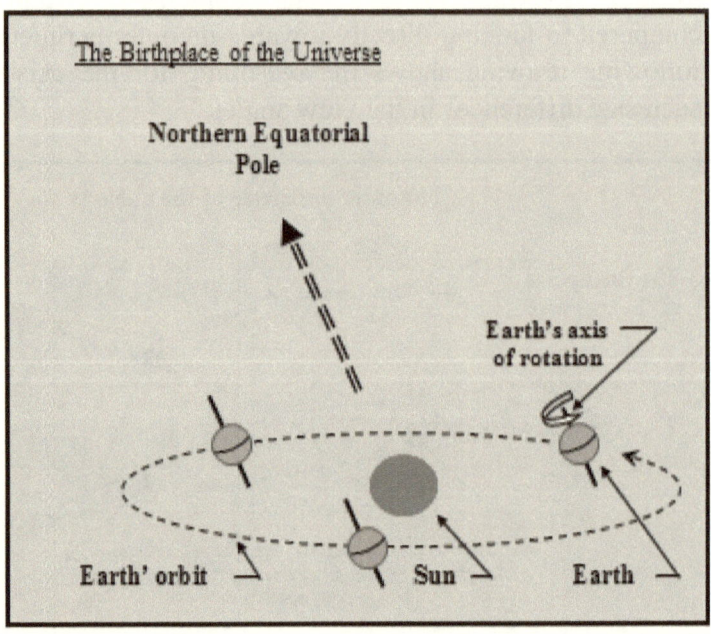

The Birthplace of the Universe

Northern Equatorial Pole

Earth's axis of rotation

Earth' orbit   Sun   Earth

The second peak is also shown to be at about 300 degrees Longitude East and about 30 degrees Latitude South (corresponding to the direction pointing towards the Southern Equatorial Pole), where it was expected to be.

The Third Method:
Using the full-faced view of the rotational planes of spiral galaxies

This method also takes advantage of the fact that, the eventual orientation of the plane of rotation of galaxies, by the time they evolve into spiral galaxies, is more or less parallel to the line of sight from the birthplace of the universe.

Therefore, spiral galaxies which happen to be along any straight line between the birthplace of the universe and its outer perimeters are expected to be viewed as edge-on or nearly edge-on to each other. In other words, <u>they are not likely to be oriented so that their planes of rotations are perpendicular to the line of sight from the birthplace of the universe.</u>

The following figure shows **the least likely orientation of rotational planes of spiral galaxies** with respect to the birthplace of this universe and its outer perimeter.

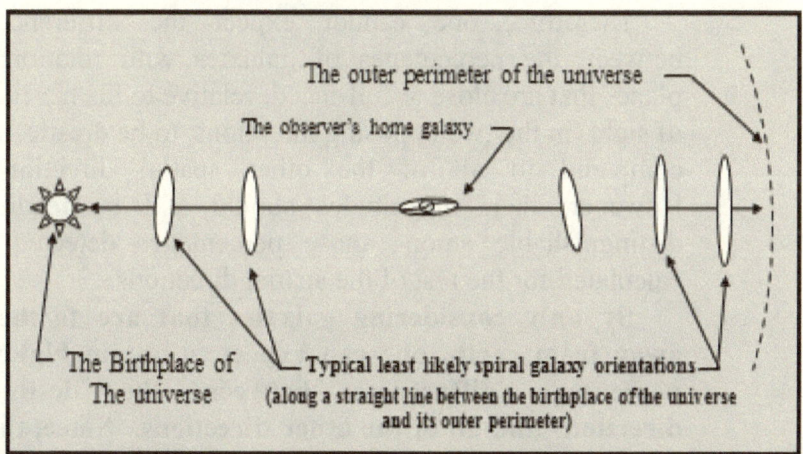

Using this method and the readily available data presented in different galaxy Catalogs, one must be able to identify the direction (Longitude and Latitude coordinates) that points towards the birthplace of this universe. Because, according to the above figure, by looking at all of the spiral type galaxies that are spread in different spatial directions,

One can expect to detect two opposing directions, hosting the lowest percentages of galaxies with rotational planes that are full-faced or nearly full-faced with respect to his/her line of sight from earth.

Note that, in this method, the overall rotational directions of the galaxies are irrelevant.

Therefore, in order to narrow down the direction pointing towards the birthplace of the universe one only needs to pay close attention to the apparent angles of the spiral galaxies' rotational planes with respect to his/her line of sight from earth.

Note that, due to the long term effects of gravitational forces that neighboring galaxies exert on each other, the general orientation of most galaxies rotational planes are no longer what they used to be, when they were first formed.

Therefore, one cannot expect the differences between the percentages of galaxies with rotational planes that are close to full-faced, relative to his/her line of sight, in the two opposing directions, to be drastic, as compared to all of the other spatial directions. However, their magnitudes should still be readily distinguishable among those percentages detected / calculated for the rest of the spatial directions.

By only considering galaxies that are farther away from earth, one can expect to obtain higher percentage differences between the desired directions and all of the other directions. Since, the gravitational forces of such distant galaxies have not had sufficient time to randomize the orientation of their rotational planes.

• Results

This method is based on identifying the two opposing directions in the sky that are hosting the lowest percentages of spiral galaxies with rotational planes that are full-faced or nearly full-faced with respect to the line of sight, as they are viewed from earth. The percentages of spiral galaxies of interest are calculated and plotted for various Longitude and Latitude grid sizes, using an Excel spreadsheet program.

The two standard galaxy Catalogs used, namely The Morphological Catalog of Galaxies and The Principal General Catalog of Bright Galaxies, are examined individually and their respective results are presented separately.

Note that, the lowest percentage values calculated are of interest for this particular method, since they indicate the grid divisions that host the least percentages of spiral galaxies which are full-faced or nearly full-faced, as seen from earth.

Therefore, in the following pages, each and every figure is followed by a complementary figure. **The top figures** are based on the actual calculated percentage values, and **the bottom figures** are based on the inverse of those calculate values (multiplied by 1000).

The bottom figures make it easier to see the coordinates that are of particular interest, since **the areas with the lowest actual calculated percentage values stand out as peaks, rather than hide as valleys.**

- ## The Morphological Catalog of Galaxies

This Catalog provides the galactic inclinations (ratio of the major axis to the minor axis) of the rotational planes of the spiral galaxies, using a scale of 1 to 5, where 1 designates galaxies that are viewed as being flat-faced (full-faced) and 5 designates those that are viewed as edge-on with respect to the line of sight from earth. This Catalog also provides the size of the major axis and minor axis for individual spiral galaxies, as well.

For this method, to isolate the galaxies which are full-faced or nearly full-faced, as viewed from earth, ratios of the major axis to the minor axis of less than 1.5 were chosen.

As shown in the following two figures, when the sky around earth is divided into small Longitude and Latitude grid sizes (equivalent to 15 degrees by 15 degrees), the percentages of the spiral galaxies that are full-faced or nearly full-faced with respect to the line of sight from earth, show almost random distribution all over the plot.

However, as the data corresponding to the larger and larger grid sizes are plotted, such as the ones shown below for grid sizes of (22.5 by 22.5), (30 by 30) and (45 by 45) degrees, more and more of the peaks and valleys combine and average out and show as more or less flat sections. In the meantime, one major valley in the top figures (which is shown as a major peak in the bottom figures) becomes more and more apparent and clearly stands out.

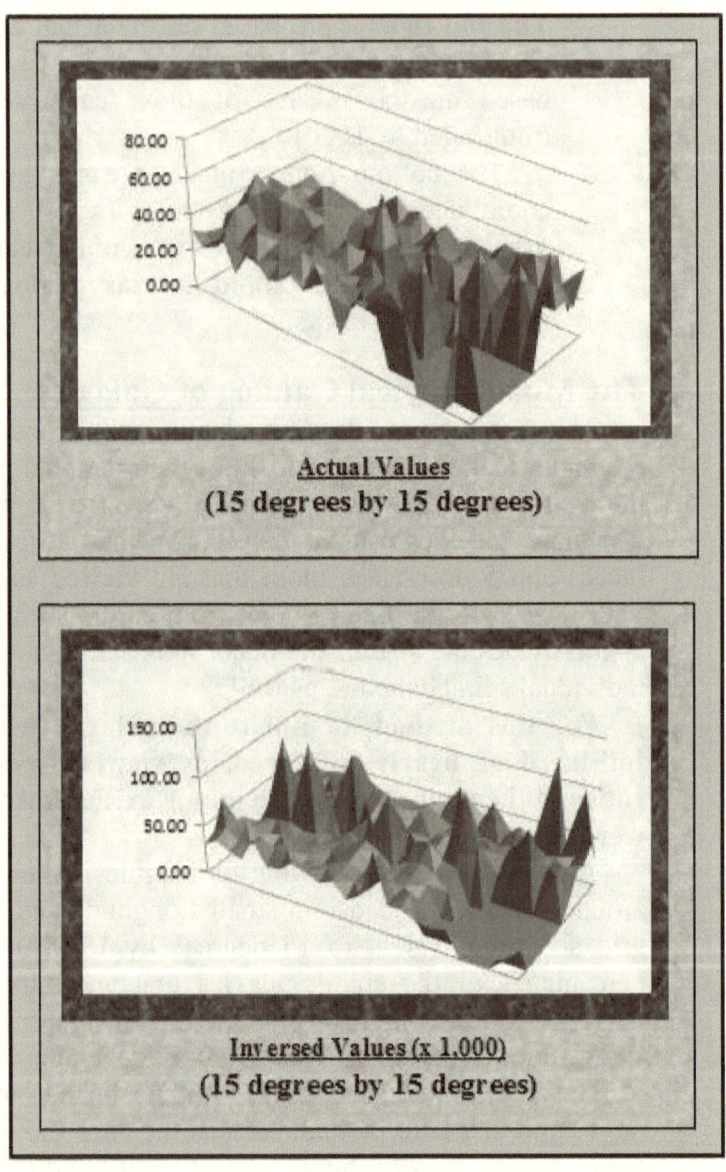

Actual Values
(15 degrees by 15 degrees)

Inversed Values (x 1,000)
(15 degrees by 15 degrees)

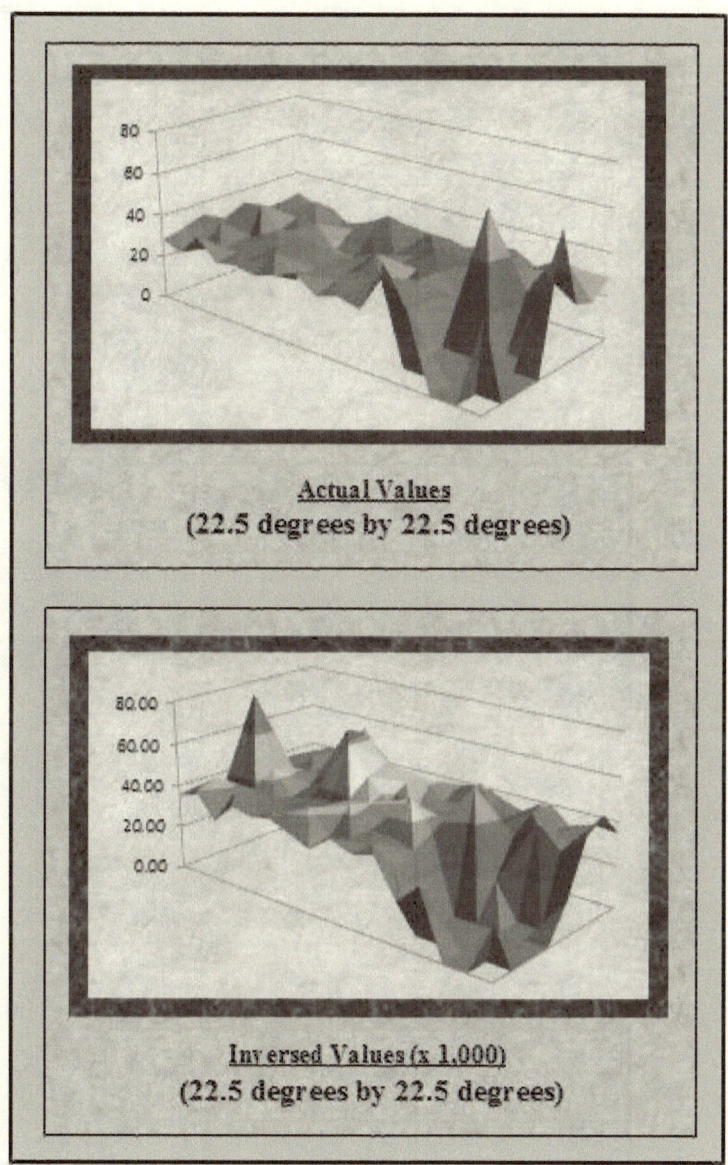

Actual Values
(22.5 degrees by 22.5 degrees)

Inversed Values (x 1,000)
(22.5 degrees by 22.5 degrees)

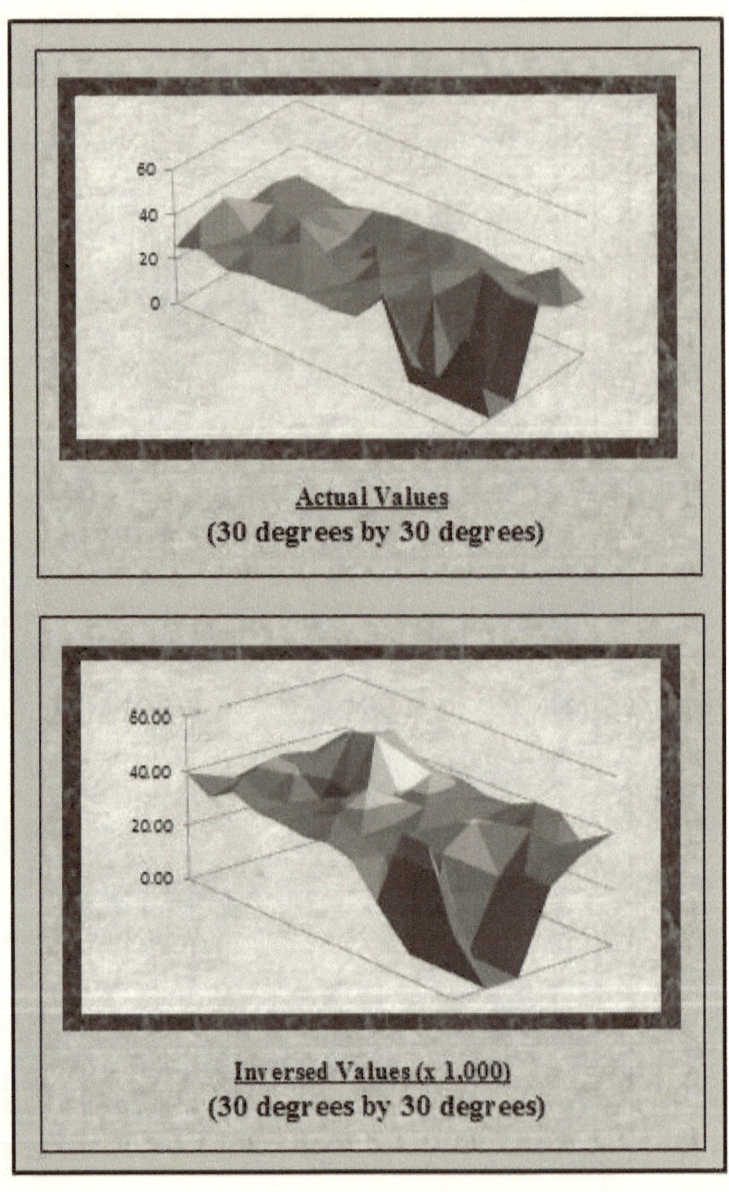

Actual Values
(30 degrees by 30 degrees)

Inversed Values (x 1,000)
(30 degrees by 30 degrees)

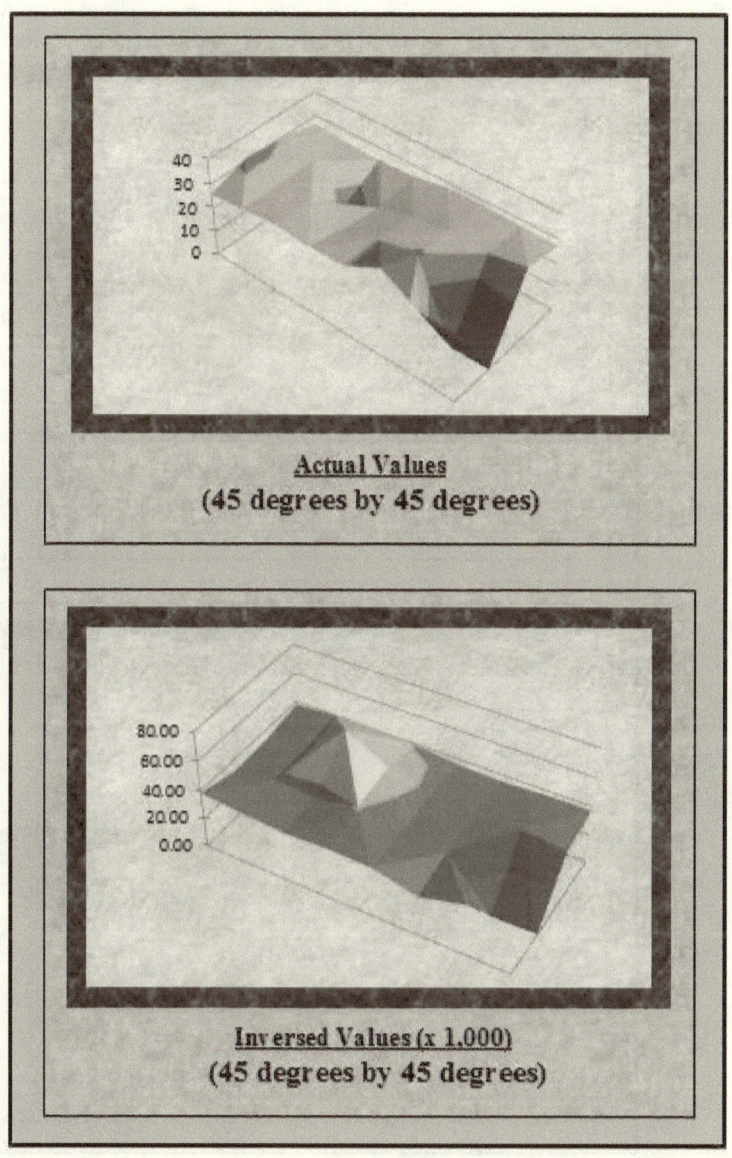

Actual Values
(45 degrees by 45 degrees)

Inversed Values (x 1,000)
(45 degrees by 45 degrees)

The following two figures show the edge view of the last two figures.

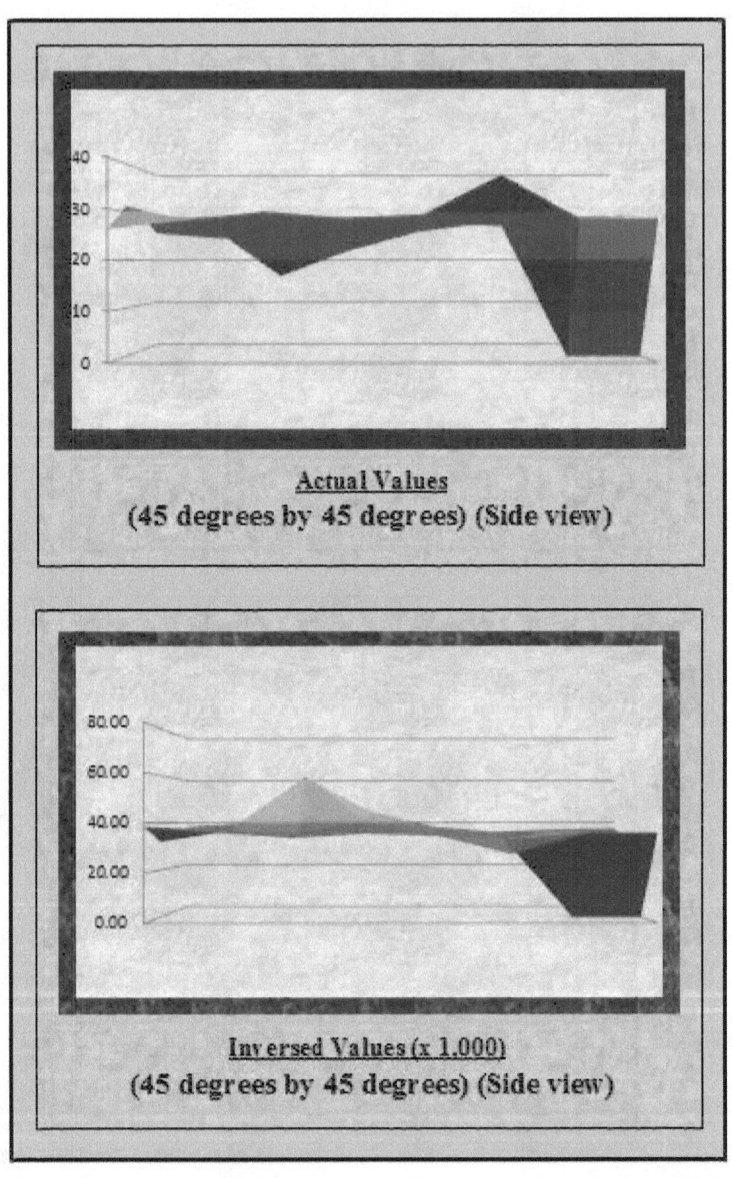

Actual Values
(45 degrees by 45 degrees) (Side view)

Inversed Values (x 1,000)
(45 degrees by 45 degrees) (Side view)

As it is clearly shown in the above two figures, for the data provided in The Morphological Catalog of Galaxies, generally the percentage of galaxies with rotational planes viewed as being full-faced or nearly full-faced or perpendicular to the line of sight from earth, average at about 28 percent. Yet, the detected lowest value (the valley shown) clearly demonstrates a value of about 17 percent, which is 40 percent less than the average values observed in the other spatial directions. This valley is clearly transformed into a peak in the bottom figure.

The approximate location of the one peak detected in this section is at about 120 degrees Longitude East and about 30 degrees Latitude North (in the galactic coordinate system) which corresponds to the direction pointing towards the Northern Equatorial Pole.

The second major peak, which is expected to be in the opposite spatial direction, is missing. This is due to the fact that, the opposite direction to the first peak happens to correspond to the Southern Equatorial Pole where it is shown as a void, due to the lack of galactic data in The Morphological Catalog of Galaxies.

Therefore, by using a Catalog that includes the data on galaxies visible from the Southern hemisphere, the second peak is expected to be found at about 300 degrees Longitude East and about 30 degrees Latitude South which corresponds to the direction pointing towards the Southern Equatorial Pole.

- **The Principal General Catalog of Bright Galaxies**

This Catalog provides the galactic inclinations by listing the sizes of the major axis and the minor axis of individual spiral galaxies.

For this method, to isolate the galaxies which are full-faced or nearly full-faced, as viewed from earth, ratios of the major axis to the minor axis of less than 1.5 were chosen.

In the previous part, one peak was detected by using the data provided in The Morphological Catalog of Galaxies. The data provided in The Principal General Catalog of Bright Galaxies is expected to confirm the existence and location of the first peak, and is also expected to show the second peak. For this part of the study, the sky is divided into Longitude and Latitude grid sizes of (15 by 15), (22.5 by 22.5), (30 by 30) and (45 by 45) degrees.

Again, on the smaller grid sizes used, particularly for the grid size of equivalent to 15 by 15 degrees (shown below), the percentages of spiral galaxies that are viewed as full-faced or nearly full-faced with respect to the line of sight from earth, show almost random distribution all over the plot.

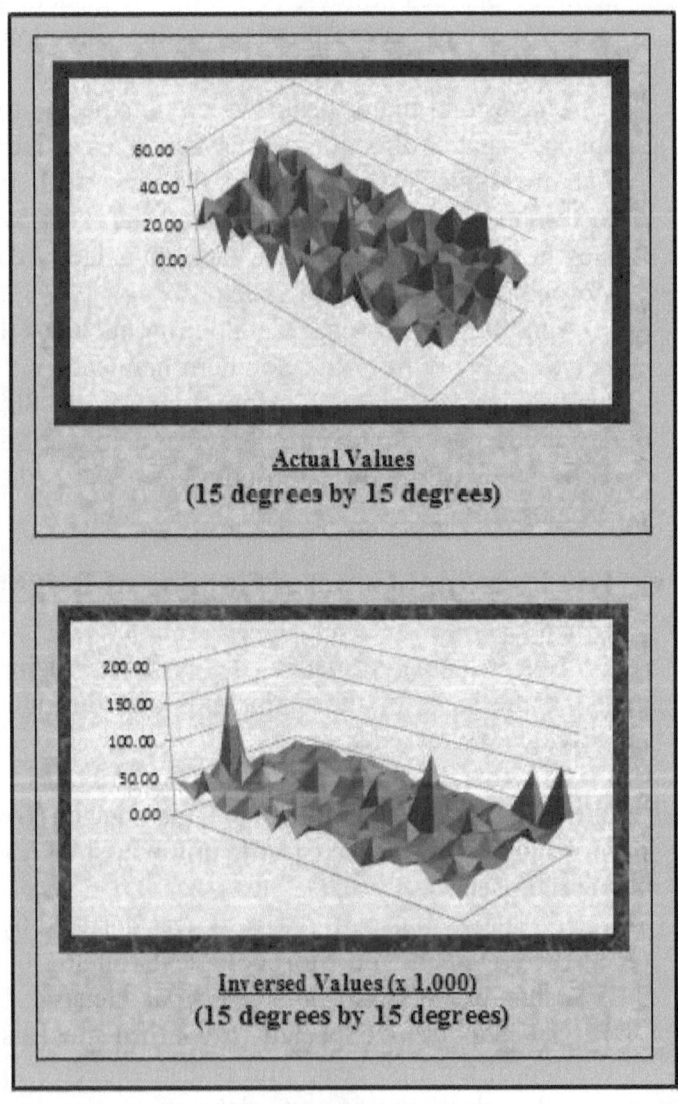

Actual Values
(15 degrees by 15 degrees)

Inversed Values (x 1,000)
(15 degrees by 15 degrees)

However, as the data corresponding to the larger and larger grid sizes are plotted, such as the ones shown below for grid sizes of (22.5 by 22.5), (30 by 30) and (45 by 45) degrees, more and more of the peaks and valleys combine and average out and show as more or less flat sections.

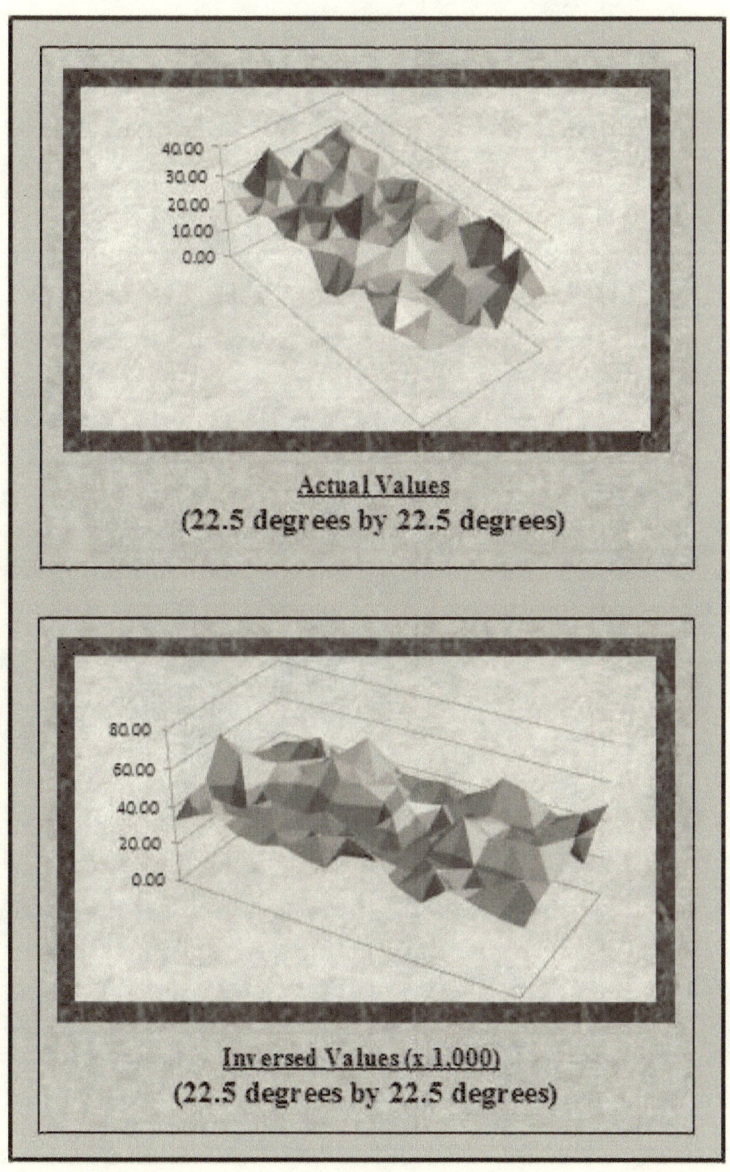

Actual Values
(22.5 degrees by 22.5 degrees)

Inversed Values (x 1,000)
(22.5 degrees by 22.5 degrees)

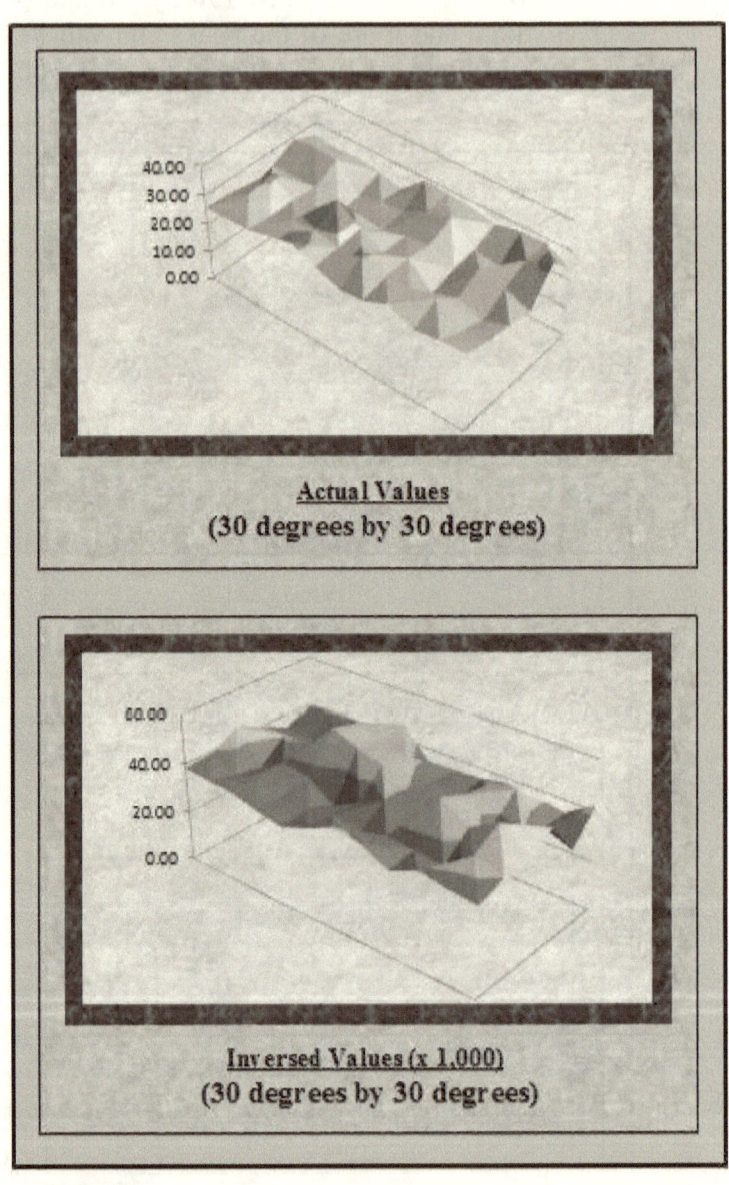

Actual Values
(30 degrees by 30 degrees)

Inversed Values (x 1,000)
(30 degrees by 30 degrees)

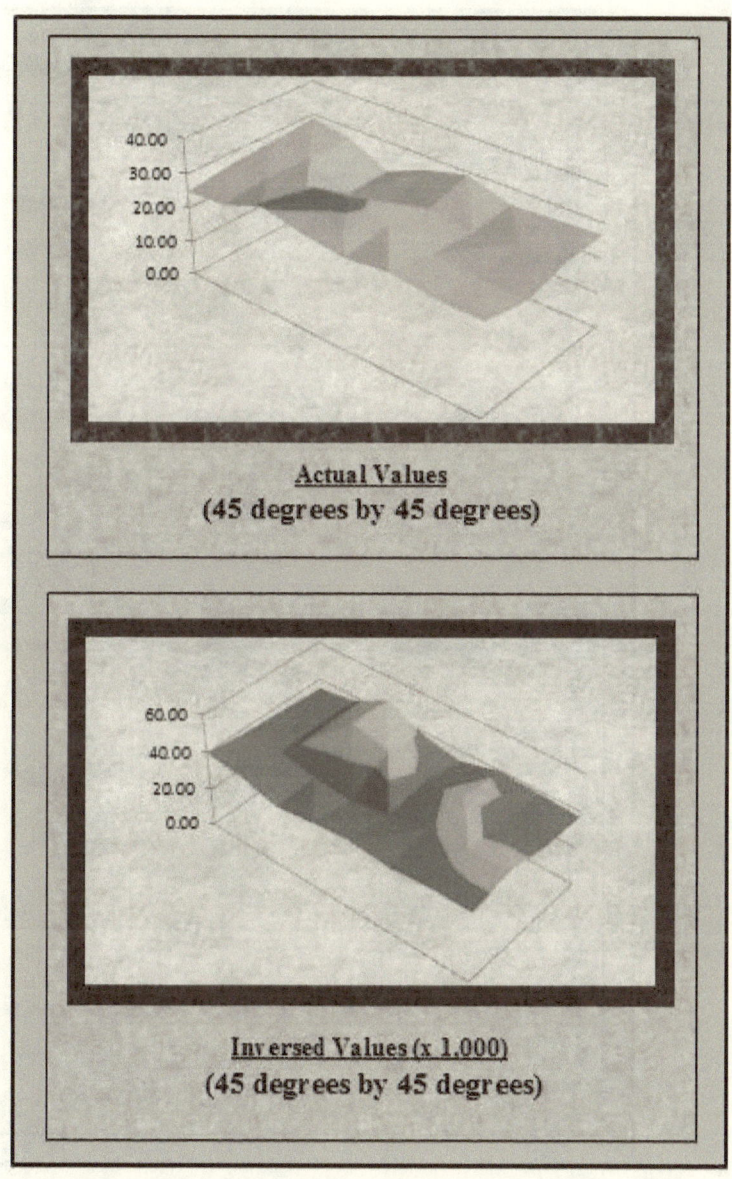

Actual Values
(45 degrees by 45 degrees)

Inversed Values (x 1,000)
(45 degrees by 45 degrees)

The following figures show the edge view of the last two figures.

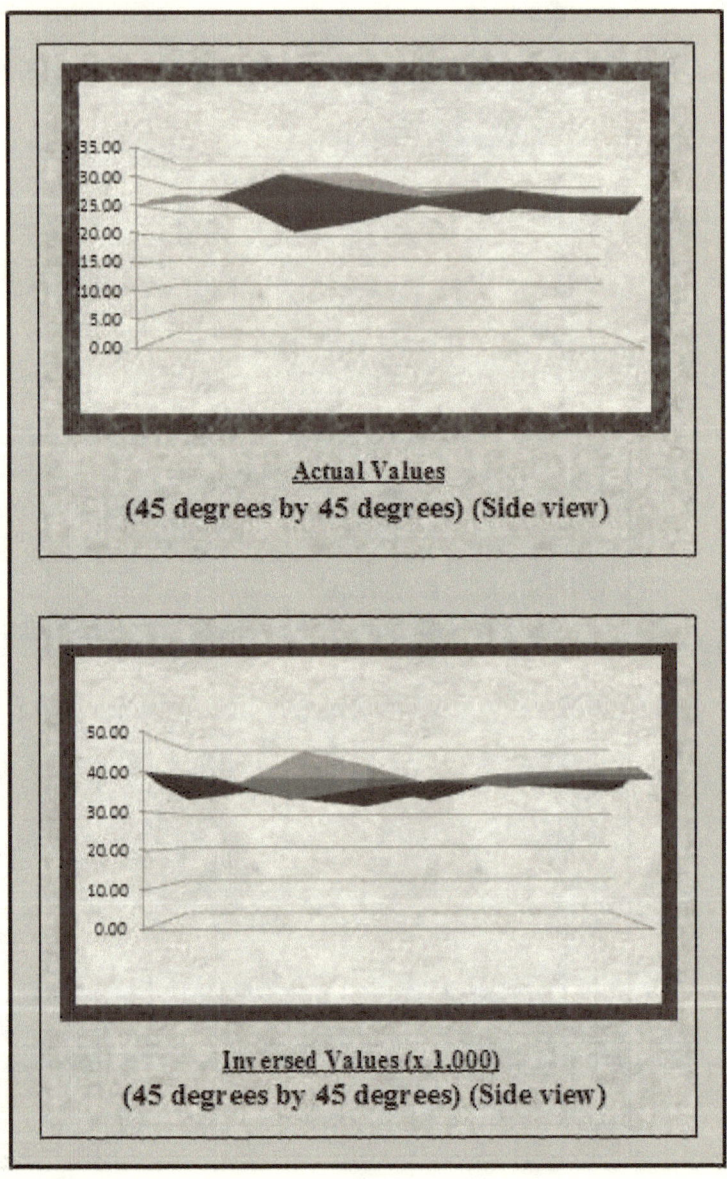

Actual Values
(45 degrees by 45 degrees) (Side view)

Inversed Values (x 1,000)
(45 degrees by 45 degrees) (Side view)

During this part of the study, as it is shown in the above two figures, the existence and location of the first peak is confirmed. Also, the second peak is demonstrated to exist where it was expected / predicted to be.

This method also has indicated that the direction pointing towards the birthplace of the universe is at about 120 degrees Longitude East and about 30 degrees Latitude North, according to the galactic coordinate system. **These coordinates also correspond to the Northern Equatorial Pole.**

The following figure shows the direction pointing towards the birthplace of this universe, in relation to earth and sun.

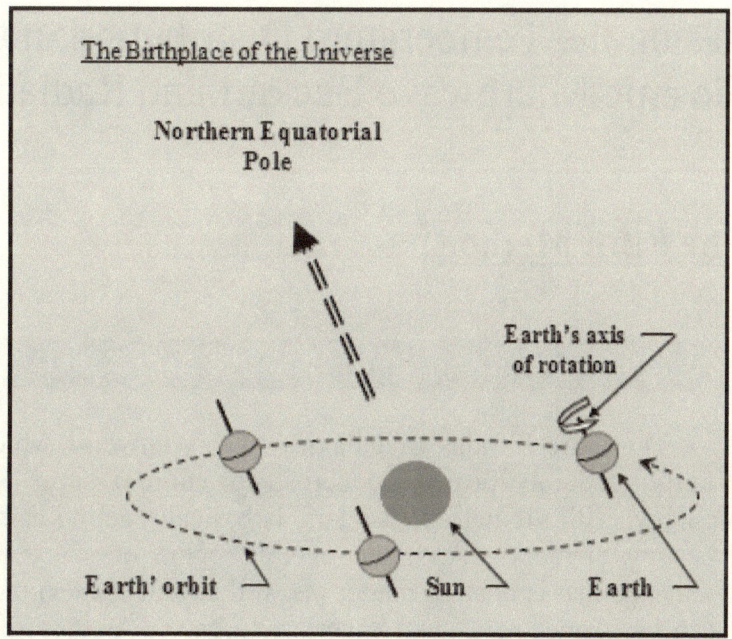

The second peak is also shown to be at about 300 degrees Longitude East and about 30 degrees Latitude South (corresponding to the direction pointing towards the Southern Equatorial Pole), where it was expected to be.

Using the Temperature Distribution in the Cosmic Microwave Background Radiation

(The Fifth Method)

- ### Necessary background information

There are billions of galaxies in this universe which are fairly randomly positioned across its entire volume. Each galaxy, with the help of its stars, is basically acting as a huge heating element. And, the overall region of space in this universe that is hosting matter (as well as anti-matter) particles can be likened to a giant spherical furnace. As these heating elements (stars in the galaxies) operate and generate heat, in the form of various types of radiation, the entire volume of this furnace will eventually experience nearly the same temperature.

However, the local energy levels (temperatures) are continuously affected by the ongoing nuclear activities within local stars. In other words,

Regions of this universe that host higher number densities of galaxies (active stars, to be more precise), automatically contribute more effectively to the overall energy level (temperature) of the background radiation in their respective vicinities. Hence, they cause the formation of hot spots in their localities.

Suppose a heat sensing satellite is placed inside this universe (the universal furnace) so that it is somewhere between its center and its outer limits.

If sensitive enough, the satellite will detect the direction pointing roughly towards the outer perimeter of the universe that is nearest to it. It will do so by detecting ONE major region of space which is the coldest as compared to all of the other regions. This is expected due to <u>net energy loss</u> through the exterior boundaries of the region of space which is hosting galaxies. In other words, the satellite will detect the direction pointing towards the region of space that is void of any heating sources, namely galaxies or even stars, and does not radiate any energy back.

> **"Once the direction pointing towards
> the coldest region of space is found, the
> opposite direction will be pointing
> towards <u>the general direction</u> of the
> birthplace of the universe."**

The opposite direction of the coldest region of space is stated as being **<u>the general direction</u>** pointing towards the birthplace of the universe because the galaxies located near the outer limits of the region of space which is hosting matter particles, are not spaced uniformly. They are more or less randomly positioned, due to the availability of matter particles that gravitated towards each other and eventually formed those galaxies.

Note that, even though the direction indicated by this method can only be a general one, it still has to be fairly close to the directions indicated by the other methods presented in this chapter.

The Cosmic Microwave Background Radiation (CMBR) actually represents this background temperature of the whole universe. The underlying waves in the cosmic microwave background radiation are due to:

- The initial phase vibrations which were introduced into this space and prompted the birth of this universe, and

- The annihilation of the matter and anti-matter particles, nearly at the end of the rapid expansion era.

However, as it was stated earlier, each and every active star contributes to and enhances (affects) the energy content (temperature level) of the background radiation in space in its immediate vicinity. This is how galaxies and clusters of galaxies, as a whole, promote the formation of hot spots in that medium's energy profile.

In short,

"To identify the direction pointing towards the birthplace of this universe, one needs to use a satellite that is capable of sensing the temperature of the background radiation in all spatial directions. Since, the direction pointing towards the birthplace of the universe is in the general opposite direction of where the coldest spot is."

Note that, in the beginning, when galaxies were not formed yet, the temperature of the cosmic microwave background radiation did not include the effects due to stars activities.

However, by following the very same methodology that is described in this section, one can identify the general direction pointing towards the birthplace of this universe. Since,

"The coldest region of space identified by the instruments will be towards the general direction of the outer layers of this universe, because at those regions, radiation that is propagating towards the outside is not

compensated for by any radiation
propagating inwards."

• Data and Results

So far a few satellites have been dedicated to this very task of mapping the temperature distribution in the cosmic microwave background radiation, in all spatial directions. They have clearly shown the hot spots, the general texture of the temperature distribution in the cosmic microwave background radiation, as well as the coldest region in space, as they surveyed the entire volume of space, both in direction and in distance.

The latest satellite which is still performing similar tasks has been / is the Planck space telescope.

The Planck space telescope has improved the resolution in the details of such a map obtained in this fashion. Particularly, it has identified a fair size region of space in the Southern hemisphere to be the coldest. The following figure shows the location of this coldest region in space.

Therefore,

According to this method, and the data collected by the Planck Space Telescope, as well as other satellites before it, <u>the birthplace of the universe is in the general direction of Northern Equatorial hemisphere.</u>

Note that, this finding is in agreement with the results obtained using the rotational planes of spiral galaxies, as presented earlier in this chapter.

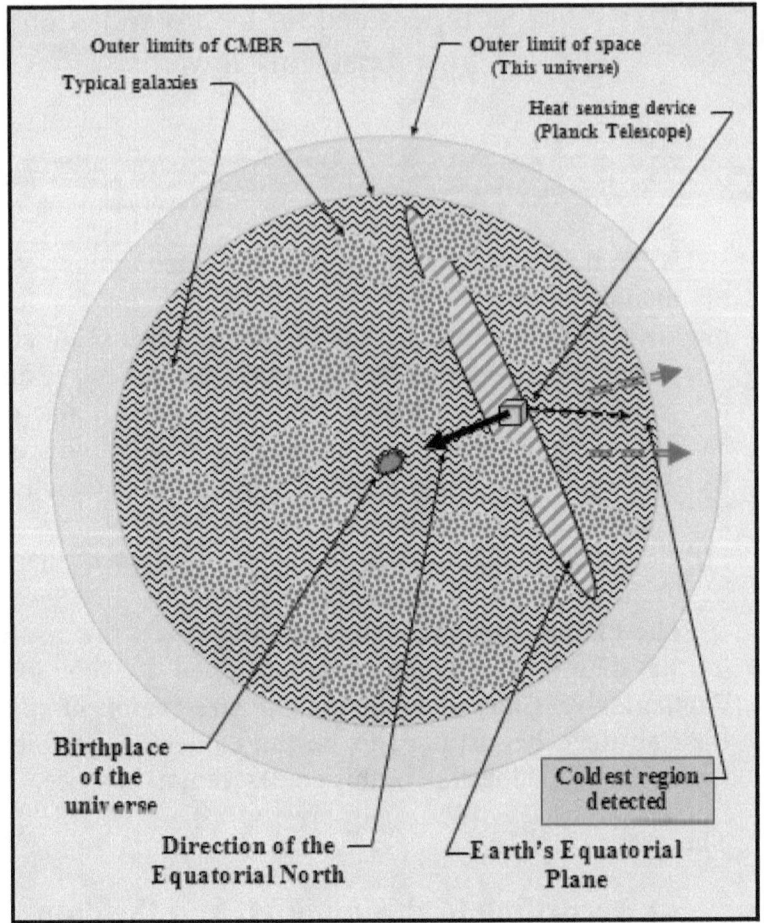

Conclusions

Even though, the first two methods presented in this chapter were looking for exactly opposite effects of the very same phenomenon, namely the conservation of momentum of the particles that had formed the galaxies, they clearly confirmed each other's findings.

Both methods concluded that the direction from earth pointing towards the birthplace of the universe is at about 120 degrees Longitude East and about 30 degrees Latitude North, according to the galactic coordinate system. **These coordinates happen to correspond to the Northern Equatorial Pole of our home planet earth.** The two methods also agreed on the location of the second peak which corresponds to the direction pointing towards the outer perimeter of this universe.

The following figure shows the direction pointing towards the birthplace of this universe, in relation to earth and sun.

The results of the last method presented here were based on the data collected by the Planck Space Telescope, indicated that the coldest region of space was in the general direction of the Southern Equatorial Hemisphere.

This finding automatically implies that the direction pointing towards the birthplace of the universe must be in the general direction of the Northern Equatorial Hemisphere. This finding is in agreement with the findings of the first two methods.

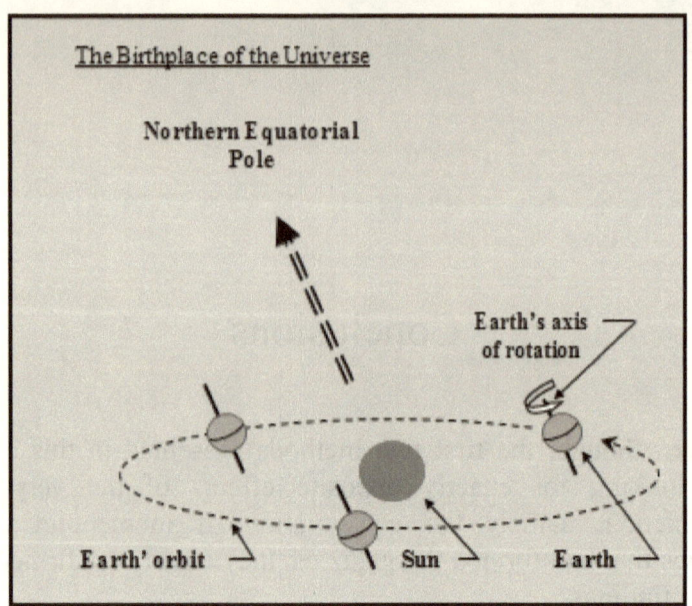

Implications

Some of the implications of the findings presented in this chapter are listed below:

1- Since there is a noticeable ratio of galaxies with rotational planes that are still oriented towards the birthplace of the universe, the following statement can be made with definite certainty,

"This universe has begun with a Big Bang."

Also, due to the very same reason,

"This universe is finite."

Since, if it were infinite, the orientations of the rotational planes of all galaxies should have been totally in random, but as it is shown in this chapter, they are not.

2- **The universe is definitely expanding**, an expansion process that is **not just WITH space, but also IN space**. In fact, it is indicated that,

"The general motion of galaxies that is taking them farther away from the birthplace of the universe is mainly IN space."

3- Analyzing the texture of the temperature distribution in the **cosmic microwave background radiation** in this universe indicates that:

- **The coldest region in deep space which is in the Southern Equatorial Hemisphere** is towards the outer perimeter (the matter frontier) of this universe. This is expected due to the lack of existence of energy sources such as galaxies (stars) further out in that direction. Such imbalance in the distribution of heat (energy) sources encourages the existence of an imbalance in the flow of energy, towards the outside of the matter frontier in this universe.

- **The temperature fluctuations / variations detected in the cosmic microwave background radiation**, as viewed from earth, must exhibit two different overall textures in the Northern Equatorial Hemisphere as compared to the Southern Equatorial Hemisphere.

4- For this particular study, all of the spiral galaxies listed (in both Catalogs) were considered for the first two methods, regardless of their distances from earth.

However, by considering only the galaxies that are located at a certain distance range (for instance, between 4.5 and 5.5 billion light years from earth) and using the relative widths of the two peaks obtained, **one can calculate the distance between the Milky Way galaxy and the birthplace of the universe.** Of course, one must keep in mind that the calculated distance will be as of 5 billion years ago, since it has taken that long for those galaxies lights to reach earth.

Note that, while applying any of the first four methods introduced in the previous chapter, **by only considering galaxies that are farther away from earth, one can expect to obtain higher percentages in the two desired directions as compared to all of the other directions.**

5- Better estimates can be made for the age of the universe.

6- Better estimates can be made for the overall physical size of the universe.

This task can be achieved by repeating the distance calculations (such as mentioned in #4, above) for different sets of galaxies that are located at 3, 5, 7, 10,... billions of light years away from earth, respectively, and correcting them for their ages, accordingly.

General Conclusions

<u>Based on the theories presented in this book, aether exists.</u> The contents of this book clearly demonstrate that, once the existence of aether is accepted and its effects are properly taken into account, the formation of this universe, along with the development of its contents, become readily comprehensible. And, the occurrences of a variety of physical phenomena become readily explicable. In fact, as it has been demonstrated throughout this book, even by pure qualitative analysis, such phenomena become readily predictable.

Aether is proposed to be a continuous medium that is freely flowing everywhere in this universe. It can be likened to a thick fog that is literally encompassing everything. It possesses somewhat different local densities / pressures, but not by much since it is always freely flowing, due to its lack of viscosity.

The ongoing expansion process of the aether medium, as a whole, as well as the gradual decrease in its overall internal pressure and density are clear indications that <u>aether is a compressible</u> medium.

The compressibility of the aether medium can be likened to that of rubber or a gaseous medium. However, aether's compressibility is not due to its being composed of individual particles which can be spaced differently, as its internal pressure is varied, but rather due to its elasticity.

The very existence of aether has given meaning to "**Space**" in this universe, as well as in its accompanying universe, just as the very existence of water has given meaning to the oceans on the surface of this planet.

Also, the very existences of phase vibrations along with their propagation in the medium of aether have given meaning to "**Time**". It is proposed here in this book that,

"The rate at which 'time' is experienced by an object (or a living being) depends on that object's speed relative to its local aether medium, as compared to the propagation speed of phase vibrations in that medium."

In the beginning, aether (in its fluid state) was the only content of this universe. However, over time, due to different circumstances, aether has manifested its many different states / effects which include various forms of matter and energy.

Matter and anti-matter particles are proposed to be bubbles that are formed as phase vibration in the medium of aether generate resonances / spikes, just like bubbles forming in a medium such as water when that medium is exposed to certain ultrasonic waves. By flowing through these bubbles, aether is escaping from this universe and entering the accompanying universe.

The very formation of matter and anti-matter particles has given birth to four major phenomena in this universe, namely:

1- **The force of gravity**, since force of gravity is the drag force which is induced by the accelerated flow of aether towards matter and anti-matter particles.

2- **The electric field**, since electric field is a one-way type of aether flow from negatively charged particles towards positively charged particles.

3- **The magnetic field**, since magnetic field is a round-trip type of aether flow induced by the motion of charged particles in that medium.

4- **The accompanying universe**, since the accompanying universe was formed as the fluid aether started going through matter and anti-matter particles and escape from this universe.

It is also proposed that, light and other electromagnetic waves are only one of many types of phase vibrations in the medium of aether. In the chapter titled "What is Light" it is clearly shown that, light is not made of any kind of particles such as photons. Even such effects as photoelectric effect and laser which were particularly emphasized on as being due to the particle nature of light were shown to be readily explicable based on light being purely a wave phenomenon.

It must be emphasized that, according to the theories introduced in this book, as an object's speed relative to its local aether medium approaches that of phase vibrations in that medium,

- Its **instantaneous speed** relative to its local aether medium dictates the variations in its **'mass'** and the rate at which the passage of **'time'** is experienced.

- Its **instantaneous acceleration** relative to its local aether medium dictates the variations in its **'weight'** and its **'length'** (along its direction of acceleration).

Aether is basically the environment in which all of the contents of this universe are literally floating. In other words,

"Everything in this universe including matter, anti-matter, dark matter, as well as all forms of energy, including dark energy and vacuum energy, are different effects / manifestations of aether, the one and only fundamental content of this physical universe."

In short,

"Aether is the only true content of this universe."

Throughout this book, it is clearly demonstrated that aether is quite a dynamic medium rather than being a stationary one, as it was assumed to be by the nineteenth century physicists. The variety of flow patterns concurrently existing in the aether medium can be readily likened to the variety of flow patterns that are experienced by air molecules in the atmosphere or by water molecules in the oceans.

It is proposed in this book that,

**"The gravitational attraction forces that
protons and neutrons exert on each other,
in the nucleus of any atom, override the
repulsion forces generated between protons
by their positive electrical charges."**

In other words,

**"The gravitational attraction forces
exchanged between protons and neutrons
ARE the forces holding the nucleons
together in the nuclei of the atoms.**

Hence, according to the aether-based theories presented in this book,

"There is no such a force as the weak nuclear force in this universe."

Also, it is proposed that,

**"The individual particles such as
electrons, protons and neutrons, as well
as their corresponding anti-particles, are
bubbles in the medium of aether, bubbles
which are not composed of any smaller
constituents."**

In other words,

"There are no such things as quarks in this universe."

As a result,

"There is no such a force as the strong nuclear force in this universe."

Therefore, according to the aether-based theories presented in this book, there are only the force of gravity and the electric field and the magnetic field that are responsible for holding the constituents of this universe together, from its most microscopic scales to its most macroscopic scales.

The force of gravity and the electric field and the magnetic field are also shown to be quite compatible with each other, since they are shown to be different types of motions of aether in this universe. In other words,

"Force of gravity and the electric field and the magnetic field are already unified."

Also, the wave nature of all matter, as well as anti-matter, particles is explained to be due to the variety of phase vibrations that are constantly propagating in the medium of aether, the very same medium that is hosting all of the particles in existence in this universe. The wavy characteristics of particles are shown to be quite analogous to the wavy motions of small boats floating on the surface of a lake. Both effects are simply due to the presence of phase vibrations in their respective mediums.

Even the double slit experiments with particles such as electrons which apparently lead to different results, depending on the experiment being watched / recorded or not, is readily explained using aether based reasoning.

One of the most important capabilities of aether-based theories, as they are presented in this book, is that they provide a variety of methods by which the birthplace of this universe can be located. Using the already collected data which are openly available to the general public through the internet, **the direction pointing**

towards the birthplace of this universe has been shown to be at about 120 degrees Longitude East and about 30 degrees Latitude North, according to the Galactic Coordinate System. These coordinates happen to correspond to the Northern Equatorial Pole of our home planet earth. In other words,

"The birthplace of this universe has been <u>DISCOVERED</u>."

Note that, since there is a noticeable ratio of spiral galaxies with rotational planes that are still oriented towards the birthplace of the universe, the following two very important statements can be made with definite certainty,

<u>"This universe has begun with a Big Bang."</u>

And,

<u>"This universe is finite."</u>

Since, if it were infinite, the orientations of the rotational planes of all galaxies would have been totally in random, but as it is shown in this book, using the data which were collected by two totally independent sources, they are not.

The following are only some of the phenomena that have been consistently explained in this book, using the variety of aether-based theories proposed. For the complete list, the reader is encouraged to refer to the "Table of Contents", or the "Appendices".

- <u>How did this universe come into existence?</u>, and how can the birthplace of this universe be identified?,

- Why did the '**Initial Rapid Expansion**' of the aether medium occur?,

- <u>Why did the initial rapid expansion slow down?</u>,

- What is '**Aether**'? and how does it relate to the variety of phenomena in this universe?,

- What is '**Space**'?, and why is it expanding?, and will its expansion ever come to a complete halt?,

- What is '**Time**'? and why is it gradually experienced at an increasingly faster pace?,

- What is '**Light**'? and why is its speed gradually increasing?,

- How were '**Matter**' and '**Anti-Matter**' particles formed?,

- Why do particles have specific **discrete physical sizes**?,

- Why are some particles **stable** while some others are not?,

- Why are some isotopes **stable** while some others are not?,

- What is '**Dark Matter**'?,

- What is '**Dark Energy**'?,

- What is '**Vacuum Energy**'?,

- What is '**Electric Charge**'?, and why is it gradually becoming weaker?,

- How is '**Electric Field**' formed?, and why is it gradually becoming weaker?,

- How is '**Magnetic field**' formed?, and why is it gradually becoming weaker?,

- What is '**Electricity**'?, and how is it formed?, and why the speed of electrical signals is gradually increasing?,

- What are '**Superconductors**?,

- How was force of '**Gravity**' formed?, and why is its strength gradually decreasing?,

- How did 'Cosmic Microwave Background Radiation' (phase vibrations) form?,

- What are 'Black Holes'? and why are their properties related to their surface areas rather than their volumes?,

- Why are relativistic effects experienced as the speed of objects approach that of light in the aether medium?,

- What happens to the motion of atoms, nuclei and electrons at absolute zero degrees temperature?

- What causes the 'Casimir Effect'?,

- What is explanation for 'Einstein-Podolsky-Rosen Paradox' of entangled particles?,

- What are 'Bose-Einstein condensates'?,

- Why do electrons form an 'electron cloud' rather than have definite orbital paths, as they go around atomic nuclei?

- Why do 'Lightning Sprites' occur?,

- What is the physical interpretation of 'Planck's Constant'?, and why is it gradually decreasing in magnitude?,

- What affects the results of different types of 'Double-Slit Experiments'?,

- How can 'Photoelectric Effect' be explained, based on light being a wave and not a particle?,

- How do 'Lasers' function, based on light being a wave and not a particle?, and how can their efficiencies be increased?,

- What is the explanation for Mr. Galileo's Experiment?,

- What is the explanation for **Mr. Michelson and Mr. Morley's Experiment,** which indicated no aether flow near earth surface**?**,

- How can **'Precession of the Perihelion of Mercury's Orbit'** be explained, <u>precisely</u>?,

- How can **'Equivalence Principle'** be explained, based on the existence of aether?,

- How are **'Magnetic Hills'** formed?, and how can artificial magnetic hills be constructed?,

- What is the explanation for **'Near-Earth Probe Flyby Anomalies'?**,

- What was the **'Science of Astrology'** based on?, and why it is no longer as accurate as it used to be?,

- Why are major **'Natural Disasters'** becoming more frequent, on a global scale?,

- Why are the **'Physical Sizes of Particles'** gradually increasing?,

- Why are the **'Masses of Particles'** gradually decreasing?,

- Why are the **Planetary Orbits** gradually becoming wider?,

- Explanation for **'Pioneer (10 and 11) anomaly'**,

- Why is the **expansion process of this universe** currently accelerating?,

- What events can be expected to occur in the future?,

And, many, many more.

<u>It must be emphasized that,</u> the approach taken in this book has been purely qualitative, and all of the proposed theories have been presented in a simple fashion so that any layperson be able to readily understand them. Since, every theory must be well examined conceptually, first. It must prove to be consistent in its explanations of various phenomena of interest. Only then, it would be wise to allocate time and resources towards its experimental verification, as well as its mathematical development. In short,

It is very crucial that, the scientific community accepts the existence of aether and takes its various effects into account.

Appendices

Appendix I: List of Nominations

Appendix II: List of Proposed Explanations

Appendix III: List of Proposed Predictions

Appendix IV: List of Proposed Experiments

Appendix V: List of 25 Fundamental Breakthroughs
          (Each one, potentially, deserving a Nobel Prize)

Appendix I:
List of Nominations

1- Esmailzadeh critical aether pressure difference
2- Scharback universal planetary cooling
3- Esmailzadeh zero gravity era
4- Esmailzadeh silent zone
5- Esmailzadeh paradox
6- Esmailzadeh orbital altitude range
7- Esmailzadeh magnetic polar orbit time anomaly
8- Esmailzadeh magnetic hill weight test
9- Esmailzadeh magnetic levitation effect
10- Esmailzadeh electric levitation effect

Appendix II:
List of Proposed Explanations

- ### On the universal scale
  1- Location of the birthplace of the universe
  2- The essence of 'Space', and why it is expanding
  3- Rapid expansion era of the universe, and why it slowed down
  4- Why the expansion of the universe is currently accelerating
  5- Sources of cosmic microwave background radiation (CMBR)
  6- Anomalies detected in CMBR by Planck space telescope
  7- Formation and development of galaxies
  8- Formation and development of star systems
  9- Formation, development and evolution of living beings
  10- Physical interpretation of the Planck's constant
  11- Reasoning for the uncertainties in measuring position and speed

- ### Aether
  1- The essence of 'Aether', and its physical properties
  2- <u>Instantaneous speed</u> relative to the local aether affects 'mass' and 'time', while <u>instantaneous acceleration</u> relative to the local aether medium affects 'length contraction' and 'weight' of objects
  4- The essence of 'dark energy'
  5- The essence of 'vacuum energy'

- ### Matter and anti-matter particles
  1- The essence of matter and anti-matter 'particles'

2- The essence of 'Mass' of particles

3- Why particles are of specific discrete sizes

4- Spontaneous appearance and disappearance of matter particles, either as individuals or as in pairs

5- Apparent production and destruction of various matter and anti-matter particles in particle accelerators

6- Stable vs. unstable particles

7- Why isolated neutrons are unstable

8- Stable vs. unstable (radioactive) isotopes

9- Relation between aether and the half-life of radioactive isotopes

10- Why particles must exhibit wave characteristics

11- Why electrons form an electron cloud around the nuclei?

12- The essence of 'dark matter'

- **Effects and paradoxes**
  1- Casimir effect
  2- Einstein-Podolsky-Rosen paradox, entangled particles
  3- Double-slit experiments (all types)

- **Time**
  1- The essence of 'Time'
  2- Which phenomena affect the rate at which 'time' progresses
  3- Why gravity induces 'Time dilation'
  4- Why acceleration does not necessarily induce 'Time dilation'

- **Light**
  1- Photoelectric effect
  2- Lasers

- **Black holes**
  1- What are 'black holes', and why their mass is proportional to their surface areas rather than their volumes
  2- What is the physical size of a black hole?
  3- 'Time' inside black holes' event horizons
  4- Why no black holes were present when the whole energy content of this universe was condensed in a very small volume of space

5- How does an object experience 'time' as it approaches a black hole and gets captured by it?

- **Gravity**
  1- Formation of the force of gravity, and the reason for its gradual weakening (Why the universal gravitational constant is decreasing)
  2- Gradual widening of planetary orbits
  3- Principle of equivalence
  4- Pioneer (10 and 11) anomaly
  5- Why science of astrology was originally more accurate

- **Magnetic field**
  1- The essence of 'Magnetic field'
  2- Near-earth probe flyby anomaly (unexpected speed variations)
  3- Magnetic hills

- **Electric field**
  1- The essence of 'Electric field'

- **Electricity**
  1- The essence of 'Electric charge' of particles
  2- The essence of 'Electricity'
  3- Why the speed of an electrical signal through a conducting Matter medium is equal to speed of light in that medium?
  4- Reason for drastic increase in the occurrences of Natural Disasters, on a global scale, particularly since 1940s

Appendix III:
List of Proposed Predictions

- **On the universal scale**
  1- Space will keep on expanding at an ever faster pace
  2- Measurements of position and speed of particles (objects) with lower uncertainties will become possible
  3- Magnitude of Planck's constant is gradually decreasing

- **Accompanying universe**
  1- The existence of the accompanying universe
  2- Currently, as compared to this universe, in the accompanying universe:
      - 'Time' is experienced at a much faster pace
      - Speed of light is much faster
      - There is no such thing as gravity, but there is what can be truly called anti-gravity
      - Particles are viewed to be larger in size
      - Black holes are seen as bright spherical light sources
      - Cosmic Microwave Background Radiation have different frequencies

- **Atomic scale**
  1- There are no such things as quarks
  2- There is no such force as Strong nuclear force
  3- There is no such force as Weak nuclear force
  4- Physical size of particles is gradually increasing
  5- Mass of particles is gradually decreasing

6- All of the particles will eventually dissolve in the aether medium

- ## Time
  1- Time is speeding up
  2- Time Anomaly onboard satellites in polar orbit

- ## Light
  1- Speed of light is gradually increasing
  2- Speed of light in a vacuum is not the maximum speed possible for objects in this universe

- ## Gravity
  1- Variable dependence of the force of gravity on distance
  2- Direction of force of gravity exchanged between two celestial bodies is not necessarily along the line connecting their centers of masses
  3- Magnitude of the force of gravity, experienced between two objects depends not only on their motions relative to each other, but also on their motions relative to the local aether medium
  4- Gravity will gradually fade away, totally

- ## Black holes
  1- Size of black holes is gradually increasing
  2- Black holes cannot have any kind of field that is due to their contents
  3- Inside a black hole's event horizon, time can be experienced only by objects and beings that are in a state of free fall directly towards the black hole but not by those following a tangential path
  4- Information that crosses a black hole's event horizon is preserved forever, in the accompanying universe
  5- All of the black holes (starting with the smaller ones) will lose their status as being black holes
  6- All of the black holes will eventually literally dissolve in space

- ## Magnetic field

1- Measuring speed of magnetic field in space and in matter mediums
2- Magnetic field is gradually weakening, and why it is so
3- Artificial magnetic hills can be built for a variety of applications
4- Magnetic propulsion system
5- Increasing laser efficiency using a magnetic field

- **Electric field**
  1- Measuring speed of electric field through a matter medium
  2- Electric field is gradually weakening, and why it is so
  3- Electric field of any and all charged particles is gradually becoming weaker
  4- Electric propulsion system
  5- Increasing laser efficiency using an electric field

- **Electricity**
  1- Electric charge of any and all charged particles is gradually becoming weaker
  2- Electricity is due to a type of phase vibration in aether which is propagating in the conducting matter medium
  3- Measuring speed of phase vibrations associated with electrical current through a matter medium
  4- Speed of electrical signals in any and all matter mediums is gradually increasing

Appendix IV:
List of Proposed Experiments

- **Aether**
  1- Measuring the current pressure difference between aether that is in this universe and aether that is in the accompanying universe
     - First Method: (Nozzle effect on gas flow)
     - Second Method: (Delay time / lag time in gravitational response of celestial bodies)
  2- Einstein-Podolsky-Rosen paradox of entangled particles (Explained): (Faces of a coin)
  3- Double-slit experiments (explained):
     - Type II: (Throwing tennis ball at a lake)
     - Type III: (Effect of camera position)
  4- Relation between age of galaxies and the percentage of their overall mass that is in the form of dark matter
  5- Causing a satellite to alter its altitude or even fall, using earth's magnetic field

- **Time**
  1- Effect of gravity on the passage of 'Time'
     - Orbital free fall vs. direct free fall
     - How does an object experience the passage of 'Time' as it <u>directly</u> approaches a black hole, in a state of free fall?
     - Esmailzadeh Paradox
     - Esmailzadeh orbital altitude Range
  2- Does acceleration induce 'Time dilation'?

- Three probes accelerating in different directions
- Two probes falling at different speeds
- Using regular satellites launched into orbit

3- Effect of magnetic field on the passage of 'Time'
- Magnetically induced 'Time Dilation' on a planetary scale
- Time anomaly experienced by clocks onboard satellites in polar orbits
- Magnetically induced 'Time Dilation' on a Laboratory scale
- Magnetic field powered time dilation machine

4- Effect of electric field on the passage of 'Time'
- Using electric field to induce 'Time Dilation'
- Electric field powered time dilation machine

- **Light**
  1- Is light a wave or a particle?
    - **Experiment:** (Is light a wave or made of photons?)
      1- Using only one detector
      2- Using too many detectors
  2- Double-slit experiments (explained):
    - Type II: (Throwing tennis ball at a lake)
    - Type III: (Effect of camera position)
  3- Direct effect of gravity on the speed of light
  4- Effects of magnetic field
    - On the speed of light
    - On the direction of light's propagation
  5- Effects of electric field
    - On the speed of light
    - On the direction of light's propagation
  6- Crossflow experiments for two monochromatic light waves or two monotonic sound waves
  7- Michelson & Morley experiment (Revised)

- **Gravity**
  1- Gravitational effects of various planets on different elements
  2- Gravitational effects of various planets on different bodily fluids

3- Relation between age of galaxies and the percentage of their overall mass that is in the form of dark matter
4- How fast variations in the strength of the force of gravity are experienced at a given distance from a star or a planet?
- Detecting cyclic variations in the direction and strength of sun's gravity
5- Direct effects of gravity on the speed of light
6- Near-earth probe flyby anomaly
- The in-line effect of aether flow on the speed of such probes
- The sideways effect of aether flow on the direction of motion of such probes
7- Effect of magnetic field on the strength of the force of gravity
- Effect of magnetic field on Mr. Galileo's experiment
- Magnetic Hill effect
- Artificial magnetic hills
8- Effect of electric field on Mr. Galileo's Experiment

- **Black Holes**
1- Does Time exist inside the event horizons of black holes?
- Orbital free fall vs. direct free fall
- How does an object experience the passage of 'Time' as it directly approaches a black hole, in a state of free fall?
2- Effect of the spin of a black hole on the in-falling objects
3- Are the electromagnetic waves reaching a black hole preserved or are they lost forever?

- **Magnetic Field**
1- Effect of magnetic field on the passage of 'Time'
- Magnetically induced 'Time Dilation' on a planetary scale
- Time anomaly experienced by clocks onboard satellites in polar orbits
- Magnetically induced 'Time Dilation' on a Laboratory scale
- Magnetic field powered time dilation machine
2- Effects of magnetic field
- On the speed of light

- On the direction of light's propagation

3- Near-earth probe flyby anomaly
- The in-line effect of aether flow on the speed of such probes
- The sideways effect of aether flow on the direction of motion of such probes

4- Effect of magnetic field on the strength of the force of gravity
- Effect of magnetic field on Mr. Galileo's experiment
- Magnetic Hill effect
- Artificial magnetic hills

5- Magnetic propulsion system
6- Measuring the speed of the aether flow manifesting as magnetic field
7- Propagation speed of magnetic field through a matter medium
8- Increasing LASER efficiency by using a magnetic field

- **Electric Field**
  1- Effect of electric field on the passage of 'Time'
  - Using electric field to induce 'Time Dilation'
  - Electric field powered time dilation machine
  2- Effects of electric field
  - On the speed of light
  - On the direction of light's propagation
  3- Effect of electric field on Mr. Galileo's Experiment
  2- Electric propulsion system
  3- Measuring the propagation speed of electric field in a matter medium
  4- Increasing LASER efficiency by using an electric field

- **Electricity**
  1- The propagation speed of phase vibrations associated with electricity in any given conducting matter medium is independent of the size of that medium's cross section or its shape
  2- Measuring the propagation speed of phase vibrations associated with electrical current in a given conducting matter medium

3- Measuring the relative speed at which phase vibrations associated with electricity propagate through different matter mediums

4- Loss of electrons due to sharp ends/edges in a conducting Medium

5- Presence of electricity throughout the whole volume of a given conducting matter medium

6- Proof that the flow of electricity is generated by a certain type of phase vibrations in the medium of aether that is occupied by a conducting matter medium

7- Propagation speed of phase vibrations associated with electrical current through a given conducting matter medium is equivalent to the speed of light in that medium

8- Do phase vibrations associated with electricity propagate through conducting matter mediums from the negative terminal of the battery towards its positive terminal or the other way around?

- **Locating the Birthplace of the Universe**

  1- Methods of locating the birthplace of the universe

  - **First method**: (Using the intersection of the rotational planes of spiral galaxies)
  - **Second method**: (Using edge-on view of the rotational planes of spiral galaxies)
  - **Third method**: (Using full-faced view of the rotational planes of spiral galaxies)
  - **Fourth method**: (Using horizontal edge-on view of the rotational planes of spiral galaxies –forming a 360° belt)
  - **Fifth method**: (Using the temperature distribution in the cosmic microwave background radiation, CMBR)

Appendix V:
List of 25 Fundamental Breakthroughs
(Each One, Potentially, Deserving a Nobel Prize)

- **Physics**:
  1- The existence of **"Aether"** in this universe, and its properties, as well as how it affects every phenomenon in this universe,

  2- The existence of the **"accompanying universe"**, and its potential application for faster communication between distant places,

  3- The essence of **"Time"**, experienced due to the speed of objects relative to the local aether being less than that of phase vibrations in that medium, and why it is gradually experienced at a faster pace,

  4- The true essence of **"Light"**, being only a wave (phase vibration in the aether medium), and why its speed is gradually increasing,

  5- The essence of **"Gravity"**, being due to accelerated flow of aether relative to objects (or acceleration of objects relative to aether), and why it is gradually weakening,

  6- The essence of **"Magnetic field"**, being a type of aether flow, and why it is gradually becoming weaker,

  7- The essence of **"Electric field"**, being a type of aether flow, and why it is gradually becoming weaker,

8- The essence of **"Electricity"**, being induced, just like photoelectric effect, by phase vibrations in the aether medium which is superimposed by the matter conductor medium, and why its signal speed is gradually increasing,

9- The essence of **"Electrical charge"**, being a type of aether flow out of negatively charged and into positively charged particles, and why it is gradually becoming weaker,

10- The essence of **"Particles"**, being bubbles, through which aether is escaping from this universe into the accompanying universe,

11- The essence of **"Mass"** of all particles, being due to the formation of a wake-like wave in the aether medium, and why it is gradually decreasing,

12- **Physical size of all particles,** being of certain discrete sizes which are gradually becoming larger, due to decreasing aether pressure in this universe

13- The essence of **"Black Holes"**, being giant bubbles, formed of unified (merged) matter particles, in fact being the largest matter particles,

14- **Planck's Constant** is proportional to the amplitude of the phase vibrations in the aether medium, and why it is gradually decreasing,

15- There is **no such force as strong nuclear force**, since there are no such things as quarks in this universe,

16- There is **no such force as weak nuclear force**, since gravity is the force responsible for holding the nucleons together,

17- Why the **expansion of this universe is accelerating,**

18- **Protons** being about 1836 times more massive than electrons, their diameter **must be about 42.85 times larger than that of electrons**, since mass ratio is equivalent to the ratio of surface areas,

19- Why **certain particles such as isolated neutrons are unstable**,

20- Why **certain isotopes are unstable**,

21- Altering the **"half-lives of isotopes / particles"**, using magnetic field and/or electric field

- **Medicine**
  1- Formation and development of **"Living Beings"** in this universe, being due to being guided and controlled by spirits residing in them,

  2- Extending lifespan of living beings, using magnetic field and/or electric field

- **Literature**:
  1- **Simplicity and clarity in presentation of new fundamental theories and concepts**

- **Peace**:
  1- **Reducing Natural Disasters**, by eliminating the 90% that are man-caused, due to using wide-spread (continental wide), synchronized Alternating Current type of electricity.

References

Most of the references listed below (other than author's own books) were sighted to gain general background information regarding different inexplicable physical phenomena. Some of the references were in fact catalogues used to access the required data to perform the analysis presented in the chapters regarding the birthplace of the universe. Also, some of the references were United Nations' studies which were used as sources for background information / data required to demonstrate the effect of Alternating Current electricity on various types of natural disasters, on a global scale.

Books:

1- J. C. Maxwell, "A Dynamical Theory of the Electromagnetic Field", Philosophical Magazine, (1864).
2- A. A. Michelson and E. M. Morley, Philosophical Magazine, (1887).
3- A. Einstein, "Relativity, the Special & the General Theory", Translated by R. W. Lawson, Methuen & Co. Ltd. London, (1920).
4- D. C. Giancoli, "Physics", Prentice Hall, Inc. Englewood, Cliffs, New Jersey, (1985).
5- E. T. Whittaker, "Theories of Aether and Electricity", Hudges, Figgis & Co. Ltd,, Dublin, (1910).
6- Kevin Krisciunas and Bill Yenne, "The Pictorial Atlas of the Universe", Bison books Ltd., London, (1989).
7- B. Esmailzadeh, "Innovative Inventions", Canada, (2012).
8- B. Esmailzadeh, "Innovative Theories", Canada, (2011).

9- B. Esmailzadeh, "Aether: Past, Present and Future of the Universe", Xlibris, USA, (2012).

10- B. Esmailzadeh, "The Evolution of Spirits", Xlibris, USA, (2012)

11- B. Esmailzadeh, "Purpose of Life in this Universe", Canada, (2012)

12- B. Esmailzadeh, "The Formation and Evolution of Physically Living Beings", Canada, (2015)

Internet:

1- http://en.wikipedia.org/wiki/Galaxy

2- http://en.wikipedia.org/wiki/Kepler%27s_laws_of_planetary_m otion

3- http://en.wikipedia.org/wiki/Newton%27s_law_of_universal_gr avitation

4- http://www.jrank.org/space/pages/2433/Lick-galaxy-catalogue.html

5- http://heasarc.nasa.gov/W3Browse/all/rc3.html

6- http://en.wikipedia.org/wiki/Fritz_Zwicky

7- http://en.wikipedia.org/wiki/Dark_matter

8- http://en.wikipedia.org/wiki/Dark_energy

9- http://metaresearch.org/cosmology/cosmology.asp

10- http://nedwww.ipac.caltech.edu/level5/Shapley_Ames/frames.h tml

11- http://en.wikipedia.org/wiki/Age_of_the_universe

12- http://en.wikipedia.org/wiki/Observable_universe

13- http://skyserver.sdss.org/dr1/en/astro/universe/universe.asp

14- http://www.phys.unsw.edu.au/einsteinlight/jw/module3_M&M .htm

15- http://en.wikipedia.org/wiki/Lorentz_transformation

16- http://en.wikipedia.org/wiki/Special_relativity

17- http://en.wikipedia.org/wiki/Black_hole

18- http://en.wikipedia.org/wiki/Inclination

19- http://nssdc.gsfc.nasa.gov/planetary/factsheet/

20- http://www.scientificamerican.com/article/the-bubbles-produced-by-u/

21- http://physics.nist.gov/cuu/Units/second.html

22- http://physics.nist.gov/cgi-bin/cuu/Info/Units/meter.html

23- http://en.wikipedia.org/wiki/Particle_accelerator

24- http://en.wikipedia.org/wiki/Photoelectric_effect

25- http://en.wikipedia.org/wiki/Casimir_effect

26- http://en.wikipedia.org/wiki/Quantum_entanglement

27- en.wikipedia.org/wiki/Pioneer_anomaly

28- http://en.wikipedia.org/wiki/Double-slit_experiment

29- http://en.wikipedia.org/wiki/Laser

30- http://en.wikipedia.org/wiki/Airbag

31- http://physics.ucr.edu/~wudka/Physics7/Notes_www/node98.ht
ml

32- http://en.wikipedia.org/wiki/Photon_sphere

33- http://heasarc.gsfc.nasa.gov/W3Browse/galaxy-
catalog/mcg.html

34- http://heasarc.nasa.gov/W3Browse/all/pgc2003.html

35- http://en.wikipedia.org/wiki/List_of_gravity_hills

36- http://www.acoustics.asn.au/conference_proceedings/AAS200
5/papers/34.pdf

37- http://en.wikipedia.org/wiki/Fuse_%28electrical%29

38- http://en.wikipedia.org/wiki/Astrology

39- http://en.wikipedia.org/wiki/Zodiac

==============================

References used in chapter titled "Reducing the Frequency and Intensity of Natural Disasters"

1- http://www.emdat.be/natural-disasters-trends

2- http://www.unep.org/geo/geo3/english/448.htm

3- http://www.munichre.com/app_pages/www/@res/pdf/NatCatSe
rvice/great_natural_catastrophes/1950-
2009_Great_natural_catastrophes_Percentage_distribution_en.p
df

4- http://en.wikipedia.org/wiki/Global_warming

5- http://en.wikipedia.org/wiki/Lightning

6- http://www.geni.org/globalenergy/library/national_energy_grid/
index.shtml

7- http://www.preventionweb.net/files/7682_2004MRNatCatSER
VICENaturalDisastersworldmapen.pdf

8- http://www.preventionweb.net/files/7681_2005MRNatCatSER
VICENaturalDisastersworldmapen.pdf

9- http://www.preventionweb.net/files/7680_2006MRNatCatSER
VICENaturalDisastersworldmapen.pdf

10- http://www.preventionweb.net/files/7679_MRNatCatSERVIC
ENaturalDisastersworldmapen.pdf

11- http://www.preventionweb.net/files/7675_20081229app2en.pdf

12- http://www.preventionweb.net/files/12216_MunichRe20091.pd
f

13- http://en.wikipedia.org/wiki/List_of_HVDC_projects

===

www.ingramcontent.com/pod-product-compliance
Lightning Source LLC
Chambersburg PA
CBHW020717180526
45163CB00001B/5